CHICAGO     LIBRARY
HAR               RARY CENTER
R

REF
QB
51
.N5

Cop 1

The New astronomy
and space science
reader

| DATE | | | |
|---|---|---|---|
| | | | |
| | | | |
| | | | |
| | | | |
| | | | |
| | | | |
| | | | |
| | | | |
| | | | |
| | | | |
| | | | |
| | | | |

REF
QB
51
.N5
Cop 1

FORM 125 M

Business/Science/Technology
Division

The Chicago Public Library

Received_____FEB 13 1979_____

© THE BAKER & TAYLOR CO.

# THE NEW ASTRONOMY
# AND SPACE SCIENCE
# READER

# THE NEW ASTRONOMY
# AND SPACE SCIENCE
# READER

Edited by

## John C. Brandt

NASA–Goddard Space Flight Center
  Greenbelt, Maryland
and University of Maryland
  College Park, Maryland

and

## Stephen P. Maran

NASA–Goddard Space Flight Center
  Greenbelt, Maryland

W. H. FREEMAN AND COMPANY
San Francisco

REF
QB
51
.N5
cop.1

**Library of Congress Cataloging in Publication Data**

Main entry under title:

The New astronomy and space science reader.

  Includes bibliographies and index.
  1. Astronomy—Addresses, essays, lectures.
2. Space sciences—Addresses, essays, lectures.
I. Brandt, John C.   II. Maran, Stephen P.
QB51.N5        520′.8        76-54316
ISBN 0-7167-0350-5
ISBN 0-7167-0349-1 pbk.

**Copyright © 1977 by W. H. Freeman and Company**

No part of this book may be reproduced
by any mechanical, photographic, or electronic process,
or in the form of a phonographic recording,
nor may it be stored in a retrieval system, transmitted,
or otherwise copied for public or private use,
without written permission from the publisher.

Printed in the United States of America

9  8  7  6  5  4  3  2

Indication of the editors' affiliation with the National Aeronautics and
Space Administration on the title page of this book does not constitute
endorsement of any of the material contained herein by NASA. The
editors are solely responsible for any views expressed.

BUSINESS/SCIENCE/TECHNOLOGY DIVISION
THE CHICAGO PUBLIC LIBRARY

FEB 13 1978

# PREFACE

Timing the falling clouds in the central region of the Andromeda galaxy, gauging the darkside temperature of Mercury by measuring its infrared radiation, exploring the landscape of Venus by radar, investigating contact of star pairs by computer, and studying the tilting of sunspots by motion-picture photography are all clearly, problems for the astronomer, but so are the identification of a priceless manuscript in the Vatican Library and the recognition of interstellar dust grains in old museum collections. These events are described in this reader, and nearly all of the articles reprinted here were first published in the 1970s. Quotations from a few of them illustrate the remarkable advances in astronomy and space research that have been made in recent years, which would have seemed fantastic to the scientists of the previous generation:

> . . . it appears that black holes of the kind predicted by theory do in fact occur in nature.
>
> S. Chandrasekhar

> Radio astronomers in Canada and the United States had observed titanic flares of energy coming from Cygnus X-3, and they wanted Conway to get the huge Jodrell Bank radio telescope on it in time to catch this unexpected and dramatic event.
>
> Ben Bova

> . . . the complex robot spacecraft *Mariner 9* fired its braking rocket and was captured in orbit around Mars, thus becoming the first man-made satellite of another planet.
>
> Bruce C. Murray

The purpose of this collection of articles is to introduce students to current topics of astronomical research, to give background information on many subjects that are touched on only briefly in formal textbooks, and to show how new technological developments in optical and radio astronomy and in space instrumentation are enabling us to advance the frontiers of astronomical knowledge. The collection includes several selections that illustrate in detail how astronomers pursue their observations and theorizing. It is intended for use as supplemental reading for astronomy and space science courses. Assuming that the student will also need to purchase a textbook, we have purposely excluded color photographs and diagrams in order to make the price of this reader as low as possible.

The selections include popular articles from *Natural History, Smithsonian,* and *Scientific American,* as well as professional papers. The latter include original research reports from *Science, Nature,* and the *Astrophysical Journal;* students will need help from instructors in working through them. Some of the selections admittedly require more knowledge of mathematics, chemistry, or other allied fields than the average student in an astronomy survey course will possess, but, by the same token, they may be of interest to science majors and others with backgrounds in those subjects. We also hope that some readers will be sufficiently inspired by the popular articles to attempt the more difficult material.

The content of survey courses varies greatly, depending on the interests of the students and the instructors. It is likely that in a given course several of the articles in this reader will be omitted because they are inapplicable, too technical, or even too simple. We would very much like to have comments from students and teachers on the content of the book and the manner in which it is used. They should be addressed to: Astronomy Publisher, W. H. Freeman and Company, 660 Market Street, San Francisco, California 94104, who will forward them to us.

John C. Brandt
Stephen P. Maran

February 1977

# CONTENTS

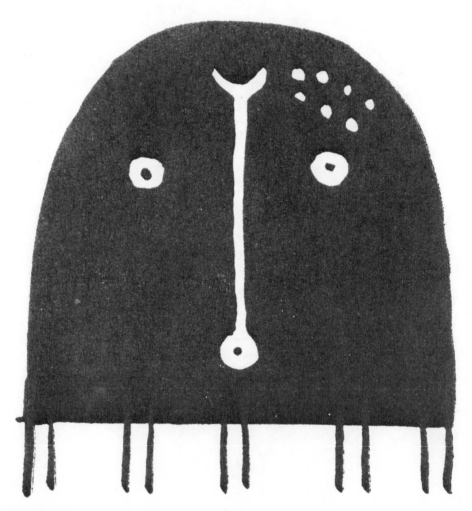

Astronomical prehistory and legends can
be as fascinating as the better known
historical developments. The figure shows
the Navaho Black God and his Pleiades.
According to Navaho tradition, he is
responsible for locating the stars that
form the more prominent constellations.
[Adapted from B. Haile, *Starlore among
the Navaho.* Santa Fe: Museum of Navaho
Ceremonial Art, 1947.]

# HISTORICAL ASTRONOMY

"The record probably implies that the eclipse was total within the boundaries of the Kingdom of Ugarit." F. R. Stephenson's account of "The Earliest Known Record of a Solar Eclipse" shows why ancient records of celestial events are usually of interest to historians. By computing the date on which an event occurred and the geographic area from which it was visible, the astronomer can enable the historian to calibrate an ancient calendar or to check on the uncertain boundary of an almost-forgotten nation. Stephenson's analysis of the eclipse of 1375 B.C. at Ugarit uses the geography of the kingdom to determine the date of the eclipse, and it is the first accurate date available to historians of this culture.

In reading the Stephenson article, it will be helpful to keep in mind that the author uses the terms "phase" and "magnitude" of eclipse interchangeably and that either term refers to the percentage of the visible disk area of the sun that was covered by the moon at the time of maximum eclipse. Thus a phase, or magnitude, of 100 refers to a total eclipse. A magnitude of $>100$ for a particular eclipse indicates that it was total and that the angular size of the moon exceeded that of the sun.

"I think my heart must have stopped for a moment when I saw the handwriting in the facsimile: I then realized that the first edition in Rome was probably also in Tycho's hand. What I had found in the Vatican Library was Tycho's original working copy, probably the most important Tycho manuscript in existence." Astronomer Owen Gingerich's tale of "Copernicus and Tycho" is at once both an essay in the history of astronomy and a personal detective story. In effect, the article comprises two parallel stories of scientific discovery, Tycho's and Gingerich's.

The history of astronomy is a scholarly field in its own right, with its own specialists, professors, students, and conferences. Its leading periodical is the *Journal for the History of Astronomy,* published by Science History Publications in Buckinghamshire, England, and replete with learned footnotes and references to books and papers written in ancient languages. Recent issues treat of such subjects as a theory of planetary motions found on an old papyrus, the authenticity of an alleged portrait of Galileo that hangs in a San Diego museum, and the rate at which observatories were established during the nineteenth century (their number doubled every 34 years). The study of ancient observations can yield information needed in modern astrophysical research. A good example is the determination of the age of the Crab Nebula and its pulsar from old oriental records, as discussed in Part XI of this book. A related subject, archaeoastronomy, deals with possible astronomical interpretations of ancient structures such as Stonehenge, with such matters as the representation of celestial phenomena in the cave paintings and other rock art of extinct Indian tribes, and with the calendar stones of vanished Central American civilizations.

# The Earliest Known Record of a Solar Eclipse

F. R. Stephenson

The excavations of ancient Ugarit, begun in 1928, have revealed numerous clay tablets of great literary interest. Their discovery has been as revolutionary in the field of biblical studies as the discovery of the Dead Sea Scrolls. From archaeological evidence the texts date from the period 1450–1200 BC. They are written in a cuneiform script which is the earliest alphabetic script as yet deciphered.

Only one of the texts so far published is of an astronomical nature; this describes a solar phenomenon. A photograph of one side of the tablet, called side *A*, is shown in Fig. 1. A recent translation of the inscription on both sides of the tablet is given by Sawyer[1] as follows:

*Side A:* The day of the new Moon in the month of Ḥiyar was put to shame. The Sun went down (in the daytime) with Rashap in attendance.

*Side B:* (This means that) the overlord will be attacked by his vassals.

The absence of any parallel to this text in the Ugaritic literature suggests that the phenomenon was rare and spectacular, and the description can hardly refer to anything other than a total eclipse of the Sun. Numerous accounts of total solar eclipses are to be found in mediaeval European and Far Eastern literature. These leave no doubt that the sudden and intense darkness, appearance of stars, and other phenomena which accompany the complete disappearance of the Sun produced a profound impression on the observers. In contrast, annular and large partial eclipses attracted little interest. Few descriptions of total eclipses survive from before the Christian era, but we should expect such records to be found in areas where a sufficiently representative bloc of literature is extant.

The archaeological evidence allows the date of the event (hereafter regarded as a solar eclipse) to be fixed only between about 1450 and 1200 BC. It cannot have occurred later than this, for the city was destroyed by invaders about the year 1200 and was never rebuilt. The inscription states that the phenomenon occurred on the day of the new Moon in the month of Ḥiyar. Ḥiyar, the second month of the year, corresponds to about April–May. Total eclipses of the Sun are rare at a given place, occurring on average about three times in a millennium; this suggests that a definite date for the event can probably be established.

Accurate computation of an ancient solar eclipse for a given place is limited by the non-uniformity of the Earth's rotation and the tidal recession of the Moon from the Earth. Irregularities in the Earth's rotation are chiefly the result of tides, but other causes are changes in sea level and electromagnetic coupling between the core and mantle of the Earth. A computation of a solar eclipse which ignores these effects may be in error by up to about 4 h in time and 50 per cent in phase near 1300 BC. The most convenient method of analysis is to assume that the Earth rotates uniformly and to make corrections to the mean longitudes of the Sun and Moon from a study of reliable records of astronomical events in history.

Figure 1. One side of the tablet with text describing a solar phenomenon.

From *Nature*, vol. 228, pp. 651–652, 1970. Reprinted with permission.

TABLE 1
*Computed circumstances for the total solar eclipses visible near Ugarit from 1450 to 1200* BC.

| Date BC | Magnitude at Ugarit | Local time (h min) | | Sun's altitude (degrees) | Zone width (km) | Total time (min s) | | Position of central zone |
|---|---|---|---|---|---|---|---|---|
| 1406 July 14 | >100 | 15 | 30 | 43 | 180 | 3 | 30 | Total at Ugarit |
| 1375 May 3 | 99 | 6 | 00 | 7 | 130 | 1 | 15 | 55 km N of Ugarit |
| 1340 Jan. 8 | 96.5 | 11 | 05 | 29 | 320 | 3 | 30 | 240 km N of Ugarit |
| 1223 Mar. 5 | 96 | 13 | 25 | 40 | 170 | 3 | 30 | 165 km N of Ugarit |

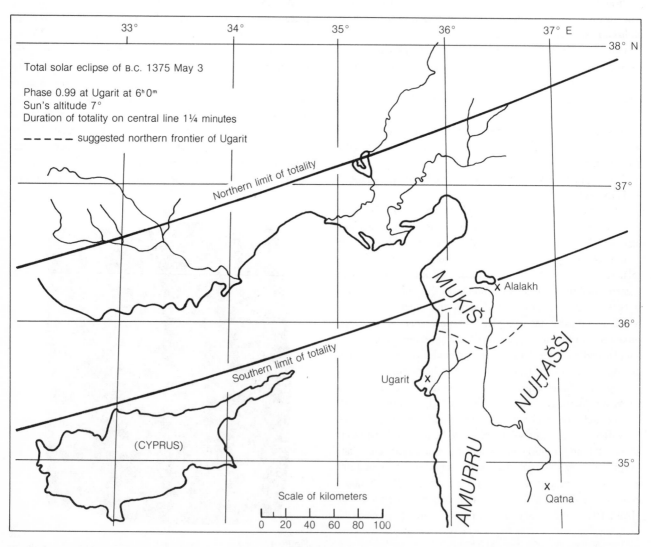

**Figure 2.** Map of the track of the eclipse in the region around Ugarit.

These observations are themselves mainly of total eclipses of the Sun, for which the production of an accurate record makes the minimum of demands on the observer. The earliest of these records dates from 709 BC (China). By extrapolation, it is expected that errors in the computation of the Ugaritic eclipse do not exceed 1 per cent in the magnitude and 5 min in the time of occurrence.

The four dates in the table were selected from a computer analysis of about 500 solar eclipses which occurred on the Earth's surface during the period covered by the texts. The program required only the input of the geographical coordinates of the city of Ugarit (35° 37′ N, 35° 47′ E) and the Julian day number for each eclipse. Apart from the four eclipses in the table, no generally total eclipse reached a phase of greater than 88 per cent at the capital. The few annular eclipses during this time would not produce appreciable darkening of the daylight.

It is clear that only the eclipse of 1375 BC May 3 could have occurred in the month of Ḥiyar. This was total near the northern boundary of the Kingdom of Ugarit (see Fig. 2). There appear to be no records of the other eclipses in the table. The loss of daylight in 1340 BC and 1223 BC would probably be too small in Ugarit to attract attention there, but the eclipse of 1406 BC must have been very spectacular. Much of Ugarit has still to be excavated, and it seems possible that a reference to this may yet turn up.

The period close to 1375 BC was one of great literary activity in Ugarit and it is not surprising to find a description of the eclipse of 1375 BC. Although the estimated uncertainty in the position of the belt of totality (about 30 km) does not make the eclipse quite complete at the capital, it would have been extremely striking in the northern part of the kingdom and may well have been total there. The reference to Rashap (=Mars) is interesting. Calculation shows that on the morning of May 3, 1375, Mars was below the horizon. Both Aldebaran and Capella, however (the latter of which was called "red" by Ptolemy and other early astronomers), were near the darkened Sun and either star may have been mistaken for Mars.

The importance of the eclipse of 1375 BC is two-fold. It is of considerable value in Ugaritic chronology, for its date is the first definite date to be established. The record probably implies that the eclipse was total within the boundaries of the Kingdom of Ugarit. As these can be fixed with a fair degree of confidence, the eclipse could prove useful in the study of the Earth's rotation.

The only known observation of a solar eclipse of comparable age to that made in Ugarit is the one recorded on one of the oracle bones found in Anyang, N. China.[2, 3] It has been deciphered as reading: "Three flames ate the Sun and a great star was seen." Liu Chao-yang has put the date of the event as the late fourteenth or early thirteenth century BC. Computation shows that between 1400 BC and 1200 solar eclipses were total near Anyang only in 1370, 1330, 1304, 1250 and 1230. If the record is authentic, and indeed describes a complete eclipse of the Sun, the observation should have been made in one of these years. Precise identification of the eclipse, however, must await further work on Shang dynasty chronology. The solar eclipse observed in Ugarit on the morning of May 3, 1375 BC, would appear to be the earliest for which there is a record.

## References

1. Sawyer, J. F. A., and Stephenson, F. R., *Bulletin of the School of Oriental and African Studies,* 33, 467 (1970).
2. Liu Chao-yang, *Yü Chou,* 15, 15 (1945).
3. Needham, J., *Science and Civilisation in China,* 3, 423 (Cambridge, 1959).

# Copernicus and Tycho

Owen Gingerich

2

In the flamboyant frontispiece to G. B. Riccioli's *Almagestum novum* (*New Almagest*), published in 1651, Urania, the goddess of the heavens, holds a balance that is weighing the Copernican world system against the Tychonic system [*see illustration on opposite page*]. In the Copernican system all the planets revolve around the sun. In the geocentric Tychonic system, which Tycho Brahe proposed several decades after Nicolaus Copernicus had advanced his heliocentric model, the sun travels around the fixed earth, carrying the rest of the planets with it. For Riccioli the geocentric system put forward by Tycho carried the most weight. Today we view Tycho's scheme as a giant step backward, but we are nonetheless disconcerted by the fact that it was proposed by the most innovative astronomical observer since antiquity. Tycho's bold plan for increasing the accuracy of observations places him far more securely in the mainstream of modern astronomy than Copernicus himself.

Just this past May I made a discovery that now puts Tycho's view in a more favorable light. Bound into an edition of Copernicus' *De revolutionibus orbium coelestium* in the Vatican Library I found the original working copy of Tycho's cosmological notes. It is perhaps unnecessary to add that this came as a

great surprise. These hitherto unknown personal notes record the germination of Tycho's conception and contain the first sketches of his planetary system.

To appreciate the full significance of the notes it is necessary to look afresh at the context of Copernican astronomy in the 16th century. When Copernicus introduced his heliocentric arrangement, the earth became one of the family of planets revolving in off-centered circles around the sun. Copernicus sought a system that was "pleasing to the mind," and he was particularly delighted with his proposal because the planets naturally ranked in order from the sun according to their speeds, with the slowest one revolving in the largest orbit. (At the time the slowest planet known was Saturn.)

In contrast, Tycho's planetary system, in which the earth regained its privileged central status, at first seems clumsily contrived. More important in the context of astronomy a century before Newton, however, is that the immobile earth of the Tychonic system fitted the accepted laws of physics where the moving-earth Copernican model did not. Tycho complained that Copernicus' system "ascribes to the earth, that hulking, lazy body, unfit for motion, a motion as quick as that of the ethereal torches, and a triple motion at that." Nevertheless, to most of Tycho's contemporaries this com-

plaint hardly mattered; the majority of 16th-century astronomers considered astronomical systems as being elaborate geometric hypotheses quite removed from physics. The gradual acceptance of a celestial physics integrated with terrestrial physics, culminating in the synthesis by Newton, is undoubtedly the most significant scientific aspect of the Copernican revolution.

The ultimate goal of a planetary model is to predict the planets' positions. In this respect the geocentric epicyclic system proposed by Ptolemy around A.D. 140 succeeded admirably. In order to be taken seriously any alternative proposals had to do as well. Copernicus realized that; thus only a few pages of *De revolutionibus* directly concern the heliocentric cosmology. The bulk of it is devoted to technical details leading to tables describing the motions of the planets. In the 16th century Copernicus was regarded as a master mathematician not for his innovative cosmology but for his ability to cope with the details of predicting the positions of the planets.

Actually the tables of *De revolutionibus* offered little improvement over those of Ptolemy. That is not surprising considering the lack of a suitable base of observations from which to compute the tables. Copernicus was forced to rely heavily on the same observations Ptolemy had recorded. Not until Tycho initiated his vast program of systematic observations in the 1570's did anyone fully realize how faulty the tables were, or whether Copernicus had offered any improvement.

Although the astronomers of Copernicus' day had only rather primitive instruments, it was possible to make certain observations quite accurately without any instruments at all. Copernicus' earliest recorded observation, in 1497,

**FRONTISPIECE TO "ALMAGESTUM NOVUM"** (*New Almagest*) by G. B. Riccioli, published in 1651, shows the earth as the bearded man at the left covered with eyes, holding a telescope and gazing up at the newly discovered marvels of the heavens: the odd shape of Saturn (due to its rings) in the top right corner, Jupiter with its four bright satellites just below, the rough appearance of the moon's surface and a comet. In the top left corner the cherubim, from top to bottom, are holding Mars, Venus, the sun and Mercury. To earth's right is the star-covered goddess of the heavens, Urania, holding an armillary sphere in her left hand and a balance in her right. The balance is testing the Copernican heliocentric system of the universe against the geocentric system of Tycho Brahe. For Riccioli the Tychonic system carried more weight. Ptolemaic system is discarded at Urania's feet; Ptolemy himself is watching over proceedings and commenting: "I erred so that I could be corrected."

From *Scientific American*, vol. 229, pp. 86–101, December 1973. Copyright ©
1973 by Scientific American, Inc. All rights reserved. Reprinted with permission.

was one of these. He was then a young man of 24, studying canon law at the University of Bologna. He had temporarily left his native Poland following his undergraduate studies in Cracow. On March 9 he watched the moon approach Aldebaran, the brightest star in the constellation Taurus, finally occulting (eclipsing) it at 11 P.M. Later he used the observation in his book as a test of the moon's varying distance from the earth.

In 1503, when Copernicus was 30, he returned to Poland to take up a lifetime post as a canon of the cathedral of Frombork, an appointment arranged through the benevolent nepotism of his uncle Lucas Watzenrode, who was the bishop. Copernicus served as administrator of the cathedral estates and as private secretary and personal physician to his uncle. His tenured position as canon gave him the time and security to pursue his astronomical work, and in 1504 he could have made a particularly interesting series of observations.

In that year all five of the visible planets (Mercury, Venus, Mars, Jupiter and Saturn), as well as the sun and the moon, moved into the constellation Cancer, affording a spectacular series of close planetary approaches. Phenomena as wonderful as these naturally attracted the close attention of astrologers; the conjunctions of Saturn and Jupiter are so rare, occurring only once every 20 years, that they were accorded the greatest astrological significance. In those days the positions of the planets were predicted on the basis of the Alfonsine Tables, which had been compiled in the 13th century by the astronomers of Alfonso X, king of León and Castile. Based on the Ptolemaic system, they predicted that Jupiter would move past Saturn on June 10, when the planets were too close to the sun for easy observation. Important conjunctions of Jupiter with Mars, however, were anticipated on January 4 and on the morning of February 22, and a conjunction of Saturn with Mars fell on March 18.

Anyone as interested in astronomy as Copernicus was could scarcely have failed to observe these phenomena. Like occultations of stars by the moon, conjunctions of the planets can be observed rather accurately without instruments. If Copernicus had watched the planets during the winter of 1503–1504, he would have found that the almanac's predictions missed the conjunctions by about 10 days. Although there is no direct record that he made such observations, Jerzy Dobrzycki of the Institute for the History of Science in Warsaw suggested to me a way by which we can be certain that the astronomer followed the planetary motions in the year of the great conjunction.

In the library of the University of Uppsala there are preserved many volumes from Copernicus' personal library, brought to Sweden in 1627 by the army of Gustavus Adolphus during the Thirty Years' War. Two leather-bound astronomical volumes decorated in the style of Cracow craftsmen of the late 15th century are of particular interest. The bindings suggest that Copernicus bought these books and began to annotate them while he was still an undergraduate and already deeply interested in astronomy. One book includes an ephemeris for 1492 to 1506; the other includes the edition of the Alfonsine Tables for 1492. Bound into the back of this second volume are 16 extra leaves on which Copernicus had added carefully written tables and miscellaneous notes. At the bottom of the last page there is written in highly abbreviated Latin:

*Mars superat numerationem plus quam gr. ij*
*Saturnus superatur a numeratione gr. 1½*

The translation is: "Mars surpasses the numbers by more than two degrees; Saturn is surpassed by the numbers by 1½ degrees."

To analyze this cryptic, undated state-

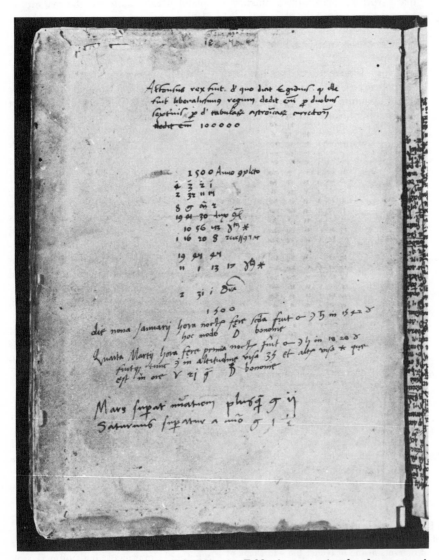

LAST LEAF of Copernicus' copy of the Alfonsine Tables for computing the planetary positions contains a cryptic note by Copernicus in his own handwriting. The note, the last two lines on the page, states that Mars was ahead of its predicted position by more than two degrees and that Saturn was behind its predicted position by 1½ degrees. If the absence of a remark about Jupiter indicates that its predictions had no appreciable error, then the only time that all of these conditions were fulfilled was during the conjunctions of the planets in February and March of 1504. Thus these two undated lines become convincing evidence that Copernicus did in fact observe the conjunctions in 1504 (*see illustration on opposite page*).

ment we can use a powerful tool made available for the first time several years ago through the use of a fast electronic computer. Bryant Tuckerman of IBM recomputed the actual positions of the planets from 601 B.C. to A.D. 1649, providing a standard with which old almanacs and ephemerides can be compared. The longitude predicted by the Alfonsine Tables for each of the superior planets (Mars, Jupiter and Saturn) shows a characteristic periodic error. In the Ptolemaic system a planet's position is predicted by computing its motion in the large orbital circle called a deferent, together with its motion in an epicycle, or secondary circle, centered on the edge of the deferent. The deferent and the epicycle each contribute their own inaccuracies, so that the resulting error pattern is distinctive and reflects a combination of the two motions. In Copernican terms the error in a planet's predicted position results from inaccuracies in the available knowledge of both the earth's motion and the planet's motion around the sun.

From graphs of the error patterns we can see that at the time of the conjunctions in February and March of 1504 there was an almost unique combination of errors: Jupiter was just about precisely on schedule but Saturn ran behind the predictions by some 1½ degrees and Mars was ahead by just over two degrees. If the absence of a remark about Jupiter indicates that its actual position showed no appreciable error from its predicted position, then these figures precisely match Copernicus' undated record. The two lines he wrote at the back of his copy of the Alfonsine Tables thus

**PREDICTED AND ACTUAL** positions of the superior planets during the winter of 1503–1504 are displayed together. The author calculated the predicted positions of the planets from the Alfonsine Tables, and Bryant Tuckerman of IBM computed the actual positions of the planets from modern tables. Each of the planets went into a westward retrograde motion during the interval, so that Mars came into conjunction with Jupiter and with Saturn on three separate occasions. The errors documented by Copernicus apply only to the conjunctions in February and March of 1504 (*band in light gray*). As can be seen, the predicted positions of Mars (*line in gray*) differ from the actual positions (*line in black*) by about two degrees; the predicted positions of Saturn (*long-dashed gray line*) differ from the actual positions (*long-dashed black line*) by 1½ degrees. The predicted positions of Jupiter (*short-dashed gray line*), however, agree well with its actual positions (*short-dashed black line*).

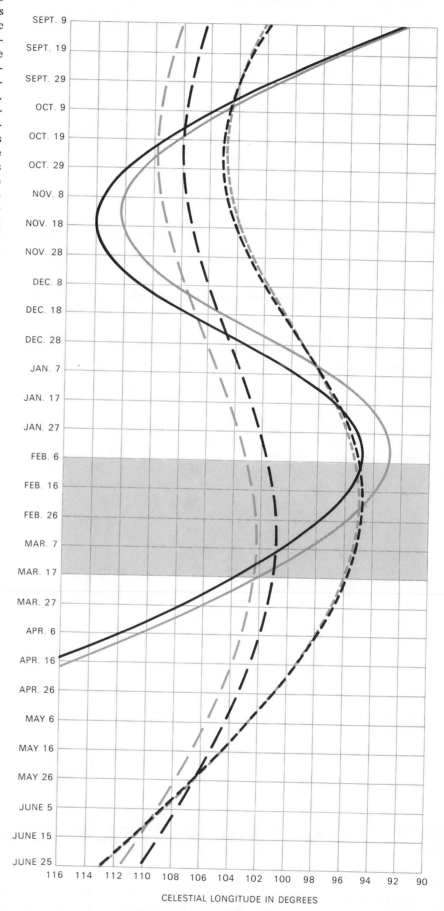

CELESTIAL LONGITUDE IN DEGREES

become convincing evidence that he did observe the conjunctions in 1504.

It would be interesting to know whether or not these celestial events crystallized Copernicus' desire to reform astronomy. If they did, there would be a remarkable parallel between Copernicus and Tycho, who as an impressionable youth three great conjunctions later resolved to devote himself to astronomy when he saw that there were still errors in the ephemerides in 1563. At that time

the Alfonsine Tables showed an even larger error for Saturn, nearly two degrees, so that the prediction of the great conjunction was a month too late. Ephemerides computed by Johannes Stadius from the Prutenic Tables, which were in turn based directly on Copernicus' work, fared much better; they were off by only a day or two, but that was still enough to shock the 16-year-old Tycho.

Even if the discrepancies between the

tables and the heavens impressed Copernicus with the need for reform, he never used the discrepancies as a major argument for his radical new cosmological system. This move was quite prudent; Copernicus knew that such errors could be corrected merely by changing the parameters of the old system. Instead he argued for his heliocentric system on philosophical and cosmological grounds, and a major consideration was its simplicity. In his *Commentariolus*, a small

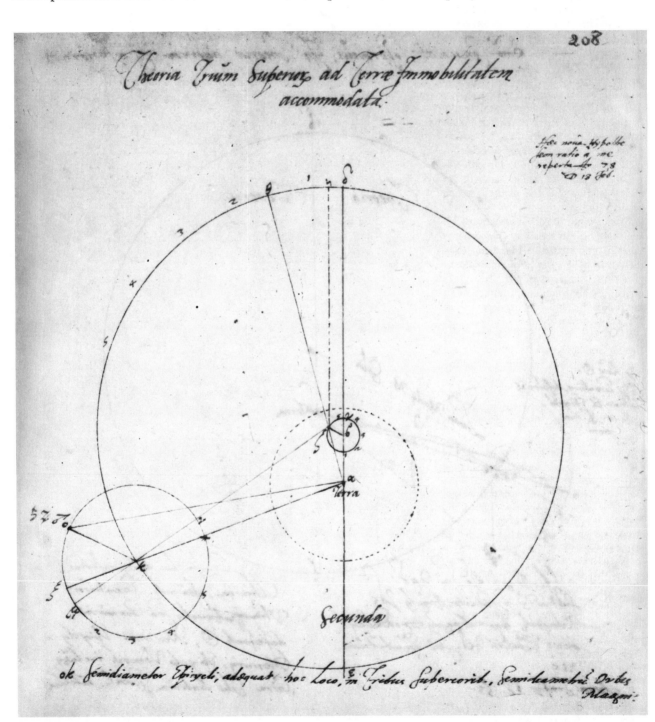

**TYCHO'S FIRST GEOCENTRIC DIAGRAM**, for the superior planets, was found by the author in the unpublished manuscript pages bound into the back of Tycho's personal copy of Copernicus' *De revolutionibus orbium coelestium*, now in the Vatican Library in Rome. It states at the upper right corner: "This new hypothesis occurred to me in the year [15]78 on the 13th day of February."

tract written about 1512 and circulated only as a manuscript, he wrote concerning the planetary motions that "eventually it came to me how this very difficult problem could be solved with fewer and much simpler instructions than were formerly used, if some assumptions were granted me." Copernicus went on to outline his new heliocentric arrangement of the planets, indicating that the apparent celestial motions could be explained by a triple movement of the earth itself: the revolution on its own axis, the orbital motion around the sun and the libration, or wobble, of its axis that accounts for the precession of the equinoxes.

In addition to putting forward the heliocentric concept, Copernicus awarded virtually the same weight to a second philosophical principle: the Platonic-Pythagorean concept that celestial movements must be composed of uniform circular motions. That principle constituted a strong criticism of one of the chief Ptolemaic devices: the equant. In order to understand the equant and the force of this argument against it, one must first notice the chief features of Ptolemy's planetary mechanism. The earth was placed near, but not exactly at, the center of a set of the large deferent circles. Each planet moved with uniform circular motion on an epicycle that was on the deferent. The epicycle produced the planet's periodic retrograde motion among the stars, in which the planet appeared to stop temporarily and then "back up" westward in the sky before stopping again and resuming its usual easterly path. In the Copernican system the epicycle is seen as the result of the earth's own orbital motion around the sun: the retrogression of a superior planet is caused by the faster-moving earth passing that planet. In the case of Mars the irregularity generated by the epicycle to account for retrograde motion could modify the position of the planet by as much as 45 degrees of arc compared with the effect of uniform circular motion on the deferent alone.

In addition to the irregularity of the planet's motion produced by the epicycle, a nonuniform motion of the center of the epicycle on the deferent was also required. To achieve this nonuniform motion Ptolemy devised the equant: an axis of uniform motion off center within the deferent. Only from that position would the planet appear to move with uniform circular motion. In addition the earth was placed equally off center in the opposite direction. In the case of Mars the equant and the off-center displacement could modify the uniform

motion by a substantial amount, up to 12 degrees of arc to the east or west. The use of the equant, the epicycle and the eccentric deferent gave Ptolemy adequate flexibility for coping with the superior planets and with Venus; with the addition of one small central circle he was able to manage Mercury as well.

Copernicus opened his *Commentariolus* with an attack on the Ptolemaic equant, which appeared to violate the principle of uniform circular motion. "A system of this sort seems neither sufficiently absolute nor sufficiently pleasing to the mind," he wrote. "Having become aware of these defects, I often considered whether there could perhaps be found a more reasonable arrangement of circles from which every apparent inequality would be derived and *in which everything would move uniformly about its proper center, as the rule of absolute motion requires.*" A major theme of Copernicus' work thus became the geometric elimination of the equant. Copernicus turned to an epicyclet, or small epicycle, to effect the substitution. The result was that, although he simplified the planetary system by removing the large epicycles, his devotion to the principle of uniform circular motion caused him to introduce new complexities.

At the end of his *Commentariolus* Copernicus remarked: "All together, therefore, thirty-four circles suffice to explain the entire structure of the universe and the entire ballet of the planets." Commentators in the 19th century used their imagination to embellish Copernicus' statement and, without checking the facts, began to propagate the story that Ptolemy's rather simple system had become overlaid with dozens of additional secondary circles by the time of Copernicus.

Alfonso X may have contributed to the legend at the end of the 13th century; he is reputed to have told his astronomers that if he had been present at the Creation, he could have given the Lord some hints. This anecdote has fed the mythology, so that a recent *Encyclopaedia Britannica* article even asserts that 40 to 60 epicycles were required for each planet! Actually even at the time of Copernicus observational astronomy was still in such a primitive state that there could have been no observational basis for such a cumbersome improvement.

The same type of computer-aided analysis that was used to decipher Copernicus' brief note on Mars and Saturn has now laid to rest this popular legend about the overwhelming complexity

of the geocentric system in the late Middle Ages. In order to test the story I carefully recomputed the Alfonsine Tables in their entirety to show that they are based on the classical and simple form of the Ptolemaic theory with only two or three minor changes of a parameter in the entire set. Next I used these 13th-century tables to compute a daily ephemeris for 300 years. When compared with Tuckerman's accurate modern computations of the planetary longitudes, the computerized Alfonsine predictions showed repetitive error patterns that were as characteristic as a fingerprint. All the old pre-Copernican ephemerides, made by the leading astronomers of the day, show precisely these same error patterns. Thus I have concluded that only the Ptolemaic scheme, with a single epicycle for each planet, was ever used for predicting planetary positions.

Is it possible that the epicycles-on-epicycles existed but just did not get to the level of almanac-making? Recent research in Islamic sources by E. S. Kennedy and his students at the American University of Beirut has revealed that elaborate models of precisely this kind were discussed by 13th- and 14th-century astronomers of the Maragha school, in particular by ibn al-Shatir at Damascus. Like Copernicus, they argued for the models on philosophical grounds, and it is unlikely that their scheme was ever used for actual planetary tables. Whether Copernicus conceived of the double-epicyclet replacement of the equant independently or whether he inherited it from the Arabs by some as yet unknown channel remains moot.

During the 1520's Copernicus worked extensively to elaborate his ideas, particularly the planetary theory, if we are to judge by the sprinkling of planetary observations recorded in *De revolutionibus*. By this time he was well settled in northern Poland, but a variety of administrative duties frequently kept him away from his celestial researches. During that time Copernicus wrestled with the writing of a monumental treatise comparable in scope to Ptolemy's *Almagest*. It was to incorporate the twin principles that the sun is at the center of the solar system and that the planetary movements are based on uniform circular motion. The heavenly movements turned out to be more complicated than he had supposed in the *Commentariolus*. In particular the lines of apsides, or the lines between the earth and the equants in the Ptolemaic system, appeared to turn slowly with respect to the stars. Only a few observations, some of them contradictory, had come down through history, and Copernicus must have toiled with some frustration trying to satisfy them all. He eventually abandoned the double epicyclets of the *Commentariolus* in favor of an off-center deferent and a single epicyclet, a scheme that more readily accommodated the shifting apsides. For Mercury, however, he retained the double epicyclet to replace not the equant but the additional central circle postulated uniquely for that

planet by Ptolemy. In the end Copernicus managed to pile up more small circles than Ptolemy.

The major part of *De revolutionibus* is devoted to a detailed analysis of the motions of the sun, the moon and the planets using combinations of large circles and small circles. This analysis, together with a mathematical commentary and tables of stars, constitutes 96 percent of the volume. Scarcely 20 pages of

*De revolutionibus* are devoted to the new heliocentric cosmology. The opening chapters review the ancient arguments for a geocentric world view along with quaintly medieval counterarguments. They give little hint of what is about to come in the chapter titled "On the Order of the Celestial Orbits." That chapter is a resounding defense of the heliocentric system based entirely on aesthetics, particularly on the principle of simplicity. In a powerful plea for

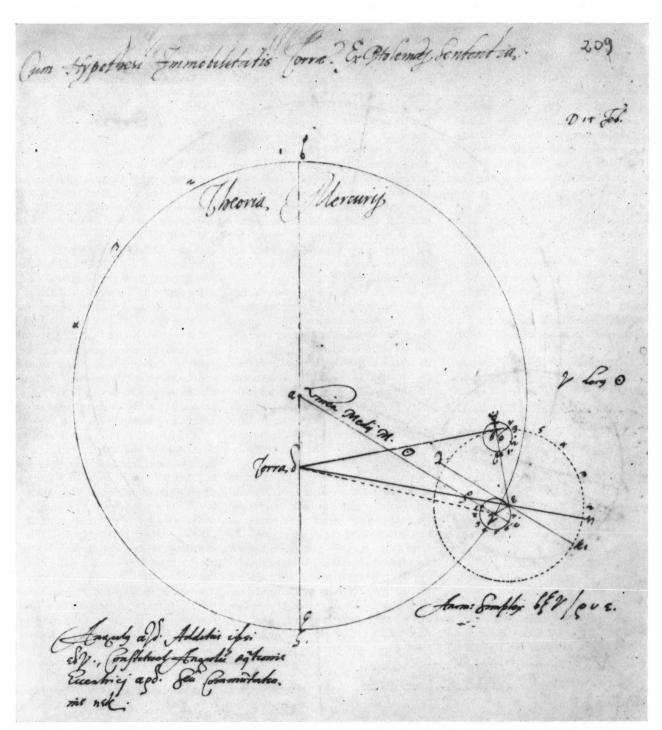

MERCURY HAS TWO EPICYCLETS (*smallest circles*) in Tycho's first attempt to account for the planet's motion in an orbit around the earth. Large circle is the deferent and the dotted one the epicycle. He notes that he drew the illustration on February 14, 1578.

the heliocentric world view Copernicus wrote: "At rest in the middle of everything is the sun. For in this most beautiful temple, who would place this lamp in another or better position than that from which it can light up the whole thing at the same time? Thus indeed as though seated on a royal throne, the sun governs the family of planets revolving around it."

The sun-centered system offered an elegant explanation for retrograde motion, including puzzling details that had no rationale in the Ptolemaic scheme. Furthermore, with the heliocentric arrangement of the planets the spacings between them are no longer arbitrary but are fully specified by the configuration. This is surely one of the most persuasive aesthetic considerations in favor of the Copernican cosmology. There is a whiff of reality here, particularly in Copernicus' resounding conclusion: "So vast without any question is this divine handiwork of the Almighty Creator." Yet very few people in the 16th century grasped the harmonious, aesthetic unity Copernicus saw in the cosmos.

Any thought that Copernicus was talking about a real system must have rapidly vanished from the minds of those few astronomers who managed to plod their way through the remainder of the treatise. Copernicus' application of the second aesthetic principle, uniform circular motion, was far from unambiguous. Like Ptolemy, he occasionally paused to mention alternative geometric arrangements. The crushing blow to any physical reality for the planetary epicyclets, if any were needed, comes in Book Six, where Copernicus was obliged to adopt details for predicting the latitudes of the planetary positions different from those he had used for predicting their longitudes. Although few readers ever managed to get that far, the hypothetical nature of the constructions was stated at the very beginning of the book—not by Copernicus but by an anonymous editor in the printshop. Most major astronomers of the ensuing decades found out that its author was a Lutheran theologian, Andreas Osiander, who probably preferred not to sign a statement in a book written by a Catholic and dedicated to the Pope. Osiander's introduction stated that the author, following his duty as an astronomer, had devised hypotheses so that the planetary positions could be calculated for any time. "But these hypotheses need not be true or even probable," he wrote. "If they provide a calculus consistent with the observations, that alone is sufficient."

The careful reader of De revolutionibus interested merely in the technical details of planetary motion might have been greatly inspired by Copernicus' adherence to the principle of uniform circular motion; he surely would have agreed with Osiander's analysis. On the other hand, a reader with a speculative turn of mind, seizing the aesthetics of the heliocentric principle, might have found himself in violent disagreement with Osiander.

Did the book actually have any careful readers? That question arose in a conversation I had three years ago with another student of Copernicus, Jerome Ravetz [see "The Origins of the Copernican Revolution," by Jerome R. Ravetz; SCIENTIFIC AMERICAN, October, 1966]. In our discussion we speculated that there are probably more people alive today who have read De revolutionibus carefully than there were in the entire 16th century. We counted on the fingers of two hands the candidates for that early era: (1) Georg Rheticus, the Wittenberg scholar who came to Poland and who persuaded Copernicus to publish his book; (2) Erasmus Reinhold, the Wittenberg professor who stayed home and who later composed the Prutenic Tables based on Copernicus' work; (3) Johannes Schöner, the Nuremberg scholar who was closely associated with the printshop and who was the man to whom Rheticus addressed his first published report on the Copernican system; (4) Tycho Brahe; (5) Christopher Clavius, the Jesuit who engineered the reform of the Gregorian calendar; (6) Michael Maestlin, Kepler's astronomy teacher, and (7) Johannes Kepler himself.

At the time I was on a sabbatical leave from the Smithsonian Astrophysical Observatory. Two days after talking with Ravetz I happened to visit the remarkable Crawford Collection of rare astronomical books at the Royal Observatory in Edinburgh. There I admired one of the prize possessions, a copy of the first edition of De revolutionibus, handsomely annotated in legible inks of several colors. As I examined the book I realized that the intelligent and thorough notations were undoubtedly made before 1551, that is, within eight years after the publication date of 1543. My speculation with Ravetz from two days earlier seemed to be knocked into a cocked hat, because it hardly seemed likely that, if intelligent readers of De revolutionibus were so rare, the very next copy of the book that I saw would be so well annotated.

Then a second thought crossed my mind: Perhaps the Crawford copy had been annotated by one of the handful of astronomers we had mentioned. The list quickly narrowed to Rheticus, Reinhold and Schöner, the only ones active before 1550. Internal evidence suggested that the reader was Erasmus Reinhold. Although his name is not in the book, I soon found the initials E R stamped into the decorated original binding. Within moments my initial enthusiasm was dampened when I made a rubbing of the cover: the letter S, previously hidden by a stain, appeared, making the initials E R S. It was not until two weeks later, when I was working at the British Museum and the Royal Astronomical Society, that I established that Reinhold always incorporated the Latinized form of his hometown Saalfeld, using the name Erasmus Reinholdus Salveldenis, in perfect agreement with the initials. Ultimately I was able to obtain additional specimens of Reinhold's distinctive handwriting, which settled the matter beyond all doubt.

One of the most interesting annotations in Reinhold's copy appears on the title page, where he had written in Latin: "The axiom of astronomy: celestial motion is circular and uniform or made up of circular and uniform parts." Reinhold was clearly fascinated by Copernicus' adherence to the principle of circular motion, whereas the paucity of annotations in the first 20 pages show that he was not particularly interested in the heliocentric principle. Apparently he accepted Osiander's statement that astronomy was based on hypotheses. Reinhold was intrigued by the model-building aspects of Copernicus' hypothesis and by the whole idea of alternative mechanisms for expressing the motions of the planets. Whenever such alternatives appeared in the book, he made conspicuous Roman-numeral enumerations in the margins.

Because Reinhold published the convenient Prutenic Tables for carrying out the Copernican planetary calculations, he is frequently listed as an early adherent of the heliocentric cosmology. The nature of the tables, however, makes them quite independent of any particular cosmological system. Moreover, although his introduction is full of praise for Copernicus, nowhere does he mention the heliocentric cosmology. With his great interest in alternative mechanisms, there is reason to suspect that Reinhold was on the verge of independently discovering the Tychonic system, but he died of the plague at an early age before he could consolidate any cosmological speculations of his own.

Flushed with the success of identifying Reinhold's copy, I resolved to examine as many other copies of the book as possible in order to establish patterns of readership and ownership, always hoping to find further interesting annotations. For three years I systematically sought out copies in such far-flung places as Budapest and Boston, Leningrad and Louisville, Copenhagen and San Juan Capistrano. In the process I saw and photographed copies annotated by Kepler (in Leipzig), by Maestlin (in Schaffhausen, Switzerland), by Tycho (a second edition, in Prague) and by Rheticus (a presentation copy in Connecticut). I also found copies annotated by astronomers we forgot to put on the original list: the ephemeris-maker Johannes Stadius (at the United States Military Academy at West Point) and by Casper Peucer, Reinhold's successor as astronomy professor at Wittenberg (at the Paris Observatory). In all, I had managed to see 101 copies of *De revolutionibus* by this past spring. The investigation confirmed that the book had rather few perceptive readers, at least among those who read with pen in hand. In spite of that, however, the book seemed to have a fairly wide circle of casual readers, much larger than has been generally supposed.

In May I had the opportunity to visit Rome, where there were seven copies of the first edition that I had not examined. My quest took me first to the Vatican Library, where I went armed with shelf-mark numbers provided by my Polish colleague Dobrzycki. Some of the books in the Vatican Library had come there with the eccentric Queen Christina of Sweden, who had abdicated her throne in 1654, abandoning her Protestant kingdom for Rome. Her father, Gustavus Adolphus, had ransacked northern Europe during the Thirty Years' War and among other things had captured most of Copernicus' personal library. Dobrzycki

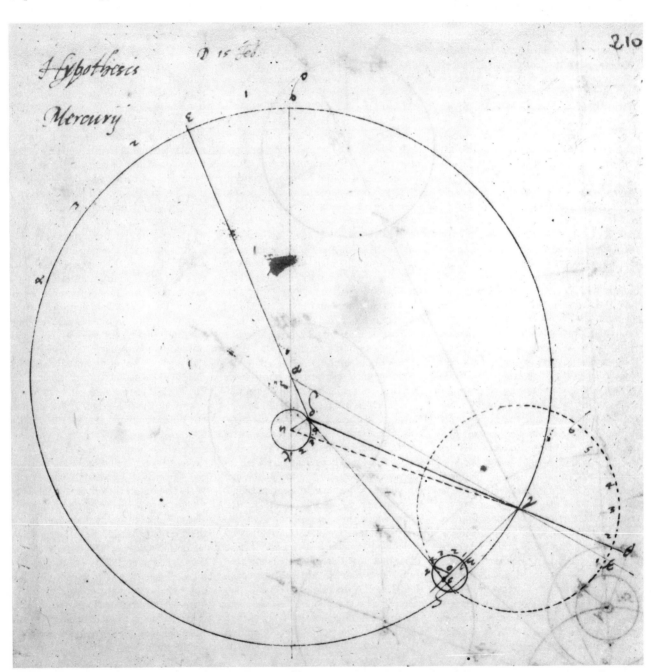

**EPICYCLETS ARE REARRANGED** in Tycho's second attempt to explain Mercury's motion in an orbit around the earth, which he drew on February 15, 1578. In a similar manner he explored alternative models that would account for motions of planet Venus.

had gone to Rome in search of Copernican materials that Queen Christina might have taken along. In the Vatican he found an unlisted copy of Copernicus' book among the manuscripts. Because the book was published just when Copernicus died Dobrzycki knew that it could not have belonged to Copernicus himself, and so he proceeded on to other materials. Fortunately for me, he gave me the number for the volume, which could not have been found in the regular Vatican card catalogues.

When I examined this copy, I realized that the extensive annotations in the margins must have been made by a highly skilled astronomer. At the end were 30 most interesting manuscript pages bound into the volume, full of diagrams made by someone working along the same lines as Tycho and dated 1578. Although there was no name anywhere on the volume, I conjectured that the annotations might have been made by the Jesuit astronomer Christopher Clavius, because his copy of *De revolutionibus* was one I had not yet found. In a state of considerable excitement I got in touch with D. J. K. O'Connell, the former director of the Vatican Observatory. With his help I obtained Xerox copies of two letters by Clavius from the Jesuit Archives. Eagerly I returned to the Vatican Library, only to have my hypothesis smashed within minutes. There was no possibility that the writing in the *De revolutionibus* was the hand of Christopher Clavius.

I left Rome in a baffled and troubled state for a conference on Copernicus in Paris. There, by a fantastic stroke of luck, I received the new facsimile of the second edition of *De revolutionibus* that is preserved in Prague, and which contains annotations by Tycho. I think my heart must have stopped for a moment when I saw the handwriting in the facsimile: I then realized that the first edition in Rome was probably also in Tycho's hand. What I had found in the Vatican Library was Tycho's original working copy, probably the most important Tycho manuscript in existence. The second edition in Prague was a derivative copy; it had been annotated by Tycho with the thought that it might possibly be published.

I rebooked my flights and went back to Rome. After I had put the Prague facsimile side by side with the Vatican copy, it took only a few minutes to prove this conjecture. Afterward the Vatican librarians traced the book to Queen Christina, who must have gained possession of it in 1648 when her troops

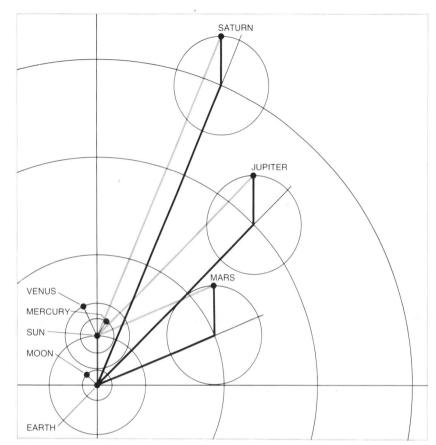

**TYCHONIC SYSTEM** can easily be obtained from Tycho's final diagram (*see illustration on following page*). Vertical line from center of each epicycle to its planet is the same length as the distance between the earth and the sun since the size of the epicycles and of the sun's orbit is the same. All that remains to finish construction of Tychonic system is to complete the parallelogram by drawing a line from sun to each superior planet (*gray*).

captured the collections formed by Rudolph II in Prague.

On the title page of Tycho's copy in the Vatican appears the very same words that Reinhold had inscribed on his copy: "The axiom of astronomy: celestial motions are uniform and circular, or composed of uniform and circular parts." I had already known that in 1575, three years before the dated annotations in this book, Tycho had visited Reinhold's son in Saalfeld and had seen Reinhold's manuscripts. Knowing this, I had guessed that Tycho's own cosmological views grew from seeds planted at Wittenberg by a tradition that honored Copernicus but followed Osiander's admonition that it is the duty of the astronomer to "conceive and devise hypotheses, since he cannot in any way attain the true causes." The newly found Tycho copy dramatically confirmed this intellectual heritage, not only through the motto on the title page but also within the book itself, where numerous annotations are copied word for word from Reinhold. In particular Tycho, like Reinhold, specifically numbered any alternative arrangements of circles indicated by

Copernicus. Like Reinhold, he must have viewed Copernicus as a builder of hypothetical geometric models.

The most exciting aspect of the Vatican volume is the sequence of 30 manuscript pages at the end. Apparently they are very preliminary working notes on Tycho's own geocentric system. The first opening is dated January 27, 1578, the day after the spectacular comet of 1577 had been seen for the last time. The diagrams on those two pages are heliocentric, and a note in the corner indicates that it was drawn according to the third hypothesis of Copernicus, corresponding to one of Tycho's enumerations marked in the margin of the printed book.

Diagrams drawn about three weeks later show a geocentric arrangement for the first time, and Tycho noted that "this new hypothesis occurred to me in the year [15]78 on the 13th day of February" [*see illustration on page 12*]. Seldom can we pinpoint the moment of a centuries-old discovery so precisely! In a series of charts Tycho explored alternative positions of the single epicyclet for Venus and the pair of epicyclets for Mercury.

Two days later, on the very next page, Tycho drew the most interesting diagram of the entire sequence [*see illustration on this page*]. It is easy to recognize as a proto-Tychonic system: the earth is at the center, circled by the moon and the sun. Tycho placed the orbits of Mercury and Venus around the sun. He still arranged the three superior planets in circles around the earth, but he made each epicycle the same size as the sun's orbit. He then drew a line from the earth to the center of each of the three epicycles and connected the center of each epicycle to its planet by a vertical line. This vertical line in each case is the same length as the distance between the earth and the sun by virtue of the fact that the size of the epicycles and the sun's orbit is the same; all that is necessary to finish the construction of the final Tychonic system is to complete the parallelogram by drawing a line from the sun to each superior planet [*see illustration on preceding page*].

Tycho was now surely within grasp of his final system. His caption, however,

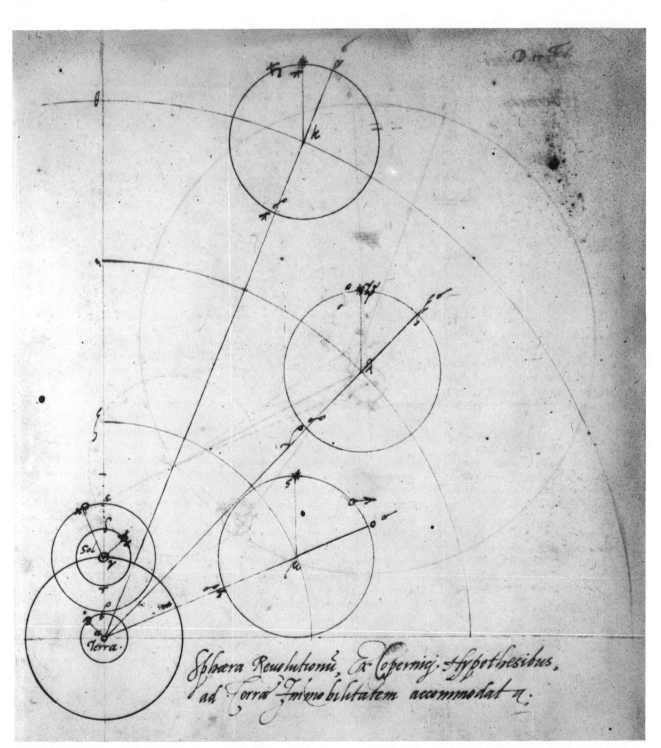

**PROTO-TYCHONIC SYSTEM** was drawn by Tycho on February 17, 1578. It is only one step removed from his final planetary system. In this diagram he shows all the planets. Venus and Mercury circle the sun; the sun circles the earth. The deferents of Mars, Jupiter and Saturn are centered on the earth; the epicycle for each planet is the same size as the orbit of the sun around the earth. Tycho then drew a line from the earth to the center of each epicycle and connected this center to its planet by a vertical line.

is worth notice: "The spheres of revolution accommodated to an immobile earth from the Copernican hypotheses." Tycho was playing the astronomical-geometry game, greatly under the influence of Copernicus, and somehow supposing that a geocentric system was compatible with the heliocentric teachings of the master.

It is curious that Tycho did not publish his new system until a decade later. Tycho was a dynamic young man of 31 when he wrote this manuscript and already well established on the island of Hven, but perhaps still uncertain where his observations for the reform of astronomy would lead him. A passage in his *De mundi aetherei recentioribus phaenomenis* implies that he did not establish the Tychonic system until around 1583, five years after he drew these diagrams. I can only suppose that those five years were an important time of maturing. In that interval Tycho must have speculated on the movement of the great comet of 1577, realizing that it would have smashed the crystalline spheres of the ancient astronomy, had they existed, as it moved through the sky. Perhaps he began to look for greater certainty in astronomy and to suppose that the observations made with the giant instruments at his Uraniborg observatory could lead beyond hypothesis to physical reality. If so, like his contemporaries in that pre-Newtonian, predynamical age, he must have viewed the physics of the sluggish, heavy earth as a most important phenomenon to be preserved. We can well imagine that Tycho believed he was taking a great step forward toward understanding the physical reality of the universe when he adopted his own geocentric system.

Curiously enough, Tycho's system retained many of the advantages offered by Copernicus: the planetary orbits were linked together in a coherent unit, and retrograde motions were "naturally" explained. Although his scheme conceded little to the special properties of the sun among the planets, it carefully preserved the time-honored uniqueness of the central earth. Yet there was one aspect, absolutely crucial for the further development of physics, lacking in the Tychonic arrangement. In the Copernican system the planets are harmoniously ordered out from the sun so that the shorter the period, the closer the distance. It is this pattern that opened the way to the mathematization and mechanization of the universe.

In retrospect the Tychonic system, which weighs so heavily in Urania's balance, looks clumsy and wrong—a monumental step backward. Thus Tycho is today rejected as a cosmographer and he is lauded as the ingenious instrument builder whose systematic observations provided the foundations for Kepler's laws. Kepler in turn is hailed as the number-juggler whose elliptical orbits finally vanquished the ancient requirement of uniform circular components for celestial motions.

Weighed against the Tychonic scheme, the heliocentric system appears neat and orderly to us living after the time of Newton. Indeed, it is precisely this elegant organization that Copernicus found pleasing to the mind, and which led to his cosmology. Early in *De revolutionibus* he wrote: "In this arrangement, therefore, we discover a marvelous symmetry of the universe, and an established harmonious linkage between the motion of the spheres and their size, such as can be found in no other way."

It was precisely in this arrangement that Kepler saw the real possibility of a celestial physics, and he took the first groping steps toward a dynamics of the heavens—dynamics that, when they were reshaped and powerfully formulated by Newton, ultimately proved to be the primary justification for the heliocentric universe. It may well be that Kepler's vision of a cosmic physics shaped the development of science far more significantly than the discovery of his three laws of planetary motion. And quite possibly Tycho's own insistence on a physically acceptable and not merely hypothetical astronomy influenced the young Kepler's views on the nature of the universe. It was in this tradition that both Kepler and Galileo taught us to use our senses to distinguish between the various hypothetical world views, allowing only the ones that were consistent with the observations.

Only in our own generation have we been able to break the terrestrial bonds; men flung out toward the moon have seen the spinning earth, a blue planet, sailing through space. Copernicus' daring concept has been vindicated. Seen by him only in the mind's eye, the concept of the moving earth became the essential first step toward a physics that embraced both the earth and the sky. Thus in reality the Copernican quinquecentennial celebrates the origins of modern science and our contemporary understanding of the universe. Fittingly we honor not only the renowned Polish astronomer but also his illustrious successors—Kepler, Galileo, Newton, and, in a new light, Tycho Brahe.

The 4-meter (158-inch) Nicholas U. Mayall
telescope at the Kitt Peak National
Observatory. Photographs taken with this
telescope are shown on pages 150, 172,
220, and 260. [Kitt Peak National
Observatory.]

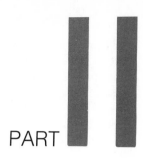

PART

# TELESCOPES AND OBSERVATORIES

"Edison planned to put a loop of telephone wires around a huge field of iron ore in New Jersey!" Gart Westerhout offers this anecdote among others in "The Early History of Radio Astronomy." He mentions that toward the end of the nineteenth century, several decades before the first successful detection of cosmic radio waves, the radio noise pollution from the city of Liverpool was sufficient to interfere with Sir Oliver Lodge's experiments. Unfortunately, technological progress has been accompanied by an increase in the number of sources of interference. In 1974, for example, astronomers complained that signals from artificial satellites interfered with some of the measurements made by sensitive radio telescopes at the National Radio Astronomy Observatory. (This observatory, located in Green Bank, West Virginia, and the Kitt Peak National Observatory near Tucson, Arizona, offer exhibits and facility tours to the public that are well worth the visit.)

". . . the two components appear to be receding from a common point of origin with an apparent velocity about three times the velocity of light." Kenneth I. Kellermann reports on a major development in the recent history of radio astronomy: the remarkable discoveries made possible by the technique of very long baseline interferometry (VLBI). International cooperation among scientists is mandatory for this research because it is necessary to coordinate simultaneous measurements by radio telescopes located on opposite sides of the earth. In "Intercontinental Radio Astronomy," Kellermann describes an astonishing finding made possible by VLBI, namely, the motions of structures in quasars at apparent speeds greater than the speed of light. He also explains several theoretical models that may resolve this apparent contradiction of Einstein's conclusion that objects cannot reach the speed of light. (Incidentally, there is nothing in the theory of relativity that forbids objects from travelling *faster* than light. It is just motion *at* the speed of light that is theoretically impossible. Several physicists are conducting experiments to search for the *tachyons*—hypothetical atomic particles that move faster than light.)

"There won't be one telescope, or two, but tens or even hundreds. . . . Most nights the telescopes will be searching in different directions, testing different ideas. Now and again, as night falls, you will see them turn as one to focus on some star or nebula at the remotest outpost of space-time." In "Telescopes for Tomorrow," Mike Disney, an astronomer who participated in the famous discovery of optical pulses from the Crab Nebula pulsar, relates some ideas and hopes for the future of astronomical observatories.

# The Early History of Radio Astronomy

Gart Westerhout

3

The organizers of this Conference have asked me to present a paper on the early history of radio astronomy, using personal experience where possible. Upon further inquiry, I found that this history should end around early 1950, which makes the task of using personal experience rather hard, as I started graduate studies in that year. The literature on the subject contains a wealth of material on the early days, however, and since the science is so young, the historian can still talk to many of the original investigators. Since most of the early history can be found in the introductions to textbooks and popular books, this paper will necessarily contain many items well known to the reader.

Contrary to popular belief, radio astronomy did not start with Jansky's discovery in 1932, but actually very shortly after Hertz produced and measured Maxwell's electromagnetic waves in 1887. And believe it or not, it was Thomas Alva Edison himself who, in 1890, proposed an experiment involving a radio telescope weighing many megatons! Edison reasoned that since disturbances were seen on the sun in visual light, they might also radiate at radio wavelengths. Edison planned to put a loop of telephone wires around a huge field of iron ore in New Jersey! The ore was magnetite, which becomes magnetized by induction. The electromagnetic disturbances from the sun might magnetize the ore; this might cause an induction current in the telephone wires, which could then be listened to. The poles for the telephone wires actually arrived on the site, but there is no record of the actual experiment having taken place. Solar radiation would not have been detected, firstly because even this huge detector would not have been sufficiently sensitive, and secondly because radiation of wavelengths long enough to be picked up by this equipment would not have penetrated the ionosphere.

Two of Hertz's associates, Wilsing and Scheiner, tried to look for radio waves from the sun around 1896, and it may well be that Hertz himself was involved. In England, Sir Oliver Lodge made a serious effort to look for solar radio waves between 1894 and 1900. He placed a very crude detector "behind a blackboard or some other opaque substance," thus shielding out any visible light. Interestingly enough, the electrical interference in the city of Liverpool, where these experiments were made, was already strong enough to make Lodge give up his early attempts. Shortly afterwards, in 1901, a Frenchman, C. Nordmann, made another attempt to detect solar radio waves but was equally unsuccessful. Nordmann applied reasoning that was well ahead of his time. He went to the top of a high glacier, to be away from man-made radio noise and as high above the atmosphere as he could get. He suggested that the radio emission from the sun might vary with the solar cycle, and as a consequence that the radio emission from the sun might well originate in sunspots. Unfortunately, he only conducted his experiment for one day (the glacier was perhaps too cold under his feet) and then gave up. If he had been more patient, and had waited for a solar maximum (1901 was a solar minimum), he might have succeeded.

The reason for the long delay in further experimentation after these initial attempts may well be due to Planck, who announced his radiation theory in 1902. If this theory were correct, it followed that the radiation from the sun should be blackbody radiation, and the radio emission from a 6000°K blackbody was, of course, undetectable at that time. Similarly, radiation from the stars should be undetectable and one could in fact calculate that the planets and the Milky Way could not possibly give any detectable signals either. This is the unfortunate type of preconceived notion that throughout the history of science has delayed progress considerably, but fortunately is more than offset by chance discoveries or the occasional appearance of a brilliant theoretician or observer. As far as I know, no one consciously tried to look for radio waves from space until after Jansky's accidental discovery in 1932.

Everybody, of course, is familiar with the famous discovery of radio waves by K. G. Jansky, who in 1930 installed a novel rotatable directional antenna system to study the characteristics of radio noise as a function of time and direction. Jansky, of course, discovered thunderstorms both locally and at large distances, but he also noted a steady radiation that quite clearly

From *Annals of the New York Academy of Sciences,* vol. 198, pp. 211–218, 1972. Reprinted with permission.

was not caused by his receiver and that varied with direction and time. It is a great tribute to Jansky's scientific mind that, knowing nothing about astronomy when he started, he first eliminated the sun as the origin of the radiation; then, after having collected several months' worth of data, he deduced that the time difference in the reception of maximum intensity of this radiation (four minutes per day) must mean it had its origin outside the solar system. A year's observation enabled him to establish the position of the maximum of the radiation with an accuracy which, despite his long wavelength (14.6 meters) and his very wide beam width, was probably within a few degrees of being right. His paper, delivered before the Washington meeting of the International Scientific Radio Union (URSI) in 1933, marked the true beginning of radio astronomy.

The discovery was well publicized by Jansky's employer, the Bell Telephone Company, and radio listeners throughout the United States were allowed to listen in to the "hiss of the Universe." Jansky continued his observations off and on for another year or so, and published his final paper[1] in 1935. He concluded that the distribution of the sources of the radiation along the Radio Milky Way was disk-like, similar to the distribution of stars in our Galaxy. He ruled out the stars as a source of radiation, as he could not detect any radiation from the nearest star, our sun. He also found that the characteristics of the noise were similar to those from the electric charges in a resistor, and suggested that the source of the radiation was in fact very hot charged particles in the interstellar medium. How right he was!

Jansky's was the first truly observational paper in radio astronomy. It was presented at the Detroit meeting of the Institute of Radio Engineers (IRE), and there were only 20 people in the audience!

Jansky's ability as a true scientist stands out even more strongly if we compare his achievements and deductions with those of several people in the United States and Japan who had noted a high degree of noise in their receivers before and after radio fadeouts. In 1938 an American paper suggested that this might be caused by charged particles from the sun hitting the ionosphere; in 1939 a Japanese paper concluded that the noise happened only during the daytime and came from high elevations. No one apparently thought of electromagnetic radiation from the sun itself.

Jansky's final proposal was that for better resolution and steerability, a 100-foot paraboloid should be constructed; preferably by one of the universities, where basic research has its place. It is sad to realize now that had the astronomical community listened, we might have advanced in our science much more rapidly than we did. We lost a great opportunity here, and we lost another because between 1932 and 1946, when radio astronomy really started, the low-frequency end of the radio window became almost entirely fogged over by the rapidly growing radio communications industry. A warning is in place here. Astronomers have to do battle for every Hertz of radio frequency; even the very shortest wavelengths penetrating through our atmosphere are threatened by our "civilization." Galileo was the first to put the optical telescope to scientific use, yet it was not really used extensively until the nineteenth century; the sunspot cycle was not discovered until

1840. Astronomy moved slowly in those days. It still moved slowly in the 1930s. Is it really moving faster in the 1970s?

Anyway, Mother Bell decided that now that we knew the radiation was there, let someone else discuss it further: Jansky must go back to the business at hand. Luckily, someone else did want to discuss it further. Kraus and Adel[2] briefly looked for radiation from the sun in 1934, but it was not until Grote Reber appeared on the scene that the subject was picked up again properly. Reber's account[3] of these early years is a gem to read. Reber reasoned that Planck's law predicts that "for radio waves at any probable temperature, the intensity per unit bandwidth is proportional to the square of the frequency"; he also knew of course that the higher the frequency, the better the resolution. Therefore, a very high frequency (or short wavelength) was indicated. With the occasional help of the village blacksmith, he constructed his backyard 30-foot diameter paraboloid in his hometown, Wheaton, Illinois, in 1937; the instrument had a surface accuracy of a few centimeters, and he had the foresight to leave a central hole for a possible future Cassegrainian or Gregorian arrangement, a scheme not used in radio astronomy until some 25 years later.

Reber started his observations at a 9-cm wavelength, without success. He promptly concluded that Planck's blackbody law was not valid for celestial radiation! So he went up to 33 cm (still a wavelength forty times smaller than Jansky's) and a sensitivity a hundred times greater. Again no success. This time he concluded that "perhaps the actual relation between intensity and frequency was opposite from Planck's Law," a conclusion which was later shown to be correct. "Being a stubborn Dutchman," says Reber, "this whetted my appetite even more," so up he went to 187 cm. Why did he choose that particular wavelength? Well, a circular wave-guide type antenna could be constructed for that wavelength out of a standard-size piece of aluminum! In spring, 1939, the equipment was installed, but the Milky Way transited during the day when the automobile interference was bad and the heat of the sun put the receiver out of balance. Reber was thus restricted to very cloudy mornings falling on weekends, when he was not at work (remember that this was a spare-time occupation) and there were few cars. He was finally successful. He located the Milky Way, and kept improving his equipment, tabulating meter readings every minute, and mapping point by point.

The first results were put before the astronomical community in the Astrophysical Journal in 1940.[4] When Otto Struve, the editor at that time, received the paper, he could find no referee for it. It was the first paper on radio astronomy submitted to an astronomical publication. The story goes that Struve finally traveled to Wheaton, inspected the telescope, talked to Reber, and accepted the paper!

A second paper, presenting a complete map, appeared in 1944.[5] In that paper Reber mentioned in passing the sun also emitted radio waves, and that if these were really of thermal origin, this would mean that the sun had a temperature of about a million degrees, "which had no meaning at the time."

In the meantime, Hey in England, working with meter-wave radar, and Southworth in the United States working at 3 to 10 cm, discovered that the sun was a powerful and highly variable radio source (but this information was of a military nature and thus was not generally available). On one day in 1942 all British radars were jammed by the enemy, so badly that some people feared a major attack. The enemy turned out to be the sun, and for many a student of very faint galactic and extragalactic sources, the sun remains an enemy, just as it is for the stellar observer, albeit for different reasons. Reber detected solar bursts in 1946, when he switched to 62-cm wavelength. The sun was strong, but on some days he also noted more automobile interference during solar transit than before or after! "Why the cars should pick this time to be particularly objectionable seemed quite mysterious," until he moved the antenna away from and back to the sun, and found that all radio waves came from the sun itself. The radio bursts from the sun, often tens to thousands of times more intense than its thermal emission, have at first sight very much the same characteristics as automobile ignition interference.

This then takes care of the early history of radio astronomy. What happened after the war? The radar and radio engineers found themselves with laboratories full of equipment and no prospective new enemy in sight. So at last they could devote their time to real basic research. And some of them turned to radio astronomy. I shall list only a few of the many discoveries that followed in rapid succession between 1945 and 1951. One man needs special mention. He is J. S. Hey, the wartime radar researcher. He was responsible for three major discoveries, maybe one third of all the major discoveries made in the subject so far. He discovered and described the radiation from the sun; he discovered that meteor trails produce radar echoes and detected new streams of meteors, thus starting a new era in meteor research; and he was the first to find a small source of radio emission, a radio "star," the famous extragalactic source Cygnus A.

This latter discovery warrants our attention for a moment, as it shows again the deductions an eminent scientist can make. During a survey of the sky at a 5-m wavelength, Hey, Parsons and Phillips[6] found short-period fluctuations (of the order of minutes or seconds) in the constellation Cygnus. The fluctuating source was smaller than two degrees, and Hey deduced that the fluctuations must come from a very small source, such as a star, rather than from the general interstellar medium which was thought to be responsible for the Milky Way emission. (Later it was found that the fluctuations were due to scintillations, the effect of the earth's ionosphere, just as small angular-size stars scintillate more violently than large planets.) Thus the first radio source was discovered.

In Australia, meanwhile, the Commonwealth Scientific and Industrial Research Organization (CSIRO) radio physics laboratory switched to peacetime research. A 1944 preliminary attempt by Kerr to observe the Milky Way was frustrated by wartime research, but in 1945 J. L. Pawsey and his colleagues started to study the sun, and for this they invented a novel instrument, the cliff interferometer. An old wartime radar antenna, mounted on a cliff overlooking the sea, was used as a Lloyds mirror interferometer, receiving both the direct rays from a rising or setting object and those reflected by the sea and thus traveling along a longer path. The interferometer provided high angular resolution and Pawsey was able in 1946 to locate some very strong disturbances over a huge sunspot that fortunately appeared a few days after the equipment had

been assembled. The Australian work on the sun led to the development of P. Wild's most sophisticated solar radio spectrometers and interferometers. Throughout the Australian development Pawsey was the leading spirit, and the leading position of Australia in radio astronomy is largely due to his enthusiasm and inspiring leadership, which continued until his early death in 1962.

In England, Martin Ryle and his group developed the two-element interferometer, and likewise studied the sun in 1946. Ryle's work, of course, led to the extensive development of the field of radio interferometry, which, with all its refinements is ultimately the only way that very high resolutions can be reached. Ryle could undoubtedly be called the father of both radio interferometry and radio cosmology. He and his group in Cambridge have set the pace for the investigation of extragalactic radio sources, which early in the 1950s led to the realization that the enormous intrinsic brightness of some types of sources could widen our observational horizon by an order of magnitude and eventually could help solve some of the burning questions of cosmology.

Although investigations of many different types were started in many countries, it is undoubtedly true that Ryle in England and Pawsey in Australia were by far the outstanding leaders in the field until the early 1950s.

Going back to the Australians for a moment, John C. Bolton and Gordon Stanley, after having seen Hey's discovery of a radio "star," put the cliff interferometer to work, found Cygnus A to be less than seven minutes of arc in diameter, and went on to discover several more of such discrete sources. They were the first to identify (in 1947) such a source with a visible object; this was the Crab nebula, the remnant of the supernova that exploded in the year 1054. This was a daring identification, as their positional accuracy was relatively good in one direction but not in the other. In order to get a different look at these sources they took their cliff interferometer to New Zealand, where instead of looking east across the ocean, they could look west. The accurate time of their rising *and* their setting allows a much better determination of the sources' coordinates.

The galactic radiation was mapped at a number of different wavelengths, confirming Jansky's and Reber's results. It was typical of that time that even though the lack of resolution of those early telescopes was well known, Oort and I in 1951 compared Bolton and Westfold's radio map with the distribution of mass in a model of the Galaxy, and found the two to be rather similar: the sources of radio emission seemed to be distributed much like the ordinary stars. A few years later it became clear that this was just a coincidence; with better resolution, the radio emission was found to be concentrated much closer to the galactic plane. The mechanism of this emission became clear around 1950 through the work of Alfvén and Herlofson and Kiepenheuer, later developed in detail by Ginzburg and Shklovski. Although the radiation at high frequencies is clearly due to thermal emission from hot gas, the low-frequency radiation is due to the synchrotron process, first detailed by Schwinger in the early 1900s. Verification of this theory was obtained in the late fifties, by the detection of the predicted linear polarization of the galactic radiation.

My final point will center around the prediction of the

21-cm line. In a symposium held in Leiden, The Netherlands, in 1944, the field of radio astronomy was discussed. Van de Hulst[7] concluded that astronomers did not pay much attention to the existing data on radio astronomy, because they were too rough; no good discussion was possible. This seems to have been the general attitude of astronomers in those days: "We'll be glad to discuss and use your data, but you radio engineers had better come up with something good first." Van de Hulst also concluded that in any case the production of radio waves in the interstellar gas was such a small portion of the energy budget that purely theoretical considerations of their production would not teach anything new! After having damned radio astronomy in this way, however, Van de Hulst produced what in essence was the first scholarly paper in the field, containing three (not one) predictions which all came true. The paper is an outstanding example of deep insight into the field of astrophysics, of how to apply that insight to a new and unexplained phenomenon, and of how to direct further observations. Expanding on Henyey and Keenan's 1940 article[8] on the spectrum of the free-free emission from the interstellar gas (Reber[4] was actually the first to point out that free-free emission is the source of the radiation) he noted the uncertainties in the theory, as most of the hot gas is in H II regions and is not smoothly distributed. He also discussed the possibility of observing the H II regions individually, and calculated the strength of the recombination lines (presently a source of all sorts of new information about H II regions). He concluded that the theoretical spectrum more or less fitted Reber's data, but that Jansky's low-frequency data were ten times too high. "It seems possible," he said, "that this is due to his poorly known directivity." The astronomer was again reluctant to accept new data that did not fit his theory.

At the suggestion of Oort, Van de Hulst made a very interesting calculation about how at radio wavelengths we can distinguish between an expanding and a static universe, because the spectrum of the radio emission is entirely different from that of the optical and ultraviolet spectra of galaxies. He could completely eliminate, on the basis of the observations then available, the theory of the static universe, where red shift is not due to the Doppler effect. Van de Hulst thus started here the field of radio cosmology!

Finally, in half a page, he showed that the spin transition should be observable in the neutral hydrogen atom at 21 cm. His discussion followed Oort's suggestion that if spectral lines could be found in the radio region, the kinematics and dynamics of the entire Galaxy could be studied. Van de Hulst's paper, therefore, was a milestone in three different subjects: the 21-cm line, radio cosmology, and the hot interstellar gas. He has never been given credit for the latter two.

As we know, it was not until 1951 that Ewen and Purcell, followed within two months by the Dutch group and the Australians, detected the line. Reber might have detected it in 1947, when he assembled a receiver for it, but because of personal reasons he "terminated his operations in Illinois" before he tried it out.

Should we go on here? I don't think so; we come to textbook material.

Some memories might be in place, such as those of the early days in Kootwyk, The Netherlands. The classical map of the spiral structure of the Galaxy depends on the data obtained between 1953 and 1955, using a 25-foot dish, moved in elevation and azimuth (to follow a point in the sky) by turning two small hand cranks every 2.5 minutes for two years (student labor!), and on data obtained from the Australian 40-foot dish, which could not be cranked at all as it was a transit instrument. I still remember the reduction of the first real data, leading to the classical 1952 paper by Van de Hulst, Oort, and Muller, outlining the rough spiral structure. All Leiden astronomy students were herded into the lecture room and were given sheets of recorder paper, and within a week we had everything reduced to intensities as a function of velocity and position.

Fred Whipple recalls the first grant given to radio astronomy by an astronomical institution: $50, given to him by H. Shapley in 1937 to buy a radio receiver. Whipple and Greenstein[9] had just finished a paper showing that radiation from interstellar dust could not possibly explain Jansky's observations. Whipple was going to build a diamond-shaped (rhombic) antenna on the top of the 61-inch dome at the Agassiz Station. The dome would provide an excellent base for a rotatable telescope!

Finally, an interesting question: why were astronomers slow to accept radio as another tool? Was it fear of electronics? Maybe, but I think Van de Hulst summed it up much better: we had the feeling that, with perhaps a few exceptions, nothing new could be learned. And then too, there was so much to do in the old established field. The mountain observatories were buzzing. Baade wrote to Bok in 1953, when Bok started radio astronomy at Harvard: "Why bother, the Dutch have done it all." Radio astronomy seemed so far removed that it was not really considered astronomy, but rather engineering; the results should be treated as entirely separate from those of "real astronomy." How many schools still teach a course called "radio-astronomy" in which the sun, galactic structure, and cosmology are all discussed? I suspect there are many. But are there courses called "photography" where these subjects are discussed? I doubt it. At the University of Maryland, the teaching of radio astronomy fits into the observatiohal astronomy cycle, but quasars fall under extragalactic astronomy, and the sun belongs under solar physics.

One man stands out as a classical astronomer who very early in the game saw the importance of radio astronomy as a tool and used it as a tool and pushed for it as a tool, from the time when he first knew of the existence of radio waves from space. Jan H. Oort organized the Dutch Symposium of 1944; he acquired old German radar dishes in 1946 before they were demolished; he started planning for a large telescope in 1950. In 1956 the Dutch 82-foot telescope was the largest radio telescope in the world (followed shortly by the Jodrell Bank instrument). Planning for the largest radio telescope now in operation, the Westerbork Array, started in 1959. I recall a note Oort wrote on top of a preprint of Ryle's first cosmology paper, around 1951: "This is the most fascinating and far-reaching paper I have ever seen." Oort realized that radio astronomy was a new tool. I think that realization was lacking in almost all the astronomical world, and it is only in the last ten years, when striking new discoveries have been made, not "just things we know already, like galaxies and H II regions," that the full impact of radio techniques on astronomy has been realized. We have finally woken up, but it took $10^{60}$-erg alarm clocks!

## Acknowledgments

A list of references to early papers would cover many pages. I have relied heavily on the excellent 1958 Radio Astronomy Issue of the Proceedings of the IRE. The introductory article by Haddock in that issue contains a wealth of references.[10] I have also relied heavily on the introductory chapters of many textbooks and popular books on radio astronomy, notably those by Kraus,[2] Smith[11] (to whom I owe the story on Edison), and Smith and Carr[12] (where I found the story on Nordmann). Because of the abundance of references in other review papers, notably Haddock's,[10] I have given very few references in this paper.

## References

1.  Jansky, K. G. 1935. A note on the source of interstellar interference. Proc. IRE 23: 1158.
2.  Kraus, J. D. 1966. Radio Astronomy. Penguin Books. Baltimore, Md.
3.  Reber, G. 1958. Early radio astronomy at Wheaton, Illinois. Proc. IRE 46: 15.
4.  Reber, G. 1940. Cosmic static. Astrophys. J. 91: 621.
5.  Reber, G. 1944. Cosmic static. Astrophys. J. 100: 279.
6.  Hey, J. S., S. J. Parsons & J. W. Phillips. 1946. Fluctuations in cosmic radiation at radio frequencies. Nature 158: 234.
7.  Van de Hulst, H. C. 1945. Radiogolven uit het wereldruim; herkomst der radiogolven. Ned. Tijdschr. Natuurk. 11:210.
8.  Henyey, L. G. & P. C. Keenan, 1940. Interstellar radiation from free electrons and hydrogen atoms. Astrophys. J. 91: 625.
9.  Whipple, F. L. & J. L. Greenstein. 1937. On the origin of interstellar radio disturbances. Proc. Nat. Acad. Sci. U.S. 23: 177.
10. Haddock, F. T. 1958. Introduction to radio astronomy. Proc. IRE 46: 3.
11. Smith, F. G. 1960. Radio Astronomy. McGraw-Hill. New York, N.Y.
12. Smith, A. G. & T. D. Carr. 1964. Radio Exploration of the Planetary System. Van Nostrand. New York, N.Y.

# 4

# Intercontinental Radio Astronomy

K. I. Kellermann

Until recently the images of celestial objects formed by radio telescopes lacked the detail of those formed by optical telescopes. The reason is that the resolution of a telescope increases with the ratio of its aperture to the wavelength of the received signal. Since radio waves are roughly a million times longer than light waves, it has been generally accepted that radio telescopes are fundamentally limited to a poor angular resolution compared with optical telescopes.

Actually this is not the case, for two reasons. First, the resolution of large optical telescopes is limited not by their size but by irregularities in the earth's atmosphere. The limit is about one second of arc—only about 100 times better than the unaided human eye. At radio frequencies the fluctuations in the length of the path of the incoming signal through the atmosphere are small compared with the length of radio waves, so that the effect of atmospheric irregularities is much less important. Second, the radio signal or the optical signal must be coherent, or in phase, over the entire dimensions of the telescope. Coherent radio waves are much easier to manipulate than coherent light signals, so that radio telescopes can operate much closer to the theoretical limit of resolution than optical telescopes.

The commonest form of radio telescope is the steerable paraboloid, which typically has an aperture, or diameter, between 10 and 100 meters. Although the resolution of a given aperture increases as the wavelength decreases, the performance begins to deteriorate significantly when the wavelength approaches the dimensions of the structural imperfections in the antenna. Since for obvious reasons the largest antennas have the least precise surfaces, the best resolution that has been obtained with paraboloidal antennas does not depend strongly on wavelength and is about one minute of arc. Although it is possible to build more precise large antennas, it does not seem feasible to achieve a resolution much better than .1 minute of arc in this way. For that reason radio astronomers have turned to interferometry, where in effect two relatively small antennas act as opposite edges of a single huge radio telescope.

Interferometers working with light waves were employed for astronomical purposes as early as 1920. At that time A. A. Michelson and F. G. Pease used an instrument with two mirrors separated by as much as 20 feet to measure the diameter of a few bright stars. In the Michelson type of interferometer a light wave is intercepted by two separated mirrors that reflect two beams to a common point, where the beams are combined [see illustration on page 30]. If the path of one beam is made slightly longer or shorter than the other, the light waves in one beam will be out of phase with the waves in the other. When the beams are combined, the two trains of waves will interfere both constructively and destructively, and if one looks at the combined beam, one will see a pattern of alternating light and dark "fringes." In a radio interferometer the advancing wave front from a celestial object simply falls on two separate radio telescopes, and the signals are carried to a common point and compared electrically.

Attempts to extend the optical interferometer to longer baselines in order to obtain higher resolution have been unsuccessful, primarily because it is difficult to keep the light waves in the system coherent and to maintain the alignment of the mirrors within a fraction of a wavelength. With a radio instrument the problems of transmitting the signals from each interferometer element to a common point are much simpler. Thus two or more radio antennas can be employed to synthesize large apertures and achieve high angular resolution. With such an instrument one can determine not only the size and shape of discrete radio sources but also their precise position in the sky.

As the spacing of radio interferometers has increased and as advances in technology have made it possible to work with shorter wavelengths, the resolution obtained by radio instruments has steadily improved. Although many radio sources are well resolved by an interferometer that has a baseline about a kilometer long and a resolution of the order of one minute of arc, it was realized by the late 1950's that further increases in resolution would be required to study the structure of these sources in more detail and to resolve the smaller sources.

It is now generally accepted that the radio emission of discrete sources such as radio galaxies and quasars is "synchrotron" radiation from electrons moving in weak cosmic magnetic fields at relativistic speeds, that is, speeds close to the velocity of light. The energy required to account for the observed radiated power is exceedingly large, and the problem of the origin of this energy and its conversion to relativistic particles has been one of the most challenging in modern astrophysics [see "The Astrophysics of Cosmic Rays," by V. L. Ginzburg; SCIENTIFIC AMERICAN, February, 1969].

From Scientific American, vol. 226, pp. 72–83, February 1972. Copyright © 1972 by Scientific American, Inc. All rights reserved. Reprinted with permission.

Soon after the discovery of galactic and extragalactic synchrotron emission it was thought that the relativistic particles might be accelerated by what is called the Fermi process, in which the electrons repeatedly bounce off moving magnetic clouds that act as magnetic mirrors. Because head-on collisions are more frequent than overtaking collisions (as driving on the wrong side of the road will quickly demonstrate) the electrons can be accelerated up to relativistic velocities.

Estimates of the characteristic time scale for the electrons to reach sufficient energy in this way to radiate at radio frequencies range from a million to 100 million years. The recent discovery of rapid variations in the intensity of some radio galaxies and quasars, however, implies that the time scale of acceleration is measured in months rather than in millions of years. That precludes the possibility of any acceleration of the type described by Fermi.

Initially the observations of radio galaxies and quasars were concentrated at relatively long wavelengths (near one meter), but as techniques at the shorter wavelengths have improved there has been increasing emphasis on the centi-meter and millimeter wavelengths, particularly in the U.S. and the U.S.S.R. Today two essentially different types of radio source are distinguished. One has a large angular extent and is strongest at the longer wavelengths. The other is relatively compact and is strongest at the shorter wavelengths. Somewhat surprisingly there is no simple relation between the angular extent of the radio emission and the optical emission from galaxies and quasars. Compact radio sources are not restricted to quasars; many are identified with the nuclei of galaxies. Moreover, many quasars are large extended radio sources.

The large radio sources have a complex distribution of radio emission that typically extends over several hundred thousand light-years of space, corresponding to angular dimensions between a few arc seconds and a few arc minutes. Many of these "extended sources" consist of two or more spatially separated components whose dimensions are of the order of half of the distance between them. Where the radio source is identified with an optical object the identified galaxy or quasar may either lie near the center of radio emission or be coincident with one of the radio components. In many instances there are bright knots of radio emission either at the center of the source or within one or both of the separated components. In at least one case—the double radio galaxy Cygnus A—each of the two major components apparently contains within itself a tiny double source.

Although it is generally agreed that the relativistic particles are ejected from an explosion at some common origin, it is difficult to understand how the smaller components can remain intact after being ejected so far from the point of origin. Some astrophysicists have speculated that what is being ejected instead are massive objects, which then produce the high-energy particles *in situ* either as the result of single or repetitive explosions or by some continuous acceleration process.

The compact radio sources are so small and their particle density is so great that, because relativistic particles absorb radiation as well as emit it, the source becomes opaque to its own radiation at long wavelengths and little energy escapes. The self-absorption cutoff frequency depends only on the flux density, on the angular size and (weakly) on the magnetic-field strength. The smaller the source, the shorter the wavelength at which it becomes opaque.

TWO RADIO TELESCOPES at the California Institute of Technology's Owens Valley Radio Observatory can be linked together and used as an interferometer to investigate the emission from celestial radio sources. The 90-foot dishes are mounted on heavy rails and can be moved as much as 1,600 feet apart. Moreover, this pair of antennas has been used with another pair at the National Radio Astronomy Observatory in Green Bank, W.Va., to form an interferometer for the simultaneous observation of two radio sources.

Since the compact sources are relatively weak at the long wavelengths where most of the early observations of extragalactic radio sources were made, they remained essentially unnoticed for many years until sensitive receivers for short wavelengths became available. At that time, as a result of the discovery of radio sources that are opaque at decimeter and even centimeter wavelengths, it was realized that for "reasonable" values of the magnetic-field strength (near $10^{-4}$ gauss) the angular size of the opaque sources must be very small indeed, perhaps as small as a hundredth or a thousandth of a second of arc and orders of magnitude beyond the resolution of any existing radio telescope.

Further indirect evidence for such small dimensions for the opaque radio sources comes from the dramatic discovery by William A. Dent in 1965 that the radio emission from the quasar 3C 273 is variable. It was widely assumed that the radiation from any object could not vary on a time scale significantly less than the time it would take light to travel across the source. Otherwise the apparent variations would be smoothed out by the differences among the travel times for signals coming from different parts of the source. Since 3C 273 showed a significant change on a time scale of

about a year, the dimensions of the variable component were estimated to be of the order of a light-year or less. The measured red shift of 3C 273 corresponds to a recessional velocity of about a sixth the velocity of light. The Hubble expansion law that relates red shift and distance gives the object a distance of about 1.5 billion light-years and thus an angular size of less than .001 second of arc.

Further studies of variable radio emission have shown that the phenomenon is not confined to quasars but occurs in the nuclei of some galaxies as well. The observed variations do not show any periodicity such as is found in many ordinary variable stars. Rather the variations appear to be generally in the form of large outbursts that may appear first at short wavelengths and then propagate with reduced amplitude toward longer wavelengths. The typical time scale for an individual outburst ranges from a few weeks to a few years, and in some sources the interval between outbursts is shorter than the duration of a single outburst. It has been concluded from the observed variations that there are repeated explosions in the nuclei of galaxies and in quasars that produce an expanding cloud of relativistic particles radiating by the synchrotron process. Each cloud appears to be produced on a

time scale of only a few months and in a volume of space less than a few light-months across. Initially the cloud is opaque at the longer radio wavelengths. As it expands the amount of energy escaping increases as the radiating surface becomes larger; at the same time the strength of the magnetic field decreases and the electrons lose energy as a result of the expansion. When the cloud becomes so tenuous that it is transparent, the intensity then decreases with time, first at the shorter wavelengths and then at the longer ones. At the time of each outburst the magnetic-field strength in the cloud appears to be of the order of one gauss, about equal to the field on the surface of the earth. The expansion of the cloud causes the field strength to decrease rapidly until it reaches about $10^{-4}$ gauss, after which the relativistic particles appear to diffuse through a fixed magnetic field. It is believed the continued production and diffusion of the relativistic particles from the compact centers of activity then lead to the establishment of the familiar extended radio sources.

The basic question, however, of the source of energy and how this energy is converted to relativistic particles remains unanswered. Some astronomers believe that here we have reached the limit of conventional physics and that only fundamentally new theories will explain the seemingly fantastic energy output of galactic nuclei and quasars. Clearly the solution to the problem lies in being able to study these incredibly tiny objects in sufficient detail to unravel the complex phenomenon that produces the intense radio emission. Until recently, however, obtaining the required angular resolution was beyond the most optimistic dreams of radio astronomers, since it called for interferometer baselines that appeared to be unreasonably long.

Longer baselines for conventional interferometers were not feasible partly because of the increased cost of the cable between the interferometer elements, and partly because of the greater difficulty of getting right-of-way across roads and private property, not to mention the natural limits imposed by rivers, mountains and ultimately oceans. A significant improvement in resolution was obtained by replacing the connecting cables with microwave radio relay links. This technique was used first by Australian and British radio astronomers to get baselines more than 100 kilometers long and a resolution of better than one second of arc. In order to resolve the compact variable sources with expected

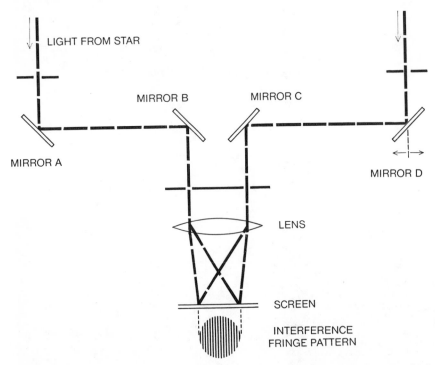

LIGHT FROM STAR

MIRROR B        MIRROR C

MIRROR A

MIRROR D

LENS

SCREEN

INTERFERENCE
FRINGE PATTERN

LIGHT INTERFEROMETER illustrates the principle of the radio interferometer. Light from a distant source is reflected from the outer mirrors (A and D) to the inner mirrors (B and C) and then combined at the projection screen. Moving one of the mirrors (D) makes one light path slightly longer than the other, shifting its light waves out of phase with respect to the waves that travel the other path. The two beams, when they are combined at the screen, will interfere with each other, creating a pattern of "fringes" (bottom).

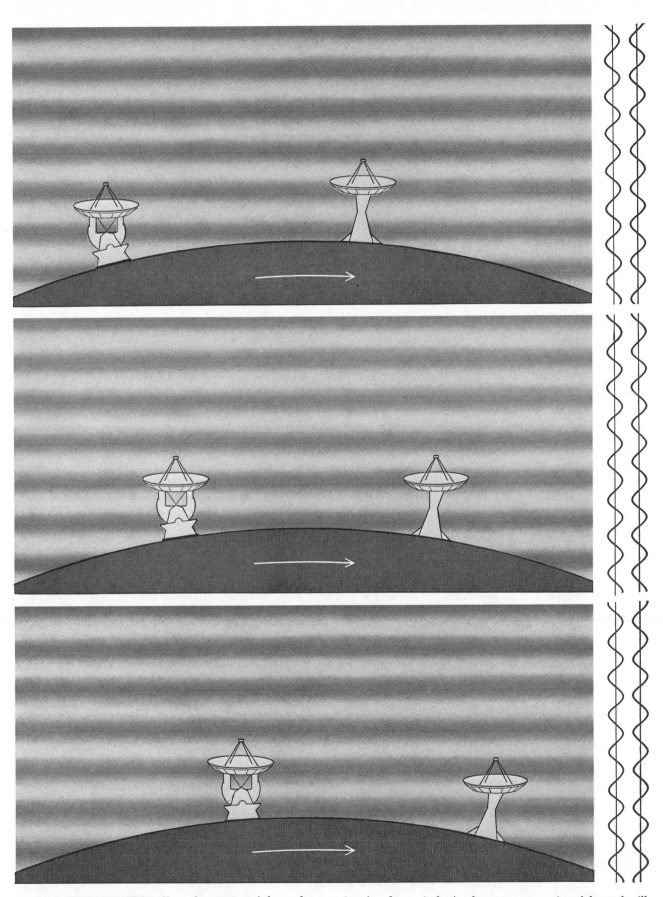

RADIO INTERFEROMETER allows the rotation of the earth to change the distance that celestial radio emission must travel to each radio telescope. At a given moment a crest of the radio wave front may be received at one telescope and a trough at the other (*top*). The waves will be out of phase with each other and will interfere destructively. At a later moment rotation of the earth will have moved the telescopes into a position such that both will simultaneously receive crests and troughs (*middle*). At a still later moment the waves will be out of phase again. Here the amount of rotation needed to change the phase relations is greatly exaggerated.

dimensions of about .001 second, however, baselines comparable to the dimensions of the earth were needed. Further large improvements in the resolution of radio-link interferometers were not practical; microwave radio links are limited to line-of-sight operation, and the installation of large numbers of repeater stations is costly and technically complex.

For many years radio astronomers had discussed the possibility of completely eliminating the direct electrical connection between interferometer elements by separately recording the signals at each end on magnetic tape and later comparing the two recordings. If this is to be done successfully, two requirements must be met. The first is that the recordings on the two tapes must be synchronized so that the time when a given wave front is received at each station is precisely known. The required precision of the time synchronization is approximately the reciprocal of the bandwidth of the recorded signal, or about one microsecond (one millionth of a second) for a typical bandwidth of one megahertz (one million cycles per second). The other requirement has to do with the fact that radio telescopes receive signals that are at too high a frequency to be recorded directly on magnetic tape. Independent local oscillators must therefore be used to "heterodyne" the radio-frequency signal, which is typically several gigahertz (billion cycles), to a much lower intermediate frequency (near one megahertz) so that it can then be recorded. If the intermediate frequency signals are to be corre-

lated, the oscillators must remain coherent over the observing time. This means that the relative phase change of the two oscillators must remain small over the observing period, so that the change in frequency is less than the reciprocal of the recording time. For example, at a frequency of one gigahertz a 100-second integration time calls for a frequency stability better than one part in 100 billion.

The possibility of independent-local-oscillator tape-recording interferometers was considered in the U.S.S.R. as early as 1961. At that time, however, the stable frequency standards and wide-band tape recorders needed for high sensitivity were not generally available. A tape-recording interferometer was first actually used in radio astronomy by a University of Florida group to study the dimensions of the radio storms on the planet Jupiter at a frequency of 18 megahertz. The Jupiter bursts are so intense that an integration time considerably less than a second and a bandwidth of about one kilohertz give adequate sensitivity. Thus a frequency stability of only one part in 100 million and a time synchronization of about one millisecond were sufficient to maintain coherence at the two ends of the interferometer, and the necessary frequency stability and time synchronization were easily provided by the time signals from the National Bureau of Standards station WWV. In this way it has been possible to determine that the dimensions of the radio-emitting regions are less than .1 second of arc, or 200 miles on the surface of Jupiter. This resolution is considerably better than the highest res-

olution in photographs made of Jupiter at optical wavelengths.

The use of tape-recording interferometers to study the much weaker radio emission of radio galaxies and quasars had to wait until stable atomic frequency standards and high-speed tape recorders were commercially available. At that time two systems for tape-recording interferometry were developed independently in the U.S. and in Canada. A system developed by a joint Canadian team from the National Research Council and the University of Toronto employed television tape recorders to record data in a four-megahertz bandwidth. (The recorders were of the same type that is used to show an "instant replay" during a televised sports event.) A system developed by a U.S. group from the National Radio Astronomy Observatory and Cornell University used a standard computer tape-drive to record digital data in a 300-kilohertz band. Although the Canadian system had the advantage of greater bandwidth and thus greater sensitivity, it required more complex special equipment to synchronize the tapes on playback. In the American system the digital data were simply fed to a large computer, which stored and correlated the two data streams. A new system has recently been developed at the National Radio Astronomy Observatory by Barry G. Clark and others that employs television tape recorders to record digital data and a special-purpose digital processor to correlate the tapes. In this way both large bandwidth and the convenience of modern digital data-processing techniques are preserved.

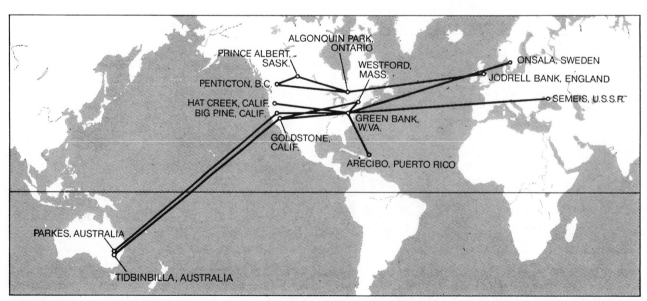

**WORLD MAP** shows some of the interferometer baselines used for high-resolution studies of radio galaxies and quasars. The long-est baseline employed so far is the one stretching across the 6,600 miles separating Goldstone, Calif., and Tidbinbilla in Australia.

In these systems atomic frequency standards are used to provide accurate time synchronization throughout the recordings and to maintain the frequency of the local oscillator. Although the atomic clocks keep sufficiently accurate time in the course of an individual recording lasting up to many hours, it is first necessary to synchronize the clocks at the different locations. This can be done by directly comparing the two clocks at the same location and then transporting one to the distant site, which is often inconvenient, or by reference to one of the continuously running atomic clocks kept at time bureaus in many countries or to the Loran C navigational signals that are available in many parts of the world.

Synchronization of the order of one microsecond is required, but the initial synchronization need not be so precise, because the data can be played back repeatedly with different time delays until the interference fringes are found. In this way relative time delays up to 100 microseconds can be readily searched; when the fringes are found, the clocks are synchronized after the fact to better than one microsecond. With larger bandwidths, or with observations made at several frequencies to synthesize a large effective bandwidth, it is possible in principle to synchronize remote clocks in this way to an accuracy of a few nanoseconds (billionths of a second).

The atomic frequency standards used in the earlier experiments were commercially available rubidium-vapor standards with a relative frequency stability of about one part in 100 billion. Hydrogen masers have a stability of about 100 to 1,000 times better, corresponding to a clock error of less than one microsecond per year. These masers, however, are expensive, complex to operate and not easy to transport. Although the rubidium standards are sufficiently stable to provide an interference pattern and a measurement of its amplitude, they do not in general have sufficient stability to allow the phase of the interference pattern to be measured. Some experiments have already been made with hydrogen masers as frequency standards to provide improved phase stability, and it is expected that as masers become more readily available they will be routinely employed for the purpose of controlling time and frequency in tape-recording interferometers.

The first long-baseline measurements with tape-recording interferometers were made in 1967 at radio wavelengths of 75, 50 and 18 centimeters on base-

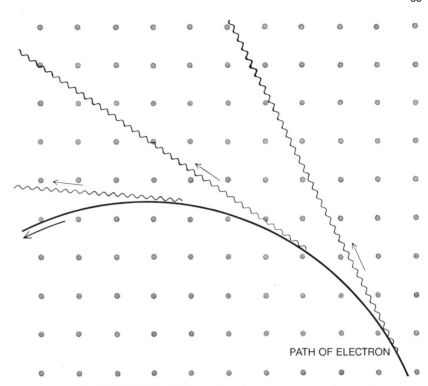

SYNCHROTRON RADIATION, which is believed to be the source of the radio emission of radio galaxies and quasars, is generated by electrons spiraling at relativistic speeds (that is, near the speed of light) in a magnetic field. Here the lines of magnetic force are perpendicular to the page (*gray dots*). The electron is traveling counterclockwise and emitting radiation (*wavy lines*) in the plane of the page. The predominant wavelength of the synchrotron radiation depends on energy of electron and strength of the magnetic field.

lines across Canada and the U.S. They quickly confirmed the expected small size of many radio galaxies and quasars. Although some of the objects studied were found to be resolved at dimensions of the order of a hundredth of a second of arc, many were still unresolved. Higher resolution was obviously needed.

The baselines were rapidly extended in a series of cooperative intercontinental experiments conducted in 1968 and 1969 by American, Swedish and Australian radio astronomers. With a wavelength of six centimeters resolutions of about .001 second of arc were obtained on the longest baselines. Still, many sources had components that appeared to be unresolved even at this extraordinary resolution. The California-to-Australia baseline was already 6,600 miles long, more than 80 percent of the earth's diameter, so that a further significant increase in the physical baseline was not feasible without the expensive procedure of setting up stations in space or on the moon.

A cheaper and simpler alternative was to observe at shorter wavelengths. Outside North America, however, only two radio telescopes were suitable for operation at short wavelengths and also large enough to provide adequate sensitivity for long-baseline interferometry. Both

of these instruments are in the U.S.S.R.

Even in the joint experiment within the Western countries logistical problems such as transporting magnetic tapes and fragile equipment and negotiating with U.S. and foreign customs authorities had proved to be as much of a challenge as technical considerations. These problems were particularly formidable for experiments between the U.S. and the U.S.S.R., since there is relatively little exchange of advanced scientific or technical equipment between the two countries.

The first joint interferometer experiment between the U.S. and the U.S.S.R. was completed late in 1969. The instruments at the two ends of the baseline were the 140-foot antenna at the National Radio Astronomy Observatory in Green Bank, W.Va., and a new 72-foot precision radio telescope located on the shores of the Black Sea in the Crimea. Last spring a second experiment was conducted involving in addition to the Green Bank and Crimea telescopes the ultrasensitive 210-foot radio telescope at Goldstone, Calif., and the "Haystack" telescope in northern Massachusetts. (The Goldstone instrument is operated by the National Aeronautics and Space Administration; the Haystack one, by the Massachusetts Institute of Technology.)

More than 20 investigators from eight institutions in both countries participated in the second experiment, which included observations of interstellar clouds of water vapor as well as of radio galaxies and quasars.

The data from the Goldstone-Crimea baseline obtained at a wavelength of 3.5 centimeters give the highest resolution obtained so far in the study of radio galaxies and quasars: approximately .0003 second of arc. This is a very small angle indeed: it is equivalent to the angle subtended by the height of these letters at a distance of about 1,500 miles. In contrast, an optical telescope operating under ideal conditions can just distinguish an object the size of a man at the same distance. The measurements of the water-vapor clouds, which were made at shorter wavelengths, give even higher resolution.

The smallest physical dimension that has been directly measured in an extragalactic radio source is in the nucleus of the nearby radio galaxy Messier 87; it has an angular size of about .001 second of arc, corresponding to a linear extent of only a quarter of a light-year. It is believed the relativistic particles that now fill an extensive radio "halo" more than 100,000 light-years in diameter have been produced or accelerated in this compact nucleus by continuing or repeated activity over a million years or more. Similar radio nuclei are found in other spiral and elliptical galaxies as well as in strong radio galaxies.

By combining data from the various long-baseline experiments we have begun to obtain crude pictures of the structure of the radio nuclei and the compact quasars with resolutions exceeding .001 second of arc. This resolution is about 1,000 times better than is possible at optical wavelengths, where quasars and the nuclei of galaxies appear as only fuzzy points of light, and it presents unprecedented opportunities for studying the violent events in these objects in considerable detail.

In conventional radio interferometers one or more antennas can be moved over distances of up to a few miles in order to obtain the number of interferometer spacings needed to reconstruct an image of the source. For very-high-resolution interferometry, where the baselines span oceans and continents, it is clearly impractical to move large antennas around. The observations have therefore been made with already existing radio telescopes at fixed locations throughout the world, restricting the range of available baselines. This gives rise to considerable ambiguity in the detailed interpretation of the data, and so only crude pictures are obtained.

For those sources that have been studied in some detail the structure has been found to be complex and generally lacking circular symmetry. Often there is a hierarchy of component sizes ranging from less than .001 second of arc to .1 second or more in a single radio galaxy or quasar. In general the smallest components are strongest at the shorter wavelengths, a result that can be predicted from the synchrotron theory. In those cases where the angular size, spectral cutoff frequency and peak flux density are observed, the magnetic-field strength can be estimated from the synchrotron theory and is typically of the order of $10^{-4}$ gauss.

Many of the compact components appear to be spatially dissected, with a region of relatively low brightness near the middle. Although generally it has not been possible to decide unambiguously between an elliptical ring structure and a double structure, in most cases the data appear to favor the double structure. It is a remarkable fact that this characteristic double structure of radio galaxies and quasars is apparent on angular scales between .001 second of arc and several minutes of arc. This is a range of about one to 100,000 in size,

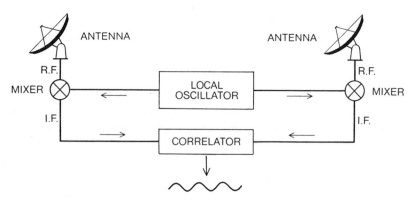

INTERFERENCE FRINGE PATTERN

**CONVENTIONAL RADIO INTERFEROMETER** employs a cable to link the two telescopes, which are separated by distances of up to a few kilometers. For purposes of analysis the high-frequency radio signal received from the celestial radio source must be converted to a lower frequency. This is done by mixing a signal from a local oscillator with the radio-frequency signal (*denoted R.F.*) to create an intermediate-frequency "beat" signal (*I.F.*).

INTERFERENCE FRINGE PATTERN

**RADIO-LINK INTERFEROMETER** joins the radio telescopes by a microwave-relay link similar to those used for long-distance telephone transmission. The local-oscillator signal is transmitted to the mixers, and the I.F. signals are returned for correlation. If the I.F. signals from each antenna are to be coherent (in phase), the local-oscillator signal at each end must also be coherent and so is derived from a common source and carried to the mixers.

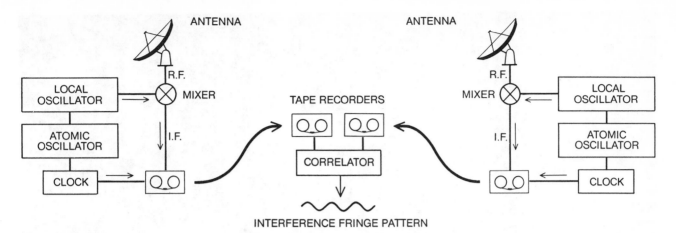

INTERFERENCE FRINGE PATTERN

**TAPE-RECORDING INTERFEROMETER** makes it possible to do interferometry with radio telescopes as far apart as opposite sides of the earth. (The microwave-relay link is practical only for antennas that are on a line of sight with respect to each other.) The intermediate-frequency signals from each antenna are separately recorded at each end of the interferometer system, and later the magnetic tapes are transported to a common location where recorded data are correlated in a large digital computer or a special correlator.

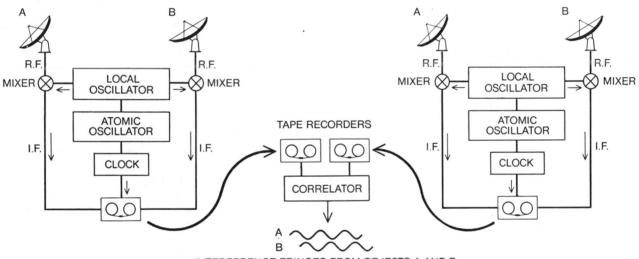

INTERFERENCE FRINGES FROM OBJECTS A AND B

**MORE SOPHISTICATED** tape-recording system employs two radio telescopes at each end of the interferometer to observe two sources simultaneously. One source serves as a point-source reference to determine the phase of the other source. The phase difference of the two interference patterns can be measured directly, independent of any phase changes in either of the local oscillators or in the atmospheric path to the two antennas. In this way it is possible to make precise measurements of the positions of radio sources and to completely reconstruct the intensity distribution by combining observations that have been made on many baselines.

which corresponds to linear dimensions measuring between less than one light-year and several hundred thousand light-years.

Although the measured angular dimensions of the individual components and those predicted from the self-absorption cutoff frequency are in good agreement with the synchrotron model, the rapid variations in intensity, and the large energies required to explain the powerful radiation observed from some quasars, are difficult to explain. For some time it was hoped that a more definitive test of the synchrotron theory would come from measuring the change of size of the variable sources, since the theory predicts a close relation between the rate of change of size and the rate of change of intensity. Now that data of this kind are available, however, the results are perplexing and raise more questions than they answer.

It has been suggested that the individual outbursts observed can be explained as being the result of a violent explosion that releases a dense cloud of relativistic electrons, which then radiate by the synchrotron process. The rapid variations in intensity that are observed in some of the radio sources require that the particles be accelerated to relativistic energies on time scales between a few months or less and a few years. If it is assumed that the dimensions of the variable source cannot exceed the distance traveled by light during the characteristic time scale of the variations, one can set an upper limit to the size of the variable source. Typically this size is of the order of a light-year. If the distance is known, one can calculate an upper limit to the angular size and the magnetic field and a lower limit to the electron energy.

For relatively nearby radio galaxies this minimum required energy is of the order of $10^{52}$ ergs of relativistic particles per outburst. This is of the same order as the energy released in a powerful supernova explosion. In the case of the quasars (assuming that their red shifts are associated with the expansion of the uni-

RADIO GALAXY CYGNUS A, the strongest radio source in the sky, appears as a fuzzy spot in the center of the photograph at left, which was made with the 200-inch telescope on Palomar Mountain.    Interferometer studies of the radio emission of the source indicate that it comes from two regions on each side of the visible object. The apparent structure of these regions is shown at the same scale

verse and that they are at cosmological distances) the energy apparently required in each outburst is as much as a million times greater, and it is very difficult to understand how so much energy can be released in such a small volume of space and in such a short time. Moreover, it has been pointed out that the intense synchrotron-radiation fields from the dense collection of relativistic electrons would in turn cause the electrons to immediately lose their energy by inverse Compton scattering rather than by synchrotron emission. Such scattering results when an energetic electron encounters a photon of lower energy. The photon gains energy and the electron loses it; the net result is that the electron energy is radiated away at shorter wavelengths such as infrared radiation or X rays rather than by synchrotron radiation at radio wavelengths.

This apparent paradox was interpreted by some investigators as evidence that either the quasars must be at closer distances than is indicated by their red shift, and hence their radiation field must be much weaker, or the radiation mechanism is more efficient than the synchrotron process. Martin J. Rees of the University of Cambridge has suggested a possible alternative: the "superlight-velocity" theory. If the radio source expands at a velocity close to the velocity of light, then since it takes a finite time for radio signals to reach an observer, the signal arriving from the receding part of the source will have originated at an earlier time, when it was closer to the point of origin, than the signals from the approaching parts. Under these conditions the *apparent* velocity of expansion may indeed exceed the velocity of light.

Thus the true dimensions of the source may be greater than what is given by the restriction that the apparent expansion velocity is less than the velocity of light, and the energy requirements, which are inversely proportional to the 10th power of the size, are much reduced.

Evidence that this effect might be important in the quasar 3C 279 was first obtained in a series of transpacific observations made between 1968 and 1970 by a joint team from Australia and the California Institute of Technology. Working with NASA tracking antennas at Goldstone in California and Tidbinbilla in Australia as two elements of a long-baseline interferometer, this group found that a component of the source that had first appeared in 1966 had reached a diameter of about .001 second of arc by the end of 1969. The corresponding linear diameter is about 12 light-years, assuming that the source's red shift of .54 is of cosmological origin.

Hence, as predicted by Rees, the apparent expansion velocity was about twice the velocity of light. More detailed measurements were made of 3C 279 in October, 1970, by a group from M.I.T., NASA and the University of Maryland using a transcontinental baseline. These observations were originally designed to measure the gravitational bending of the radio signals from 3C 279 as it approached the sun on October 8, but they showed clearly that the compact source in 3C 279 was complex and appeared to have at least two components separated by .00155 second of arc, or about 20 light-years. In February, 1971, this source was reobserved by the same group and also by workers from

the National Radio Astronomy Observatory, Cal Tech and Cornell group using the same baseline and techniques. There had been a distinct change in 3C 279 in only four months. Again the source appeared double, but the separation was greater by two light-years: 10 percent more than previously. Thus the two components appear to be receding from a common point of origin with an apparent velocity about three times the velocity of light.

Although this would seem to confirm the predictions made by Rees, a serious problem remains. The difference in the Doppler shift of the two components should cause the approaching component to appear much more intense than the receding component. Yet the observed intensities of two components are equal to within a few percent, so that the effect of the Doppler shift must be exactly canceled by the difference in the intrinsic luminosity of the two components. This would seem to be a somewhat remarkable coincidence.

On the other hand, the apparent motions may be a "searchlight" effect caused by the excitation of stationary material by a moving shock front. The point of contact between the shock front and the material may appear to move with almost unlimited velocity, like the beam of a rotating searchlight seen at a great distance. It is also possible that the apparent change in size observed in the structure of 3C 279 may not be the result of component motions at all; it may merely reflect a change in the relative intensity of one or more stationary components. For example, a ring with a point source at the center or a triple source would appear to expand if

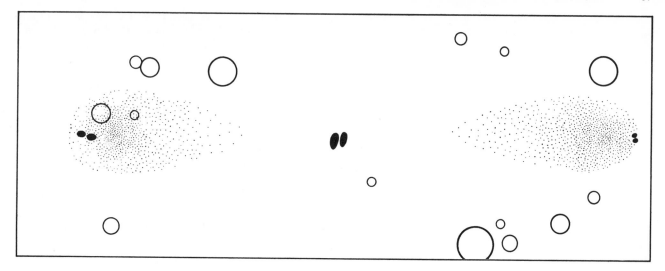

in the drawing at right. Each of the two major components contains within itself a tiny double source. The radio picture is based on observations made with a one-mile radio interferometer at the University of Cambridge, a 2.7-kilometer interferometer at the National Radio Astronomy Observatory and a 35-kilometer microwave-link interferometer also at the National Radio Astronomy Observatory.

the intensity of the inner component decreased or the intensity of the outer parts increased. Although this is the most straightforward and least spectacular interpretation of the data, it still leaves unexplained the problem of the huge energy requirement, unless the distance of 3C 279 is very much less than is deduced from its red shift.

The present measurements do not distinguish between these various models, although it may be expected that future detailed observations will demonstrate whether or not there is actual motion of the individual components. Even if the large apparent velocities are confirmed, however, there will still be more questions than answers. Those who believe that the quasar red shifts are cosmological can cite the super-light-velocity theory to reduce the enormous energy requirements, which have been one of the main arguments against the cosmological interpretation of the quasar red shifts [see "The Evolution of Quasars," by Maarten Schmidt and Francis Bello; SCIENTIFIC AMERICAN, May, 1971].° Those who believe the quasars are local emphasize that the apparent super-light-velocity may be only an artifact of the assumption that the quasars are at great distances. They point out that the apparent angular expansion rate, the observed flux densities and the rates of variation are similar for the nearby galaxies and the supposedly more distant quasars, and that the apparent large differences in the corresponding intrinsic properties are merely the result of the assumption that the quasar red shifts are cosmological.

Although measurements of the apparent change in the structure of compact radio sources may not help to clarify this problem, they will determine (1) whether the successive outbursts that are observed occur in exactly the same volume of space or whether they are spatially separated, (2) the kinematics, or mechanics, of the expansion and (3) how the magnetic field and total energy vary with time. With this information we may hope to reach a better understanding of the process by which violent events lead to the creation of intense radio sources, in particular a better understanding of the source of energy and how this energy is converted into relativistic particles.

Although the sensitive long-baseline interferometer systems were initially developed to study the compact extragalactic radio sources, the technique has also been used to study the radio emission from interstellar clouds of hydroxyl radicals (OH) and water vapor. These clouds, which radiate like giant interstellar masers, emit such powerful radio signals that high-resolution interferometer measurements of their size can easily be made with standard high-fidelity audio-frequency tape recorders. All the observations made so far of these interstellar masers, however, have been made with the more sophisticated wide-band techniques that allow the simultaneous measurement of several clouds (which, because of their high velocities in various directions, are Doppler-shifted and radiate over a wide range of frequencies). Typically the clouds are dispersed over a volume of the order of several light-years across, although the individual components are as small as one astronomical unit (the distance between the sun and the earth).

One of the main restrictions of tape-recording interferometry so far has been the lack of sufficient oscillator stability to determine the phase of the interference pattern. We have noted the improvement to be expected from the use of hydrogen masers as frequency standards. Even with infinitely stable oscillators, however, there are still problems created by fluctuations in the path length through the atmosphere at the two widely separated observing sites. Experiments are now being conducted with a total of four antennas to overcome this problem and to eliminate the effect of local-oscillator instabilities. In these double-interferometer experiments two antennas located at Green Bank, W.Va. (at the National Radio Astronomy Observatory), and two at the Cal Tech Owens Valley Observatory in California are used to observe two sources simultaneously and to determine their relative positions.

If phase-stable interferometry could be done on baselines comparable to dimensions of the earth, angular positions could in principle be established with an accuracy approaching .0001 second of arc. Such precision would make it possible to measure the actual motion of nearby galactic objects such as pulsars, interstellar molecular masers and those stars that show strong radio emission. The possibility of applying this technique to determining small departures from the uniform motion of stars that might be caused by the gravitational attraction of planets is an exciting one; it may offer the only direct way of detecting other planetary systems in our galaxy. In addition, the detection of small apparent motions in pulsars or interstellar molecular masers caused by the

°In Part XI of this reader.

motion of the earth around the sun (parallax) would allow for the first time a direct determination of their distance.

Another particularly interesting application of tape-recorder interferometry is the testing of the prediction of Einstein's general theory of relativity that an electromagnetic wave such as a radio or light wave that passes close to a massive body is deflected by the gravitational field of the body. The expected deflection for rays from a distant celestial source that pass close to the edge of the sun is 1.76 seconds of arc, and many optical measurements have been made of the apparent positions of stars near the sun at the time of a solar eclipse. Although these measurements have shown a displacement of about the magnitude predicted by the general theory, the accuracy of the measurements is at best of the order of 10 percent of the deflection. This is not good enough to distinguish between the predictions made by the general theory and competing gravitational theories.

The ability to determine relative angular positions of celestial objects with an accuracy of .001 second of arc by radio interferometry offers a potential accuracy considerably better than 1 percent. The circumstances for such an experiment are favorable each October, when the compact quasar 3C 279 is eclipsed by the sun. A second strong compact quasar, 3C 273, is located only 10 degrees away in the sky and provides a convenient position reference for measuring the daily deflection of 3C 279 as it passes closer and closer to the sun.

Richard Sramek has conducted an experiment of this type with a conventional interferometer at the National Radio Astronomy Observatory that has a baseline of 2.7 kilometers. His results show a slight discrepancy from the result predicted by the general theory of relativity and agree better with an alternative theory: the Brans-Dicke theory of gravity. The accuracy, however, is not sufficient to definitely exclude the general theory. Nevertheless, this is one of only a few experimental results in the more than 60 years since Einstein's theory was published that even suggest a possible discrepancy, and the current opportunity to improve the accuracy by one or two orders of magnitude is an exciting prospect for fundamental physics.

An M.I.T.-NASA group has conducted extensive studies of the October eclipse of 3C 279 with transcontinental tape-recording interferometers, but so far no conclusive results have been obtained. The joint program of the National Radio Astronomy Observatory and Cal Tech utilizing two antennas at Green Bank and two at Owens Valley to measure the simultaneous difference in the positions of 3C 279 and 3C 273 is ex-

pected to overcome many of the experimental difficulties and may well yield a very precise test of the general theory of relativity.

Even when the instrumental problems of controlling local-oscillator stability and eliminating the effect of the atmosphere are overcome, the measurement of angular positions to an accuracy of better than .001 second of arc will not be simple. At that level of precision significant errors are introduced by uncertainties in the rate of rotation of the earth itself, in the direction of the earth's axis of rotation at any one moment and in the relative coordinates of the antennas on the earth's surface. The errors in phase produced by uncertainties in these quantities are all related, and repeated observations of a suitably selected grid of sources can sort out these effects with techniques commonly applied to optical astrometric measurements. When this is done, it will be possible to determine (a) positions in the sky to within .001 second of arc, (b) the length of the day to within .0001 second of time, (c) transcontinental and intercontinental baseline distances to within a few centimeters and (d) global clock synchronization to within a few nanoseconds. Such precise data will open up to radio astronomers an entire class of geophysical experiments, including the direct measurement of tides in the solid earth, of continental drift, of variations in the rate of rotation of the earth and of the wobble of the earth's axis.

Is there a limit to the resolution of long-baseline interferometers, or will the baselines continue to extend into space, to the moon and beyond? Just as turbulence in the atmosphere affects optical astronomy, so at short wavelengths will it affect radio astronomy. Even more important is the effect of the scattering of radio waves by electrons in the solar "wind" and in the interstellar medium, which become important for measurements made at long radio wavelengths. The present evidence indicates that only for wavelengths less than about 10 centimeters can baselines much greater than the diameter of the earth be effectively used. Thus hydroxyl emission regions, which radiate at 18 centimeters, and pulsars, which radiate most strongly at meter wavelengths, are not likely to be targets for a space interferometer.

Where the synchrotron sources are concerned there is a more fundamental limit to the maximum baseline. As we have noted, if the dimensions of the synchrotron system are below a critical size, the relativistic particles quickly lose

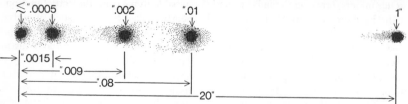

QUASAR 3C 273 is seen in the photograph at top, made with the 200-inch telescope. Extending to the right of the central image is a peculiar "jet." The drawing at bottom, which is not to the same scale as the photograph, shows the radio structure of the object. The numbers below the drawing give the distance in seconds of arc from the center of the object on a logarithmic scale. The numbers above the drawing give the size of the components in seconds. Wavelength of the radio emission is progressively longer with distance from center.

all their energy by inverse Compton scattering and so do not have time to radiate radio energy. This critical angular size is proportional to the wavelength at which the flux density is the highest. Since the resolution of a fixed-length interferometer is inversely proportional to the wavelength of observation, the two effects cancel each other if the observations are made near the wavelength of maximum intensity (as they usually are to obtain the highest sensitivity). The maximum baseline needed to resolve synchrotron sources is hence nearly independent of the wavelength of operation; in fact, it is comparable to the diameter of the earth for the stronger radio galaxies and quasars. Baselines in space or on the moon will therefore probably not be necessary to study even the smallest radio nuclei and quasars.

The simple extension of interferometer baselines is only part of the story, however. Many baselines of intermediate length are also required to give a complete picture of the complex radio structure observed for the radio galaxies, quasars and interstellar molecular masers. Today the simple two-element radio interferometer of the 1950's has grown into multielement synthesis arrays. Radio astronomers look forward to the time when many radio telescopes throughout the world can be linked together to form an ultra-high-resolution array.

Although radio telescopes currently exist that are suitable for this purpose, they are not at optimum locations. Moreover, in view of the great variety of other important problems being studied by radio astronomers throughout the world and the wide range of instrumentation being employed at different observatories, it is virtually impossible to schedule a large number of radio telescopes for observing the same object at a common wavelength. In order to obtain a sufficient range of interferometer spacings, it may be necessary to construct a special global long-baseline interferometer network.

The task of collecting all the tape recordings at a common location and correlating all possible pairs of telescopes will be a formidable one. An attractive solution will be to dispense with the tape recordings entirely and telemeter the data from each antenna to a common center by means of synchronously orbiting satellites. Such a very-long-baseline array would require a truly international effort, but the opportunities it would provide for both astrophysics and geophysics are enormous.

# 5

# Telescopes for Tomorrow

Mike Disney

Modern astronomy had its birth when astronomers fitted cameras to their telescopes and began to photograph the heavens. By exposing their plates for long hours through the nights they were able to discover objects far too faint and distant for the human eye to perceive. Not only could they see farther and in more detail, but they had permanent records which they could show other men, and from which systematic measurements, the basis of any science, could be made. Astrophysics was born. Within 50 years men had fathomed why stars shone, discovered that they inhabit an expanding universe, and stood in awe of the new-found depths of space and time.

The photographic process has served astronomical man well. Even today, the observing astronomer collects most of his light with its aid. But even as he used it the astronomer came to realize its defects. To begin with it is very slow and inefficient. Several minutes are needed to detect what the human eye can see at a glance. Of every 100 photons that fall on a plate 99 are undetected or wasted. Only in its ability to integrate over long exposure times does it come to surpass the eye. Then again it has a non-linear response; that is to say, if one star is twice as bright as another its image is not necessarily twice as deep on the plate. Figure 1 shows the characteristic response of the plate to different levels or intensities of light. If a star is so faint that its intensity lies below the threshold ($I_0$ in Figure 1) the plate will not respond to it at all. No matter how long the exposure is continued the star will remain undetected. On the other hand, if the general sky intensity is sufficiently high, and the exposure sufficiently long, the whole plate will be equally blackened, or saturated, so that nothing can be distinguished on it at all. This saturation effect is particularly serious, since nowhere, even the darkest part of the heavens, is the sky completely black. Faint airglow from the atmosphere, background light from distant galaxies, diffuse galactic light from unresolved stars and interstellar dust and, most serious of all, the light from our ever-growing cities fog up our plates.

As they have come to know the drawbacks to photography, astronomers have wrestled to overcome them as best they can. The response curve tells us that if we are to detect the faintest stars we must raise the intensity of their light above the threshold value of our plates. We need the biggest telescopes we can afford to collect the most light and then, with the aid of lenses or fast camera systems, we need to condense that light down onto the minimum area of plate. The history of optical astronomy over the last 100 years is largely an account of our struggle to do just that. Led by the great American astronomer, George Ellery Hale, formidable technical and financial obstacles have been overcome. As a monument to his, and to other men's efforts the 200-inch Hale telescope at Palomar can photograph stars ten million times fainter than we can see with the naked eye, and 20 times fainter than the darkest parts of the night sky. Beyond that we cannot see. Yet we know, if only we could see farther, a host of fascinating answers and

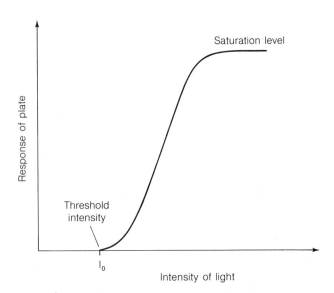

**Figure 1.** How a plate responds (relative blackening) to different intensities of light.

From *New Scientist*, vol. 58, pp. 147–149, 1973. Reprinted with permission.

intriguing problems would lie within our grasp. How *are* we going to see farther—for see we must?

## Limits to Big Telescopes

The obvious answer is to build an even larger telescope. But we must beware, for in astronomy, as in most sciences, there comes a point of diminishing returns. I hope to demonstrate that the large telescope may well have over-stepped that point already.

To begin with there's cost. The price of a telescope rises as the cube of the mirror diameter whereas its light-gathering capacity rises only with the square. Twice as big a mirror collects four times the light but at eight times the cost. The 200-inch would cost £10 million today so a really big telescope could cost as much as the CERN accelerator.

Cost aside, however, there are fundamental scientific problems. A very faint star is fainter than the night sky. To detect it what we really have to do is detect the contrast between one piece of night sky and another bit which is maybe a few per cent brighter because the light from our star is superimposed on it. It is the *contrast* which is important. Now collecting more light will help, but the extra light from the star is accompanied by more unwanted light from the sky too. Thus as the size and cost of the telescope rises sharply, the contrast, and with it the performance of the telescope, improves only slowly. We say the observation is "sky-limited" to distinguish it from observations of bright stars which are "starlight-limited."

So far I have concentrated on direct photography but nowadays big telescopes spend the larger part of their time doing spectroscopy. A spectrogram contains far more information about the star or galaxy we are observing than a straight photographic plate. It is virtually identical with a direct photo except that before the light is focused onto the plate it is split up with a prism or grating into its constituent colours. Instead of a single image the spectrogram now contains a large number of images side by side each formed in a different coloured light. If the star preferentially emits in certain colours the images in those particular colours will be especially strong and will be obvious on the spectrogram. If the star is weak in certain colours this will be obvious too on the spectrogram for the images will be correspondingly weak. Now the various colours of light are emitted and absorbed by different atoms in particular states of excitation. Thus the spectrogram contains information about the chemical composition, temperature, distance, rotation-speed, density, and gravity of the star being observed. Much of this information can be obtained by no other means and it was by studying spectrograms of galaxies that Hubble and others discovered the expansion of the universe.

Figure 2 shows a portion of a spectrogram and illustrates a further serious problem which besets large telescopes. Big telescopes produce larger images on the spectrogram, but the separation between the images in different colours is produced entirely by the grating (or prism) in the spectrograph. Unless the larger telescope can be fitted with a larger grating the stellar images in different colours begin to overlap one another and the usefulness of the spectrogram is severely degraded. We say the observation is "grating-limited." It could be that one

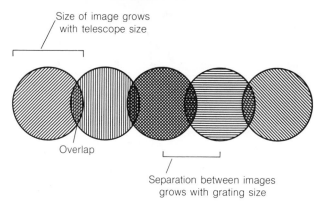

Size of image grows with telescope size

Overlap

Separation between images grows with grating size

**Figure 2.** Portion of a spectrogram. The images in different colours overlap.

day we'll be able to build gratings much larger than we can at present. But the technical problems are truly formidable. We can probably say it would be easier to construct a 400-inch telescope than to build the gratings that would be needed to make use of its extra light.

All telescopes, then, are equally limited by the sky background for faint observations; but the large telescope is further handicapped by its disproportionate cost and by the lack of suitable gratings.

For these reasons astronomers have looked wistfully, from time to time, at the idea of using several smaller telescopes as a cooperative array to simulate the effect of a single giant instrument. If, say, the array telescopes were to photograph the same star-field at once, the plates could be superimposed later in the laboratory to simulate a single and deeper plate. This attractive idea has been abandoned for two main reasons. First there's the non-linear threshold problem. If no single array telescope can record a stellar image then N times nothing is still nothing. Then there's a more practical problem. It may take days to superimpose the plates so that if the results proved unsatisfactory it might be too late to repeat the experiment.

In principle, much the best approach would be to put our telescopes outside the atmosphere. If we could do so, most of our difficulties would disappear. A wider spectrum is available and the sky background is darker. Most important, however, with no blurring by the atmosphere, the stellar images will shrink into sharp points of light. That vastly increases the intensity contrast between star and sky and eliminates the grating problem. But the costs are so staggering that it is unlikely we'll see more than one or two medium-sized telescopes in space this century. The main thrust of the advance must still come from the ground.

If photographic detection were the only tool available to ground-based astronomy, then the prospect of extending our present optical horizon would look altogether bleak. Fortunately, and in the nick of time, technology heralds an exciting new way ahead.

## Television to the Rescue

In laboratories and observatories across the world, a quiet revolution is going ahead which will replace the photographic

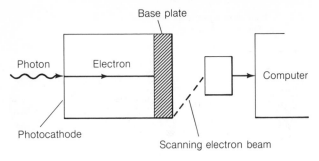

**Figure 3.**  The basic components of an integrating system.

plate with something much better. We call it the integrating television system. Figure 3 illustrates, in a schematic way, how it works. Photons of light striking the photocathode eject electrons which are accelerated through a high potential and focused onto the base-plate. An amplified charge distribution builds up on the base-plate which becomes a sort of electrical photograph of the light image on the cathode. Before the base-plate can saturate, an electronic beam is scanned across it which neutralises and measures the charge distribution. To each point on the plate there corresponds a memory address in the computer. Every time the point is scanned the amount of charge residing on it is turned into a number and the number is added to the appropriate computer address. Thus a numerical picture is built up in the computer memory which corresponds exactly to the light picture on the cathode. At any time the astronomer can order the computer to display its "picture" in a suitable form. It could, for example, produce a highly realistic television picture of the starfield or spectrum under observation. When the astronomer is satisfied he has the depth and detail required, he can discontinue the integration and start a fresh observation.

I must emphasise that Figure 3 is purely schematic. Moreover there are several different television-type systems under development. It is not possible, at this stage, to pick the one which holds out the greatest promise. They all have this in common, however: photons of light are converted into electric charges, the charges into numbers, and the numbers are added in a computer. And they all promise colossal advances over the photographic plate. To begin with they're far more sensitive, 20 to 50 times more sensitive at a given colour. Second they're sensitive over a much wider colour range than a plate. Third they're *linear*, and the problems of threshold and saturation do not apply. Fourth, they afford a "real time output." That is to say, the astronomer can monitor his results as they come in; he does not have to wait until after his plate is developed. Last, but not least, they now make the array concept a feasible proposition.

### The Revolution to Come

Of course the TV system has its problems. The photocathodes are rather small. Stray electrical noise in the system can degrade the picture. Large and very fast computer systems may be needed to keep up with the showers of individual photons which can now be detected. Nevertheless, scientists at the Princeton and Lick observatories, using integrating television,

have already made observations of faint quasars which could not have been made in any other way. One of the most promising systems is already working at University College, London. There can be no doubt that, given modest development funds, integrating television can become an every-day, or every-night, tool in the average observatory well within a decade. A 40-inch telescope fitted with such a system would have the same capacities that the 200-inch possesses today.

Over the past 100 years telescope design has largely been dictated by the properties of the only detector available, the photographic plate. Quite rightly, despite cost and grating problems, astronomers have fought to get the biggest telescopes society could afford. So long as the threshold problem existed the array has been rightly left to languish as an attractive but impractical idea.

Integrating television is going to change all that. While the special handicaps of a big telescope remain, those of the array have vanished almost overnight. The day for building big telescopes could well be over. Independently of one another several astonomers, among them myself, have concluded that the most efficient telescope is somewhere between 60 and 80 inches in diameter. For almost any observation an array of such telescopes, all fitted with TV, and all adding their results together in real-time in a central computer, will be more efficient (faster) than a single large telescope built for the equivalent cost (see Figure 4). Indeed, in its submission to the US government for funds in the 1970s, the American astronomical community *as a whole* has reiterated the need for a greater light gathering capacity. But in preference to another large telescope they urgently stress the need to build an array.

The advantages of an array are many. It is faster. It can be built more easily and quickly than a monster. It is not dependent on problematical advances in grating technology. There is no limit to its effective size because more units can be added as funds become available. It will be more reliable, for mal-

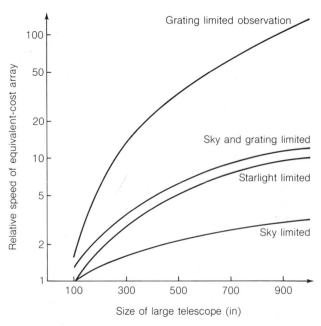

**Figure 4.**  For nearly all types of observation, an array of smaller telescopes will be faster than a single large telescope of the same cost.

function in a single unit will not ruin an observation, only marginally slow it down. It can do experiments denied its bigger brother. It provides an ideal cooperative basis for astronomers from different institutions or nations who can use their telescope units independently, or in cooperation, as they wish. Last, and most important, it will provide the opportunity for more astronomers to try out original ideas than is possible on a single big telescope. The universe is so large, so complex and so mystifying that the main problem is not how to observe but *what* to observe, and how to interpret what you see. Time and again advance has slowed, even stopped, when a fashionable hypothesis (dogma) has stood in the way of a fresh idea. I remember the great astronomer Martin Schwarzschild getting up at conference and saying "For God's sake will everyone stop trying to agree with everyone else." Skeptical, imaginative minds are needed at the forefront of astronomy. The more of them the better.

How then do I see the observatory of the future? It will be high on some mountain peak remote from the encroaching lights of city man where the skies are clear and the seeing is good. There won't be one telescope, or two, but tens or even hundreds. It won't be a national observatory but an international one. Astronomers will come and go from all corners of the Earth. Much of the control will be remote. Most nights the telescopes will be searching in different directions, testing different ideas. Now and again, as night falls, you will see them turn as one to focus on some star or nebula at the remotest outpost of space-time. Computers will hum and men will doze, and think, and search. And what will they be searching for? Who knows?

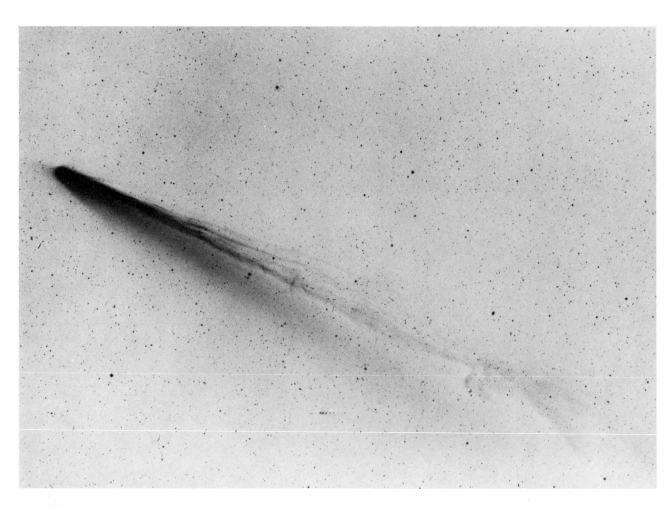

Comet Kohoutek on January 11, 1974
(negative print). The photograph shows
the anti-tail, dust tail, ion tail, and huge
plasma cloud in the ion tail located about
15 million kilometers from the nucleus.
[Joint Observatory for Cometary Research:
NASA-Goddard Space Flight Center and
New Mexico Institute of Mining and
Technology.]

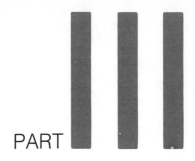

PART

# ASTEROIDS, METEORS, AND COMETS

"The missing planet might conceivably have strayed within the Roche limit of Jupiter." William McD. Napier and Richard J. Dodd give a short review of theories on the formation of the asteroids in "The Missing Planet." They use a few simple formulas in their analysis and conclude that an ordinary (chemical) explosion could not account for the breakup of a planet to make the asteroids. Likewise, they reject a breakup due to tidal force in the event that the hypothetical planet passed close by Jupiter, and they describe objections to the idea that the missing planet was disrupted by radioactive and nuclear effects. Napier and Dodd mention Kuiper's theory that the asteroids are the debris of collisions among small planets and Alfvén's theory that, on the contrary, the asteroids are slowly *building up a planet* by collisions! Here are some terms that Napier and Dodd employ: m.y. means million years; $M_\oplus$ and $R_\oplus$ denote units of the Earth mass and Earth radius, respectively; Roche limit is the distance from Jupiter at which an approaching large object such as another planet would be torn apart by tidal force—that is, by the difference between Jupiter's gravitational pull on the near side of the object and its pull on the far side; the Poynting-Robertson effect is a process that causes certain particles in interplanetary space to spiral inward toward the sun.

". . . some 100 tons of nickel-iron crashed to earth after fragmenting in the atmosphere." Peter M. Millman, in "The Meteoritic Hazard of Interplanetary Travel," examines the danger to life that meteorites pose. Although this has been a common theme in science fiction accounts of space travel, we have yet to lose an astronaut as a result of a meteoroid collision, and, indeed, Millman calculates that there is little to worry about.

Beware of scientists' predictions! Fellow scientists greet each new theory or major prediction of their colleagues with healthy skepticism, for they know the track record, on the average, of those who would attempt to second-guess Nature. This matter was brought out again when the widely heralded Comet Kohoutek disappointed an expectant public in January, 1974. The story is told by one of the editors in the selection "A Funny Thing Happened to Comet Kohoutek."

# The Missing Planet

W. McD. Napier and R. J. Dodd

The idea that the asteroids are fragments of an exploded planet was first put forward by Olbers more than 150 yr ago and has since been widely accepted. Recently Ovenden,[1] in an attempt to account for Bode's Law, has argued persuasively that a planet of 90 $M_\oplus$ vanished between Mars and Jupiter some 16 m.y. ago. According to this point of view the asteroids are remnants of this event.

Other possibilities have been suggested, however, for the origin of the asteroids. For example, Kuiper[2] suggested that they are remnants of successive collisions which occurred between a few primaeval planetoids; Alfvén[3] and others regard the asteroids as being in the process of forming a single planet by accretion. The difficulty of disrupting a planet of 90 $M_\oplus$ was mentioned by Ovenden. Here we eliminate mechanisms which are incapable of providing the energy needed both to break up the hypothetical planet and to remove most of its mass beyond the Solar System. The mechanisms for disrupting a planet might be chemical, gravitational or nuclear in nature.

In the chemical case, to dissipate a planet of mass $M$ and radius $R$, an energy $E \sim GM^2/R$ is required, where $G$ represents the gravitational constant. For a mean density comparable to that of the Earth, $R \sim 4.5\ R_\oplus$ and $E \sim 8 \times 10^{42}$ erg, or $\sim 1.5 \times 10^{13}$ erg g$^{-1}$. This seems to exclude the possibility of a chemical mechanism: the detonation energy of TNT, for example, is $\sim 5 \times 10^{10}$ erg g$^{-1}$.

Gravitational tidal forces due to the Sun or Jupiter are negligible at a heliocentric distance of 2.5 a.u. The missing planet might conceivably have strayed within the Roche limit of Jupiter. Such a close passage would greatly disturb, if not disrupt, the Galilean satellite system, which, according to Ovenden's theory, required $2 \times 10^9$ yr to settle into the observed resonances. Consequently the missing planet could not have strayed so close a mere $1.6 \times 10^7$ yr ago.

Nuclear effects might be the heat released by radioactive decay somehow converted to kinetic energy, fusion, or a chain reaction. Radioactive decay is a slow process and if an explosion is to result the energy generated must be stored. Thus a high pressure core must develop, constrained by a solid mantle until bursting occurs. The crushing strength of rock is $\sim 10^9$ erg g$^{-1}$, so even if the entire planetary mass were involved in containing the central force, breaking point would be reached long before the required $1.5 \times 10^{13}$ erg g$^{-1}$ was available. Consequently any explosion would lack the energy required by a factor $\sim 10^4$.

A similar objection applies to the hydrogen fusion reactions with the exception of $^2\mathrm{D} + {}^1\mathrm{H} \rightarrow {}^3\mathrm{He} + 5.5$ MeV which has a reaction time of only $\sim 6$ s. But this reaction is only significant at temperatures above $5.4 \times 10^5$ K and this is unattainable in planetary interiors. More generally, masses less than $2.4 \times 10^4$ $M_\oplus$ are too small to become hot enough for nuclear reactions.

Assuming a relative $^{235}\mathrm{U}$ abundance throughout the planet equal to the terrestrial crustal abundance, there is just enough energy to disperse the planet if the $^{235}\mathrm{U}$ could be assembled into a super-critical mass. In addition a chain reaction has the advantage that the necessary force could be developed in less than the travel time of shock waves through the planet. But it is clearly improbable that, in nature, enormous numbers of sub-critical masses of $^{235}\mathrm{U}$ could be assembled and brought together simultaneously within the planet.

There remains the possibility, mentioned by Ovenden, that the 90 $M_\oplus$ might have been in the form of a ring. For the ring to dissipate, the particles must either leave the Solar System or be absorbed into the Sun.

In the first situation the energy difficulties are as severe as in the planetary case ($1.9 \times 10^{42}$ erg required) with the additional problem that a suitable mechanism for dispersal is even more difficult to conceive. In the second case the ring particles are required to lose virtually all their angular momentum over a very short time scale (say, $<10^6$ yr). The only effective mechanism for systematic angular momentum loss is the Poynting-Robertson[4] effect. To be removed in $10^6$ yr or less, the particles must have been $<0.02$ cm in diameter; but for the ring to have persisted for the previous $4.5 \times 10^9$ yr, they

From *Nature*, vol. 242, pp. 250–251, 1973. Reprinted with permission.

must have been $\gg 600$ cm in diameter. No mechanism seems to exist whereby bodies of this size could exist for so long and yet suddenly disintegrate on a time scale $< 10^6$ yr.

We conclude that the destruction of a planet or ring in the recent past, with the mass required by Ovenden's theory, is physically improbable: one must look elsewhere for an explanation of Bode's Law and the existence of the asteroids.

## References

1. Ovenden, M. W., *Nature*, 239, 508 (1972).
2. Kuiper, G. P., *Astron. J.*, 55, 164 (1950).
3. Alfvén, H., *Icarus*, 3, 52 and 57 (1964).
4. Robertson, H. P., *Mon. Not. Roy. Astron. Soc.*, 97, 423 (1937).

# 7

# The Meteoritic Hazard of Interplanetary Travel

Peter M. Millman

Some 30,000 or more years ago, when the ancestors of Folsom Man inhabited what is now the Arizona desert, there was an earth-shattering explosion that hurled massive boulders weighing many tons for distances up to 10 km and left a hole in the desert over 220 m deep and 1.25 km across. A 25-m chunk of nickel-iron had collided with the earth at an impact speed of 15 km per second, and the explosive energy of this event produced such catastrophic results that the scattered boulders and the great crater are as visible today as they were in past millennia, long before the dawn of history (Nininger 1956). The Barringer Crater, near Flagstaff, Arizona, is one of the most interesting natural wonders in the world.

It was a bright sunny morning on February 12, 1947, in far eastern Siberia when, at 10:38 A.M., woodman Ashlaban noticed strange dancing shadows in the trees. Quickly looking up he saw a ball of fire exceeding the sun in brightness. Lineman Yefteyev, up a telegraph pole, felt a distinct shock, and astronomy instructor Mizerov, in the village of Ivanovichi, observed an unusual lighting effect on the plaster wall of a room. Running outside he saw a vertical pillar, bright white at the top and gradually shading to hues of rose and dark blue near the horizon. Later he heard detonations like those from a powerful motor. The first flash had been more brilliant than the sun, and a shock wave had broken windows, forced open doors, and knocked down plaster. In a heavily forested, remote area of the Sikhote-Alin Mountains north of Vladivostok some 100 tons of nickel-iron crashed to earth after fragmenting in the atmosphere. The result was over 100 craters among the trees, crater sizes ranging from diameters of half a meter up to 30 m (Krinov 1966). The biggest single piece of iron recovered weighed nearly two tons, and thousands of smaller fragments were collected over an elliptical area $1 \times 2$ km in extent.

In western Canada, just north of Edmonton, Alberta, a bright fireball crossed the sky from west to east at 1:06 A.M., MST, in the early morning hours of March 4, 1960. This was followed two or three minutes later by a thundering noise, and

windows rattled. Next morning farmers near the little town of Bruderheim found black stones of various sizes scattered over the snow-covered ground (Folinsbee and Bayrock, 1961). Over 300 kg total weight of these stones was recovered from an oval area $3 \times 5$ km in extent. Half a dozen stones were 30 cm or more across, and the rest ranged down to little pellets not much larger than a grain of wheat. All were covered with a dull black fusion crust, but inside the stones were light gray with small particles of metallic nickel-iron scattered throughout their substance. The bulk of the material consisted of various minerals, whose major constituents were oxygen, magnesium, silicon, and iron.

The three events described above are typical examples of collisions between "spaceship earth" and some of the solid debris that is scattered throughout interplanetary space. Such encounters immediately raise the question: What is the hazard to our interplanetary spacecraft of damage from similar collisions?

These accounts involve relatively rare events, where the colliding material was massive enough to reach the ground in some quantity. In the great majority of cases the small solid particles that collide with the earth are completely consumed in our upper atmosphere, which acts as a protective blanket. All that is left of the original particle is a bit of dust and vapor, and the event appears to the eye as a meteor or shooting star.

Man's literary records from distant centuries bear testimony to the fact that meteors have ever been a feature of the night sky. A Chinese classic, *Ch'un-ch'iu,* reported that on March 23, 687 B.C., "stars fell like a shower," and a Japanese work, the *Shinzan Shū;* reported that on November 8, A.D. 1698, "meteors fell like the weaving shuttle" (Imoto and Hasegawa 1958). Shakespeare, in *Richard II,* act 2, gives Salisbury this line: "I see thy glory, like a shooting star, fall to the base earth from the firmament"; and Alfred Lord Tennyson, in *The Princess,* part 7, describes the phenomenon perfectly: "Now slides the silent meteor on, and leaves a shining furrow."

It is apparent that, throughout the history of our planet, we have encountered this interplanetary debris, both large and

From *American Scientist,* vol. 59, pp. 700–705, 1971. Reprinted with permission.

small. It has been termed the meteoritic complex of the solar system, and the individual particles are called meteoroids. To assess their hazard to space travel we must have quantitative information concerning the total number of meteoroids in a given volume of space, the relative numbers of different sizes or masses, and the general nature of their motions around the sun.

The scientific study of the meteoritic complex started only a little over a century ago. At first, practically all observations were made visually, and all data were subject to the errors inherent in this form of record. Over the last few decades, instrumental programs of meteor observation have become more common, and now we have relatively reliable quantitative values for the most significant physical parameters of the complex.

## Obtaining Data

For particles in the meteoritic complex heavier than one kilogram, most of the information about size distribution and total numbers comes from a study of the frequency of meteorite falls on earth and the statistical analysis of the lunar impact craters. These heavier meteoroids are so uncommon that impacts between them and any spacecraft would be extremely

rare, and they are not particularly relevant to the subject under discussion. When we come to objects weighing 100 grams and less, we are dealing with much greater numbers per unit volume of space. Now we are in the range of the meteoroids that produce the normal visual meteor. As a very rough average figure one could say that a meteoroid of one gram produces a meteor as bright as the better known stars, like Arcturus, Vega, Regulus, or Spica, while a meteoroid one-hundredth of a gram in weight would result in a shooting star near the lower limit of normal vision.

There are various techniques by which we can obtain a figure for the numbers of meteoroids in a given volume of space. Teams of visual observers, such as those active near Ottawa, Canada (Fig. 1), make counts of the total numbers of meteors seen per hour over the whole sky for various nights in the year. Similar results are obtained through the use of fast cameras, specially designed to photograph faint meteors over a relatively large field (Hawkins and Upton 1958). Meteors too faint for detection by visual or photographic techniques must be recorded by radar. The small meteoroid, on entering the upper atmosphere at high velocity, leaves a trail of electrons (negatively charged particles) along its path and this acts as a radar target. At the Springhill Meteor Observatory near Ottawa, a meteor patrol radar was in use 24 hours a day for the period

Figure 1.  A team of nine visual meteor observers on duty at the Springhill Meteor Observatory, National Research Council of Canada, Ottawa, Ontario. These observers work in conjunction with photographic and radar programs, carried out simultaneously at the same site. (Photo by Bill Lingard—Photo Features.)

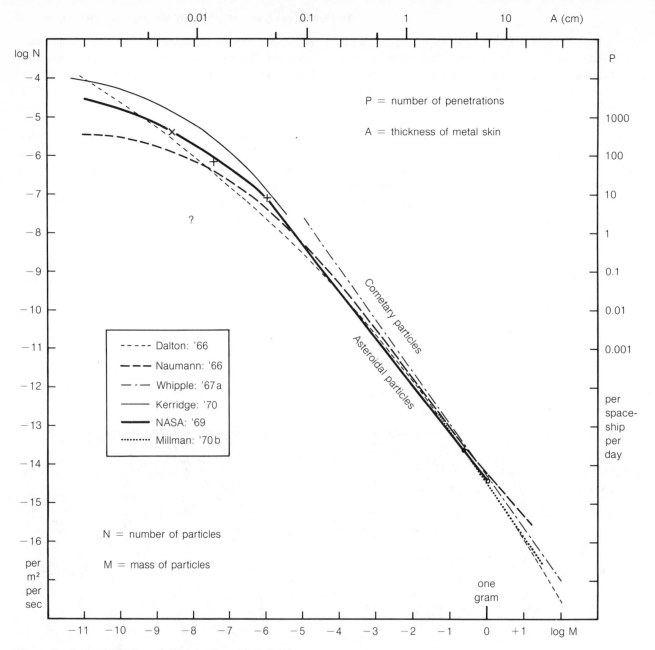

**Figure 2.** A log plot of the number of meteoroids that will impact on a square meter in space near the earth per second, N, against the lower limit of mass in grams, M, to which N is counted. Also a log plot of the number of penetrations per spaceship per terrestrial day, P, against the thickness of the spaceship skin in cm, A. The two plus signs are points found from penetrations of aluminum 0.4 and 0.2 mm thick on Pegasus vehicles, the X is determined from experiments on Explorers 16 and 23.

from 1957 to 1968 (Millman and McIntosh 1964, 1966).

Results from this program, and others like it in different parts of the world, are used to compile an average value for the numbers of small meteoroids populating interplanetary space near the earth's orbit.

Meteoroids in the size range of one-millionth to one-billionth of a gram are too small to be detected easily by radar, and here the study is extended through the use of recording instruments placed on upper-air rockets or satellites. These react to the impact energy when a small meteoroid strikes a sensitized surface of the vehicle. The most reliable data of this type have been collected by the Pegasus series of satellites (Clifton and Naumann 1966). Three of these vehicles were flown in 1965 and, when the 29-m wings were extended, each

satellite exposed over 210 sq m of detecting surface to the impact of the space dust. Parallel thin sheets of aluminum in the Pegasus wings were separated by insulators and formed large flat capacitors. These were discharged when the aluminum was penetrated by an impacting meteoroid. Three different thicknesses of metal were used, 0.04, 0.2, and 0.4 mm; thus penetration rates were obtained down to three different thresholds. The number-statistics of this experiment were good, since over 4,000 penetrations were recorded and were available for computing the impact rates. These were confirmed by the records obtained on the satellites Explorer 16 and Explorer 23, which gave the best results for penetrations of metal less than 0.1 mm in thickness (O'Neal 1968).

In summary, then, meteoroid counting has been carried out

by four major observational techniques—visual, photographic, radar, and recording via satellite. After careful analysis it is found that the basic results derived from the best observational programs are in essential agreement. In the size range considered here, that is, for masses less than one gram, certain orbital and physical information suggests that most of the meteoroids encountered by the earth are cometary fragments. This material is probably more fragile and of lower bulk density than that which falls to the earth as meteorites. The denser stone and iron material of meteorites seems to be debris from small planets not more than a few hundred kilometers in diameter, and is categorized as asteroidal. It probably accounts for less than one percent of the meteors seen and recorded.

## Size and Distribution of Meteoroids

Let us first take a look at the average size distribution that has been found from observations of the cometary meteoroids. In broad general terms we can describe this as follows. Suppose we take a volume of space where, counting down to a lower mass limit of one gram, we find on the average one meteoroid. In the same volume of space, counting down to a limit of one-tenth gram, we will find 20 particles. If we choose a smaller volume of space, where there is only one meteoroid greater than or equal to 0.001 gram, we find here 20 meteoroids greater than or equal to 0.0001 gram. In other words the same mass distribution law still holds. But the picture changes when we move to the still smaller spacedust. Suppose we count in a volume of space where there is only one particle equal to or greater than 0.000,001 gram. If now we count to a limit of 0.000,000,1 gram, we find only 5 meteoroids. The mass distribution law has changed and the increase in numbers is much less rapid.

This property of the meteoritic complex can be illustrated in another way. In Figure 2 I have plotted the numbers of meteoroids against mass limit on a logarithmic scale. The mass unit is the gram, and against this I have used the number of meteoroids that will impact on one square meter of surface per second. Recent compilations of diversified observational data by several different authorities have been plotted separately to show the general range of uncertainty that still exists. The average values adopted by the National Aeronautics and Space Administration (NASA) in 1969 are also plotted (Cour-Palais 1969), and a reduction of 23 years of visual observation near Ottawa has been added (Millman 1970b). Where the mass distribution law remains constant, the log plot is a straight line. Where this changes, as it does near a millionth of a gram, the line starts to bend over.

It is estimated that the greatest total mass of meteoric material lies in the mass interval at 0.000,01 gram (Whipple 1967b), a direct result of the change in the mass distribution law. The greatest uncertainty in the absolute numbers exists at the upper end of the relation, corresponding to the smallest particles.

It must be realized that Figure 2 is a simplified representation of a complicated picture. All data have been plotted against mass. However, in the case of the visual and photographic programs, only radiation energy is observed; the radar detects the quantity of ionization, while the satellite instruments are activated by penetration energy or other parameters. Reductions to mass must take into account a number of factors, the most important of which is impact velocity. Space does not permit here a detailed discussion of the various methods by which meteor impact velocities have been measured.

Figure 3 is a histogram (Cour-Palais 1969) showing the relative frequency of various velocity ranges in any given

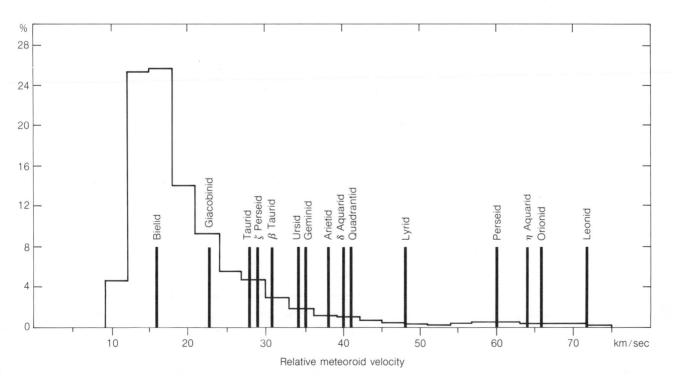

**Figure 3.** A histogram, in 3 km/sec steps, illustrating the distribution of velocities for impacting sporadic meteoroids, and vertical lines marking the mean observed velocities at which 15 of the major meteor showers meet the earth.

number of impacts on a body moving like the earth through interplanetary space around the sun. This plot refers to the average meteoritic background—what we call the sporadic meteors. At certain specific times of year we run into groups of meteoroids traveling along fixed orbits. These are called meteor showers, and all members of any one shower impact at nearly the same velocity. The shower velocities (Millman and McKinley 1963) have been indicated in the figure by heavy vertical lines. Since impact energy varies as the square of the velocity, it is obvious that, for a given mass, a Leonid or a Perseid meteoroid will have very much more effect on impact than the average sporadic meteoroid.

### Estimates of Meteoroid Penetrations

In spite of the above complications we are fairly certain that the shape of the number-vs-mass curve is essentially correct. Since we have tied it directly to the penetration experimental results, we can now use it to give a rough estimate of the penetrations to be expected on the skin of a spaceship. We can assume a surface area for a standard spacecraft in the general range of 1,000–1,500 sq m, which corresponds to a proposed twelve-man space station or an aerodynamic Mars lander. On this basis I have added two additional logarithmic scales to Figure 2, one for thickness of metal skin along the top, and the other for penetrations per spaceship per terrestrial day along the right side. Because there is no simple relation between skin thickness and penetrating mass (Cour-Palais 1969), the log scale at the top is nonuniform. Reading from these scales we see that a ship with a metal skin 0.4 mm thick would experience, on the average, ten penetrations a day. If we have a skin 5 mm thick there would be only one penetration a year, while aluminum 7 cm thick would be penetrated only once in 10,000 years.

As already noted, in the neighborhood of the earth's orbit the cometary meteoroids account for 98 or 99 percent of the

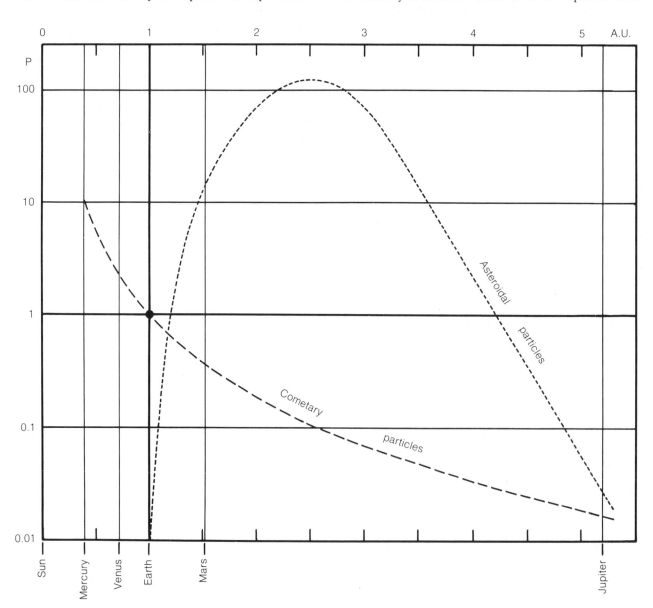

Figure 4.  Relative penetration frequencies, P, on a log scale against distance from the sun in astronomical units, a.u., on a natural scale. The mean distances of the planets Mercury, Venus, Earth, Mars, and Jupiter are indicated by vertical lines.

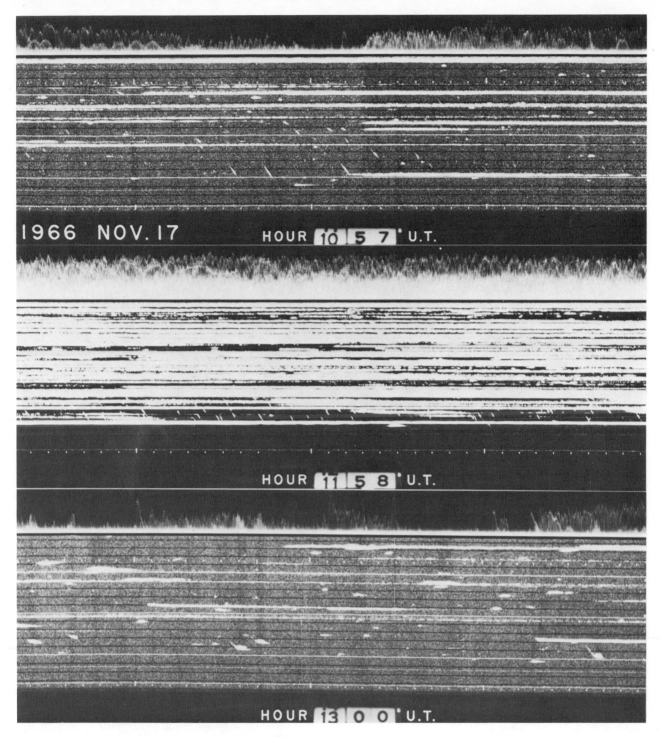

**1966 NOV. 17    HOUR 10 5 7 U.T.**

**HOUR 11 5 8 U.T.**

**HOUR 13 0 0 U.T.**

**Figure 5.** Three records from the high-power meteor radar at the Springhill Meteor Observatory, near Ottawa, covering a period of about two hours during the passage of the earth through the intense Leonid shower of 1966. Range in kilometers is represented as ordinates against time in seconds as abscissa. Echo amplitudes on a log scale appear at the top of each record. Near the peak of the shower the display tube was saturated with long-duration echoes which hid most of the head echoes, the short slanting lines. (Figure from McIntosh and Millman 1970.)

particles. In Figure 2 an attempt has been made to indicate the relative numbers of asteroidal meteoroids as compared to cometary meteoroids, but the information in this size range is very uncertain. Still more uncertain is the variation in numbers of meteoroids as we move out from the sun and the orbits of Mercury and Venus toward the orbits of Mars and Jupiter. There is considerable evidence from the zodiacal light obser-

vations and theoretical calculations that the small cometary particles decrease in numbers as we move out from the sun (Beard 1959). On the other hand it is logical to assume that the density of the asteroidal particles peaks somewhere between the orbits of Mars and Jupiter (Parkinson 1965), where most of the asteroid orbits are situated. Taking the conditions near the earth to be represented by a ratio of 100 between the com-

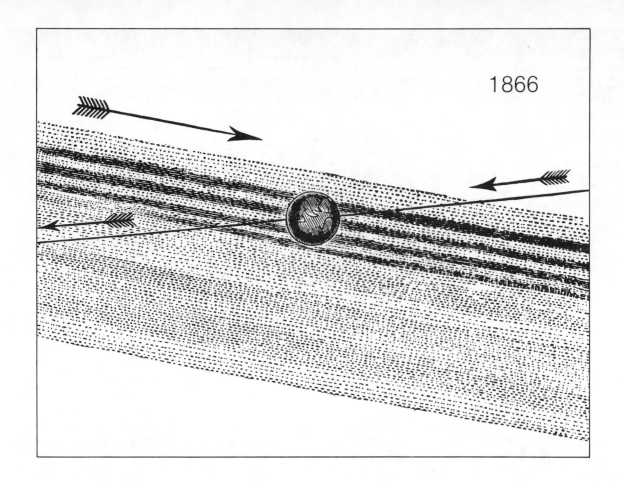

1866

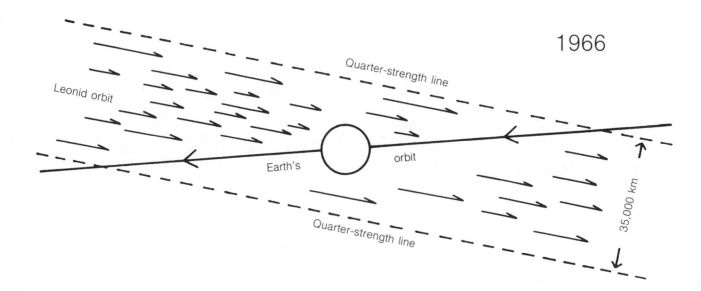

1966

Leonid orbit

Quarter-strength line

Earth's          orbit

Quarter-strength line

35,000 km

**Figure 6.** Two diagrams to scale, illustrating the passage of the earth through the strong Leonid showers of 1866 and 1966. The quarter-strength line in the 1966 plot indicates the position where the number of meteoroids per unit volume is one quarter that at the center of the stream. A few months after publishing the lower diagram (Millman 1967) the author found the upper woodcut in a popular scientific journal of the last century (Ward 1867), which shows the aptness of the Biblical quotation, "There is nothing new under the sun."

etary and asteroidal particles, Figure 4 illustrates possible density distributions of meteoroids at various distances from the sun. This plot should be regarded only as an educated guess at the present time. It is supported, at least qualitatively, by some measures of low statistical number-value made from Mariner 4 on its way to Mars (Alexander et al. 1965).

Up to the present we have been talking about the average background of the meteoritic complex found in the inner core of the solar system. Impressed on this background we have the orbits of the major meteor showers that have already been referred to. These are like broad or narrow highways in space along which the shower meteoroids travel, adding their numbers to the general background. Many of these orbits have been identified as almost identical with certain comet orbits, hence the belief that the shower meteoroids are cometary fragments. Apparently the bulk of the general background flux of meteoroids encountered by the earth consists of particles that once belonged to specific showers but which, over the centuries, have been blended into the background by orbital perturbations.

Meteor showers exhibit a wide variety in width and density (Millman 1967). In general the narrower widths have the higher space density of meteoroids. In extreme cases the numbers of meteors encountered at the peak of a shower may be 100 or 1,000 times that of the sporadic background (Fig. 5), or even higher if we are considering meteoroids heavier than a gram. As a general rule the size distribution of the shower meteors is less steep than that plotted on the right of Figure 2 (Millman 1970a). In other words, for a given number of one-gram meteoroids there are fewer of the very small meteoroids in the showers and the number–mass slope approaches that at the upper left of the figure. This means that as we count to smaller and smaller mass limits the showers become less and less prominent and eventually they are lost in the background flux.

Since it is relatively easy to protect spacecraft against penetration by small meteoroids, and impacts by meteoroids of mass greater than 0.1 gram are extremely rare, the showers will not pose a serious threat to the safety of the astronauts. It would be wise to avoid the very concentrated streams, such as the Leonids encountered by the earth in 1866 (Ward 1867) and 1966 (Millman 1967), whenever this is possible (Fig. 6). Unfortunately, we have no knowledge of the general complex of meteoroid showers in space, as we can observe only those that intersect the earth's orbit. We can assume that any comet will have meteoroids moving along its orbit, and thus all known comet orbits should also be avoided by manned spaceships. This is not always possible since often the cometary debris tends to be spread into a flat, thin distribution along the comet orbit plane (Jacchia 1963).

There is a technique by which the protection of a given weight of metal in the skin of a spaceship may be increased by roughly an order of magnitude. This is the meteor-bumper concept, first suggested by F. L. Whipple (1946, 1952). A thin metal skin is supported a short distance outside the main wall of the spacecraft. An impact on the outer skin results in a fragmentation of the meteoroid and, although all the fragments may penetrate, when they reach the thicker inner skin the total kinetic energy is reduced, the impacting fragments are spread

over a much wider area, and they are thus much less effective in penetrating the main spaceship wall. As an example, an outer skin 0.5 mm thick, placed 20 mm outside a main skin 5 mm thick, will increase the protection by a factor of about 10, and we may expect only one penetration per spaceship every ten years of flight.

Although the hazard of penetration by meteoroid impact is never of zero probability, a suitable design of the spacecraft and the space suits can minimize the probability of serious damage to such an extent that this hazard becomes relatively small compared to the other problems of interplanetary travel.

## References

Alexander, W. M., C. W. McCracken, and J. L. Bohn. 1965. Zodiacal dust: Measurements by Mariner IV. *Science* 149:1240-41.

Beard, D. B. 1959. Interplanetary dust distribution. *Astrophys. J.* 129: 496-506.

Clifton, Stuart, and Robert Naumann. 1966. Pegasus satellite measurements of meteoroid penetration. *NASA Technical Memo.* NASA TM X-1316:1-30.

Cour-Palais, B. G. 1969. Meteoroid environment model—1969. *NASA Space Vehicle Design Criteria* NASA SP-8013:1-31.

Dalton, Charles C. 1966. Effects of recent NASA-arc hypervelocity impact results on meteoroid flux and puncture models. *NASA Technical Memo.* NASA TM X-53512:1-36.

Folinsbee, R. E., and L. A. Bayrock. 1961. The Bruderheim meteorite-fall and recovery. *J. Roy. Astron. Soc. Canada* 55:218-28.

Hawkins, Gerald S., and Edward K. L. Upton. 1958. The influx rate of meteors in the earth's atmosphere. *Astrophys. J.* 128: 727-35.

Imoto, Susumu, and Ichiro Hasegawa. 1958. Historical records of meteor showers in China, Korea, and Japan. *Smithsonian Contr. Astrophys.* 2:131-44.

Jacchia, Luigi G. 1963. Meteors, meteorites, and comets: Interrelations. In Barbara M. Middlehurst and Gerard P. Kuiper, eds., *The Moon, Meteorites, and Comets.* Chicago: Univ. Chicago Press, pp. 774-98.

Kerridge, J. F. 1970. Micrometeorite environment at the earth's orbit. *Nature* 228:616-19.

Krinov, E. L. 1966. The Sikhote-Aline iron meteorite shower. In *Giant Meteorites.* New York: Pergamon Press, pp. 266-376.

McIntosh, Bruce A., and Peter M. Millman. 1970. The Leonids by radar—1957 to 1968. *Meteoritics* 5:1-18.

Millman, Peter M. 1967. Observational evidence of the meteoritic complex. In J. L. Weinberg, ed., *The Zodiacal Light and the Interplanetary Medium.* Washington: NASA SP-150, pp. 399-407.

Millman, Peter M. 1970a. Meteor showers and interplanetary dust. In *COSPAR, Space Research X.* Amsterdam: North-Holland Pub. Co., pp. 260-65.

Millman, Peter M. 1970b. Meteoritic flux determined from visual observations. *J. Roy. Astron. Soc. Canada* 64: 187-190.

Millman, Peter M., and Bruce A. McIntosh. 1964. Meteor radar statistics I. *Canadian J. Phys.* 42:1730-42.

Millman, Peter M., and Bruce A. McIntosh. 1966. Meteor radar statistics II. *Canadian J. Phys.* 44:1593-1602.

Millman, Peter M., and D. W. R. McKinley, 1963. Meteors. In Middlehurst and Kuiper, eds., *The Moon, Meteorites, and Comets,* pp. 674-773.

Naumann, Robert J. 1966. The near-earth meteoroid environment. *NASA Technical Note* NASA TN D-3717:1-38.

Nininger, H. H. 1956. *Arizona's Meteorite Crater.* Sedona, Arizona: American Meteorite Museum.

O'Neal, R. L. 1968. The Explorer XXIII micrometeoroid satellite. *NASA Technical Note* NASA TN D-4284:1-111.

Parkinson, J. 1965. Space environment criteria. *George C. Marshall Space Flight Center Contract* NAS8-11285, Aerojet-General, Rept. no. 2979.

Ward, the Hon. Mrs. 1867. The November shooting stars. *The Intellectual Observer* (London) 10:449-58.

Whipple, F. L. 1946. Possible hazards to satellite vehicle from meteorites. In Project Rand. Douglas Aircraft Co. Inc., Santa Monica, Calif.: Rept. no. SM-11827, chap. 11, appendix III.

Whipple, F. L. 1952. Meteoritic phenomena and meteorites. In Clayton S. White and Otis O. Benson, Jr., eds., *Physics and Medicine of the Upper Atmosphere.* Albuquerque: Univ. New Mexico Press, pp. 137-70.

Whipple, F. L. 1967a. The meteoritic environment of the moon. *Proc. Roy. Soc., A* 296:304-15.

Whipple, F. L. 1967b. On maintaining the meteoritic complex. In Weinberg, ed., *The Zodiacal Light and the Interplanetary Medium,* pp. 409-26.

# A Funny Thing Happened to Comet Kohoutek

Stephen P. Maran

A thrown egg, which missed, was the prompt reply when astronomer Edward P. Ney recently asked a University of Minnesota audience its opinion of Comet Kohoutek. Eagerly awaited as the "comet of the century," the celestial visitor has amply rewarded researchers, but it has been a sore disappointment to the viewing public. Kohoutek has been derided by the *Washington Star-News,* as "The Comet of the Month," and satirized by *New York Times* columnist Russell Baker as "The Comet That Couldn't." Its modest glow, which belied our great expectations, has prompted a variety of scientific explanations.

Calculations made last April indicated that Kohoutek ought to outshine the planet Venus and might even rival the brightness of a half moon. Interest focused, of course, on the latter possibility. Somewhat lost in the publicity buildup was the point that maximum luminosity would be attained only as the comet neared perihelion, its closest approach to the sun, on December 28 [1973]. However, in that proximity, the sun's dazzle would preclude most comet viewing from the earth.

By October, astronomers realized that Kohoutek was not living up to expectations, and that it probably would not exceed the brightness of Venus. This assessment was consistent with the findings of Brian Marsden, director of the Central Bureau for Astronomical Telegrams at the Smithsonian Astrophysical Observatory in Cambridge, Massachusetts, who had analyzed the comet's orbit and concluded that Kohoutek was making its first close approach to the sun. In 1951, the Dutch astronomers Jan Oort and Maarten Schmidt discovered that such "new" comets tend to brighten less strongly than other comets as they approach the sun. Nevertheless, a good show still seemed likely for much of January. Having by then rounded the sun, Kohoutek would head back out into the solar system, coming closest to the earth on January 15. Calculations by William Liller, professor of astronomy at Harvard University, suggested that on that date the visible tail length might reach 18 degrees, or thirty-six times the apparent diameter of the full moon. On December 27, when I faced a Washington press conference with the comet's discoverer, Lubos Kohoutek, I was still confident that his comet would become conspicuous in the ensuing weeks.

Even before perihelion, the comet was a scientific success, thanks to the detection of naturally generated radio signals emanating from its head. These emissions revealed that molecules of methyl cyanide ($CH_3CN$) and hydrogen cyanide (HCN) were present there. These substances, never before found in a comet, are known to occur in dense clouds in the interstellar space of our Milky Way galaxy, where new stars and solar systems are thought to be forming. Thus, Kohoutek might truly be a sample of the primordial material from which our sun and the earth itself were born.

Measurements made by NASA's Orbiting Solar Observatory-7 satellite on December 27 and analyzed by Martin Koomen, a physicist at the Naval Research Laboratory in Washington, D.C., showed that the comet, now very close to the sun, was brighter than Mercury. During the next few days, Kohoutek did become a genuine spectacle, but only for the Skylab astronauts. They watched it brighten and on December 29 discovered a brilliant sunward-pointing spike, or "antitail."

Most of the treasure trove of comet data collected by the Skylab crew was in the form of photographic film, which could only be developed and examined after the astronauts returned to earth. But a relatively crude, video-transmitted "coronagram"—a picture of Kohoutek passing behind the sun's outer atmosphere, or corona, and actually seen through it—showed that the comet was at least comparable to the brighter stars and might even have exceeded magnitude −3, close to Venus's −4 rating. (The brightness of celestial objects is ranked by magnitudes. The brightest objects are classified as being of negative value, the dimmest are of positive value. The greater the magnitude number, the dimmer the object.) Thus, a celestial show still seemed in the offing for January. If the comet did surpass magnitude −3, it had brightened by a factor of more than forty million since its discovery. Considering the vast

From *Natural History* Magazine, March 1974. Copyright © The American Museum of Natural History, 1974. Reprinted with permission.

range of brightness over which the predictions had to extend, it was not unreasonable to miss by a few magnitudes.

On the evening of December 30, astronomer Robert Roosen, then at the new Joint Observatory for Cometary Research, 10,600 feet up on South Baldy Mountain, near Socorro, New Mexico, spotted the comet. He was apparently the first person on earth to see it after its perihelion passage. "How bright is it?" I asked in a telephone conversation later that night. Roosen didn't know. He had glimpsed Kohoutek through a small, accurately pointed "finder telescope," but he viewed it through clouds that prevented comparisons with stars and planets of known brightness. Cloud cover also hampered observations at many other locations the next evening. Then, on January 1, word came from comet watcher and amateur astronomer Clay Sherrod in North Little Rock, Arkansas, that the entire comet—head and tail—was visible to the naked eye low on the horizon at twilight. Sherrod judged its magnitude to be −1.5, about equal to that of Jupiter or of Sirius, the brightest star. The circumstances of Sherrod's observation rendered it somewhat uncertain, but the magnitude value was smack in the middle of the range, −1.2 to −2.2, predicted by Donald Yeomans, an astronomer at the Computer Sciences Corporation in Silver Spring, Maryland.

Yeomans's computations indicated that on January 4, when the Goddard Space Flight Center was scheduled to launch a rocket payload to study the comet, Kohoutek's magnitude would be between +0.1 and −0.9, comparable to the very bright stars Rigel and Canopus, respectively. But on January 2, physicist Thomas Heinsheimer of Palos Verdes, California, could discern only the comet's head with his unaided eye. A report from Madison, Wisconsin, on January 3 put the comet's brightness at +1, but on that same night a different magnitude was noted at Kitt Peak National Observatory in Tucson, Arizona. Norman Tolk, a physicist and the leader of a Bell Laboratories team studying Kohoutek with the large McMath Telescope at Kitt Peak, called me to say that the comet, whose tail could be clearly seen through the clear Arizona air, was only +2 magnitude, no brighter than a Big Dipper star. Sherrod's estimate seemed to verify Yeomans's predictions, but only a few days later, the comet was much too dim. On January 4, we had word from the Skylab crew that the comet was now only comparable to the adjacent star Beta Capricorni, a third-magnitude object. (A difference of one magnitude corresponds to a brightness ratio of slightly more than a factor of 2.5.) Was the comet really fading that fast? Comet expert Elizabeth Roemer, of the University of Arizona in Tucson, points out that judging the magnitude of a comet is tricky at best. It was especially difficult for earth-bound Kohoutek observers, who, because of the low elevation of the comet after sunset, saw it through a great deal of atmosphere.

My first postperihelion look at Kohoutek came on the evening of January 4. Flying over Pennsylvania aboard a NASA research aircraft, I could perceive with the unaided eye a 5-degree length of yellow tail, about equal to the separation of the "pointer stars" in the bowl of the Big Dipper. The astronauts' estimate seemed to me to have been a little on the dim side, unless the comet was fluctuating in intensity, a not-impossible situation. But as Tolk had reported, Kohoutek was clearly of only second magnitude. I had seen three earlier comets from the ground that were more impressive than Kohoutek looked that night from the air. It was apparent that for the general public the viewing prospects, especially in light-polluted and smoggy metropolitan areas, would be very limited. At a few choice sites the comet did become a superb sight in mid-January as it moved up higher into a darker sky and became visible later in the evening than in the first week of the month. But the rave reviews came from South Baldy Mountain and from Mount Haleakala in Hawaii, which is nearly as high. Scattered reports of naked-eye sightings were received from around the country, some even from cities, but for most would-be viewers, Kohoutek was the comet that wasn't there.

What went wrong? It is possible that, as already implied, Kohoutek faded very rapidly after perihelion, as did Comet 1897 I, another new comet. Several theories have been proposed to explain such fading. Thornton Page, an astronomer, and Donald Packer, a physicist, both at the Naval Research Laboratory, speculate that the sun's heat may have stimulated chemical reactions in the comet, producing a sticky substance that deterred the outflow of sunlight-reflecting dust particles, thus diminishing Kohoutek's luminosity. Other astronomers think that the low level of activity on the sun in late December and early January left the comet insufficiently stimulated by solar radiation and particles to shine brightly. I don't find either of these ideas compelling. Maarten Schmidt, now at the California Institute of Technology in Pasadena, even doubts that the comet faded rapidly. How accurate are the near-perihelion brightness values? he asks. On the other hand, Luigi Jacchia, an astronomer at the Smithsonian Astrophysical Observatory, working on the assumption that these data are indeed valid, concludes that the comet actually flared briefly in brightness near perihelion when only the astronauts could see it. To check on this, we will have to make a careful analysis of the Skylab photographs, probably the only good records available for that time.

Whether or not Kohoutek dimmed abruptly in early January, its relative brightness when discovered in March, 1973, remains to be explained. It was that luminosity that led to what turned out to be overly optimistic magnitude predictions for the end of the year. One idea, advanced by astronomer E. J. Öpik of Northern Ireland's Armagh Observatory, is that in March, Kohoutek was surrounded by a halo of small chunks of frozen water known as ice grains. These would reflect sunlight very effectively and could account for the comet's brightness on its discovery 442 million miles away from the sun. Then, as the comet approached the sun, the ice sublimated, that is, was converted to water vapor, and Kohoutek became largely dependent on its dust particles to reflect sunlight. Observations made this winter, however, show that Kohoutek was just not dusty enough to become very bright.

A variation on Öpik's hypothesis, proposed informally by William G. Fastie, a physicist at Johns Hopkins University, suggests that the comet's initial halo was composed of dust rather than ice and that it was denser than is usual for a comet so far from the sun. At distances of several hundred million miles, far beyond the orbit of Mars, a comet's brightness is often due primarily to reflection by its solid nucleus; the dust cloud only assumes prominence later, as the comet approaches the sun. If Kohoutek is a new comet, Fastie theorizes, it was never before substantially heated by the sun. It may thus have approached that body with an intact surface layer of frozen hydrogen molecules ($H_2$). Only a slight warming would be required to sublimate this rare solid. If dust grains were embedded in the $H_2$, they would have formed a bright cloud, or coma, when the comet was far from the sun. Solar radiation pressure is very weak at great distances from the sun, and accordingly would not have pushed the dust out from the coma very rapidly. None of this applies to "old" comets, such as Halley's, which have come near the sun many times. Because their surface layers of $H_2$ or other volatile molecules have been eliminated by sublimation, such comets do not brighten as much as new ones when far from the sun.

In conclusion Comet Kohoutek may have faded rapidly because of an as yet unexplained disturbance when it came around the sun. Or it may have misled us by looking too bright, too soon, as measured by old-comet standards. Neither or, perhaps, both of these propositions may be true. We will only have a better idea of what happened after all the data are in from the space and ground observatories. The only thing we can be certain of now is that comets are probably the most unpredictable and, at least sometimes, the most impressive phenomena of our solar system.

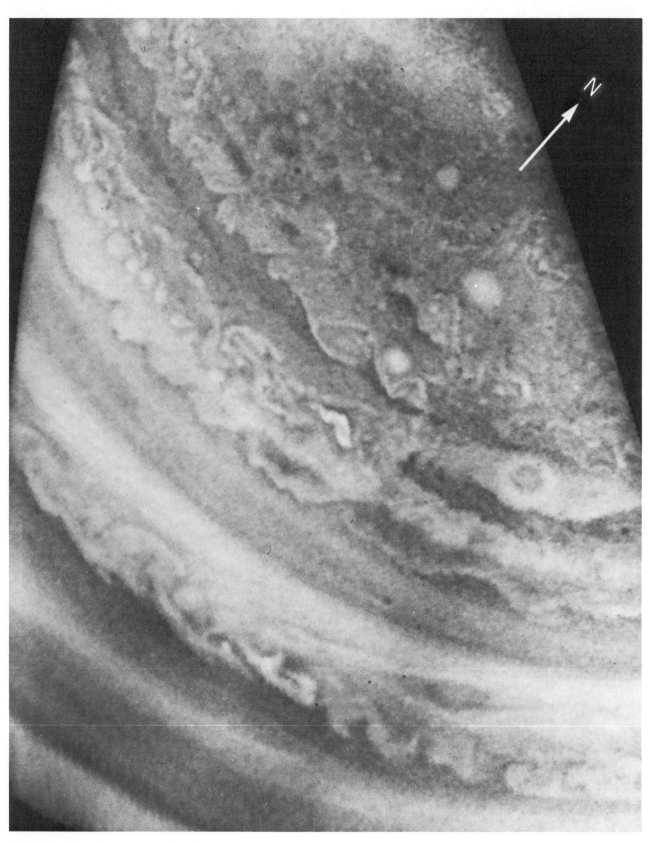

Part of the planet Jupiter as seen from the spacecraft *Pioneer 11* in December 1974, detailing the dark belts and bright zones toward Jupiter's north pole. [National Aeronautics and Space Administration.]

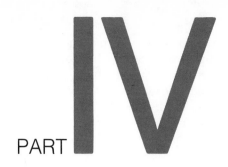

PART **IV**

# THE MOON AND PLANETS

In "Mercury: The Dark-Side Temperature," infrared astronomers Thomas L. Murdock and Edward P. Ney tell how they determined an average temperature of 111 K ($-162°$C or $-260°$F) for the dark side of Mercury—a temperature below the freezing points of the rare gases krypton, radon, and xenon. (The noontime temperature on the bright side is 620 K (347°C or 657°F), hot enough for lead and tin to exist in a molten state.) They used the *beam-switching* technique, in which the beam or field-of-view of the photometer is moved on and off the target object in a cyclic fashion (usually by means of a tilting mirror in the telescope) so that the contribution of infrared emission from the Earth's atmosphere, called "sky radiation," can be eliminated from the measurement of the radiation from the target. Murdock and Ney conclude that a parameter called Mercury's "thermal inertia," which characterizes the length of time that mercurian soil requires to adjust to a change in heat input from the sun, is comparable to the corresponding parameter for the moon. Thus, the mercurian and lunar soil layers may be similar. In 1974, television pictures transmitted from the *Mariner 10* spacecraft showed that Mercury is also heavily cratered, as are the moon and Mars.

Instruments on *Mariner 10* observed both Venus and Mercury in ultraviolet, visible, and infrared light and sampled the cosmic rays, plasma, and magnetic fields in their vicinities. In "Venus: Atmospheric Motion and Structure from Mariner 10 Pictures," a scientific team led by Bruce C. Murray describes the first results gleaned from the close-up views of Venus. The pictures show the striking cloud features of Venus's atmosphere, and demonstrate that there is a pronounced east-west or "zonal" circulation pattern. There is clear evidence in the time-sequences of the pictures that cloud patterns make a complete circuit of the planet at the equator in four days, whereas patterns at higher latitudes rotate even faster (only two days at latitude 50°). This leads to the conclusion that "angular velocity increases with latitude." The existence of widespread vertical motions, indicative of convection, was discovered in the region near the subsolar point on Venus. Remarkable curved wave patterns were also recorded. Murray and his colleagues searched for meridional (north-south) circulation, but didn't find any.

In the description of the two television cameras on *Mariner 10*, the abbreviation "nm," for *nanometer*, is used (Figure 3). A nanometer is a unit of length equal to $10^{-9}$ meter (i.e., one-billionth of a meter), or 10 angstroms. The intensities recorded by the television cameras were converted into a series of numbers for transmission to Earth by the spacecraft telemetry system. The phrase "encoded to 256 discrete levels" indicates that the exact intensities were not telemetered. Rather, each measured value was approximated by the nearest of the 256 levels on the scale of the telemetry system; "8 bits" is computer slang for the fact that 256 equals $2^8$. Other terms used in this selection are *Venera 8*, which is the name of an earlier Venus space probe that was launched by the Soviet Union, and *radio occultation*, which refers to the method by which the atmosphere of Venus was studied through its effect on radio signals transmitted between *Mariner 10* and the Earth as the spacecraft passed behind Venus. The section on atmospheric circulation is quite technical and can be omitted because the broader observational conclusions are sufficient to illustrate the new findings about the atmosphere of Venus. The next article, "Venus: Topography Revealed by Radar," examines the planetary surface that is concealed by this cloudy atmosphere.

"Higher, sharper peaks may also be present but would not be visible because the data . . . represent averages over large regions." In a pioneering exploration of the surface of Venus by means of radar, Donald B. Campbell and four colleagues discovered a huge elevated region. As implied by the preceding quote, they found tantalizing hints of individual mountain(?) ranges, but the observations were not sufficiently detailed to verify them. Current and future radar observation should improve our knowledge of Venus. We may describe a planet, in the simplest approximation, as a perfect sphere, or more accurately as an oblate (flattened at the poles) spheroid. If we want to give a more detailed description that represents the presence, for example, of elevated and depressed regions, more complicated mathematical expressions are used to represent the planet's shape. The authors mention two such methods, the *spherical-harmonic expansion* and the *double Fourier series*. Some quantities are quoted in microseconds—$\mu$sec, one-millionth of a second—an indication of the precision with which the radar echoes were timed. If we observed the Earth *from* Venus with a comparable radar system, only the very gross surface features of the Earth would be distinguished. What guesses about the surface of Earth might we make in such a case?

The *Apollo* moon missions resulted in a wide variety of findings concerning the physical nature of the moon. A number of these results are summarized by science writer Allen L. Hammond in "Lunar Science: Analyzing the Apollo Legacy." Hammond uses the acronym KREEP, which refers to lunar rocks of a certain chemical composition: K is the chemical symbol for potassium; REE stands for rare earth elements and P is for phosphorus. Other discoveries made possible by travel to the moon are described by Palmer Dyal and Curtis W. Parkin in "The Magnetism of the Moon."

"After burning 900 pounds of retro-rocket fuel, which it had transported 287 million miles in 167 days, *Mariner 9* weighed 1,350 pounds when it finally went into orbit around Mars." Bruce C. Murray describes the exploration of "Mars from Mariner 9." Murray, who is now the Director of the Jet Propulsion Laboratory in Pasadena, California, tells of the giant martian volcano Nix Olympica and of sand dunes, canyons, and other newly-discovered features of the red planet's landscape. Additional *Mariner 9* findings are reported by William D. Metz in "Update on Mars: Clues about the Early Solar System." He mentions a curious parallel between the present martian surface and the ancient Earth continent of Gondwanaland, which existed in the remote past, before continental drift caused the present arrangement of the Earth's surface.

The discoveries of *Mariner 9* laid the groundwork for *Viking*'s search for life on Mars in 1976.

Two brief reports on some results of *Pioneer 10*, the first space probe to reach the vicinity of Jupiter, are given by Metz in "Moons of Jupiter: Io Seems to Play an Important Role" and by Darrell L. Judge and Robert W. Carlson in "Pioneer 10 Observations of the Ultraviolet Glow in the Vicinity of Jupiter." *Rayleigh,* a term used by Judge and Carlson, is a unit that describes the intensity of light emitted along the line of sight through a three-dimensional luminous region, such as a cloud of aurora borealis or a region of some planet's atmosphere. For some reason, this unit is often misunderstood by practicing scientists and used incorrectly. (No doubt some instructors will tell students that we have described the rayleigh incorrectly in this paragraph!) As an example, the emission from a bright green display of the aurora in the Earth's atmosphere may reach a value of a few hundred rayleighs. *Photolysis* means the breakup of a molecule by absorption of energy from a photon of light.

# Mercury: The Dark-Side Temperature

T. L. Murdock and E. P. Ney

Abstract. *The planet Mercury was observed before, during, and after the inferior conjunctions of 29 September 1969 and 9 May 1970 at wavelengths of 3.75, 4.75, 8.6, and 12 microns. The average dark-side temperature is $111° \pm 3°K$. The thermal inertia of the surface required to fit this temperature is close to that for the moon and indicates that Mercury and the moon have very similar top surface layers.*

Radar measurements have shown that the mercurian day is two orbital periods long. When Mercury is new, the center of the dark hemisphere presented to the earth is at the midnight temperature, and 88 days have elapsed since noon for this point. The noontime temperature at the mean distance of the planet from the sun is about 620°K. A measurement of the midnight temperature will establish the average temperature of the planet, as well as the thermal inertia of the surface. In order to establish the dark-side temperature it is necessary either to observe with sufficient resolution to put one photometer beam on the cold disk without including any of the crescent, or to fill the photometer beam with the planet when the planet is near inferior conjunction and the contribution of scattered light and thermal emission from the crescent can be reduced indefinitely. Murray attempted to measure the dark-side temperature by the first method (*1*), using the 200-inch (508-cm) telescope at Mount Palomar. No signal was observed, and Murray concluded from this that the temperature must be less than 150°K.

We used the second method of observing the entire planet near its new phase. This method has the disadvantage of requiring observations at very small elongation angles from the sun. Fortunately, this can be accomplished much more easily in the infrared region than in the visible, and in September 1969 we observed down to an elongation angle of 3°. Solar heating of the mirror resulting in defocusing can cause a reduction in signal from the planet when observations are made with

From *Science*, vol. 170, pp. 535–537, 30 October 1970. Copyright 1970 by the American Association for the Advancement of Science. Reprinted with permission.

beam sizes only slightly larger than the apparent angular diameter of the planet. As shown below, it appears that this effect was small although not negligible.

Except for very small planetocentric phase angles, the infrared radiation in the 3.5- to 12-$\mu$ region is produced by thermal emission from the illuminated part of the disk, and thermal emission exceeds scattered sunlight by a large factor, even at the shortest wavelength. As Mercury approaches its new phase, the effective brightness temperature of the crescent drops and sunlight scattered from the crescent finally exceeds the thermal emission from the crescent at 3.75 $\mu$. At the longer wavelengths, the thermal emission from the dark side of Mercury exceeds the thermal radiation from the crescent when the fraction illuminated ($k$) becomes less than 0.01.

Mercury was observed six times between 21 September and 3 October 1969 and three times between 4 May and 17 May 1970, including the two days nearest inferior conjunction in September. The 1969 observations were made with a four-filter, broadband infrared system which had effective wavelengths of 3.75, 4.75, 8.6, and 12 $\mu$. The 1970 observations were made with an eight-filter, broadband system with effective wavelengths of 2.3, 3.6, 3.75, 4.75, 8.6, 10.6, 12.2, and 20 $\mu$. The angular diameter of beam projected on the sky by the 30-inch telescope at the O'Brien Observatory, University of Minnesota, was 12 seconds of arc with the four-filter system and 26 seconds of arc with the eight-filter system. The apparent angular diameter of Mercury at the September conjunction was 10 seconds of arc so that the planet nearly filled the beam, thus producing the best possible signal-to-noise ratio. Complete sky cancellation was obtained with both systems by switching the beam to a position one beam diameter away at 10 cycle/sec. The noise level was determined at the short wavelengths by the bolometer detector and at the long wavelengths by sky noise.

A comparison of the 3.75-$\mu$ intensities taken during the September conjunction with intensities measured in the conjunctions of June 1968 and October 1968 with a beam diameter of 83 arc seconds, and in the conjunctions of May and June

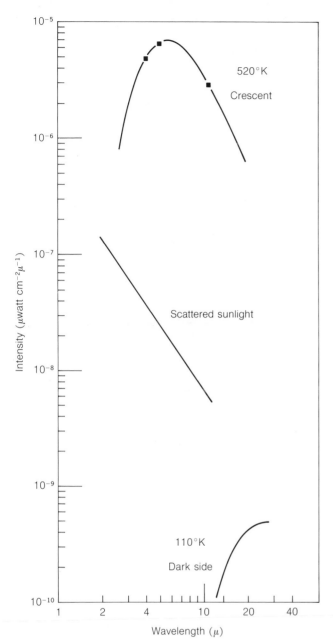

**Figure 1.** Measured intensities from Mercury at 3.75, 4.8, and 11.3 μ on 8 August 1969. At this time k was 0.853.

1969 and May 1970 with a beam diameter of 26 seconds shows that defocusing by sunlight heating caused a reduction by a factor of 1.5 in the signal from the planet. This effect also appears at the longer wavelengths. Therefore, all intensities measured during the September 1969 conjunction have been multiplied by a factor of 1.5, with the result that there is a change of 2°K in the average dark-side effective temperature. The mirror was again in equilibrium at the time the calibration measurements were taken.

Figure 1 shows a plot of the surface brightness of Mercury on August 8.8. The infrared radiation was produced by thermal emission from the crescent at a color temperature of 522°K. The solid angle required to make the brightness temperature equal to the color temperature is the solid angle of the illuminated crescent. Also plotted in Figure 1 are the expected contributions of scattered sunlight and thermal emission from the dark disk.

Figure 2 shows a plot of the surface brightness of Mercury on September 29.8. The infrared radiation was produced by thermal emission from the dark side at 12 μ, thermal emission from both the dark side and the crescent at 8.6 μ, and scattered sunlight at the two shortest wavelengths. We determined temperatures by first assuming that all of the 12-μ radiation was thermal emission from the dark side and that all of the 8.6-μ radiation was thermal emission from the crescent. A method of successive approximation was then applied. Using intensities uncorrected for mirror defocusing leads to best fit temperatures of 109°K for the dark side and 194°K for the crescent. The corrected average dark-side temperature is 111°K, and the corrected average crescent temperature is 205°K. The dark-side temperature is much better determined than the effective temperature of the crescent. Our value of the dark-side temperature is in agreement with the prediction of Morrison and Sagan (2).

The curve for scattered sunlight is drawn through the experimental points (Fig. 2) and gives an albedo for Mercury that is a factor of 2 less than that expected on the basis of an observed visual magnitude of +3.4 and the sun's color index (3). This effect could be due either to a real change in Mercury's albedo between visible and infrared wavelengths or to the fact that we were beam-switching against the zodiacal light which should have the same color index as the sunlight scattered from Mercury. Whereas our other measurements were made with a signal-to-noise ratio of up to 5 to 1 (see

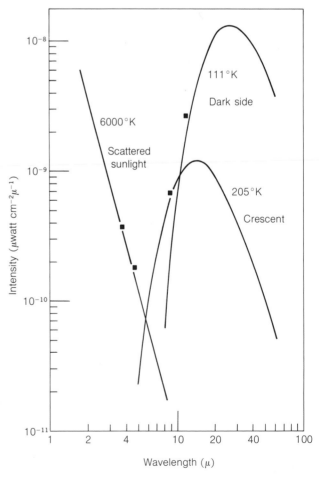

**Figure 2.** Corrected intensities from Mercury at 3.9, 4.8, 8.6, and 11.8 μ on 29 September 1969. At this time k was 0.004.

66

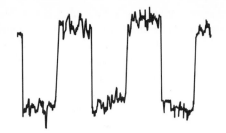

**Figure 3.** A reproduction of part of the chart record from 29 September 1969 showing signal and noise at 11.8 $\mu$.

Fig. 3), our 4.8-$\mu$ measurement was made with a signal-to-noise ratio of 1 to 1 and can properly be called only an upper limit. At an elongation angle of 3° coronal brightness is 1.2 $\times$ 10$^{-10}$ the sun's surface brightness (Fig. 4) (4). Therefore, for our beam diameter of 13 seconds the intensity of the F corona or zodiacal light at 5 $\mu$ is 9 $\times$ 10$^{-12}$ $\mu$watt cm$^{-2}$ $\mu^{-1}$ or about 1/20 of our measured 4.8-$\mu$ intensity (5). Under more favorable conditions it may be possible to detect the zodiacal light as an "antisignal."

According to our measurements, the average dark-side temperature of Mercury is 111° $\pm$ 3°K. Comparison of this average temperature with theoretical models requires that the thermal inertia, $(\kappa\rho c)^{1/2}$, (where $\kappa$ is the thermal conductivity, $\rho$ is the density, and $c$ is the specific heat), of the surface be 0.0014. Mercury and the moon therefore appear to have very similar top surface layers (6).

### References and Notes

1. B. C. Murray, "Infrared Radiation from the Daytime and Nighttime Surfaces of Mercury," *Trans. Amer. Geophys. Union* 48, 148 (1967).
2. D. Morrison and C. Sagan, *Astrophys. J.* 150, 1105 (1967).
3. H. L. Johnson, *Commun. Lunar Planet. Lab.* 3, No. 53 (1965).
4. T. J. Pepin, *Astrophys. J.* 159, 1067 (1970).
5. C. W. Allen, *Astrophysical Quantities* (Athlone Press, London, ed. 2, 1963), p. 172.
6. W. M. Sinton, in *Physics and Astronomy of the Moon*, Z. Kopal, Ed. (Academic Press, New York, 1962), pp. 407–415.
7. We thank R. Maas, D. Allen, and J. Hackwell for assistance in making the observations with the 30-inch telescope at O'Brien Observatory, University of Minnesota. We thank G. Burnett for communicating to us in advance of publication his results on thermal inertia versus average dark-side temperature for Mercury. Work supported by NASA grant NGL-24-005-008 and substantially assisted under ONR contract N00014-67-A-0115-0004.

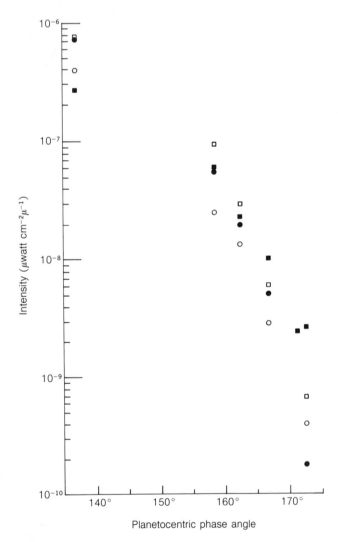

**Figure 4.** Corrected intensities at the four infrared wavelengths as a function of planetocentric phase angles for the 1969 conjunction. Open circles, 3.7 $\mu$; solid circles, 4.8 $\mu$; open squares, 8.5 $\mu$; solid squares, 11.5 $\mu$.

# Venus: Atmospheric Motion and Structure from Mariner 10 Pictures

Bruce C. Murray, Michael J. S. Belton, G. Edward Danielson,
Merton E. Davies, Donald Gault, Bruce Hapke, Brian
O'Leary, Robert G. Strom, Verner Suomi, and Newell Trask

Abstract. *The Mariner 10 television cameras imaged the planet Venus in the visible and near ultraviolet for a period of 8 days at resolutions ranging from 100 meters to 130 kilometers. The general pattern of the atmospheric circulation in the upper tropospheric/lower stratospheric region is displayed in the pictures. Atmospheric flow is symmetrical between north and south hemispheres. The equatorial motions are zonal (east-west) at approximately 100 meters per second, consistent with the previously inferred 4-day retrograde rotation. Angular velocity increases with latitude. The subsolar region, and the region downwind from it, show evidence of large-scale convection that persists in spite of the main zonal motion. Dynamical interaction between the zonal motion and the relatively stationary region of convection is evidenced by bowlike waves.*

On 5 February 1974, Mariner 10, carrying two television cameras, crossed the Venus terminator from the dark side, swinging around the planet on a hyperbolic trajectory on its way to Mercury (*1*). The cameras were designed mainly to observe the surface details of Mercury (*2*); however, the optical design incorporated special filters, coatings, and transmitting glass in order to image Venus in the ultraviolet (UV).

Faint UV markings (Fig. 1) were discovered on Venus in 1926 by Ross (*3*). Decades of subsequent UV observations from the earth suggest a retrograde equatorial motion with a 4-day period (approximately 100 m/sec) (*4*). At least one feature (which takes the form of a dark horizontal Y) appears to be quasi-permanent (*5*) or recurrent (*6*). Spectroscopically observed Doppler shifts in reflected sunlight (*7*) suggest retrograde equatorial motions of approximately 100 m/sec (at pressure levels near ~200 mbar). In situ measurements lower in the atmosphere, from Venera 8, are also consistent with these inferred motions (*8*). From studies of optical phenomena near inferior conjunction, Goody (*9*) places the tops of the clouds at altitudes lower than about the 10-mbar pressure level. Rayleigh scattering alone due to the primary atmospheric constituent, $CO_2$, would limit visibility of features in the UV to altitudes well above the 400-mbar level (*10*). Hence, the UV markings probably (i) originate in the same general region of the atmosphere as the spectroscopically observed Doppler shifts and (ii) reflect mainly mass motion in the atmosphere rather than propagating waves.

The television images returned from Mariner 10 cover the global development of the UV markings over an 8-day period, approximately two rotation periods of the troposphere/lower stratosphere region. Our sample of the dynamical regime on Venus is thus limited vertically and temporally. Nevertheless, the Mariner 10 pictures contain a surprising amount of information about the general circulation of this part of the atmosphere, which will enhance the value of ground-based observations as well as establish a specific scientific framework for future entry probes and orbiters.

In brief, the pictures display highly symmetrical motions relative to the rotational axis encompassing both north and south hemispheres; angular velocity increases with latitude. Zonal flow near the equator is consistent with an approximately 4-day retrograde rotation period. An unexpected equatorial disturbance continually develops near the subsolar point, within which cellular structures suggestive of convection are exhibited. There is dynamical interaction between this solar oriented equatorial disturbance and the main zonal flow. Bright jetlike streams spiral around the planet to merge into a conspicuous circumpolar band.

*Experiment Description.* The Mariner 10 television hardware and operations are similar to the Mariner 9 (Mars, 1971) system (*11*). Several significant improvements, however, have been made as a result of past experience and the unique requirements of this mission. Besides the extension of spectral response into the UV, the most important change for the study of Venus is a dramatic increase in the communications band width from 16 to 117.6 kbit/sec. As a result, tape storage on the spacecraft could be bypassed. Near the planet every frame

From *Science*, vol. 183, pp. 1307–1315, 29 March 1974. Copyright 1974 by the American Association for the Advancement of Science. Reprinted with permission.

acquired was transmitted in real time, making possible high-resolution time lapse studies.

To take full advantage of the Mercury trajectory, new optics with a 1500-mm focal length were developed with extended blue and UV response. The field of view is 0.36° by 0.48°. Table 1 gives the characteristics of the filters which were included with each camera, and a schematic of the system is presented in Figure 2. Figure 3 illustrates the relative spectral response of the filter/camera combination for the primary filters used in the Venus sequence.

Each television frame consists of 700 scan lines, each of which is sampled 832 times. These samples are encoded to 256 discrete levels (8 bits). Table 2 briefly describes basic

TABLE 1
*Filter characteristics, calculated for a sample selenium vidicon using the spectral radiance of Venus.*

| Filter | Effective wavelength (Å) |
| --- | --- |
| Ultraviolet | 3550 |
| Blue | 4740 |
| Orange | 5780 |
| Clear | 4820 |
| Minus ultraviolet | 5120 |
| Ultraviolet polarizing | 3580 |

Venus as photographed from Mariner 10 on March 19, 1974.
[National Aeronautics and Space Administration.]

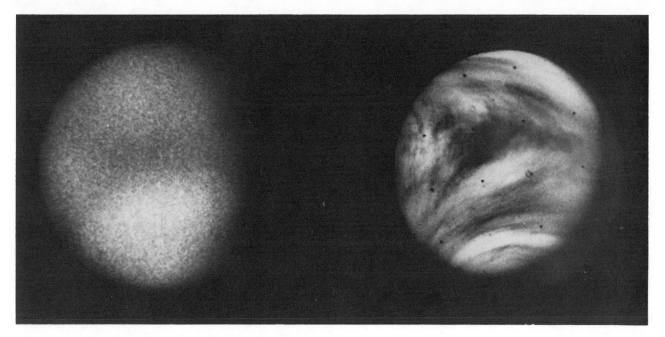

**Figure 1.** A Y-shaped feature can be seen in UV light. The picture at the left was taken at the Pic du Midi Observatory, France (04:47 U.T., 24 July 1966); it has a resolution of about 500 km. By contrast, the Mariner 10 picture at the right was taken from 3,300,000 km (03:57 U.T., 10 February 1974); it has a resolution of 65 km.

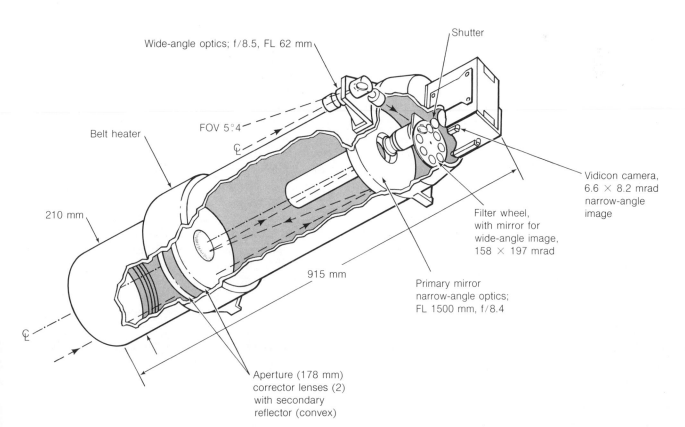

**Figure 2.** Schematic view of Mariner 10 television camera.

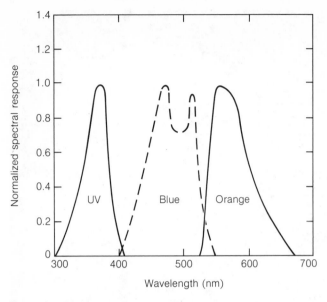

**Figure 3**. Relative spectral response curves of the filter/camera combination used in the Venus sequence. Each curve is independently normalized.

characteristics of the overall system. Two cameras are employed, with one camera reading out its image while the other is being prepared. Residual image (a low-level signal remaining from an earlier image) has been virtually eliminated by vidicon faceplate light flooding between each read and erase cycle. Changes to the high-voltage power supply and deflection/focus coil assemblies have reduced electronic noise and readout distortion.

TABLE 2
*Television performance characteristics.*

| Characteristic | Value |
| --- | --- |
| Focal length | 1500 mm |
| f/number | f/8.4 |
| Field of view | 0.36° × 0.48° |
| Sensor dimension | 9.6 × 12.35 mm |
| Format in pixels | 700 × 832 |
| Encoding | 8 bits |
| Frame time | 42 seconds |
| Resolution per television line | $9.5 \times 10^{-6}$ radians |

The flat-field signal-to-noise ratio of the cameras is better than 200 to 1. At the limiting spatial resolution, 4.5 arc seconds, the signal-to-noise ratio is better than 20 to 1. As a result, extremely low contrast scenes can be resolved through appropriate computer spatial filtering and contrast enhancing of the digital data. Discernible tonal variations in pictures accompanying this report sometimes reflect intrinsic brightness variations of less than 1 percent. As a consequence of this unparalleled discriminability, faint blemishes introduced into the pictures by the imaging system itself are made visible through computer processing. Figure 4 shows sample images which have been greatly enhanced to illustrate the artifacts. The small black squares are reseau marks to permit computer restoration of geometric integrity to the images.

Figure 5 illustrates the Venus encounter sequence through the first 20 hours past encounter, during which continuous

picture-taking was carried out. From 1 to 4 days after encounter, mosaics in the UV were obtained at 2-hour intervals. Between 4 and 6 days the images were taken at 8-hour centers, and on the 7th and 8th days 12-hour intervals were used (*12*). The spatial resolution at cessation of photography was 130 km, which is about twice the best Earth-based resolution.

About 3400 useful frames were acquired, but fewer than half of these have been processed and studied so far. Figure 6 illustrates the "footprints" of the frames on the planet for the highest-resolution mosaic, which has not yet been reconstructed. Most of the preliminary results included in this report have been obtained from mosaics of frames taken 1 day out and later.

In the description of atmospheric motions it is often useful to refer to the points of the compass and also to draw analogy with motions on the earth. On Venus, however, rotation is in a retrograde sense. We refer here to north as in the direction of the north ecliptic pole. Thus we must accept a left-hand rotation convention. All pictures here are printed with north at the top. Rotation is from right to left, and the right-hand edge of the disk is the morning terminator.

**Description of the Observed Markings.** Venus has been observed through all of the filters. In the blue and orange, very faint global scale markings may be present occasionally, but these frames have not yet been studied in detail. Hence, we restrict discussion to UV pictures and to certain pictures of the planetary limb taken through the orange filter.

The UV images have three general characteristics: (i) a mottled appearance full of small-scale (100 to 500 km) features found in the equatorial zone surrounding the subsolar point (Fig. 7, top; Fig. 8); (ii) streaky and banded structures in the higher latitudes of both hemispheres (Fig. 7, bottom; Figs. 9 and 10); and (iii) a strongly divergent flow pattern around the subsolar point and symmetrical about the equator (Figs. 9, 10, and 11). These patterns are evident in all of the global pictures and mosaics. Figure 12 illustrates the global aspect of the planet from 1 to 8 days after encounter.

The major light and dark markings on the surface of the planet, which have dimensions of the order of 1000 km, are found to be composed of a wealth of smaller-scale features with dimensions down to about 10 km. The maximum contrast detected so far between the major light and dark UV regions is about 30 percent, consistent with Earth-based observations (*13*). Brightness differences of 5 to 10 percent are found in the mottled areas and in the streaks with dimensions of more than several hundred kilometers. The smallest resolvable dark streaks, 10 to 20 km wide, differ by only 2 to 5 percent from the surrounding background.

The lifetimes of the various light and dark markings between ±50° latitude are variable. Preliminary estimates are illustrated in Figures 8 and 10. Both large and small features (~200 to 1200 km) can retain their basic geometrical configuration during passage across the disk almost from terminator to limb, implying lifetimes in excess of 12 hours. Features a few hundred kilometers in diameter can persist for a period of 2 hours in some cases or, alternatively, can become unrecognizable over the same time period. Many fine-scale (50 to 100 km) cellular features in the vicinity of the subsolar point became unrecognizable in a period at least as short as 2 hours,

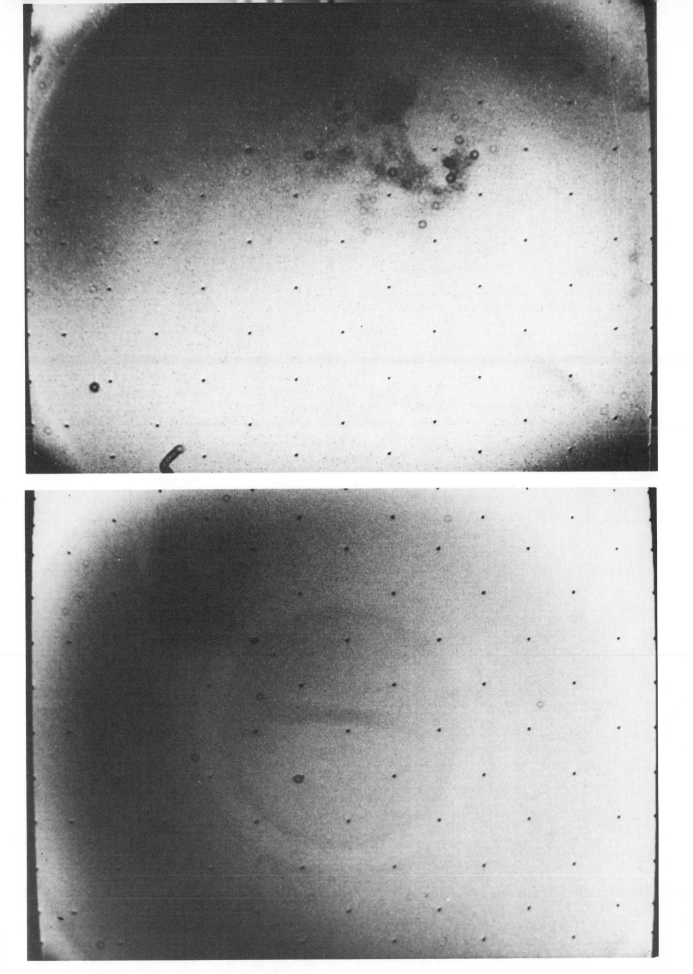

**Figure 4.** Flat-field images from the "A" camera (top) and "B" camera (bottom), contrast-enhanced to map and illustrate artifacts and blemishes that are present in the pictures.

72

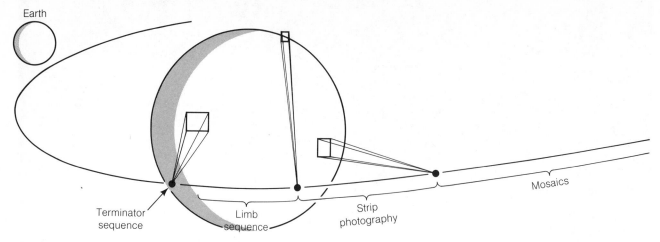

**Figure 5.** Picture sequences at near encounter were designed to obtain data concerning specific questions and later to monitor the entire planet.

the time interval between consecutive mosaics; however, overlapping frames within the same mosaic exhibit lifetimes greater than 15 minutes in some cases.

**Characteristics of the Observed Circulation.** We are confident from the nature of the spatial patterns and temporal variations of the UV markings that global atmospheric motions are being recorded. Analysis of vertical structure (discussed later) suggests that the markings probably originate within the visible cloud region encompassing the stratosphere and upper troposphere of Venus. The sharp spectral dependence of the markings suggests variation in absorptivity rather than particle

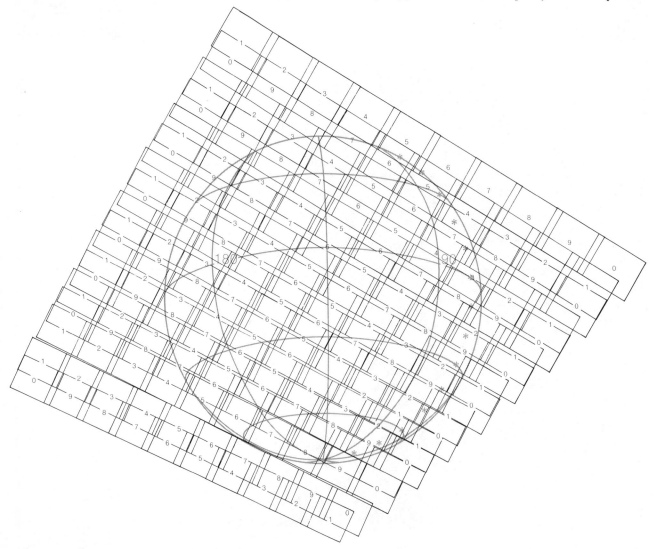

**Figure 6.** "Footprints" of a multiframe mosaic in the Venus sequence, taken about 6½ hours after encounter.

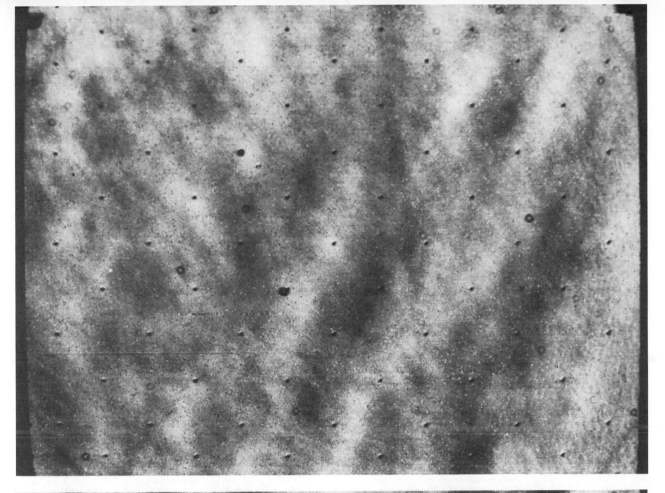

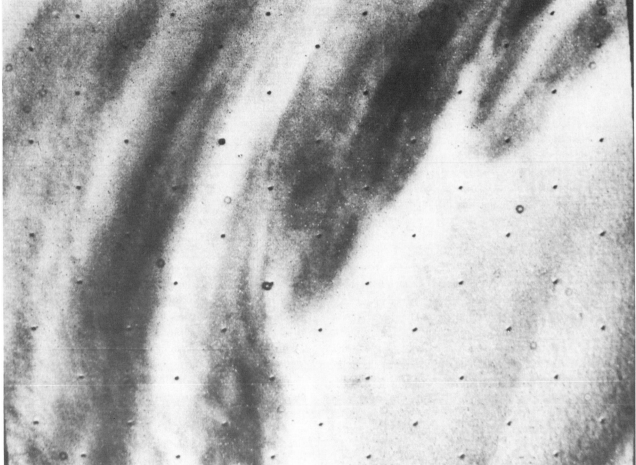

Figure 7. Venus has a mottled appearance in some areas of the equatorial region (top), while streaks and whorls are seen at high latitudes (bottom).

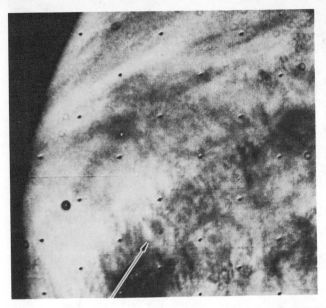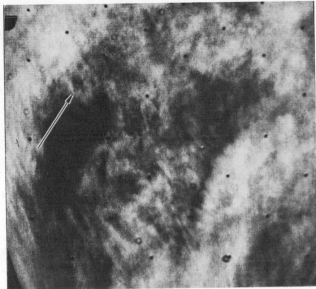

Scale: 5000 km

**Figure 8.** High-resolution views taken 2 hours apart in the vicinity of the subsolar region of Venus, showing the persistence of the large cell (280 km) indicated by the arrows and transitory nature of the smaller cells (170 km) to the right (east).

size as the primary optical process involved. The temporal behavior of the markings suggests the formation and disappearance of condensate clouds rather than solely dust or photochemical products.

Nevertheless, we are unable to specify the probable constituents and their distribution within the clouds responsible for the observed markings. Hence, the inferences concerning atmospheric motion presented here must be regarded as tentative, subject to reinterpretation.

In order to provide a basic description of the observed motions, we introduce the nomenclature which is illustrated schematically in Figure 9.

**Subsolar Disturbance.** The obliquity of Venus is observed to be sufficiently small that no seasonal effects are anticipated (*14*). The subsolar portion of the equatorial region is characterized by the presence of cellular features, as shown in Figure 8. In high-resolution mosaics, the larger (500 km), less distinct cells are bounded by dark edges. Some are polygonal in shape. Preliminary comparison of successive mosaics suggests lifetimes of a few hours at most. The interior of these cells is highly structured. A tenuous network of this type of cell has been traced over about 5 percent of the area of the planet near the subsolar point. Slightly smaller (~200 km) cells bounded by light material (as identified in Fig. 8) are also found which move with the wind and change markedly over a 2-hour interval. The subsolar disturbance extends at its widest over ±20° [in latitude] and at least 80° in longitude. The actual extent in longitude is not known, as the region extends beyond the limb. The subsolar point itself is located near the eastern extremity. The subsolar disturbance is locked to the sun-Venus line; it is continually being regenerated there, presumably in response to maximal solar heating.

**Mid-latitude Streamlined Flow Region.** Conspicuous streaks originate in the equatorial region and spiral up to higher latitudes. In high-resolution mosaics, they are bor-

dered on the equatorial side by finer streaks and, occasionally, whorls, suggestive of horizontal shear and turbulence. There are at least two major systems of jetlike spiral streaks in each hemisphere; one set appears to be symmetrical across the equator. The spiral features appear to be most prominent in mid-latitudes (~ ±30°). They merge into a bright polar ring at 50° latitude after progressing some 200° to 300° of longitude around the planet. We have not observed any evidence of instability on a global scale in these features, nor evidence of structures similar to large-scale cyclonic eddies.

Figure 11 is a preliminary attempt to display the temporal appearance of the mid-latitude regions. The patterns are complex. Prominent dark, poleward markings sometimes appear fully formed when first viewed on the morning terminator. Other patterns seem to originate in the equatorial region.

Preliminary measurements of small-scale features as they move across the disk of the planet provide crude estimates of zonal motions. In the equatorial zone (outside of the subsolar disturbance) large discrete areas of light and dark material move primarily zonally at approximately 100 m/sec westward relative to the fixed planet, corresponding to an apparent equatorial retrograde rotation period of about 4 days. Smaller-scale features, however, do not always share this motion. Zonal components of about 80 to 90 m/sec in some cases are suggested. At higher latitudes motion is also primarily zonal.

The angular velocity of the markings increases with latitude although angular momentum diminishes somewhat with latitude. The suggestion of shear in the polar ring could mean an even higher zonal velocity. At 50° latitude the rotation period of the upper atmosphere could be as short as 2 days.

We have not detected any measurable meridional motions in low-latitude regions. At higher latitudes (30° to 50°), particularly associated with the spiral streaks, poleward flow of the order of 10 m/sec is possibly indicated by the data, but must be substantiated by further analysis.

**Polar Region.** The southern polar ring is perhaps the

most distinctive and stable feature in the light markings. It encompasses a band of latitude 10° to 15° wide, with its equatorial side at 50° latitude. There are strong indications that a similar ring exists in the northern hemisphere, but the viewing geometry is unfavorable for a definite conclusion. Our preliminary impression is that the pole of atmospheric rotation is on the terminator (*14*). There is an indication of vortex structure in the streaks emanating from the poleward side of the edge of the polar region. Indeed, the entire polar region may be a vortex fed by meridional flow from the equatorial regions. A major systematic analysis of all the picture data is planned especially to elucidate the magnitude of possible meridional motions and the variation of angular momentum with latitude.

***Interaction Features.*** Very faint circumequatorial belts appear on some of the mosaics. Often three or four appear at one time between latitudes ±20° and are parallel to latitude circles. The belts are less than 100 km in width, and appear to be moving rapidly around the globe in the same direction as the general motion and also drifting across latitude circles.

Dynamic interaction between the strong zonal flow and the solar-locked subsolar disturbance is evident. In some mosaics, we have noted the presence of darker features suggestive of bow waves generated by interaction with a "soft" obstacle (Fig. 9). These features move relative to the obstacle (unlike true bow waves). They are symmetrical about the equator, extending to at least ±30° latitude and give the impression of being present in pairs. In one of the best examples they are

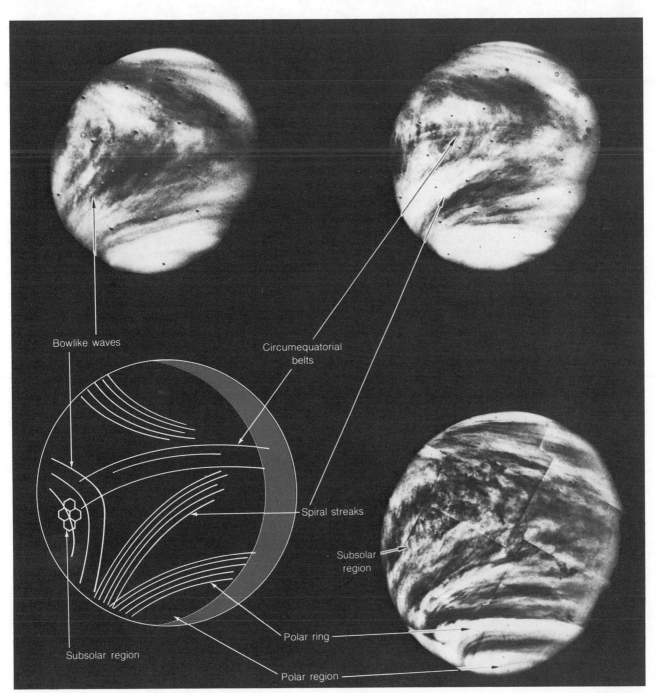

**Figure 9.**  The major features of the circulation pattern are identified in the sketch and are seen in most of the global pictures.

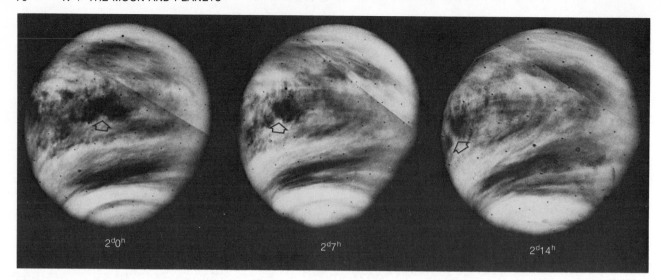

**Figure 10.** Series of mosaics at 7-hour intervals showing the persistence of large light and dark markings over a 14-hour period. The size of the feature indicated is about 1000 km.

separated by about 1000 km, each one being roughly 250 to 350 km in width. It is our impression that these features form at irregular intervals a few tens of degrees upwind of the subsolar disturbance and then propagate through (or over) that disturbance at roughly 80 percent of the average rotation speed.

***Relation to Ground-based UV Photographs.*** The outstanding characteristic of the Mariner 10 pictures is a diverging pattern centered on the equator and opening in the direction of rotation, which is present throughout the entire 8-day period. Four days after encounter, a pronounced dark horizontal Y appeared on the equator, suggestive of the Y-shaped feature observed from the earth (Fig. 1). This feature was observed to rotate from morning terminator to limb at a rate consistent with an approximate 4-day rotation. The Y morphology recurred again 8 days after encounter.

Earth-based UV photographs often show a reversed C pattern in the evening terminator region (6). The same kind

of morphology can be seen in a Mariner 10 picture after it was projected onto a globe and then rephotographed to give an unforeshortened view of regions near the evening terminator (Fig. 13). The reversed C is evidently associated with the Y feature and the bowlike wave structures to the west of the subsolar region. It is interesting to note that, in the numerous Earth-based UV photographs of Venus taken over the years, the Y and C features always open in the direction of rotation (15). Very long term stability is suggested for the diverging patterns so apparent in the 8 days of Mariner 10 observations. However, these preliminary results do not provide an "explanation" of the markings.

***Characteristics of the Limb, Terminator, and Cusp.*** Photographs of the cusps, terminator, and limb were acquired at very high spatial resolution (Fig. 14). The cusp appears devoid of small-scale structure, indicating a very homogeneously stratified medium. There is no evidence of shadowing or other horizontal brightness variation in the terminator

Equatorial longitude (based on 4$^d$0 rotation at equator)

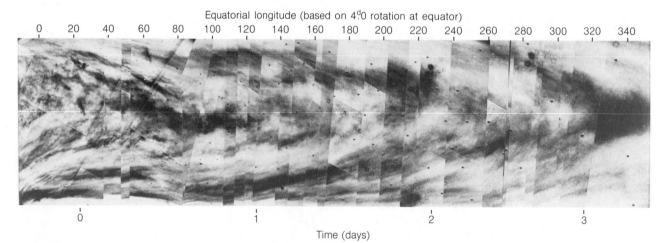

**Figure 11.** Temporal display of the UV markings between +40° latitude and −50° latitude. The subsolar point moves from left to right in the picture as the planet rotates. The map contains considerable geometric distortion as the pictures were not rectified; however, it has been useful in studying the circulative pattern.

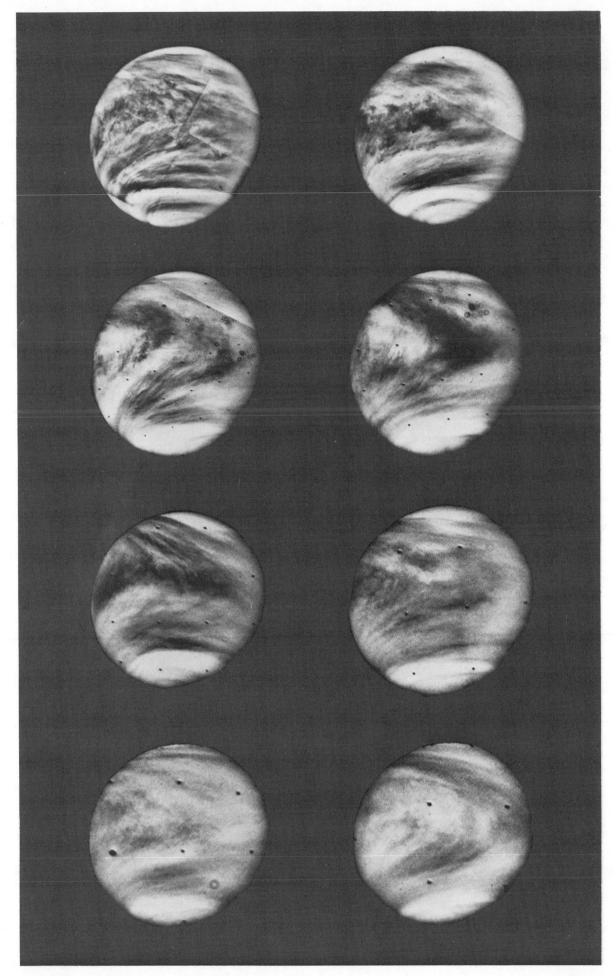

**Figure 12.** Montage of global views on 1-day centers. Time progresses from top left to right, then down the page. The first frame (identical to the picture on page 68) is a mosaic of frames taken at encounter plus 1 day.

**Figure 13.** (Left) Earth-based UV photograph of a reverse C feature on the evening terminator of Venus on 24 May 1967, 01:35 U.T. (courtesy of New Mexico State University Observatory). (Center) Mariner 10 picture 4 days after encounter, projected on a globe and rephotographed to give an unforeshortened view of regions near the evening terminator. (Right) the same Mariner 10 picture viewed from the direction of the spacecraft.

region. At the low sun angles involved, this observation implies the absence of opaque shadowcasting clouds with vertical relief greater than about 20 m relative to any overlying clear atmosphere.

A sequence of 45 pictures of the limb of Venus in UV and orange filters was acquired from equator to polar latitudes at resolutions better than 1 km. Of these, ten have been investigated for analysis of any limb haze structure. Figure 15 shows four high-resolution views of the limb near the equator. In the orange frames, highly stratified layers ~1 km thick are found. There is a suggestion that the vertical structure is different at locations ~1000 km apart. Some UV frames also show evidence of layering.

***Inferred Vertical Structure.*** The presence of cellular structures in the subsolar region suggests to us the presence of large-scale convection. In that region at least, we may be seeing down to the 100-mbar pressure level or deeper because the Mariner 10 radio occultation suggests a temperature inversion at ~100 mbar with a steep lapse rate below (*16*). The limb photographs apparently refer to a region much higher in the atmosphere. As a lower limit we take the UV and orange limbs to be defined by a slant optical path of unity in rayleigh scattering. We find they correspond to regions in the atmosphere no lower than the 10-mbar and 90-mbar pressure levels, respectively. The existence of stratification in some UV limb pictures indicate that haze cloud particles are located up to the 10-mbar level and perhaps higher, consistent with the visual transit measurements (*9*).

Preliminary measurements of the curvature of the limb in selected frames indicate a haze layer radius of about 6130 km in both UV and orange light. Agreement between four differ-

ent frames is within ±2 km, but it is difficult at this time to estimate possible systematic errors. Comparing these data with the radio occultation results (*16*), we infer that both the UV and orange limbs are defined by a level in the stratosphere approximately 15 km above the tropopause and near the 10-mbar level. There is also a preliminary indication of similar haze structure in orange and UV frames taken near a latitude of 22° north (Fig. 15, c and d), again suggesting the presence of a particulate haze at or above the 10-mbar level in both colors.

Very thin, highly stratified limb hazes are indicative of the great stability in the atmosphere at these levels. Presumably the vertical eddy diffusivity is similar to, or less than, that found in the earth's stratosphere where similar phenomena occur. The resolution of the cusp pictures is rather low (~15 km). However, we can infer from these pictures that gross vertical separation into layers in the stratosphere more than 15 km thick does not occur.

The lack of shadowing in the terminator frames is consistent with the presence of an enveloping thin haze in the stratosphere. Any layers of cloud with well-defined vertical relief must be deep enough in the atmosphere so that at low sun angles the sunlight has been diffused enough by the overlying haze that shadows are effectively absent. At present, we see no basic conflict with recent models of the cloud structure (*17*) which require a stratospheric haze layer near the 20- to 50-mbar level with a vertical optical depth of ~2 above an optically dense second deck near the 200-mbar level.

***Discussion of Atmospheric Circulation.*** Despite the preliminary nature of the data, the patterns and motions of the UV markings suggest a spectrum of models for planet-wide

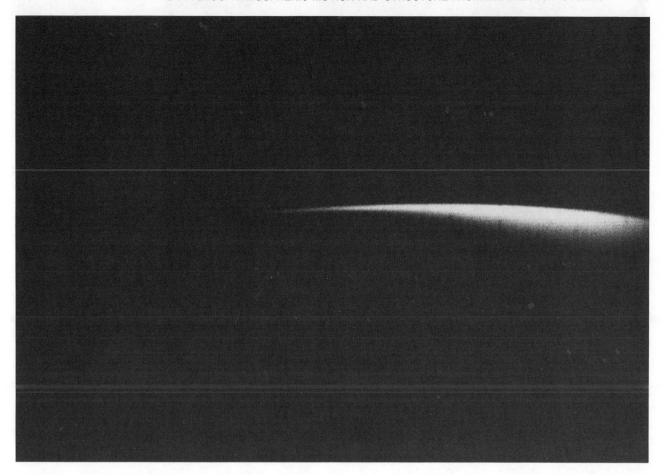

**Figure 14.** View of the cusp region. The dark markings in the cusp are artifacts in the imaging system.

atmospheric circulation which are useful to discuss in order to develop a frame of reference for subsequent data analysis and future missions.

Some time ago, ground-based observations raised a fundamental question about atmospheric circulation on Venus: What is the source of angular momentum for rapidly moving UV clouds? A non-axisymmetric flow mechanism is needed. Schubert and Whitehead (*18*), followed by Gierasch (*19*) and Malkus (*20*), developed the "moving flame" model for Venus. Periodic solar heating of the top of the atmosphere results in cellular motion ultimately leading to uniform zonal flow. Non-axisymmetric motion takes place in vertical planes. Meridional motions are neglected.

To reconcile such two-dimensional concepts with the observed patterns of markings, it would be necessary to regard the spiral markings primarily as wavelike disturbances (not streamlines) which move with the mean zonal flow. The markings obviously are disrupted by the subsolar disturbance zone, but this is not taken to be any indication of divergence in flow. Figure 11, with its herringbone pattern, would be interpreted as somewhat like the result of rotating a cylinder with painted, or at least recurrent, markings (like a rotating barber's pole in which zonal motion gives the illusion of meridional motion).

The three-dimensional implication of the moving flame

concept was considered very briefly by Malkus (*20*); he noted the possibility that a weak Hadley cell could develop near the top of the atmosphere and, by virtue of the equator-to-pole temperature contrast, transfer energy and angular momentum to higher latitudes. To apply this kind of interpretation, the spiral streaks of the mid-latitude streamline flow region would be inferred to be streamlines in this Hadley cell in a coordinate frame rotating with the mean equatorial motion. The spiral streaks presumably are clouds associated with disturbances induced in the Hadley flow by interaction between the equatorial flow and the subsolar disturbance. Both Malkus (*20*) and Gierasch *et al.* (*21*) allude to the possibility of gravity waves arising from the interaction of zonal flow and the subsolar region. The bowlike waves observed in the subsolar disturbance indeed are suggestive of some such interaction.

Carrying the impression of meridional motion further, the spiral streaks and general divergent pattern can be taken as evidence of an unexpectedly large influence of the subsolar region on global circulation. Great significance would be attached to the kinetic energy generated by velocity divergence in the subsolar high-pressure area; convection cells are interpreted as indicating a higher temperature and therefore a higher pressure. The resultant local meridional pressure gradients would imply velocity divergence and strong cross-isobaric flow to accelerate the zonal wind toward the poles;

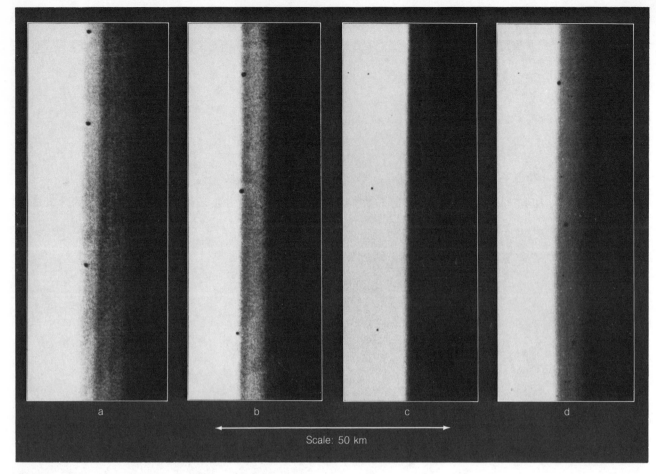

Scale: 50 km

**Figure 15.** Four views of the limb. Pictures (a) and (b) were taken through the orange filter near the equator; the two to the right are (c) orange and (d) UV photographs taken at 22°N latitude. All four pictures show structure indicating the presence of particulate matter in the stratosphere.

the spiral streaks are interpreted as associated jet streams. The kinetic energy sink is in the polar vortices due to velocity convergence at low pressure. As a further speculation, the bright polar ring would represent excess condensation associated with a kinetic energy maximum there. The bowlike waves are taken to be clear evidence of the imbalance between the pressure excess in the subsolar area and the mean zonal flow. Angular momentum conservation requires that poleward flow be deflected right in one hemisphere and left in the other. Thus, both flows would combine to add momentum in the same zonal direction. Equatorward return flow and also weak zonal counterflow at deeper levels are required to maintain the planetary angular momentum balance.

Thus, one extreme interpretation minimizes the importance of suggestions in the pictures of departures from uniform zonal flow. They are assigned at most to a superficial Hadley cell at the top of the atmosphere. The other extreme interpretation maximizes the implications of possible divergence of zonal flow and convergence at the pole, which could result from a persistent sun-locked high-pressure anomaly. Deep stirring of the atmosphere return flow is implied. At one extreme, the subsolar disturbance mainly generates cloud patterns; non-axisymmetric motion takes place only in vertical planes. At the other, the subsolar disturbance is a primary element of the global circulation system and non-axisymmetric flow is partly meridional as well.

To proceed further in interpretation, we will have to fully exploit the Mariner 10 television data, especially regarding evidence of meridional motions and variation of angular momentum with latitude.

***Implications for Future Studies.*** Such disparate interpretations imply significantly different vertical temperature and wind profiles in equatorial and polar regions. Direct measurements of these and other atmospheric parameters from carefully targeted entry probes can provide a clear-cut choice— or point toward presently unimagined possibilities. In addition, in situ measurement of the cloud particle composition high in the atmosphere will probably be required for a clear understanding of the detailed origin of the UV markings. Certainly, additional UV imaging of Venus from an orbiter can now be regarded as a powerful tool for atmospheric research. Hence, even very preliminary assessment of the television pictures from Mariner 10 carries implications for the Venus Pioneer program of the United States scheduled for 1977 and 1978. In addition, the Soviet Union may plan future Venera systems with much larger spacecraft, comparable to those used at Mars in 1971–1972 and 1974. An exciting era of exploration of our nearest planetary neighbor is emerging in which imaging can play a significant scientific role.

## References and Notes

1.  The gravitational field of Venus was used to reduce the heliocentric velocity of the spacecraft, following a suggestion made in 1963 by M. A. Minovitch, who was then a graduate student at the University of California, Los Angeles.

2.  B. C. Murray *et al., Icarus* 15, 153 (1971).

3.  F. E. Ross, *Astrophys. J.* 68, 57 (1928).

4.  C. Boyer and H. Camichel, *Ann. Astrophys.* 24, 53 (1961).

5.  C. Boyer and P. Guerin, *Icarus* 11, 338 (1969).

6.  A. H. Scott and E. J. Reese, *ibid.* 17, 589 (1972).

7.  B. Guinot and M. Feisael, *J. Observ.* 51, 13 (1968); W. Traub and N. Carleton, paper presented at the NATO Advanced Study Institute, Istanbul, Turkey (1972). These authors have informed us that the spectroscopic data may indicate a more complicated situation than a simple equatorial mass motion of approximately 100 m/sec (private communication).

8.  M. Y. Marov, V. S. Avduevsky, V. V. Kerzhanovich, M. K. Rozhdestvensky, N. F. Borodin, O. L. Ryabov, *J. Atmos. Sci.* 30, 1210 (1973).

9.  R. M. Goody, *Planet. Space Sci.* 15, 1817 (1967).

10. This corresponds to four optical depths, an extreme estimate of maximum depth of visibility.

11. H. Masursky *et al., Icarus* 12, 10 (1970).

12. The original plans called for picture-taking for 16 days, but with an extended gap at 3 days. As it turned out, we achieved 8 days of almost uniform coverage instead.

13. D. L. Coffeen, in *Planetary Atmospheres,* C. Sagan, T. C. Owen, H. J. Smith, Eds. (Reidel, Dordrecht, Netherlands, 1971), p. 84.

14. The atmospheric rotation pole appears to be oriented parallel to the planetary rotational pole. However, detailed processing and measurements will be necessary to determine the relationship precisely. Any differences would be highly significant for a dynamical understanding of the atmosphere.

15. B. A. Smith, private communication.

16. H. T. Howard *et al., Science* 183, 1297 (1974).

17. W. A. Traub and N. Carleton, *Bull. Am. Astron. Soc.* 3, 278 (1971); J. E. Hansen and A. Arking, *Science* 171, 669 (1971).

18. G. Schubert and J. A. Whitehead, *Science* 163, 7172 (1969).

19. P. Gierasch, *Icarus* 13, 25 (1970).

20. W. V. R. Malkus, *J. Atmos. Sci.* 27, 529 (1970).

21. P. J. Gierasch, A. P. Ingersoll, R. T. Williams, *Icarus* 19, 473 (1973).

22. We acknowledge the ingenuity, diligence, and dedication of the many engineers and scientists of the Jet Propulsion Laboratory who have contributed to the Mariner 10 television experiment. In particular the skilled staff of the Space Photography Section, the Image Processing Laboratory, the Mission Test Computer, and the Mission Test Information System have played a major role in our work. James Anderson and Michael Malin of the California Institute of Technology, Robert Krauss of the University of Wisconsin, and Ken Klaasen and Robert Toombs of the Jet Propulsion Laboratory made major contributions. It has been a special privilege to work with Gene Giberson, manager of the Mariner-Venus-Mercury Project, and his able staff. Andrew Ingersoll of the California Institute of Technology supplied valuable criticism. Contribution No. 2458 of the Division of Geological and Planetary Sciences, California Institute of Technology, Pasadena 91109. Research sponsored in part by NASA contract NAS 7-100. Kitt Peak National Observatory is operated by AURA, Inc., under contract to the National Science Foundation.

# 11

# Venus: Topography Revealed by Radar Data

D. B. Campbell, R. B. Dyce, R. P. Ingalls, G. H. Pettengill, and I. I. Shapiro

Abstract. *Surface height variations over the entire equatorial region on Venus have been estimated from extended series of measurements of interplanetary radar echo delays. Most notable is a mountainous section of about 3-kilometer peak height located at a longitude of 100 degrees (International Astronomical Union coordinate system). The eastern edge has an average inclination of about 0.5 degrees, which is unusually steep for a large-scale slope on Venus. The resolution of the radar measurements along the surface of Venus varied between about 200 and 400 kilometers with a repeatability in altitude determination generally between 200 and 500 meters. The mean equatorial radius was found to be 6050.0 ± 0.5 kilometers.*

The surface of Venus can be studied from afar only with radio waves. For the past decade interplanetary radar measurements of ever-increasing precision have been made of the round-trip echo delays of signals transmitted from the earth toward Venus. During the most recent inferior conjunction in November 1970 the errors in the measurements of delay were no more than 1 $\mu$sec—about a thousandfold improvement compared with the earliest such data obtained in 1961. From the newer observations we have been able to extract values for surface height variations over the entire equatorial region on Venus. These results form the basis for this report.

In principle, surface heights are simple to determine. Given the orbits of both the earth and Venus, a time-delay measurement can be interpreted directly in terms of the average altitude of the reflecting area that contributes to the echo. The size of the area depends on the effective extent of the pulses of radio energy and on the scattering law of the surface (*1*). This approach is complicated by two factors: (i) the orbits of the earth and Venus must be determined from the very data that are to be used for the inference of surface heights; and, relatedly, (ii) the spin of Venus is synchronous, or nearly so, with the relative orbital motion of the earth and Venus (*2*). The latter fact makes the separation of surface and orbital

effects more difficult than, for example, with radar observations of Mars (*3*). Nonetheless, since Venus completes four rotations on its axis, as seen by an observer on the earth, between successive inferior conjunctions, topographic effects, which nearly repeat for each rotation, tend to be distinguishable from orbital ones. The distinction is not completely straightforward because the latitude of the subearth point does not repeat exactly with the longitude, particularly near inferior conjunction where the greatest precision is possible. Variations of surface height with small changes in latitude therefore distort the interpretation. The nearly commensurate 13:8 ratio of the orbital periods of the earth and Venus insures that in each 8-year cycle the subradar point traces out virtually the same path on Venus's surface. This near periodicity is also a complicating factor. But, with patience, a complete separation of topographic from orbital effects will be possible since, presumably, the time scale for physical change in surface structure is usefully measured in units far longer than decades.

To achieve this separation with the present limited data span, we utilized several techniques. First, we simply ignored topography and used the earth-Venus and other radar data simultaneously to solve for the maximum-likelihood estimates of the relevant orbital initial conditions of the four inner planets, a single average radius for each of these planets (except the earth), and several other necessary astronomical constants (*4*). All told, the values for 23 parameters were obtained. The data were from both the Massachusetts Institute of Technology's Haystack Observatory and Cornell's Arecibo Observatory (now the National Astronomy and Ionosphere Center). The post-fit residuals from the earth-Venus data, as a function of the longitude and latitude of the subradar point, were taken to represent the surface height variations on Venus with respect to its mean equatorial radius.

As a second approach, we used a spherical-harmonic expansion to represent the topography of Venus and the other target planets and estimated the low order (primarily sectorial) coefficients. In a variation on this technique, we substituted a double Fourier series to represent the planetary surface-height variations (*5*).

The results from these two analyses and minor variants

From *Science*, vol. 175, pp. 514–516, 4 February 1972. Copyright 1972 by the American Association for the Advancement of Science. Reprinted with permission.

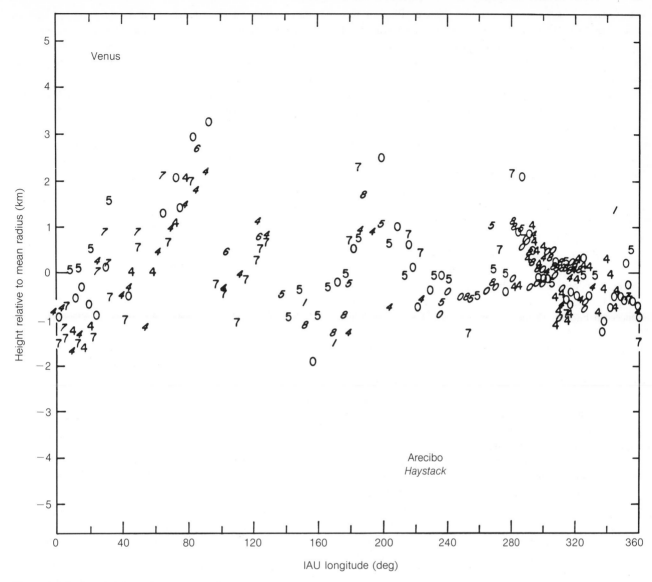

**Figure 1.** Surface-height variations over the equatorial region on Venus inferred from radar echo-delay measurements. The data were obtained at radar frequencies of 430 and 7840 Mhz at the Arecibo and Haystack observatories, respectively. The longitude scale is based on the IAU system (8). The times of the observations are coded by the number of axial rotations made by Venus as seen from the earth; the epoch is the late-August 1967 inferior conjunction, and the final data are from the November 1970 inferior conjunction. During this period the latitudes of the subearth points on Venus varied from about $-9°$ to $+10°$. All the data points have standard deviations under 7 $\mu$sec; the most recent (eighth revolution) data have typical standard errors of 1 $\mu$sec. The reference level is the mean equatorial radius, estimated to be 6050.0 $\pm$ 0.5 km after correction for the small effect of retardation in Venus's atmosphere.

thereof were very similar. Good consistency was obtained between Haystack and Arecibo data (6) and for both sets between one conjunction and the next. The estimates for the mean equatorial radius all fell within the limits 6050.0 ± 0.5 km, in excellent agreement with our previous result (7), which was based on fewer observations. For the 23-parameter solution, the surface heights relative to this radius are shown in Fig. 1 as a function of the International Astronomical Union

(IAU) longitude (8) of the subearth point. The data are distinguished by synodic rotation number—the number of rotations of Venus as seen by an observer on the earth—starting with zero at the inferior conjunction of late August 1967 and continuing through the corresponding conjunction of November 1970. The standard error associated with each measurement was omitted so as to prevent obscuration of the similarities and differences among the residuals. All of these errors were under 7 μsec, with the Haystack data having un-certainties in delay of no more than 1 μsec (equivalent to 150 m uncertainty in surface height) near the last inferior conjunction. The regions on the surface to which the average heights apply range in size from about 200 to 400 km (*1*). The repeatability of the results for different rotations is an indica-tion of their reliability and a measure of the success of our separation of the topographic from orbital effects (*9*). This internal consistency for most longitude regions appears to be within about 200 to 500 m. Because of the variations in the latitude on Venus to which data points from successive rota-tions refer, and because of the decreased accuracy associated with the measurements further from the inferior conjunctions, it is difficult to place more precise bounds on the uncertainties of the surface heights shown in Fig. 1.

The latitudes of the subearth points on Venus exhibit the largest excursions near inferior conjunctions. For the three conjunctions represented in Fig. 1, these latitudes were, chro-nologically, about 10°, −9°, and 5°. Some latitude depen-dence can be seen in the topography, but the most striking feature is the 3-km-high peak at about 100° longitude. This feature seems to rise gently from the west (smaller longitudes) with a slope of about 0.04 degree, but it then drops precipi-tously to the east with a slope perhaps as large as 0.5 degree. More closely spaced observations in the longitude region between about 90° and 110° are required in order to determine the true shape of the eastern slope. The extent of this elevated region is at least 500 km in latitude and about 6000 km in longitude. Its existence was even apparent in the data at much lower resolution reported earlier (*10*). Other, shallower peaks and valleys are clearly evident in Fig. 1 (*11*). Higher, sharper peaks may also be present but would not be visible because the data in the figure generally represent averages over larger regions. One very narrow peak, about 1.5 km in altitude, has in fact already been observed by a different method (*10*).

What are the prospects for improvement in the surface resolution and in the accuracy of the altitude determinations? Near inferior conjunction, the main limitation is placed not by the signal-to-noise ratio but by the effective pulse length, $\Delta t$ (*1*). For Haystack, and similarly for Arecibo, relatively straightforward modifications would make it possible for phase codes with $\Delta t$'s of 1 μsec to be transmitted and the echoes analyzed. The corresponding surface-area resolution cell would be about 50 km in radius; the height resolution for strong signals might be as fine as 30 m. As the surface resolu-tion improves, the frequency of obtaining closure observations (that is, multiple measurements of the same surface cell but from different relative orbital positions) will perforce de-crease—until the surface heights, on a scale corresponding to the precision of the echo-delay determination, begin to be well correlated over distances that exceed the surface resolu-tion of the measurements. Although this improvement reduces the number of strict closure points obtained in a given ob-servation interval, the weight of those available may more than compensate in the separation of orbital and topographic effects. An even more promising solution is to use methods in which a relatively large portion of the planet's topography is deter-mined simultaneously with little loss in resolution. One such technique (*12*) provides from 1 day's measurements fine reso-lution along an extended portion of the apparent Doppler equator. The newer method of delay-Doppler interferometry (*13*) will allow the topographic determinations to be extended away from the equatorial regions but with increasingly poorer resolution. The latter technique utilizes the delay, Doppler, and fringe-phase measurements to provide three-dimensional maps of a planetary surface. For useful accuracies, more sensi-tive radar systems such as the new Goldstone radar (*14*) and the soon-to-be-improved Arecibo radar (*15*) must be used.

The present results, although limited, show that Venus has a rich, varied, and durable topography, its high surface tem-perature of 800°K notwithstanding. The degree to which isostatic compensation may take place, however, cannot be de-termined reliably until detailed gravity data become available.

## References and Notes

1.   In all current time-delay measurements, a phase-coded waveform is transmitted; in the present work, the constant-phase intervals (or baud lengths) varied from 6 to 60 μsec. The extent in time of the code element, $\Delta t$, determines the effective resolution area, $\Delta A$, on the surface through the simple formula

$$\Delta A \simeq \pi R c \Delta t$$

where $c$ is the speed of light and $R$ is the planet's radius. This result depends on the surface being nearly spherical and on the scattering law being nearly con-stant in the vicinity of the subradar point. Under these assumptions, the resolution cell on the surface of Venus ($R \simeq 6050$ km) will have a diameter of about 300 km for $\Delta t \simeq 10$ μsec. If the echo power falls off significantly with increased separation of the reflecting region from the subradar point, as is normally the case, then the effective resolution at the subearth point is perforce improved. The precision of the determina-tion of average altitude is also proportional to $\Delta t$ and, in most cases of interest, is nearly inversely propor-tional to the signal-to-noise ratio. Further details on the applications of phase-coded waveforms to radar astronomy and the role of the scattering law are pro-vided by G. H. Pettengill, in *Radar Handbook*, M. Skolnik, Ed. (McGraw-Hill, New York, 1970), chap. 33.
2.   See, for example, R. M. Goldstein, in *Moon and Planets*, A. Dollfus, Ed. (North-Holland, Amsterdam, 1967), p. 126; I. I. Shapiro, *Science* 157, 423 (1967); R. F. Jurgens, *Radio Sci.* 5, 435 (1970); R. L. Carpenter, *Astron. J.* 75, 61 (1970).
3.   G. H. Pettengill, A. E. E. Rogers, I. I. Shapiro, *Science* 174, 1321 (1971); A. E. E. Rogers, M. E. Ash, C. C. Counselman, I. I. Shapiro, G. H. Pettengill, *Radio Sci.* 5, 465 (1970); R. M. Goldstein, W. G. Melbourne, G. A. Morris, G. S. Downs, D. A. O'Handley, *ibid.*, p. 475; G. H. Pettengill, C. C. Counselman, L. P. Rainville, I. I. Shapiro, *Astron. J.* 74, 461 (1969).
4.   The radar data are relatively insensitive to the values of the "out-of-plane" orbital parameters, which can therefore be fixed in accord with the results from optical observations of the planets without the con-comitant errors introducing a significant distortion into the altitude determination. A detailed descrip-tion of the methods used to obtain the estimates of the other parameters can be found in M. E. Ash, I. I. Shapiro, W. B. Smith, *Astron. J.* 72, 338 (1967). See also I. I. Shapiro, M. E. Ash, R. P. Ingalls, W. B.

Smith, D. B. Campbell, R. B. Dyce, R. F. Jurgens, G. H. Pettengill, *Phys. Rev. Lett.* 26, 1132 (1971).

5.  Another technique, currently being tested, involves the use of only data corresponding to longitude-latitude resolution cells that were observed from widely separated orbital positions. By the addition of one radius parameter for each such cell, the separation of orbital and topographic effects can be achieved. With the orbits thus obtained, all of the data can be interpreted directly in terms of surface heights. Difficulties with this approach include the effects of (i) omission of a large fraction of the data in the orbit-determination process and (ii) differences in the exact subearth point for data assigned to the same resolution cell.

6.  In fact, in some of the computer experiments a parameter representing the "bias" of the Arecibo data relative to Haystack's was added; the estimate for this parameter was always under 1 $\mu$sec, which is equivalent to a bias in radius of under 150 m.

7.  M. E. Ash, D. B. Campbell, R. B. Dyce, R. P. Ingalls, R. Jurgens, G. H. Pettengill, I. I. Shapiro, M. A. Slade, W. B. Smith, T. W. Thompson, *Science* 160, 985 (1968).

8.  At its last general assembly in 1970, the IAU proposed a standard value for the rotation vector of Venus that is consistent with current knowledge. The precise definition is given in *Proceedings of the 14th General Assembly of the International Astronomical Union* (Reidel, Dordrecht, 1971), p. 128. The central meridian on Venus, as seen from the center of the earth at 0 hour, universal time, on 20 June 1964, is 320° in this system and increases with time. The assumed inertial spin period is 243.0 days (retrograde), which is slightly less than the synchronous period of 243.16 days. Partly because of this difference, the data in Fig. 1 are not plotted in exact accord with the IAU system; they are displaced toward smaller longitudes, but nowhere by more than a few degrees.

9.  The number of orbital degrees of freedom is not overwhelmingly large. For Venus, only the four in-plane parameters are relevant [see (*4*)]; for the earth, the corresponding orbital elements are also constrained by the radar observations of Mercury and Mars. Thus, of the 23 parameters involved in the solution, relatively few are more than minimally correlated with the topography estimates. This conclusion is verified by the results from solutions in which topography parameters were estimated explicitly (see text).

10. W. B. Smith, R. P. Ingalls, I. I. Shapiro, M. E. Ash, *Radio Sci.* 5, 411 (1970).

11. For example, note the modest (1 km) near-equatorial rise at 280° longitude.

12. R. P. Ingalls and L. P. Rainville, in preparation; see also (*10*).

13. I. I. Shapiro, M. A. Slade, A. E. E. Rogers, S. H. Zisk, T. W. Thompson, in preparation.

14. The sensitivity of this radar (R. M. Goldstein, personal communication) is about 50 times that of Arecibo, and of Haystack when the latter is observing Venus. Venus's thick $CO_2$ atmosphere absorbs Haystack's X-band radiation and thereby reduces the echo power by a factor of about 6 relative to the echoes from the Goldstone and Arecibo signals, which are at substantially lower radio frequencies.

15. F. D. Drake, personal communication.

16. We thank M. E. Ash, R. Cappallo, R. F. Jurgens, and W. B. Smith for their vital contributions to earlier phases of this study and the staffs of the Haystack Observatory and the National Astronomy and Ionosphere Center for their aid in performing the radar experiments. Research at the Haystack Observatory is supported by NSF grant GP-25865 and NASA grant NGR22-174-003; contract NAS9-7830. The National Astronomy and Ionosphere Center is operated by Cornell University under a contract with the National Science Foundation.

# 12

# Lunar Science: Analyzing the Apollo Legacy

Allen L. Hammond

Man's knowledge of the moon has been dramatically transformed during the brief $3\frac{1}{2}$ years between the first and last Apollo landing. One of the more remarkable aspects of this transformation is how rapidly data from surface experiments, orbital observations, and analyses of lunar samples have become available. In contrast to past voyages of discovery, where the detailed record often did not appear until 20 or more years later, coordination between the space agency and the burgeoning lunar science community has produced preliminary Apollo results almost before the dust from the departing lunar module has settled. A truly impressive array of reports, including early evidence from Apollo 17, was presented at last week's Fourth Lunar Science Conference in Houston, Texas. While it was clear that detailed study of the Apollo legacy has only begun, the meeting marked the end of the adventure itself and the first attempt at a summary of what has been gained.

Among the highlights of the Apollo 17 data was the confirmation by several investigators that the widely reported orange-colored soil found near Shorty crater was not rust but rather a red-tinted glassy substance with a composition similar to that of basalt (Fig. 1). Dating established that the glass beads were formed about 3.7 billion years ago, and while their origin remains obscure, with some proposing volcanic extrusion and others suggesting meteoritic impact processes, the notion that the soil might indicate a recent volcanic vent has been dispatched. A rock found by Apollo 16, however, did turn out to contain rust in considerable abundance. But it also was shown to contain zinc compounds and other volatile minerals uncommon on the moon, thus raising the question of whether the rust is due to indigenous water or, as many investigators believe, possibly came from a comet.

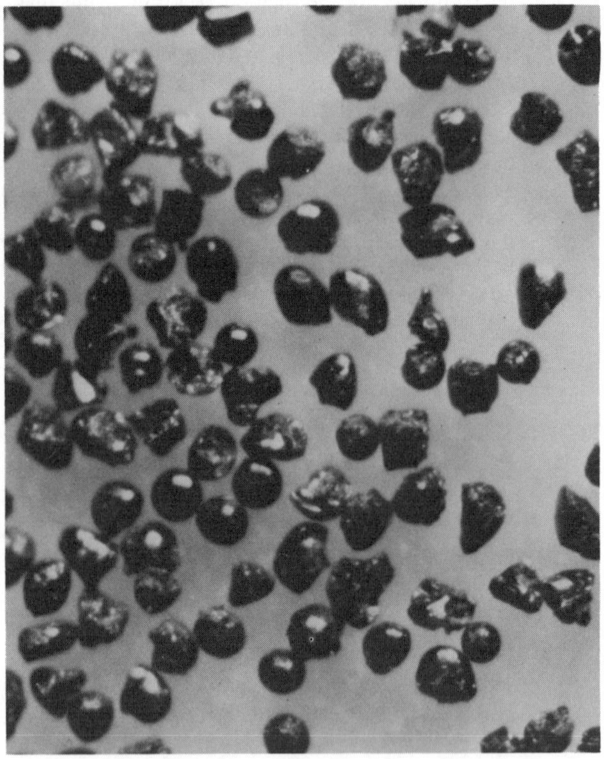

The Apollo 17 results and the more detailed analyses of data from earlier missions (including the Soviet Union's Luna 16 and Luna 20) reported at the Houston meeting are part of a virtual flood of empirical results pouring out of lunar labora-

Figure 1. Photomicrograph of glass spheres and fragments from the orange-colored soil found by the Apollo 17 astronauts. The samples shown here are in the 150- to 250-micrometer range. [Source: National Aeronautics and Space Administration]

tories. Emerging more slowly and still the subject of intense debate is a model of planetary evolution that can order and help explain the data. Many of the major constraints that such a model must meet are beginning to be clear, however, and the evidence is now overwhelming that the moon differs greatly from the earth in internal structure, in chemistry, and hence probably in the nature of its birth.

From *Science*, vol. 179, pp. 1313–1315, 30 March 1973. Copyright 1973 by the American Association for the Advancement of Science. Reprinted with permission.

Before Apollo, virtually nothing was known of the interior of the moon except that its density was singularly less than that of any other terrestrial planet, and that it appeared (from its moment of inertia) to be a homogeneous body. The shape of the moon, as determined by laser altimeter data from Apollo 15 and Apollo 16, is such that the center of figure is displaced away from the earth by about 2.5 kilometers from the center of mass. But this discrepancy may be explained, according to W. M. Kaula of the University of California at Los Angeles, by the presence of the low maria on the near side of the moon and their absence on the far side. The mean altitude of the maria below the highlands is about 3.5 kilometers. If the crust is substantially thicker on the far side of the moon, as J. A. Wood of the Smithsonian Astrophysical Observatory in Cambridge, Massachusetts, and others have proposed, it may have inhibited generation or extrusion of the maria and hence explain why they occur only on the moon's near side. Revised measurements of the moon's moment of inertia do not allow a homogeneous internal structure, but, according to Kaula, do seem to indicate that gravitational isostasy prevails.

### Few Quakes in a Quiet Moon

Evidence from the four seismometers that comprise the lunar seismic network indicates that moonquakes are rare and weak events compared to earthquakes. According to G. Latham of the University of Texas in Galveston, 43 active moonquake zones have been identified. Moonquakes in each zone occur at monthly invervals and show longer term variations that also correlate well with lunar tides.

A more puzzling finding concerns the spatial pattern of the moonquakes. Latham finds that the quakes occur at depths between 800 and 1100 kilometers, much deeper than any seismic activity on earth. Their focal centers seem to lie along two distinct belts that arc around the moon for about 1000 kilometers, one running roughly north to south and the other northeast to southwest. The pattern does not correlate with the rims of the lunar maria, as had been suggested earlier, but seismologists were at a loss to explain just what might give rise to the belts.

The depth of the moonquakes seems consistent with a model based on seismic data of the moon's interior structure. According to the model, the moon has a thick crust (for so small a body), a rigid mantle, and possibly a partially molten core. Measurements of the travel times of seismic signals generated when spent Saturn rocket stages are crashed into the moon had earlier shown the presence, at a depth of about 60 kilometers, of a sharp discontinuity in the velocity with which the signals propagated. The discontinuity is inferred to be the lower edge of the lunar crust. Data reported by M. N. Toksoz of M.I.T. from the last two Apollo missions and from the fortuitous impact of a large meteorite seem to confirm this view. The crust is thought to consist of two layers in the mare regions, of which the uppermost is inferred, by comparison with laboratory velocity measurements, to be basalt, and the second is thought to be rocks similar to the aluminous basalts and anorthositic gabbros found in the lunar highlands. Other investigators believe that the upper 20 kilometers of the lunar surface may be largely crushed and broken rock.

Below the crust, seismic velocities are higher and relatively constant to depths of about 1000 kilometers, indicating the presence of more rigid material with constant elastic properties. Seismic shear waves from events on the far side of the moon seem to be damped out when they pass through the deep interior, an indication, according to Latham, that below 1000 kilometers the moon contains a partially molten core. If this model is correct, and if the core material is silicate rock, temperatures of about 1500°C would be required. The model makes plausible the explanation that moonquakes occur and that tidal energy is released primarily in a zone between the rigid lower mantle and the less-rigid core material.

Heat flow measurements at the Apollo 17 site, reported by M. G. Langseth of Lamont-Doherty Geological Observatory, bear out the surprisingly high values reported on Apollo 15. The installation of the experiment on the lunar surface, which according to Langseth's description was "as difficult as plowing a straight furrow in New Hampshire," went off without a hitch. But the reported flux, about 2.8 microwatts per square centimeter, seems to indicate that there is more radioactive material and hence a hotter moon than had been expected. The heat flow corresponds to a lunar abundance of uranium between 0.05 and 0.075 parts per million. Langseth raised the possibility of regional bias in the heat flow measurements, since the two data points were both obtained on the margins of large mare basins and also suggested that corrections for the topographic effect of the narrow valley at Taurus-Littrow might lower the Apollo 17 result slightly. Nonetheless the heat flow measurements exceed those predicted by many geochemical models based on a chondritic composition and appear to put a strong constraint on calculations of the moon's thermal history.

The history appears to have been a complicated one, including such events as volcanic flooding of the mare basins between 3.1 and 3.9 billion years ago. Toksoz and his colleagues have computed models of the moon's thermal history that provide for the mare volcanism and also meet the high values for heat flow. The resulting theoretical picture requires complete melting of the outer few hundred kilometers of the moon at the time of formation, then gradual cooling of the exterior while the interior heats up because of radioactive decay. The heating progresses downward, possibly melting each deeper zone of the moon in turn. At present, the model predicts a warm lunar interior with partial melting below about 800 kilometers, consistent with the deep seismic evidence.

The major objection to a moon with an initially cold core has come from investigators concerned with magnetic phenomena. The remanent magnetic field of the moon varies from one location to another, but it seems inescapable that the mare lava flows crystallized in a magnetic field with a strength of a few thousand gammas, much stronger than the present one. Speculation has centered on a self-generated field, excited by convection in a molten iron core that later cooled, turning off the field. This explanation, however, would require an initially hot moon and thus runs contrary to the prevailing consensus.

A second hypothesis, proposed by S. K. Runcorn of the University of Newcastle-upon-Tyne and H. C. Urey of the University of California at San Diego, is that the moon became magnetized early in its history. In the form advanced at the

Houston meeting by D. W. Strangeway of the National Aeronautics and Space Administration's (NASA) renamed Johnson Space Center (JSC), this model is based on the assumption that the moon's interior was initially below 780°C, the curie point of iron, and was briefly exposed to an external magnetic field as high as 20 gauss. As the magnetized moon evolved, the iron-bearing rocks near the moon's surface cooled and acquired some residual magnetism from the internal field. Later the moon's interior heated up, destroying the moon's original magnetic field, but leaving the remanent magnetism at the surface. This model agrees much more closely with the other evidence for the moon's thermal history, but the source of the magnetizing field—whether an interplanetary field of some kind or, as H. Alfvén of the University of California at San Diego has suggested, an amplified geomagnetic field—is still a matter of speculation.

The chemistry of the moon was also an enigma prior to Apollo. Data from the diverse and often complex surface materials returned from the moon have demonstrated that compared to the earth the moon is enriched in refractory elements and depleted in volatile elements. Although there is general agreement that the mare basaltic rocks were produced by internal melting, the composition and origin of the lunar highlands is more complicated. Both the samples and orbital geochemistry experiments indicate that the three most common rocks on the lunar surface are iron-rich mare basalts, plagioclase or aluminum-rich anorthosites (more than 25 percent $Al_2O_3$), and uranium- and thorium-rich basalts that are also enriched in potassium, rare-earth elements, and phosphorus (KREEP basalts, with 15 to 20 percent $Al_2O_3$). A fourth major rock type proposed by P. W. Gast of JSC is a class of very high aluminum (VHA) materials (20 to 25 percent $Al_2O_3$) that may represent basalts (Fig. 2). In addition to their bulk composition, the pattern of rare earths varies greatly from one rock type to another—a characteristic depletion of europium in the KREEP and VHA basalts, for example, as compared to mare basalts. Unambiguous explanations for the chemical and mineralogical properties of most of these rocks are not yet available. The distribution of these rocks on the lunar surface—in particular the concentration of KREEP basalts around Mare Imbrium—is also very baffling.

Despite the petrological puzzles of the major rock types and the variety of minor rock types found on the moon, the preponderance of evidence is that the outer parts of the moon were formed by igneous processes and then modified by the effects of massive bombardment with meteorites. Major collisions such as those that resulted in the mare craters are thought to have had ample kinetic energy to melt large quantities of surface rock. According to Gast, the combined thermal and mechanical effects of the moon's collisional history have largely obliterated the structural and textural characteristics of the ancient lunar rocks. From their chemistry, however, he believes that early melting of the outer surface of the moon led to the formation of an anorthositic crust and that all subsequent magmatic activity was the result of internal melting caused by radioactive heating—a model that accords well with the physical evidence. A consequence of the model, however, is that the moon was chemically zoned in its original

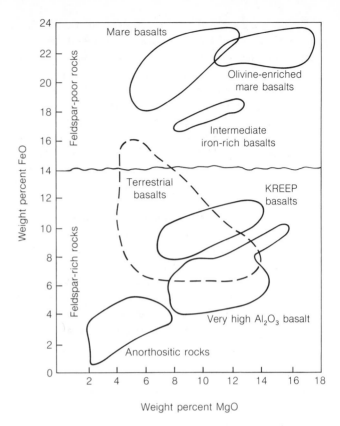

**Figure 2.** Iron oxide, magnesium oxide, and aluminum oxide content of different lunar rock types. [Source: Paul Gast, Johnson Space Center, Houston, Texas.]

state—rich in calcium and aluminum in its outer layer and richer in iron in deeper layers. Because of the implications for the process by which the moon was formed, not everyone agrees with the concept of an initially heterogeneous, largely cold moon.

There is also disagreement about the processes that led to formation of highland rocks. The evidence from Apollo 16 seems to indicate that the Cayley formation, earlier thought to be an example of highlands volcanism, is a breccia deposit—rocks composed of previously crystallized fragments that have been shocked and transformed by meteorite impacts. The lesson, according to N. Hinners of NASA, is that impact processes are capable of yielding materials whose surface morphology looks like that of volcanic terrains. As a result, new emphasis is being given to the study of breccias, whose complex history makes their ages and origins difficult to unravel, and to the chemistry of impact processes.

The bombardment of the lunar surface by projectiles as large as tens of kilometers in diameter has also apparently had the effect of resetting the isotopic clocks used to determine the absolute age of rocks in most of the samples brought back so far. Thus the chronology of lunar events, particularly the filling of the mare basins, is reasonably well determined up to about 4.0 billion years ago, but is unknown between that date and the formation of the moon perhaps 600 million years earlier. Dates obtained by the group headed by G. J. Wasserburg of the California Institute of Technology for highland rocks from several different locations on the moon (including data from Luna 20) all seem to indicate that the rocks have been formed or have undergone metamorphic changes around 3.95 billion years ago. Because his dates cluster so narrowly around 3.9 to 4.0 billion years, Wasserburg proposes that the moon underwent a major cataclysm at this time. An alternative hypothesis is that the number of impacts was so high as to steadily reset the isotopic clocks throughout the moon's early history. The impact which formed the huge Imbrium basin is thought to have occurred about 4.0 billion years ago and may have been enough of a super-impact to scatter a blanket of debris over much of the moon, or the cataclysm may have been a series of events which terminated very abruptly about 3.9 billion years ago. In either case the collisions must have been, as Wasserburg put it, "one hell of a good show."

Not everyone agrees with the Wasserburg hypothesis, and there was controversy at the Houston meeting over dates of about 4.2 billion years reported for some breccia fragments from Apollo 16 by L. Husain and his colleagues at the State University of New York at Stony Brook. If these earlier dates are confirmed, the rocks would antedate the hypothesized cataclysm. Other investigators such as C. Meyer, Jr., of JSC believe that the Imbrium impact may have occurred much earlier than 4.0 billion years ago. Meyer put forward the hypothesis that the distribution of KREEP basalts on the lunar surface may be explainable as an ejecta pattern of material formed in the Imbrium basin and scattered by later impacts, if the Imbrium impact itself occurred about 4.4 billion years ago. But the cataclysm hypothesis clearly seemed to be the favorite of most investigators.

Wasserburg also believes that no significant amount of volcanism occurred on the lunar surface after 3.1 billion years ago, when the youngest maria are believed to have been formed. This view, held by many sample analysts, contrasts with the view of those investigators who study the orbital lunar photographs and who have attempted to apply the traditional methods of field geology to the moon. The debate centers around dark deposits on the lunar surface that seem to form a mantle over or cover large areas of the preexisting topography. According to F. El-Baz of the Smithsonian Institution in Washington, D.C., these deposits are probably volcanic and as recent as 2.5 billion years ago. But if this material was sampled at the Apollo 17 site, it has not yet turned up among the soils that have been dated, and many lunar scientists are consequently inclined to be skeptical of photogeological interpretations.

One of the more interesting models of lunar evolution presented at the Houston conference was that described by D. L. Anderson of Caltech. He proposes that, as the earth and moon [formed in] the solar nebula, the larger and earlier-formed body grew at a more rapid rate and trapped more of the late condensing, volatile elements. Hence the refractory-rich composition of the moon is a consequence of its delayed condensation and of its subsequent competition with the earth, in this model. The high temperatures at which most of the moon would have condensed and the high concentration of radioactive elements might lead to early and extensive geochemical differentiation. Among the consequences of the model is that the deep interior of the moon is virtually barren of iron. The model is thus consistent with those explanations for the moon's remanent magnetism that do not require a molten iron core. However, D. H. Green of the Australian National University in Canberra finds that Anderson's model implies a lunar interior more dense than it is thought to be.

Understanding of how the moon was formed is still evolving, and at a rapid pace. Renewed emphasis on the study of brecciated rocks as the key to the surface history may help to clear up a number of unresolved puzzles such as why the soils on the lunar surface appear to be so much older than the rocks from which they presumably came. Where the meteorites or moonlets that caused the lunar cataclysm came from is also unknown. The answers to these and other outstanding questions may with luck yet be found among the nearly 400 kilograms of lunar samples that, in the view of lunar scientists, amount to a treasure house of incalculable scientific worth. What they hope to gain from the Apollo legacy is not only insight into the origins of the moon, but of the earth and the solar system as well.

*APOLLO 12* SURFACE MAGNETOMETER was photographed by astronaut Alan L. Bean just after he finished deploying it near the *Apollo 12* landing site in Oceanus Procellarum. Astronaut Charles Conrad, Jr., can be seen in the background aiming the central-station *S*-band antenna toward the earth. The magnetometer was isolated from other scientific instruments in order to avoid measuring their magnetic fields. The device operates continuously during lunar day and night on power from a nearby nuclear generator.

# The Magnetism of the Moon

Palmer Dyal and Curtis W. Parkin

Magnetometers placed on the moon by the astronauts of *Apollo 12* and *Apollo 14* have measured two types of lunar magnetic field: permanent fields due to fossil magnetic material and transient fields due to electric currents generated deep in the interior of the moon. These magnetic measurements yield unique information about the history and present physical state of the moon. The origin of the fossil magnetism is unknown and points to a mysterious epoch in lunar history. The transient fields are induced in the moon by changes in the magnetic field associated with the "solar wind": the stream of electrically charged particles ejected by the sun. The transient fields provide data for computing the electrical resistance and the temperature of the lunar interior. The magnetic measurements demonstrate a promising method for the future scientific exploration and study of bodies in the solar system, such as Mars, that resemble the moon.

Although the permanent fields measured on the moon so far are less than 1 percent as strong as the magnetic field of the earth, the lunar fields are much stronger than was expected on the basis of magnetometer measurements made earlier by American and Russian spacecraft that either flew past the moon or went into orbit around it. The permanent fields measured by the Apollo instruments vary from place to place on the lunar surface and do not combine to form an overall dipole pattern similar to that of the earth. This indicates that the measured fields are due to local sources. Moreover, the high field strengths indicate that these sources were magnetized by an ambient field much stronger than the one that now exists at the moon. Evidently at some time in the past the moon either possessed a strong magnetizing field or was immersed in one. We are presented with some fascinating questions about lunar history: Did the

ancient moon have an earthlike field, generated as the earth's is thought to be by an internal "dynamo"? Was the moon once in an orbit closer to the earth and within the strong terrestrial field? Was the moon magnetized in another part of the solar system and later captured by the earth? The answer to these questions may be recorded in the fossil magnetism

of the lunar material. The fact that the moon has changed little for billions of years offers opportunities to investigate magnetic records of the ancient solar system. Similar magnetic information recorded in the earth's crust would have been erased long ago by crustal activity and exchange of magnetic surface material with the earth's molten interior.

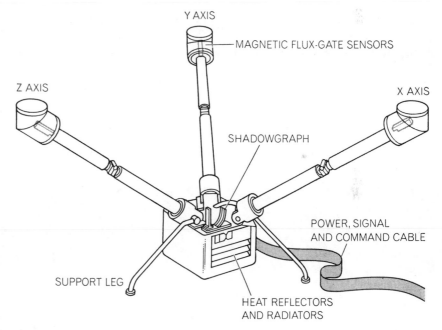

LUNAR SURFACE MAGNETOMETER was designed by workers at the Ames Research Center of the National Aeronautics and Space Administration to be emplaced at a single fixed location on the moon, where it can measure three components of the lunar magnetic field continually through the lunar day and night. Three "flux gate" sensors (*gray*), each directed at right angles with respect to the other two, are installed at the top ends of the booms about 30 inches above the surface. By remote command from the earth the sensors can be individually flipped 90 degrees or 180 degrees or rotated around their axes by a pulley drive (*not shown*), allowing all three sensors to be aligned simultaneously along any one boom direction. Motors and data-processing equipment are located inside the box, and the entire assembly is encased in a thermal blanket. Heat rejection (during day) and retention (during night) are controlled by an array of parabolic reflectors on two sides of the electronics box. A bubble level and an azimuthal shadowgraph on top of the box enable the astronauts to orient the magnetometer accurately. Power, digital signals and commands are conveyed through a ribbon cable connected to the central-station receiver/transmitter.

From *Scientific American*, vol. 225, pp. 63–73, August 1971. Copyright © 1971 by Scientific American, Inc. All rights reserved. Reprinted with permission.

APOLLO 14 PORTABLE MAGNETOMETER, a self-contained battery-operated device, was designed to be transported by the astronauts along a surface traverse for the purpose of recording measurements of the permanent lunar magnetic field at different locations. In the photograph above, made by astronaut Edgar D. Mitchell, Jr., the lunar portable magnetometer is shown stowed aboard the two-wheeled mobile equipment transporter. Astronaut Alan B. Shepard, Jr., can be seen working on another piece of equipment behind the transporter. In the photograph below the sensor-tripod portion of the magnetometer is shown deployed about 50 feet from the electronics box, which remains on the transporter during the measurements. After completing all the readings and transmitting them to the earth, the astronaut stows the sensor-tripod assembly, reels up the ribbon cable and continues on the surface traverse.

The transient fields generated by electric currents flowing in the lunar interior are associated with the entire moon rather than with any one region. The fields wax and wane rapidly in response to changes in the solar wind. The characteristics of the induced lunar fields depend on the electrical resistance of the moon's interior, and this resistance is related in turn to the temperature of the material. Therefore the magnetometer can be used as an indirect "resistance thermometer" to determine the internal temperature of the moon.

Analysis of the transient fields shows that the temperature of the moon is only about 1,000 degrees Celsius at a depth halfway to the center, or some 3,400 degrees cooler than the earth is halfway to its center. Lunar temperature determinations are consistent with the low seismic activity reported by the seismometers that have been placed on the moon. These experimental results indicate that the moon is not hot and molten inside but is relatively cool, and that its surface has been modified more by the impact of meteorites than by volcanic activity. Additional magnetic-field measurements should yield a more accurate temperature profile of the lunar interior and also shed light on the nature of the internal heat source. Through this knowledge fundamental questions about the origin and evolution of the moon will be answered.

The lunar magnetometer experiment can be regarded as the modern counterpart of the first planetary magnetic experiment conducted by William Gilbert in 1600. Observing that a balanced magnetized needle dips toward the earth at an angle, Gilbert inferred that the earth itself acts as a large magnet. Since then the geometrical and time-varying characteristics of the earth's field have been intensively analyzed to determine the electrical resistance and the temperature of the earth's crust and mantle. Long-term measurements also indicate that the earth's molten interior is rotating at a slightly lower speed than the crust. This differential rotation is thought to produce the dynamo action needed to account for the earth's magnetic field.

With the development of rocket technology it became possible to place magnetometers on earth-orbiting spacecraft, to map the terrestrial magnetic field in nearby space and to determine its interaction with the solar wind. Other magnetometer experiments, conducted by Edward John Smith of the Jet Propulsion Laboratory, were flown on board

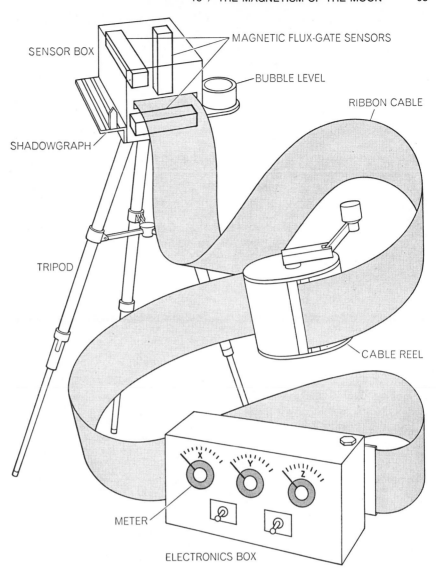

LUNAR PORTABLE MAGNETOMETER contains three mutually perpendicular flux-gate sensors, identical with those used in the lunar surface magnetometer, that are housed inside a cubical sensor block and mounted on top of a tripod. The sensor-tripod assembly is connected by means of a 50-foot ribbon cable to the electronics box, which contains a battery pack, electronics equipment and three meters. At each surface site the three magnetic-field vector components are read aloud from the meters by an astronaut and radioed to the earth.

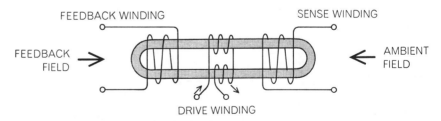

FLUX-GATE SENSOR is used in both types of lunar magnetometer to detect the ambient magnetic field. The sensor consists of an easily magnetized, toroidal Permalloy core that is driven to saturation by an alternating signal at a certain frequency. A sense winding detects the superposition of the drive-winding magnetic field and the ambient magnetic field; as a result a signal of twice the frequency of the driving frequency is generated in the sense winding with a magnitude that is proportional to the strength of the ambient field. The phase of this second harmonic signal with respect to the drive signal indicates the direction of the ambient field with respect to the axis of the sensor. The harmonic signal is electronically processed and sent through the feedback winding to "null out," or cancel, the ambient field within the sensor. Operating at null increases the thermal stability of the sensor.

Mariner spacecraft to the vicinity of Venus and Mars, measuring the interplanetary solar fields en route and determining how these planets interact with the solar wind. No magnetic fields intrinsic to either Venus or Mars were detected.

Magnetic fields associated with the sun and celestial objects outside the solar system have been measured by telescopic observation. As early as 1908 George Ellery Hale observed that the magnetic field of the sun splits lines in the solar spectrum (the Zeeman effect). Subsequently Horace W. Babcock and others used Hale's technique to measure the magnetism of other stars and entire galaxies. Recently James C. Kemp of the University of Oregon has observed an extremely strong field (100 million times stronger than the field of the earth) in a dwarf star in the constellation Draco.

Magnetic studies related to the moon were begun in 1850, when Karl Kreil discovered that the earth's magnetic field varied in a systematic way with the phase of the moon. Later it was shown that the effect was created by the moon's gravitational pull on the earth's atmosphere and ionosphere and had nothing to do with the presence or absence of a lunar magnetic field. It was not until spacecraft able to reach the vicinity of the moon were developed that direct measurements of the moon's field finally became possible.

Since no rigorous theory has evolved that satisfactorily explains the earth's permanent magnetic field, it is not surprising that no one predicted the magnitude of the lunar permanent field. Investigations of other lunar properties, such as optical measurements of changes in the moon's surface temperature and studies of the moon's shape and dynamic behavior, have indicated, however, that the moon does not have an internal dynamo.

Theoretical predictions of other lunar magnetic properties were made before the manned lunar missions by several investigators, including F. Curtis Michel of Rice University, Thomas Gold of Cornell University and John R. Spreiter of Stanford University. They attempted to predict how the moon would interact with high-speed particles in the solar wind if, at one extreme, the moon was completely nonmagnetic and if, at the other, it had a magnetic field as strong as the earth's. (The earth's field at the Equator is about 30,000 gammas.) At the nonmagnetic extreme the moon would act simply as an opaque sphere,

intercepting only the charged solar particles that happened to strike it and allowing the others to pass undeflected [see illustration at right]. If the moon had a significant overall field, however, the particles of the solar wind would be deflected around the lunar sphere in a complex manner. Moreover, a shock front would be created on the upwind side of the moon, in rough analogy to the shock front created ahead of a supersonic aircraft.

Several possibilities lying between these extremes were considered. If the moon did not have a global field of at least 40 gammas (as was later found to be the case), time-varying magnetic fields induced inside the moon could still substantially perturb the flow of the solar wind. Two basic types of induced lunar field are possible. The first is a type of field that would be induced in the moon by the motion of the solar-wind field past the moon. This process was mentioned by Gold and was developed theoretically by David S. Colburn and Charles P. Sonett of the Ames Research Center of the National Aeronautics and Space Administration. The second possibility is a time-varying dipolar field, which would be induced in the moon by temporal changes in the solar-wind field [see illustration on page 99]. In this process eddy currents are set up inside the moon by time-varying solar-wind fields in much the same way that currents are generated in the secondary coil of a transformer by changing fields in the primary coil [see illustration on page 100]. This latter mechanism was suggested by one of us (Dyal) in 1966 and was developed theoretically by Joel Blank and William Sill of Bellcomm, Inc., in 1969. Magnetometer measurements made on the surface of the moon have subsequently proved that the dipolar eddy-current field is indeed the dominant lunar induced field.

The measurement of magnetic fields in the vicinity of the moon began in January, 1959, when the Russian spacecraft Luna 1 carried a magnetometer to within several hundred miles of the moon. In September, 1959, Luna 2, also equipped with a magnetometer, crashed into the moon. The instrument aboard Luna 2 set an upper limit of 100 gammas for a possible lunar field at an altitude of 30 miles above the moon's surface. In April, 1966, Luna 10, carrying a magnetometer 10 times more sensitive than Luna 2's, was successfully placed in a lunar orbit that came to within 220

miles of the moon. The Luna 10 magnetometer measured a magnetic field varying between 24 and 40 gammas in the neighborhood of the moon. This field, which was correlated with changes in magnetic activity at the earth's surface, was interpreted by the Russian investigator L. N. Zhuzgov as indicating the existence of a weak lunar magnetosphere.

A year later (July, 1967) the U.S. placed Explorer 35 in orbit around the

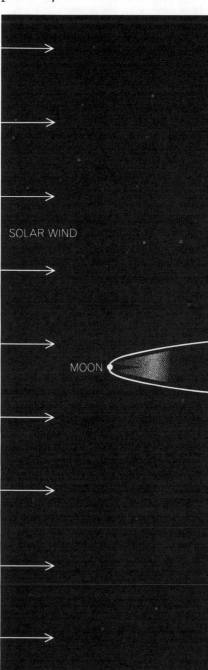

MAGNETIC ENVIRONMENT of the moon varies considerably depending on the moon's orbital position. The "solar wind," made up of particles blown outward from the sun at supersonic speeds, sweeps the earth's

moon with two magnetometers aboard. The spacecraft is still operating in a trajectory that takes it to within 520 miles of the moon's surface. With instruments whose sensitivity is ±.2 gamma *Explorer 35* successfully measured magnetic properties of the solar-wind cavity downstream from the moon. *Explorer 35* failed, however, to detect the lunar magnetosphere indicated by *Luna 10* measurements or the shock front and induced-field configuration suggested by

Gold. In an analysis of the *Explorer 35* results Sonett concluded that if a permanent lunar field existed at all, its magnitude would be less than two gammas, a disappointingly low value, at an altitude of 520 miles.

This was the situation, with all signs pointing to a magnetically dead moon, when we finished testing the magnetometer package scheduled to be carried to the Ocean of Storms by *Apollo*

*12* in November, 1969. One can imagine our surprise and delight when the instrument measured a permanent magnetic field of 38 gammas at the *Apollo 12* landing site. Then in February, 1970, Alan B. Shepard, Jr., and Edgar D. Mitchell, Jr., of *Apollo 14*, working with a portable magnetometer, measured magnetic fields of 43 gammas and 103 gammas at two different locations near their landing site, some 110 miles east of where *Apollo 12* had landed.

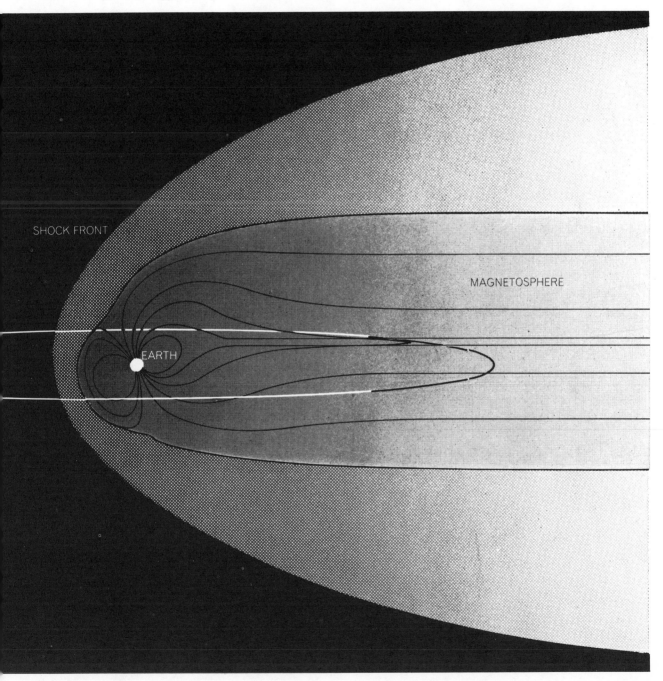

magnetic field into a tubular shape known as the magnetosphere (*gray area*). The magnetosphere in turn acts as a barrier to the solar wind, causing a bow-shaped shock front to form ahead of the earth. When the moon is immersed in the magnetosphere, the moon's magnetic environment is dominated by the comparatively steady magnetic field

of the earth. In the free-streaming solar wind the moon is subjected not only to the solar magnetic field but also to plasma waves traveling from the sun. The intermediate magnetosheath region (*shaded from dark to light*) is characterized by erratic solar-particle flow and the most turbulent fields of the lunar orbit.

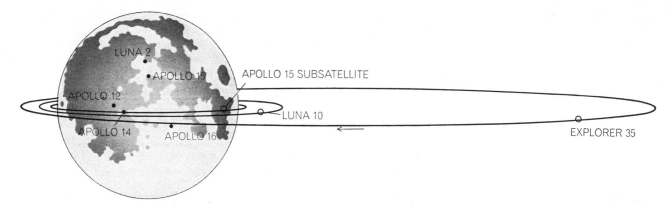

LOCATIONS of past lunar magnetometer experiments and those scheduled by the U.S. through 1972 are indicated on this map of the front half of the moon. The Russian *Luna 2* spacecraft, which crashed into the moon in 1961, and the *Luna 10* orbiter, launched in 1966, made the earliest magnetic measurements of the lunar environment but were not sensitive enough to detect either the steady or the transient lunar fields. The U.S. *Explorer 35* orbiter, placed in lunar orbit in 1967, has made possible detailed study of the effects of the moon on the solar wind, but it is too distant from the moon to measure the lunar fields. The magnetometers placed on the moon at the *Apollo 12* and *Apollo 14* sites respectively in 1969 and 1971 have found the moon to be much more active magnetically than was anticipated. Magnetometers scheduled for future Apollo missions to the moon should provide more detailed information.

The magnetometers placed on the moon by the astronauts of *Apollo 12* and *Apollo 14* were designed at the Ames Research Center. Sonett originated the *Apollo 12* experiment and one of us (Dyal) extended this concept to a network of stationary and portable magnetometers for a series of subsequent Apollo missions. The instrument used in the *Apollo 12* experiment was designed and fabricated by the Philco-Ford Corporation under the direction of John Keeler of the Ames Research Center. It was integrated into an array with four other experiments, collectively called the Apollo Lunar Surface Experiments Package (ALSEP), by the Bendix Corporation. The *Apollo 14* portable magnetometer was fabricated by an Ames Research Center team led by Carle Privette and was integrated into the Apollo spacecraft by the Manned Spacecraft Center of NASA.

The design of the *Apollo 12* instrument took into account the magnetometer readings provided by *Explorer 35* and *Luna 2*. In the Apollo instrument three components of the vector magnetic field are detected by three "flux gate" sensors located at the ends of three mutually perpendicular booms [*see illustration on page 91*]. The sensors are separated so that the uniformity of the local magnetic field can be determined by measuring field values at each sensor location. Each sensor weighs less than an ounce and operates on 15 milliwatts of power.

A flux-gate sensor consists of an easily magnetized core (Permalloy) with three windings: one to drive the core to saturation in the presence of the ambient magnetic field, another to sense the sum of the driving field and the ambient field, and a third to cancel out the ambient field inside the sensor. A signal with twice the frequency of the driving frequency is induced in the sense winding; the magnitude of this signal is proportional to the ambient field [*see bottom illustration on page 93*]. Each sensor has a frequency response ranging from zero to three hertz (cycles per second) and an angular response that is proportional to the cosine of the angle between the magnetic-field vector and the sensor axis. The instrument can be set at ranges of ±100 gammas, ±200 gammas or ±400 gammas, with an accuracy of .2 percent in each range.

The electronics subsystem is self-contained except for power supplied by a radioactive-isotope thermoelectric generator and timing and telemetry supplied by the ALSEP central station. The subsystem weighs six pounds, occupies a volume of 300 cubic inches and has some 6,400 electronic parts. The electronic circuits drive the flux-gate sen-

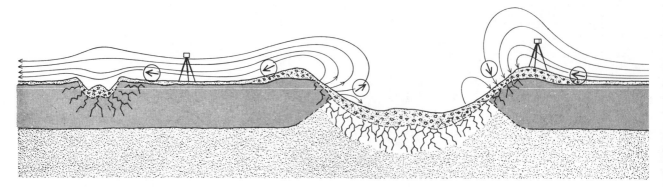

NONUNIFORM STEADY MAGNETIC FIELD was measured by the *Apollo 14* astronauts at two locations separated by about three-quarters of a mile, indicating that strongly magnetized material exists in that vicinity. The nonuniformity could be explained by the existence of a subsurface slab of material (*dark gray area*) that was uniformly magnetized at one time but has subsequently been altered by local processes such as shock demagnetization from meteorite impacts. The moon has little or no global dipole field; it appears rather that the permanent lunar magnetic field is dominated by local features such as the one illustrated here and that it varies from place to place. Therefore a compass (*large arrows*) would probably not be useful for navigation on the moon.

sors, measure the orientation of the instrument, drive three motors for sensor orientation and process the field data before radio transmission to the earth. The instrument is wrapped with thermal insulation and has directional heat reflectors to radiate heat out into space during the lunar day; resistance heaters supply heat during the lunar night. In this way the temperature of the instrument is maintained between −25 degrees C. and 75 degrees C. even though the temperature of the lunar surface on which it sits varies from about −150 degrees to 120 degrees.

During the *Apollo 12* mission the magnetometer was placed in operating position by astronaut Alan L. Bean at 23.35 degrees lunar west longitude and 2.97 degrees south latitude. Bean leveled the instrument with a bubble level and aligned it in azimuth by adjusting a shadowgraph to read within marked preset values. The instrument was then activated by radio command from the Manned Spacecraft Center at Houston, and measurements transmitted back to the earth showed that the instrument was operating successfully.

The success of the *Apollo 12* magnetometer experiment lead to the design of a portable magnetometer to measure the local permanent fields at the *Apollo 14* landing site. The *Apollo 14* magnetometer was designed to be entirely self-contained, so that the astronauts could measure the steady magnetic field at different points along their traverse. This design makes it possible to measure the changes in magnetic field over distances on the order of a mile rather than, as in the case of *Apollo 12*, 4.5 feet. Three mutually perpendicular flux-gate sensors, identical with those in the *Apollo 12* instrument, are mounted on top of a tripod; the sensor-tripod assembly is connected by a 50-foot cable to an electronics box that contains a battery pack, the electronics and three meters [*see top illustration on page 93*]. The astronauts read the meters and report the needle positions back to the earth by radio. The 50-foot separation minimizes the effect of the magnetic fields associated with the electronics and the astronauts' backpack. The instrument weighs 11 pounds, operates on 1.5 watts of power and functions over the temperature range from zero degrees C. to 50 degrees. It is designed to filter out high-frequency fluctuations in the solar-wind field. Two magnetic-field measurements were made by Shepard and Mitchell during their second exploratory excursion on the moon. The first measurement was obtained 350 yards to the

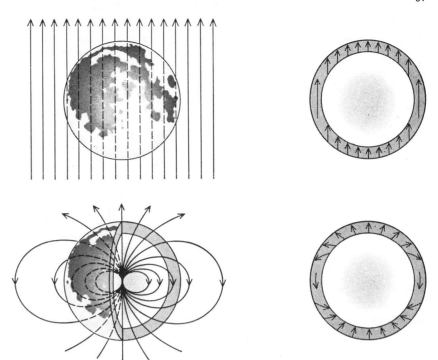

MAGNETIC MEASUREMENTS SUGGEST that at one time in the lunar past ambient fields much stronger than those currently observed existed over much or all of the lunar surface. One explanation is that the moon may have taken a "magnetic snapshot" of an early stage in the evolution of the solar system. If the crust of the entire moon was magnetized at that time, different global patterns of magnetized material should exist depending on whether the ancient ambient field originated outside or inside the moon. An external field (*top left*) that was stronger than the solar or terrestrial fields that exist at present would tend to magnetize the crustal material unidirectionally (*top right*), whereas an internal field (*bottom left*) would result in a varied pattern of magnetization (*bottom right*).

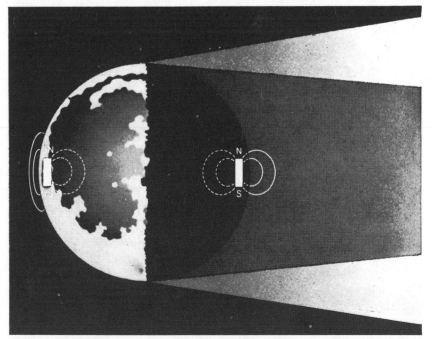

SOLAR WIND COULD COMPRESS a steady lunar magnetic field, forming a bow-shaped shock front, provided that the steady field is strong enough and its source is large enough. A global lunar dipole field, if it exists at all, would be too weak to create a shock front like the shock front of the earth. Instead the solar wind impinges directly on the lunar surface on the daytime side and forms a turbulent "wake" downstream (*light-gray area*), which bounds a "cavity" essentially empty of solar-wind particles (*dark-gray area*). A localized steady field of sufficient strength and extent, although unaffected while in the cavity, would be compressed while on the daytime side of the moon. Such a steady-field compression has been observed at the *Apollo 12* site for data gathered during a solar magnetic storm.

east of the lunar module; the second, .7 mile farther east near the rim of a crater.

Measurements returned from the Apollo magnetometers on the lunar surface have yielded much more information than was anticipated. Not only were steady magnetic fields measured that were up to 25 times stronger than had been predicted but also well-defined lunar induction fields were observed that enabled us to calculate values for the electrical resistance and temperature of the moon. The variation in readings from less than 40 gammas to more than 100 gammas told us that our instruments were measuring fields due to highly magnetized local sources rather than an overall lunar dipole field.

The steady-field value of approximately 38 gammas—some 10 times the maximum predicted value—radioed back to the earth just 40 minutes after Bean had unfolded the *Apollo 12* magnetometer was so unexpected that we hesitated to believe it until we had calibrated the instrument by sending radio commands from Houston back to the experiment package. Three days after the magnetometer was turned on, the moon passed into the tail of the earth's magnetosphere, where the solar wind is excluded by the earth's magnetic field. During the time when the moon was inside this region of very steady magnetic fields we sent commands to the instrument that enabled it to function as a gradiometer. In this mode the sensors

were rotated by motors so that all three vector-field components could be measured sequentially at each of the three sensor locations. The three measurements showed that the field varies less than the instrument's maximum sensitivity of .2 gamma over the 4.5-foot distance between any two sensors.

The spatial uniformity of the field indicates that if the highly magnetized source were a single meteorite buried near the surface, it would have to be more than 200 yards away from the instrument. On the other hand, the upper limit placed on the permanent field by *Explorer 35* required that the source be within 125 miles of the *Apollo 12* instrument. One can compute the minimum and maximum dimensions of the source if it is assumed to be a sphere with the same remanent magnetization as that of lunar samples brought back to the earth. We calculated that if the hypothetical source lay 200 yards from the magnetometer, it would be 50 yards in diameter, and if it were 125 miles away, it would be 30 miles in diameter.

Another observation that sheds light on the dimensions of the steady field is the compression of the field by the solar wind. Aaron Barnes and Patrick Cassen of the Ames Research Center predicted that the 38-gamma permanent magnetic field would be compressed if its effective extent above the surface were more than about six miles. Such a compression of the 38-gamma field at the *Apollo 12* site has been observed during times when the moon is showered by a high density of solar-wind particles [*see bottom illustration on preceding page*]. The solar-wind density was measured by Conway Snyder and Douglas Clay of the Jet Propulsion Laboratory.

The magnetic-field measurements made at the *Apollo 12* and *Apollo 14* sites, together with the high magnetic remanence found in rock samples returned from all the Apollo sites (including the *Apollo 11* site some 900 miles to the east of *Apollo 12*), demonstrate that the moon has been magnetized in widely dispersed regions. John Mihalov of the Ames Research Center has reexamined the magnetometer and charged-particle data from *Explorer 35* and has concluded that several magnetized areas may exist on both the near and the far side of the moon. Thus we have strong evidence that much of the lunar surface—perhaps even a crustal shell around the entire moon—was magnetized at some time in the past. Evidently ambient magnetic fields much stronger than those observed today existed over much or all of the lunar surface. It may be that

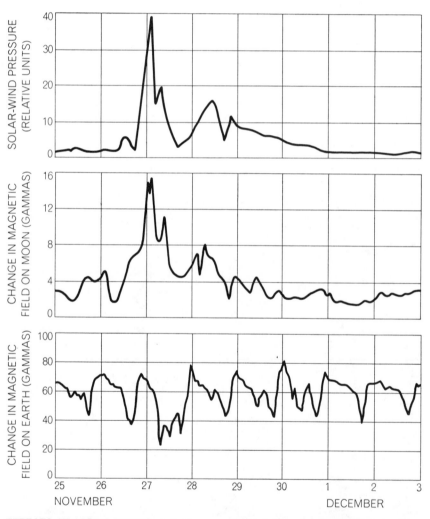

**EFFECTS OF SOLAR STORM** on the lunar magnetic field at the *Apollo 12* site and on the terrestrial magnetic field at a site on the surface of the earth are contrasted in this illustration. The top curve shows the rise in solar-wind pressure that accompanies the arrival of solar-storm particles in the vicinity of the earth-moon system. The middle curve shows that the horizontal component of the *Apollo 12* steady field rises in direct proportion to the solar-wind pressure. The bottom curve shows that the terrestrial field intensity, in contrast, exhibits a corresponding decrease during the solar storm. This decrease in the earth's total field is believed to be caused by the formation of earth-encircling ring currents involving charged particles trapped in the earth's magnetosphere; no known mechanism exists for the formation of an analogous ring current around the moon.

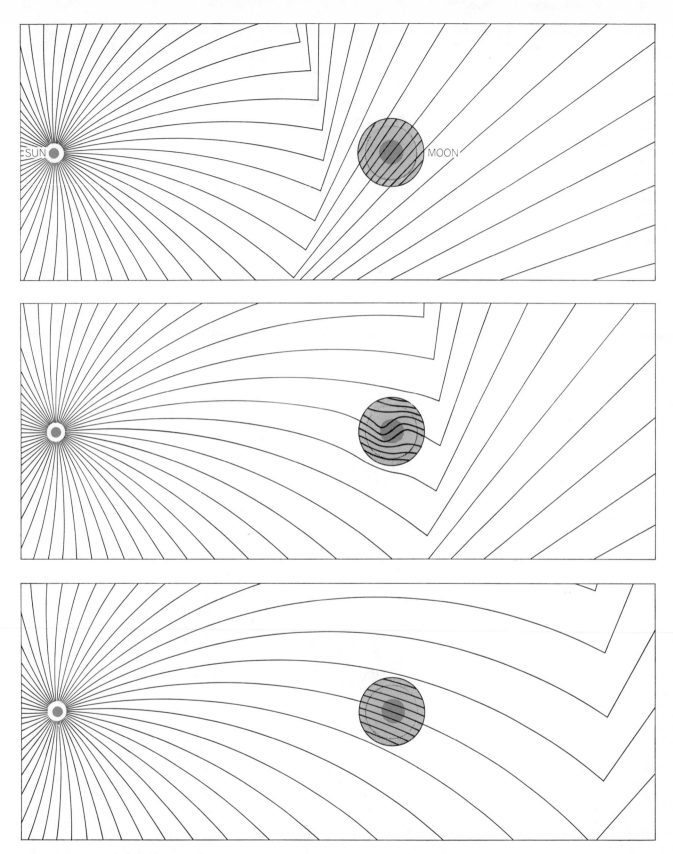

**TRANSIENT MAGNETIC RESPONSE** of a hypothetical three-layered model of the moon is represented for a case in which a directional discontinuity in the solar magnetic field travels outward past the moon at a supersonic velocity. The three layers are characterized by successively decreasing electrical resistance with distance into the moon. The top drawing shows the moon immersed in a steady solar field before the arrival of the discontinuity. The middle drawing shows the magnetic field through the moon immediately after the passage of the discontinuity. Eddy currents have been induced in the moon according to Lenz's law, which states that in an electrically conducting medium, currents and fields will be induced that tend to oppose any change in the original ambient magnetic field. Such eddy currents persist longest in regions of low resistance, so that the original magnetic-field orientation is maintained longest in the core. Eventually all the eddy currents decay and distortions of the total field disappear (*bottom drawing*).

the moon has taken a magnetic snapshot of an early evolutionary phase of the solar system.

The magnetic-field measurements made at the *Apollo 12* and *Apollo 14* sites are strikingly similar: all the vectors point down and toward the south, and their magnitudes correspond to within a factor of three. This suggests that the two *Apollo 14* sites and possibly the *Apollo 12* site are located over a slab of material that was uniformly magnetized at one time. Subsequently the magnetism of the slab could have been altered by local processes, for example by tectonic activity, by fracturing or by demagnetizing shock of meteorite impacts [*see bottom illustration on page 96*].

The material near the lunar surface was probably magnetized at the time of the moon's crustal solidification some 3.7 billion years ago. The remanent magnetization in the samples from *Apollo 11* and *Apollo 12* would have required an external magnetizing field stronger than 1,000 gammas; ambient fields of this magnitude have not been measured in space near the moon. The source of the ancient ambient field could have been external to the moon (produced by the sun or the earth) or inside the moon (produced by dynamo action or internal electric currents).

The earth's magnetic field could have magnetized the lunar material if the earth's field was much stronger in the past or if the moon's orbit was once much closer to the earth. If the terrestrial field was never stronger than it is today, the moon would have had to approach to within two or three earth radii in order for the moon to be subjected to a magnetic field of 1,000 gammas; this would be close to the "Roche limit," where tidal forces would break up the moon. For the required magnetizing field to have been produced by an intrinsic lunar dynamo, the moon would have had to possess both a hot core and a fairly high rate of spin at the time the surface material cooled below its Curie temperature (the temperature where magnetization is "frozen" into the material). This hypothesis requires some mechanisms, unknown at present, for lowering the temperature of the lunar interior and slowing the spin rate to their present values. Our knowledge of the moon is still too limited to allow a choice among these and other hypotheses. One hopes that further mapping of the moon's steady fields during future surface and orbital missions will elucidate the moon's "magnetic epoch" and solve one of the more interesting puzzles of lunar history.

Whatever models may be devised to explain this epoch, they will have to take account of the moon's low magnetic permeability. There has been speculation that portions of the lunar interior might consist of a high percentage of iron, in which case the moon could exhibit substantial permeability. Our analysis of magnetometer data from *Explorer 35* and *Apollo 12* shows otherwise. If the moon were significantly permeable, it should measurably distort the lines of magnetic force in the lunar environment. Little or no distortion is observed; the moon's relative permeability appears to be about as low as that of free space. We conclude that the moon does not as a whole possess the properties of a large magnet.

Those of us engaged in developing the magnetometer experiments had hoped that measurements made at the *Apollo 12* site might provide clues to the electrical conductivity of the moon's interior from which we might be able to infer the moon's internal temperature. We could not predict in advance, however, whether readings obtained during the lunar day or the lunar night would be most useful for the purpose. It turned out that the most easily interpreted measurements were those made during the lunar night. During the day the solar wind compresses the 38-gamma permanent field, and this compression is proportional to the highly variable density of the solar wind. Thus if the daytime data are to be accurately analyzed, simultaneous solar-wind effects must be taken into account. During the night the moon can be treated as an electrically conducting sphere in a vacuum.

EDDY CURRENTS AND MAGNETIC FIELDS can be induced in the moon in a manner analogous to the way currents and fields are set up in a passive resistive-inductive secondary of a transformer (*top*). The primary of the transformer has a power source that can drive a current in the coil and change its magnetic field. Field lines from the primary thread the secondary coil, and any change in the primary field results in the formation of a dipole field in the secondary. The time necessary for the decay of the secondary current and secondary field depends on the inductance and resistance of the secondary circuit. By analogy the sun acts as a primary in the sense that it is a source of a magnetic field and has the capacity to change that field. A sudden change in the solar field accordingly results in the induction of a corresponding current and a dipole field in the moon (*bottom*). The decay time of the lunar current and field is a function of both the electrical resistance and the size of the moon. The temperature of the moon's interior can be estimated from calculations of the electrical resistance for various assumed lunar compositions. For example, on the assumption that the moon is composed entirely of pure olivine, the temperature of the moon increases with depth to about 1,000 degrees Celsius in the central region.

Because the sun and the moon are electromagnetically coupled by the solar wind, the sun acts as the primary coil of a transformer and the moon acts as a secondary coil. Any sudden change in the strength of the primary magnetic field causes a change in the electric current and magnetic field induced in the secondary. The time required for the secondary current and field to decay is a function of the electrical resistance and size of the secondary, that is, of the moon itself.

The magnetometer on the surface of the moon at night and the magnetometer aboard *Explorer 35* give us simultaneous readings that can be compared whenever there is a "step" (sudden) transient in the strength of the magnetic field carried by the solar wind [*see illustration at right*]. In general the surface fields that point vertically out of the moon respond slowly to rapid changes in the field external to the moon, whereas fields that point along the surface respond rapidly and are amplified compared with simultaneous values recorded in nearby space.

We have examined the magnetometer readings produced by more than 100 step transients in the solar wind. The analysis clearly shows that the entire moon responds to each transient and that the induced currents start immediately near the surface and diffuse slowly into the deep interior. Our calculations show that the outer portion (about 40 percent of the distance to the center) of the moon has a resistance of approximately 10,000 ohm-meters and that the inner region, or core, has a resistance of only 100 ohm-meters. Comparable resistance values for the earth are respectively .1 ohm-meter and .00001 ohm-meter.

The calculated resistance values of the moon can be used to estimate the temperature at various depths in the moon. For this one uses laboratory measurements of resistance as a function of temperature (and pressure) for the most likely lunar constituents. Because such measurements are made with pure minerals they can at best only approximate the moon's actual composition. If one assumes that the chief constituent of the moon is peridotite (a common mineral in the mantle of the earth), the observed electrical resistance of the moon corresponds to a maximum temperature of between 600 degrees and 1,000 degrees C. for the bulk of the moon's interior. These temperatures are only approximate, but they support the hypothesis that for most of its history the moon has not had a large molten core.

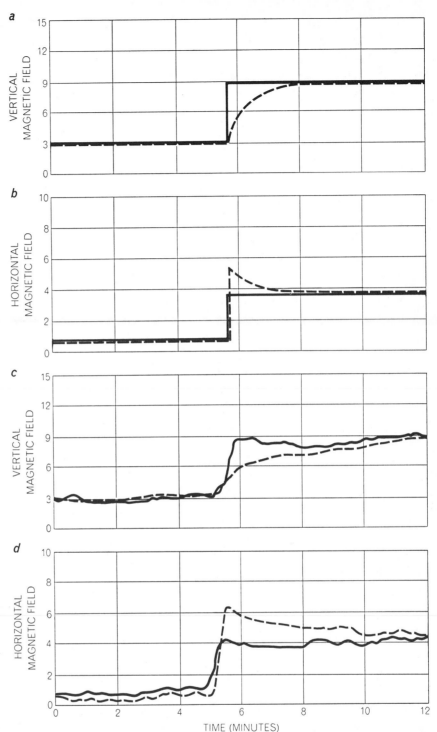

MAGNETOMETER DATA, recorded as a sharp solar-field directional discontinuity passed by the moon, were used to calculate the moon's internal electrical resistance. The data were gathered while the surface magnetometer was on the dark side of the moon, so that the compressive effects of the solar wind did not have to be considered. Theoretical solutions for dark-side surface fields on a moon of homogeneous internal resistance are shown in top two graphs. The solid curves show the "step" solar-field change in two external magnetic-field components (vertical and horizontal to the lunar surface). The dashed curves show the theoretical total surface-field values characteristic of field components that are vertical (*a*) and horizontal (*b*). The bottom two graphs show actual magnetometer data samples. The solid curves represent measurements of a solar-field change recorded by the *Explorer 35* lunar orbiter. The dashed curves are simultaneous vertical (*c*), and horizontal (*d*) components of lunar surface fields measured by the *Apollo 12* magnetometer. The deviations of the data from the theory are believed to be caused by a core of lower resistance than that of the outer layers of the moon. (The deep resistance can be calculated from the magnitude of the deviation.) The distinct difference between vertical and horizontal components illustrates that the moon is responding as an entire body to changes in the solar magnetic field.

# Mars from Mariner 9

Bruce C. Murray

A little more than a year ago, in November, 1971, the complex robot spacecraft *Mariner 9* fired its braking rocket and was captured in orbit around Mars, thus becoming the first man-made satellite of another planet. From its orbital station, ranging between 1,650 and 17,100 kilometers (1,025 and 10,610 miles) above the surface, *Mariner 9* started sending back to the earth a steady stream of pictures and scientific information that was to continue for nearly a year. By the time its instruments had been turned off *Mariner 9* had provided about 100 times the amount of information accumulated by all previous flights to Mars. It had also decisively changed man's view of the planet that generations of astronomers and fiction writers had thought most closely resembled the earth. As a result of the *Mariner 9* mission it is now possible to make plausible conjectures about the geology of Mars, conjectures comparable, say, to those made about the moon in the early 1960's.

It will be recalled that the first close-up pictures of Mars, made in 1965 by *Mariner 4*, revealed a planetary surface whose principal features were large craters reminiscent of the bleak surface of the moon. Four years later the pictures sent back by *Mariner 6* and *Mariner 7* showed that the Martian surface was not uniformly cratered but had large areas of chaotic terrain unlike anything ever seen on either the earth or the moon. In addition a vast bowl-shaped area, long known to earthbound astronomers as the "desert" Hellas, turned out to be nearly devoid of features down to the resolving power of the Mariners' cameras. None of the pictures returned by the first three Mariners showed any evidence of volcanic activity, leading to the view that Mars was tectonically inactive.

This view has had to be drastically revised in the light of the photographs sent back by *Mariner 9*. The new evidence emerged slowly as the clouds of dust that had shrouded the planet for weeks settled. It revealed, among other things, four large volcanic mountains larger than any such volcanic features on the earth. The *Mariner 9* pictures also show a vast system of canyons, tributary gullies and sinuous channels that look at first glance as if they had been created by flowing water. Elsewhere on the planet's surface there is no suggestion of water erosion. That is probably the major mystery presented by the highly successful mission of *Mariner 9*.

Designed and built by the Jet Propulsion Laboratory of the California Institute of Technology, as were the earlier Mariners, *Mariner 9* was crammed with instruments and electronic gear. After burning 900 pounds of retro-rocket fuel, which it had transported 287 million miles in 167 days, *Mariner 9* weighed 1,350 pounds when it finally went into orbit around Mars. The cameras and instruments it carried were designed by several groups of investigators from Government laboratories and more than a score of universities. The television team, to which I belonged, was headed by Harold Masursky of the U.S. Geological Survey and had nearly 30 members. Somewhat smaller groups were responsible for designing and analyzing data from the ultraviolet spectrometer, the infrared radiometer and the infrared interferometric spectrome-ter. Other groups had the task of analyzing the trajectory data (which have provided information about the gravitational anomalies of Mars) and the data provided by nearly 100 occultations of the spacecraft's radio signals (which have yielded new knowledge of the planet's atmosphere and surface).

## The Old Mars

There were strong reasons for the traditional belief in the resemblance between Mars and the earth. Mars rotates once every 24½ hours and its axis is tipped from the plane of its orbit by almost exactly the same amount as the axis of the earth, thus providing the same basis for the seasonal changes in the amount of solar radiation received by the planet's two hemispheres. Mars has white polar caps, originally thought to be composed of water, that alternate from one hemisphere to the other once every Martian year (687 earth days). The planet also exhibits dark and light markings that change on a seasonal basis.

Early astronomers speculated that the dark markings might be vegetation. Later and more cautious workers still found it plausible that Mars had had an early history similar to the earth's, which implied the existence of oceans and an atmosphere with enough water vapor to precipitate and erode the surface. Because of Mars's small mass (a tenth the mass of the earth) and low gravity, such an aqueous atmosphere was assumed ultimately to have escaped, leaving the planet in its present arid state. This view of an earthlike Mars strongly influenced

From *Scientific American*, vol. 228, pp. 49–69, January 1973. Copyright © 1973 by Scientific American, Inc. All rights reserved. Reprinted with permission.

**CLEAR PICTURE OF MARS** was taken by *Mariner 7* at a distance of 395,249 kilometers as it approached the planet in August, 1969. Since it was then winter in Mars's southern hemisphere the south polar cap was at its maximum size. The north pole was shrouded in haze. The bull's-eye-shaped feature in the upper right quadrant, known as Nix Olympica ("Snows of Olympus"), was thought to be a huge crater. Subsequent closeup photographs taken by *Mariner 9* revealed it to be a gigantic volcanic mountain (*see illustration on page 112*).

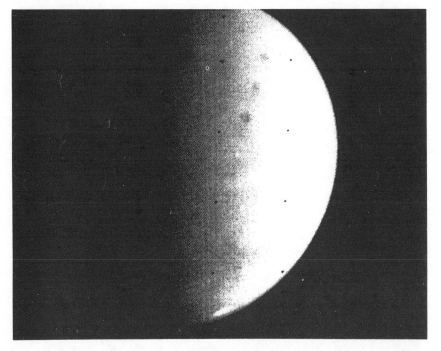

**MARS SHROUDED BY DUST** was photographed by *Mariner 9* at a distance of 400,000 kilometers, a day and a half before reaching the planet on November 13, 1971. The greatest dust storm in more than a century had obliterated all surface features except for the south polar cap, rapidly shrinking with the approach of spring in the southern hemisphere, and four dark spots in the upper right quadrant. The spot nearest the shadow line is the top of the volcanic mountain Nix Olympica. The other three spots are also volcanic peaks.

proposals for the biological exploration of the planet at the beginning of the space age. It seemed reasonable to suppose that life could have originated on Mars much as it had on the earth, presumably as the result of high concentrations of suitable precursor molecules in primitive oceans. Once life had appeared on Mars, microorganisms, at least, could very well have been able to adapt to changing environmental conditions and so could have survived for discovery and analysis by robot devices launched from the earth.

Such expectations were dampened by the findings of *Mariner 4*. Not only did Mars appear bleak and moonlike but also it was found to lack a magnetic field, which could have shielded its surface against energetic charged particles from the sun. Moreover, Mars's atmospheric pressure was found to be less than 1 percent of the earth's, lower by a factor of at least 10 than had previously been estimated. Since the force of gravity at the surface of Mars is more than a third the force of gravity at the surface of the earth, Mars should have easily been able to hold an atmosphere whose pressure at the surface was a tenth the pressure of the earth's atmosphere at the surface.

*Mariner 6* and *Mariner 7* extended these observations. They confirmed that the polar caps are composed of very pure solid carbon dioxide—"dry ice" rather than water ice. The pictures revealing a chaotic terrain suggested that parts of Mars's surface had collapsed and that there had been a certain amount of internal activity. As a result some investigators speculated that the planet might just now be heating up, a circumstance suggested independently by thermal models of the interior. The preponderant view of the Mariner experimenters, however, was still that Mars was basically more like the moon than like the earth. By then the light and dark markings on Mars seen through telescopes were generally attributed to some kind of atmospheric interaction with dust. Indications that the interaction was controlled by local topography were seen in the second set of Mariner photographs, but no general explanation was deduced. Even so some investigators held to the belief that the markings might instead reflect variations in soil moisture.

The 1971 mission to Mars was originally designed to employ two spacecraft, *Mariner 8* and *Mariner 9*, both of which were to be placed in orbit around the planet. The purpose of the two orbiters was to map most, if not all, of the planet's surface at a resolution high enough to reveal both external and internal processes, to study transient phenomena on the surface and in the atmosphere and to provide reconnaissance over a long enough period (from nine months to a year) to observe seasonal changes in surface markings in the hope of clarifying their origins. When *Mariner 8* was lost during launching, the complementary missions of the two spacecraft had to be combined.

When *Mariner 9* reached Mars on November 13, 1971, the greatest dust storm in more than a century was raging on the planet, almost totally obscuring its surface. The first views from a distance of several hundred thousand miles

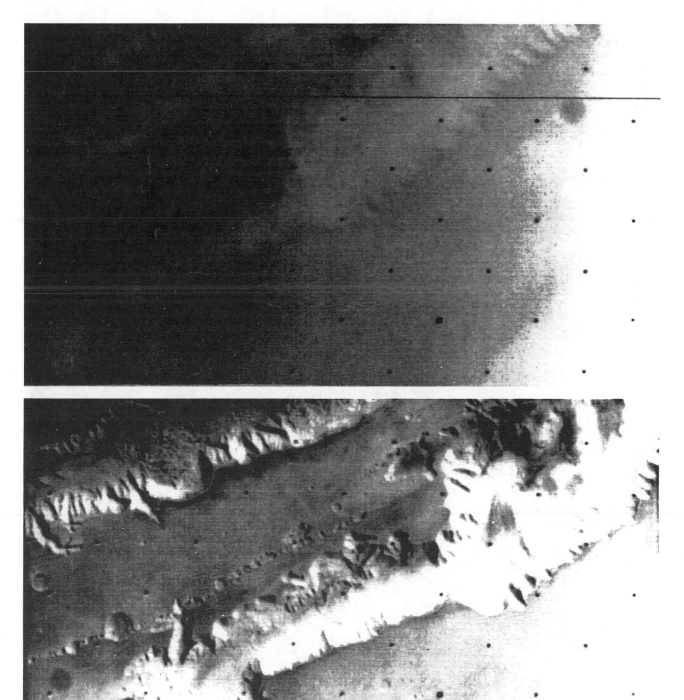

CLEARING OF DUST STORM is shown in these two views of a region in Coprates near Mars's equator. The picture at top, taken when *Mariner 9* had been in orbit 41 days, gives little or no hint of the rugged canyon that was revealed in the bottom picture, taken on *Mariner 9*'s 80th day in orbit. Coprates had long been recognized by astronomers as a feature that changes unpredictably in brightness. It now appears that the canyon looks brighter than the surrounding region toward the end of a dust storm, when the canyon atmosphere is still filled with light-reflecting particles and atmosphere above surrounding plateau has become largely dust-free.

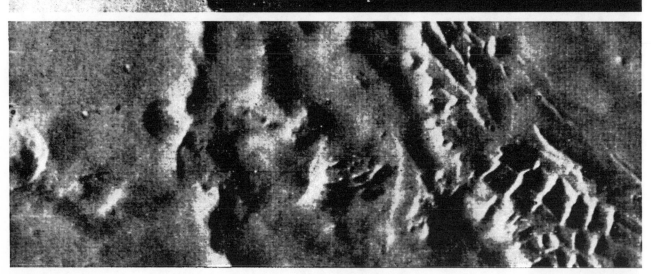

RESULTS OF COMPUTER PROCESSING are demonstrated in these three displays of the same picture showing a region near Mars's south pole. The processing, done within the space of five minutes by J.P.L.'s "mission test computer," was carried out on every picture as it was received from *Mariner 9*. Each image was transmitted as a coded radio signal in which the brightness of each point in the picture was represented by a sequence of nine binary digits rather than as an intensity level (AM signal) or as a frequency tone (FM signal). Each picture frame consisted of 700 lines made up of 832 picture elements per line. The received signal was recorded on magnetic tape and simultaneously displayed on a cathode ray tube. In the initial presentation (*top*) the picture looks rather gray and featureless because the actual contrast on the surface of Mars is quite low. The first stage of computer processing yields the shading-corrected version (*middle*), which indicates the relative brightness in the scene along with some enhancement. The right-hand portion of the picture is dark because it is beyond the terminator and hidden in shadow. A second level of processing (*bottom*) suppresses large-scale differences in brightness in order to enhance preferentially the topographic detail in both the light and the dark areas. The picture is now seen to reveal a complex of transecting ridges two to five kilometers apart that the Mariner television group nicknamed Inca City. The ridges seem to be composed of a resistant material that has filled cross-cutting fractures and that later has been exposed by erosion of the surrounding material. *Mariner 9* took the picture at a distance of 2,937 kilometers.

revealed essentially no detail except a glimpse of the south polar cap [*see bottom illustration on page 104*]. The dust storm delighted the investigators who wanted to study the planet's atmosphere, since it promised to reveal how particles were transported by such a thin medium, but it was a disappointment to the investigators concerned with surface features. For example, there had been plans to take a sequence of far-encounter pictures, ultimately to be printed in color, showing the planet getting larger and larger, thereby providing a visual bridge between the level of detail seen through telescopes from the earth and the detail eventually visible in pictures taken from orbit around Mars. A limited effort to produce far-encounter pictures showing Mars in natural color had been made with images taken by *Mariner 7* through separate red, green and blue filters.

### The Great Volcanoes

The dust storm delayed the systematic mapping of the Martian surface for nearly three months. Even during the storm, however, four dark spots in the equatorial area were repeatedly seen in the early pictures taken from orbit. The spots clearly represented permanent surface features high enough to stick up through the dust. Presumably they looked dark simply because their surface reflectivity was lower than that of the bright, dusty atmosphere.

One of the four spots corresponded to the location of Nix Olympica ("Snows of Olympus"), so named because it was normally visible from the earth as a bright feature and also a variable one. When this dark spot was observed with the high-resolution, or narrow-angle, camera on *Mariner 9*, the image that emerged was breathtaking. What one saw was the characteristic pattern of coalesced craters that constitute a volcanic caldera. Such calderas are not uncommon on the earth, for example in the Hawaiian Islands. The Martian caldera, however, was 30 times larger in diameter than any in the Hawaiian chain. When the dust had settled, Nix Olympica was seen in full to be an enormous volcanic mountain more than 500 kilometers in diameter at the base, much larger than any similar feature on the earth; the caldera was only the summit [*see illustration on page 112*]. Atmospheric-pressure maps made later with the aid of the ultraviolet spectrometer and other techniques show that Nix Olympica is at least 15 kilometers high and possibly 30. For purposes of com-

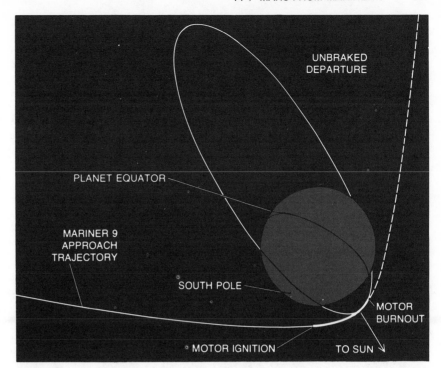

CAPTURE OF *MARINER* 9 BY MARS took place on November 13, 1971, after the spacecraft's retro-rocket had generated a decelerating thrust of about 300 pounds for 15 minutes 15.6 seconds. The 287-million-mile voyage from the earth had taken 167 days. *Mariner 9* approached Mars from below and went into an orbit inclined at an angle of about 64 degrees to the planet's equator. The initial orbit ranged from 1,385 kilometers to 17,300 above the planet's surface. A subsequent correction reduced the orbit's high point to 17,100 kilometers and raised the low point to 1,650, achieving the desired orbital period of 11.98 hours.

parison, Mauna Loa, the tallest volcanic cone in the Hawaiian Islands, rises less than 10 kilometers from the floor of the Pacific. High-resolution photography revealed that the other three dark spots were also volcanoes, somewhat smaller than Nix Olympica, strung together to form a long volcanic ridge. Following the traditional name for that area, it is now called Tharsis Ridge [*see illustration on the following two pages*].

The first recognizable features photo-

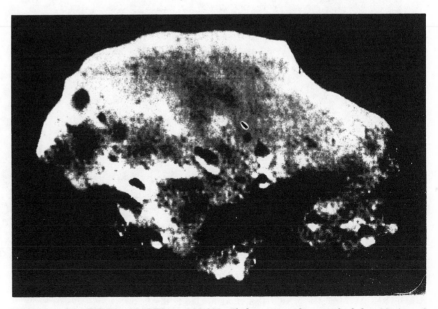

HEAVILY CRATERED MARTIAN MOON, Phobos, was photographed by *Mariner 9* from a distance of 5,500 kilometers. The inner of the two Martian satellites, Phobos is about 25 kilometers long and 20 kilometers wide. Deimos, the outer satellite, is about half the size of Phobos. The picture of Phobos has been greatly improved by computer processing.

graphed by *Mariner 9* presented a fascinating question: How can one explain why one entire hemisphere of the planet, the hemisphere observed by three earlier Mariners, shows scarcely any evidence of internal activity, whereas the first area to be investigated in detail on the opposite side of the planet has four enormous volcanoes? The explanation apparently is that Mars is just beginning to "boil" inside and produce surface igneous activity. Presumably this process is now well advanced in the Nix Olympica–Tharsis Ridge area but has not yet spread to the planet as a whole. We may be witnessing on Mars a phase similar to one the earth probably went through early in its history, a phase whose record has been totally erased by subsequent igneous and sedimentary processes.

The rate at which a planet's interior heats up depends on a number of factors, chiefly the amount of radioactive material in its original accreted mass and the total mass, which determines the pressure in the interior and the degree of in-

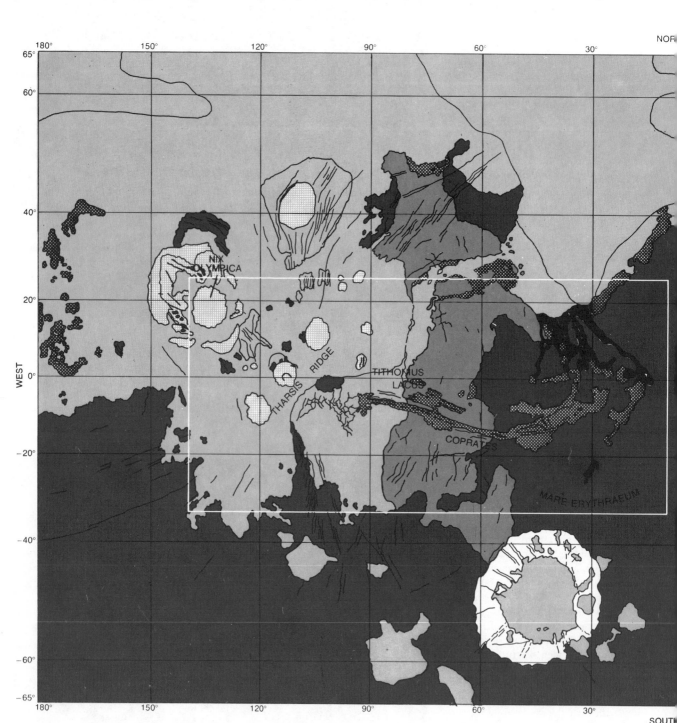

**GEOLOGICAL MAP OF MARS** is an effort to classify the features that compose the surface terrain. The white rectangle encloses the region depicted in the mosaic on the next two pages. The areas in light gray are smooth plains; areas in medium gray are cratered plains; dark-gray areas are old cratered terrain. White areas are mountainous terrain. Areas in light cross-hatching are volcanic, embracing Nix Olympica and the three volcanoes that form Tharsis Ridge. Areas in dark cross-hatching identify terrain that has been modified in some fashion. Areas in black are channel deposits. Short hatch lines are inferred faults. A number of features referred to in the text or depicted

sulation. In very general terms, if Mars had the same original composition as the earth, one would expect it to heat up more slowly because it has only a tenth the mass of the earth. The sheer size of Nix Olympica suggests the possibility that deep convection currents are churning, a process that conceivably could lead some hundreds of millions of years

hence to the kind of plate-tectonic phenomena responsible for the slow drift of continents on the earth.

Immediately to the east of the volcanic province is a highly fractured area, and beyond that another extraordinary topographical feature was discovered: a series of huge canyons stretching east and west along the equator [see illustra-

tion on pages 110 and 111]. These canyons, 80 to 120 kilometers wide and five to six and a half kilometers deep, are much larger than any found on the earth. Again we must assume that their origin is due to fairly recent internal activity. Presumably large-scale east-west faulting has exposed underlying layers of the planet whose composition could con-

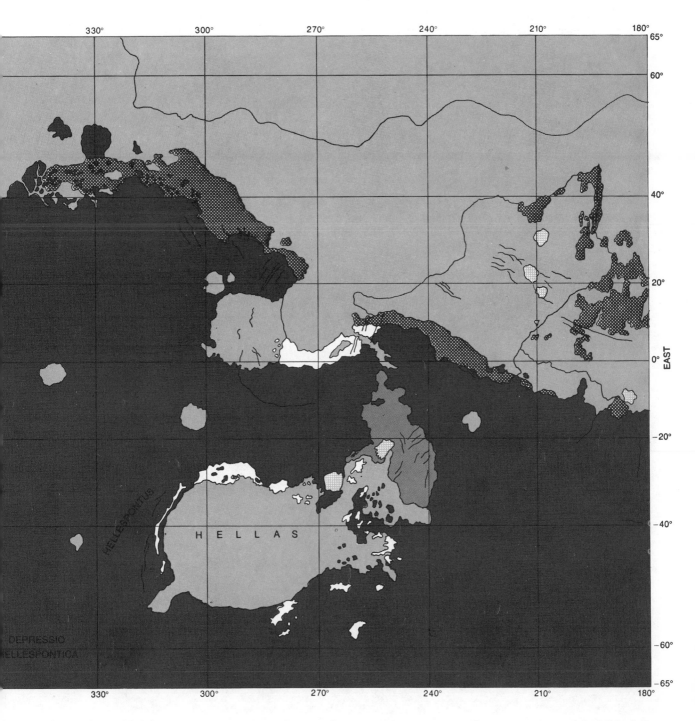

in photographs are labeled. A canyon in Coprates is shown in the illustration on page 105. A canyon in Tithonius Lacus appears at the top of page 113 and a crater in Hellespontus at the bottom of the same page. A sinuous valley in Mare Erythraeum is shown in the photograph on page 115. Hellas is a nearly featureless bowl more than

1,600 kilometers across. The map was prepared by Michael Carr, John F. McCauley, Daniel Milton and Don Wilhelms of the U.S. Geological Survey in cooperation with several members of television-experiment team of Mariner 9 project.

ceivably trigger an erosion process of some kind.

One speculation is that deep permafrost is involved, associated perhaps with the arrival near the surface of juvenile water preceding and accompanying the rise of molten rock near the surface of the planet during the volcanic episode apparent to the west. Mars is everywhere below freezing just a short distance below the surface. Once permafrost was exposed to the atmosphere, its water content would sublimate, making avail-

able a loose, friable material sufficiently mobile to serve as an eroding agent in a mass-transport process. We must then ask where the material went. One possibility is that the winds of Mars have transported it as dust to other localities. (Although the Martian atmosphere is thin, its winds may blow at several hundred miles per hour.) Alternatively, the missing material may yet be discovered somewhere to the east of the canyons. Still a third possibility is that it may even have disappeared into the planet's

interior in a complex exchange process.

The largest of the canyons corresponds to a feature long known as Coprates, whose appearance sometimes changes with the seasons. By observing this canyon as the dust storm ended we were able to gain insight into its variable appearance. The canyon is so deep that considerable dust persists in the atmosphere between the canyon walls after the atmosphere above the surrounding region is comparatively dust-free [see illustration on page 105]. The

MOSAIC PANORAMA was made from several dozen *Mariner 9* wide-angle photographs specially computer-processed and matched by Raymond Batson of the U.S. Geological Survey at Flagstaff, Ariz. The mosaic depicts a region extending some 7,000 kilometers along the Martian equator. The area covered by the panorama is indicated on the geological map of Mars on the preceding two pages. The superposed outline map of the U.S. shows vividly the dimensions of the great canyons that parallel the equator for 4,000 kilo-

dust-filled atmosphere makes the canyon look brighter than the surrounding landscape. Once the canyon atmosphere clears up, there is little contrast between the interior of the canyon and the surrounding area. Hence the variable "surface" markings associated with Coprates probably have nothing to do with the surface at all.

Similar atmospheric processes may well explain some of the other variable markings formerly attributed to seasonal changes on the surface. Other kinds of variation are not so simply explained, but evidently they always involve the interaction of dust, topography and atmosphere.

Like the volcanoes, the great canyons of Mars suggest a fairly recent episode in the history of the planet characterized by large-scale events. On the earth one often finds a reasonably steady state between processes of erosion and processes of restoration; thus one sees a range of surface morphologies from youthful to mature. In the case of the Martian canyons erosion does not seem to be balanced by a corresponding restoration; we do not see old degraded canyons with a mature form.

## The Channels

The eastern extremity of the canyons joins a large area of chaotic terrain, a small portion of which was glimpsed by *Mariner 6*. The appearance of the chaotic terrain strongly suggests that it is the result of some kind of collapse and

meters. Much larger than anything like them on the earth, the canyons average 100 kilometers in width and reach a depth of more than six kilometers. At the extreme left of the mosaic one can see the giant volcanic mountain Nix Olympica and immediately to the right the series of somewhat smaller volcanoes that form Tharsis Ridge. The sun was generally shining from the lower left when the pictures were taken, so that the shadows fall toward the upper right. For some, relief will be stronger if mosaic is turned upside down.

that the collapse is genetically related to the canyons to the west. Extending out from the chaotic terrain in a north-westerly direction are some extraordinary channels, which are also found in a number of other localities on the planet. It is hard to look at these channels without considering the possibility that they were cut by flowing water. Indeed, some of my colleagues think that is the only reasonable explanation.

One can estimate the age of the channels by noting the size-to-frequency relation of impact craters on their floors. The channels are clearly younger than the crater-pocked terrain seen over much of the planet, yet they are by no means the youngest features of the Martian landscape.

The discovery of the channels has revived speculation that there may have been an earthlike epoch in the history of Mars. According to this view Mars may once have had a much denser atmo-

**CLOSE-UP OF NIX OLYMPICA** is shown in a mosaic of *Mariner 9* photographs that have been specially computer-processed and matched. The picture is printed with north at the right, so that the sunlight seems to come from the top of the page; the feature is thus seen immediately as a cone and not as a depression. The vol-canic mountain, rising from a great plain, is 500 kilometers across at the base, or much larger than similar volcanic mountains on the earth. Pressure mapping of the depth of the atmosphere indicates that Nix Olympica is 25 kilometers high. The main crater, a complex volcanic vent known as a caldera, is 65 kilometers in diameter.

sphere and water vapor in such abundance that rain could fall. Given rainfall, the channels could be easily explained. Less easily explained is why channels have survived in only a few areas and why older topography shows no evidence of water erosion. It would seem difficult to explain how the primitive Martian atmosphere, probably dry and reducing (in the chemical sense), could have evolved into a dense, wet one and then have been transformed again into the present thin, dry atmosphere consisting almost entirely of carbon dioxide. Moreover, the present atmosphere is strongly stabilized by the large amounts of solid carbon dioxide in the polar regions. If the channels were created by rainfall, it would seem that one must postulate two miracles in series: one to create the earthlike atmosphere for a relatively brief epoch and another to destroy it.

An alternative hypothesis presents at least as many difficulties. It is suggested that liquid water accumulated in underground reservoirs following entrapment and melting of permafrost. Hypothetically the reservoirs were abruptly breached, allowing the released water to create the channels. The observed channels are so large and deep, however, that a great volume of water must have been involved in their formation. Therefore it would seem even more difficult to ascribe the channels to a "one shot" open-cycle process than to a closed-cycle process such as rainfall.

The origin of the canyons and channels is one of the primary enigmas that has emerged from the *Mariner 9* mission. Because of the importance of liquid water to life as we know it, the possible role of water in creating the canyons and channels has attracted particular interest.

Finally, there are a few areas on Mars to which the term "basins" seems appropriate. The most prominent is the large circular feature Hellas, more than 1,600 kilometers in diameter. Hellas has been observed from the earth for more than two centuries. Sometimes it rivals the polar caps in brightness. *Mariner 7* demonstrated that Hellas is indeed a low-lying basin virtually devoid of features. It has been deduced from closeup photographs that the surface of Hellas has probably been smoothed by the influx of large amounts of dust carried into the basin by wind. *Mariner 9*, however, revealed that Hellas exhibited a few faint topographical features just as the planetwide dust storm was ending. This suggests that variations in the brightness of

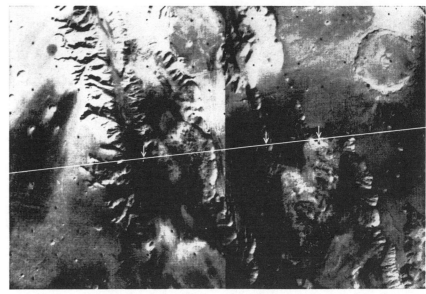

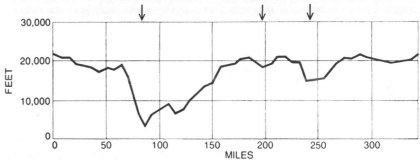

CHASM IN TITHONIUS LACUS is in the extreme western part of the canyon system that runs east and west near Mars's equator. An ultraviolet spectrometer on *Mariner 9* measured the atmospheric pressure at the planet's surface along the track indicated by the white line. The pressure reading was then converted into the jagged depth profile that appears below the pair of photographs. The difference between the highest and lowest points is more than 6,000 meters, making the chasm four times as deep as the Grand Canyon in Arizona. Tithonius Lacus is immediately south of the Martian equator at about 85 degrees west longitude.

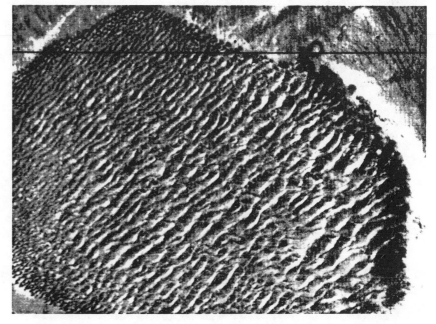

DUNE FIELD some 70 kilometers across was discovered by *Mariner 9*'s narrow-angle camera inside a crater 150 kilometers wide in the region known as Hellespontus. Ridges are one and a half kilometers apart. Dune fields as spectacular as this one appear rare on Mars.

Hellas may be due to frequent dust storms of a more local nature, a view originally adduced on meteorological grounds by Carl Sagan of Cornell University and his co-workers. Thus Hellas probably acts as a long-term collection basin for dust but may also serve as a source of dust when the Martian winds blow particularly hard. The observation by *Mariner 9* of small-scale dust storms and the recognition that they can alter the brightness of local areas give us further insight into some of the variable features that have been observed over the years from the earth.

One of the crowning achievements of *Mariner 7* was the study at high resolutions of the very large polar cap present during the southern winter of 1969. Measurements of reflectivity and temperature provided by an infrared spectrometer and an infrared radiometer on the spacecraft proved conclusively that the south polar cap was composed of very pure solid carbon dioxide, as had been predicted some years earlier. The photographs showed that the frost cover was thin (probably less than a few meters on the average) and that a variety of un-

usual surface features were also present in the vicinity of the south pole.

When *Mariner 9* reached Mars, it was late spring in the southern hemisphere, an ideal time for monitoring the wasting of the dry-ice cap and for examining in detail the unusual surface features that should have been further revealed. The disappearance of the south polar cap started out as expected but then clearly showed anomalous behavior. Curiously, the general outline of the shrunken cap persisted throughout the late summer, when the sublimation of the carbon dioxide should have been at a maximum [*see top illustration on page 117*]. This suggested to me that after the large annual cap of carbon dioxide has sublimated, it exposes a residual cap of ordinary water ice. Ordinary ice, of course, has a much lower evaporation rate than carbon dioxide, and traces of water vapor are present in the Martian atmosphere.

*Mariner 9*'s pictures also disclosed a most peculiar terrain in the south polar area, which we named laminated terrain. Although its outline is not symmetrical, it covers much of the south

polar region up to about 70 degrees south latitude. The laminated terrain is composed of very thin layers, alternately light and dark, whose gently sloping faces exhibit a certain amount of texture, or relief [*see bottom illustration on page 117*].

The thin laminas appear to be collected in units of 20 or 30 or more to constitute plates perhaps half a kilometer or more in thickness and up to 200 kilometers across. The plates have outward-facing slopes in which a banded structure can be seen. The laminar deposits have been found only in the polar regions, where carbon dioxide forms an annual deposit of frost. This suggests that the laminations are associated in some way with the coming and going of volatile substances and that they may even retain some solid carbon dioxide or water ice. Since the laminations are marked by very few impact craters, one can deduce that they are a recent development in the history of Mars.

### The North Pole

The north polar region of Mars finally

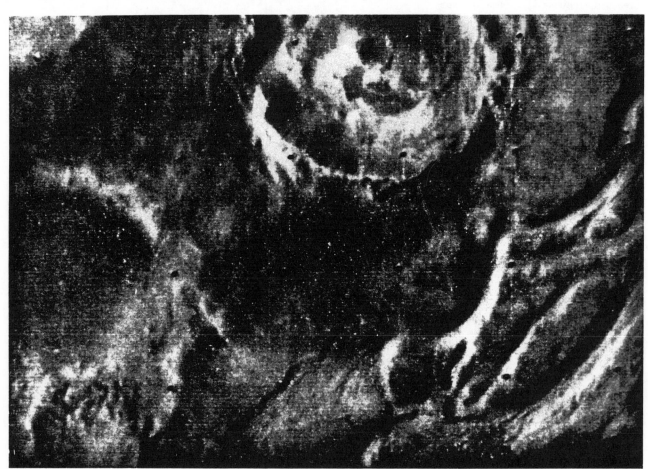

**BRAIDED CHANNEL** adjacent to an impact crater 20 kilometers in diameter is an example of the type of feature that suggests there has been fluid erosion of some kind on Mars. If the agent was actu-ally water, it is hard to understand why eroded terrain is confined to only a tiny fraction of the planet's total surface. The region shown is at six degrees south latitude 150 degrees west longitude.

became available for observation by *Mariner 9* rather late in the mission as a result of gradual changes in lighting, associated with the change in season and with the lifting of the haze that characteristically develops in the fall over each pole. The quasicircular structures characteristic of laminated terrain were found to be even more abundant around the north pole than around the south pole. One can see 20 or 30 individual plates arranged in a pattern reminiscent of fallen stacks of poker chips. The existence of the laminated terrain and circular-plate structures in the north polar regions as well as in the south polar ones indicates beyond any reasonable doubt that their formation must be associated in some way with the periodic deposition and evaporation of volatile material.

Michael C. Malin, a graduate student at the California Institute of Technology, and I have speculated that the distribution of the circular plates and their overlapping arrangement can be explained by changes in the tilt of Mars's axis. We posit that the rotational axis of the planet has been displaced over the

past tens of millions of years as a result of convection currents deep in the mantle, currents that are probably associated with the production of volcanoes in the equatorial areas. As the spin axis has shifted, the laminated plates have formed concentrically around each successive position of the poles.

This speculation is at least consistent with information about Mars's gravity distribution deduced from changes in *Mariner 9*'s orbit. The planet exhibits gravitational anomalies suggestive of deep density differences of the kind that could be associated with deep convection. Moreover, there is a strong correlation between the gravitational anomalies and the location of the equatorial volcanoes.

The regular appearance of the laminas and of the plates themselves suggests that they are also associated in some way with periodic alternations of the climate of Mars. In collaboration with two other graduate students, William Ward and Sze Yeung, I have investigated the theoretical variations in the orbit of Mars over a period of time. We find that perturbations in the orbit

caused by other planets, analyzed a number of years ago by Dirk Brouwer and G. M. Clemence, alter the orbit's eccentricity in a way that turns out to be quite favorable for our hypothesis. The eccentricity of Mars's orbit varies from .004, or nearly circular, to .141. Its present value is .09 [*see top illustration on page 120*].

The consequence of this variation in eccentricity is a variation in the yearly average amount of sunlight reaching the poles of the planet, together with a much stronger variation in the maximum solar flux when the planet is closest to the sun [*see bottom illustration on page 120*]. Although the variation in average radiant input at the poles is only a few percent, it is sufficient under some circumstances to cause a cyclical variation in the growth and sublimation of permanent carbon dioxide frost caps. Assuming that dust storms regularly deposit dust during the sublimation phases, thin laminas of the type observed could be produced. The plates themselves would therefore correspond to a periodicity of the order of two million years. Hence the laminated terrain seems to closely

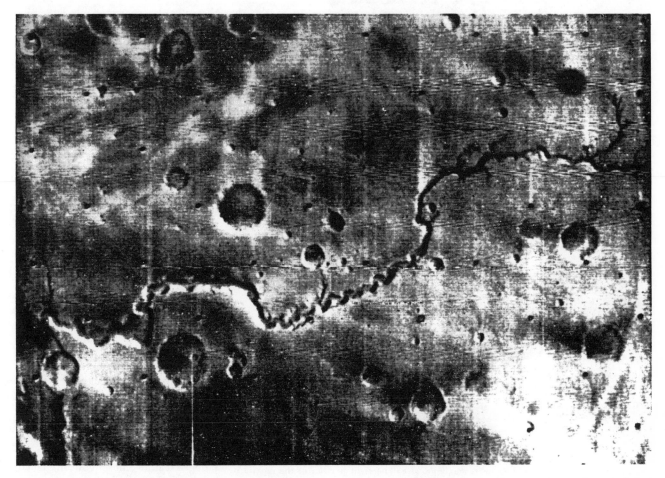

**SINUOUS VALLEY** 400 kilometers long and up to six kilometers wide is located at 29 degrees south latitude 40 degrees west longitude in the region known as Mare Erythraeum. The feature resembles the outline of a meandering river or one of the lunar rilles, which may have been created by flowing lava. The origin of such features on Mars has not been satisfactorily explained.

reflect the fluctuations in average radiant flux reaching the planet both in the short run (roughly 90,000 years) and in the long run (two million years). Inasmuch as a total of 20 or 30 plates are visible in the northern hemisphere the laminated terrain constitutes a record reaching back something like 100 million years. An alternative view regards the origin of the laminated terrain as being primarily erosional rather than constructional.

## Evolution of the Atmosphere

If all the polar laminations accumulated in only a few hundred million years at most, representing no more than the past 5 percent of the history of Mars, what happened earlier? One encounters a basic difficulty in understanding Mars if one tries to apply the famous dictum of the 18th-century geologist James Hutton: "The present is the key to the past." Whether one looks at the volcanic terrain, the canyon lands, the channels or the polar laminations, all seem to record a remarkable degree of activity and change during the most recent part of Mars's geological past. I was led by these considerations to wonder if it is possible that the atmosphere of Mars as we know it may be a fairly recent acquisition. Malin and I are presenting this view as a "contentious speculation." It may be that Mars had no atmosphere at all, or only a very thin, unimportant atmosphere throughout the middle period of its history, lasting perhaps several billion years. Presumably there was an initial primitive atmosphere associated with the accretion of the planet, but this atmosphere may have been lost quite early, particularly if it consisted chiefly of hydrogen and methane.

We think a significant fraction of the mass of the present Martian atmosphere was released during the formation of Nix Olympica and the other three volcanoes in the Tharsis Ridge area. The existence of widespread blankets of rock material and other evidence of somewhat earlier and more extensive volcanism and sedimentation suggest that rather large volumes of volatile substances have issued from the planet's interior in the later geological episodes. Thus it may be that as Mars matured enough to boil it simultaneously began to produce an enduring atmosphere. The atmosphere in turn has produced the laminated terrain and provided the wind-transport and erosion mechanisms to form the channels and the great canyons. On this hypothesis Mars is still far from reaching a steady state in which erosion would be balanced by modification, with a resulting development of a spectrum of morphological features. As part of our contentious speculation one might even imagine that in the early stages of the development of the present Martian atmosphere, before the polar cold traps for carbon dioxide were well established, enough water might have been brought to the surface to have flowed down the channels under peculiar, nonrecurring conditions. At least this possibility avoids the problem of positing two miracles in series and lets us settle for just one if ultimately liquid water is really required to explain the genesis of the channels.

The young-atmosphere hypothesis may also help to explain why "permanent dark areas" (for example the two-pronged feature Sinus Meridiani) should survive in the face of the frequent planet-wide dust storms. Again Mars some-

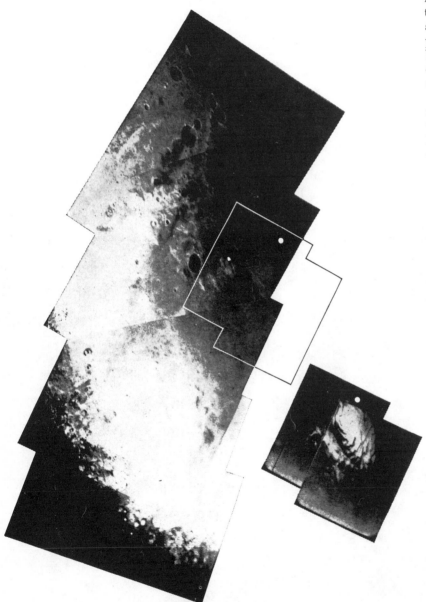

**SOUTH POLE IN DIFFERENT SEASONS** can be compared in these views taken by *Mariner 7* in August, 1969 (*mosaic at left*), and by *Mariner 9* in November, 1971 (*right*). The area covered by the two *Mariner 9* pictures is indicated by the shape superposed on the *Mariner 7* mosaic. The geometric south pole is shown by a white dot in both views. In August, 1969, it was winter in the southern hemisphere of Mars and the frost cap of dry ice (solid carbon dioxide) was close to its maximum size. The right portion of the mosaic is dark because it is in shadow. When the *Mariner 9* pictures were taken 27 months later, the frost cap was shrinking rapidly with the approach of spring in Mars's southern hemisphere.

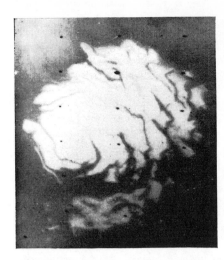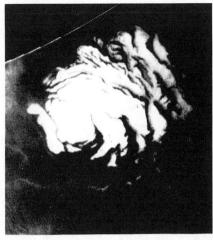

SHRINKING OF SOUTH POLAR CAP is depicted in these three views taken by *Mariner 9.* The original images have been enhanced, stereographically rectified and printed at the same scale. Viewed in the usual order, the pictures show the appearance of the frost cap after the spacecraft had been in orbit 14, 36 and 94 days. The first picture is still somewhat hazy because of the planet-wide dust storm. It can be seen, nevertheless, that the outline of the surface frost changed significantly in the three weeks between the first two views but then changed surprisingly little in the next eight weeks, when sublimation of the frost cap should have been at a maximum.

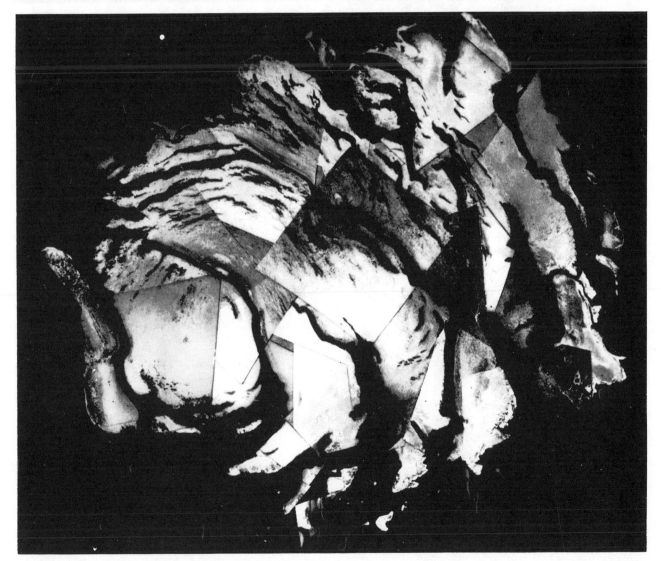

MOSAIC OF RESIDUAL SOUTH POLAR CAP was assembled (by Laurence A. Soderblom of the U.S. Geological Survey Center of Astrogeology in Flagstaff, Ariz.) from closeup pictures taken between the 58th and 94th day of orbital flight. The residual cap, which measures about 300 by 350 kilometers, is centered about 200 kilometers from true south pole (*white dot*) at about 45 degrees west longitude. Shades of gray have been optimized to bring out detail in frost-covered areas, hence frost-free surface appears black.

**LAMINATED TERRAIN NEAR NORTH POLE** resembles that in the south polar region. The author speculates that the distribution of circular plates and their overlapping arrangement may be evidence that the tilt of the axis of Mars has shifted over the past 100 million years. Such a shift could have resulted from convection currents deep in mantle of planet.

how does not seem to us to be in a steady state, although others do not share our viewpoint. According to our hypothesis the dark markings may be the site of older surface materials not yet affected by the chemical weathering associated with the new atmosphere. In fact, there seems to be some correlation between the permanent dark markings and the terrain populated by the oldest craters.

Other *Mariner* 9 investigators, such as Sagan and W. K. Hartmann, have developed a quite different view of Mars's history. The nature of the old cratered terrains suggests to them that a long period of atmospheric erosion preceded the spectacular events of the more recent past. Thus the concept of an earthlike Mars is not by any means dead. Nonetheless, concepts of the geological history of Mars are changing rapidly in the light of *Mariner* 9's highly successful mission. Perhaps ultimately some intermediate interpretation will fit the observations best.

### Is There Life on Mars?

The present *Mariner* 9 results suggest to me, however, a view very different from that of the early astronomers who thought that Mars was once earthlike and is now a dried-up fossil. I would argue that Mars is probably just now starting to become earthlike with the development of a durable atmosphere. "Just now" is hard to pin down quantitatively because the dates assigned to the craters on the basis of meteor flux rates are still highly uncertain. My guess now would be that the atmospheric "event," if it really happened, took place within the past quarter of Mars's history and certainly within the past half. If this contentious speculation should become widely accepted, it would necessarily imply pessimism about the possibility that past conditions were favorable for the appearance of simple forms of life on Mars. If Mars indeed was like the moon and lacked a significant atmosphere for much of its history, and if the maximum amount of water on the surface has been at most enough to create a few channels, it seems highly unlikely that there has ever been a sufficient accumulation of liquid water in the surface layers of Mars to allow the accidental development of life from prebiological organic materials. On the other hand, life-on-Mars enthusiasts argue otherwise and emphasize that if water has been available at all in surface layers, it would provide a favorable environment for the development of life.

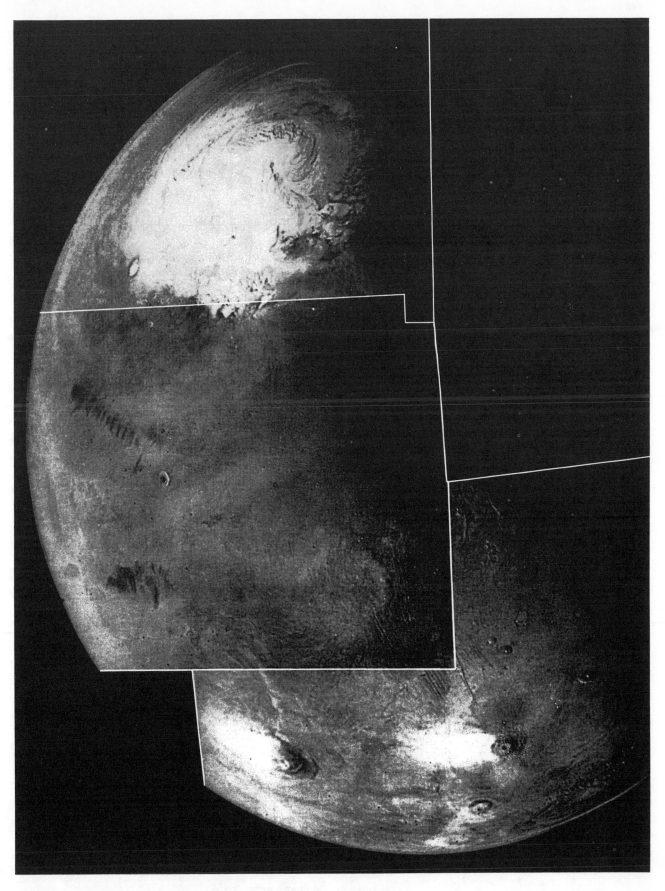

**NORTHERN HEMISPHERE OF MARS** was photographed in three frames taken only 84 seconds apart by swiveling the wide-angle camera aboard *Mariner 9*. The bottom frame clearly shows Nix Olympica, the volcanoes of Tharsis Ridge and at the lower right the huge canyon that lies just below the equator. The pictures, taken on August 7, 1972, at an altitude of 13,700 kilometers, were among the last of the 7,273 produced by *Mariner 9*'s two cameras. Clouds of water ice or of carbon dioxide crystals obscured the planet north of the 50th parallel until the final weeks of the mission. The north polar cap is shrinking during the late Martian spring.

120

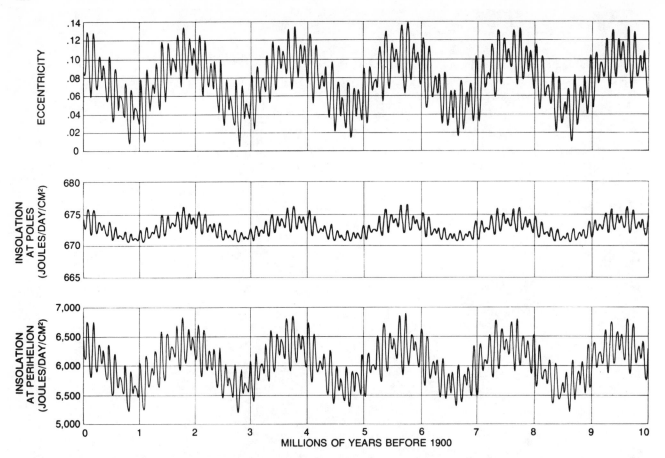

CHANGES IN ORBIT OF MARS over the past 10 million years may account in part for the peculiar circular plates and laminations observed in the planet's polar regions, according to a hypothesis developed by the author and his students. The eccentricity of Mars's orbit (*top curve*) has varied from .004, or nearly circular, to .141. This would lead to changes in the average amount of solar energy reaching the poles (*middle curve*) and in the peak energy reaching the planet when it was closest to the sun (*bottom curve*).

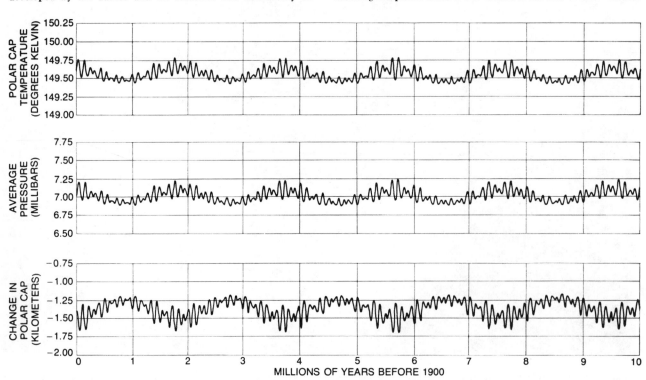

CHANGES IN SOLAR INPUT TO MARS, due to the changing eccentricity of the planet's orbit, could produce cyclical variations in the growth and sublimation of the carbon dioxide frost caps. The top curve shows the long-term changes in the temperature of solid carbon dioxide at the poles. The middle curve shows variations in average atmospheric pressure at the planet's surface that might have resulted from the variations in insolation. The bottom curve shows changes in the height of the permanent frost cap.

Obviously such a debate cannot be settled by the kinds of information collected by *Mariner 9*. The answer must wait for sophisticated chemical and mineralogical analysis of the surface soil itself.

Simultaneous with the flight of *Mariner 9* the U.S.S.R. undertook an ambitious mission whose objective was to land a capsule on Mars and conduct some analyses of the surface. Unfortunately the Russian lander *Mars 3* failed shortly after reaching the surface and apparently transmitted no useful information. I expect the U.S.S.R. to repeat this kind of mission late in 1973, and I look forward to seeing pictures from the surface and probably the results of some simple chemical analyses.

One hopes that by 1976 we shall be getting information back from a complex U.S. lander being developed in the Viking program and possibly from a second-generation Russian lander. The Viking capsule is being designed not only to look for organic compounds directly but also to perform some simple but important determinations of the basic inorganic composition of surface minerals. Such measurements will provide an important clue to the past chemical evolution of the surface minerals, including whether or not they have reacted chemically with water.

Will the Viking lander finally tell us if life now exists or ever did exist on Mars? I personally rather doubt it. I think the difficulty of obtaining an unambiguous yes or no is so great that it is probably beyond the grasp of even an investigation as ambitious as the Viking mission. My own view is that the final answer to the great search for life on Mars may have to await the return, probably by unmanned means, of a sample of Martian soil for sophisticated analysis in terrestrial laboratories. My guess is that the U.S.S.R., having demonstrated the ability to return unmanned samples from the moon, should be in a position to repeat this feat for Mars around 1980 (assuming that it continues to give unmanned space exploration the same priority that it has had in the past). The U.S. has no ambitious plans for exploring Mars beyond Viking. Thus *Mariner 9* will long be remembered as one of the high points in the American exploration of Mars.

# 15

# Update on Mars: Clues about the Early Solar System

William D. Metz

With the widening exploration of the solar system, geology is becoming a science of many more planets than the earth. For several years the term "lunar science" could adequately encompass extraterrestrial geology. But since the spacecraft Mariner 9 produced thousands of detailed pictures of Mars in 1971 and 1972, scientists have new information on a full-sized planet to consider [*Science* 179, 463 (1973)]. At a recent international colloquium (*1*) in Pasadena, California, most of the scientists who have studied Mars, as well as many lunar and earth scientists, discussed what is now known about Mars and tried to formulate the outstanding questions, as well as the best ways to answer them.

Mariner 9 has shown that many features on Mars are strikingly different from features on the earth. The volcanic structure, Nix Olympica, is far larger than any similar feature found on the earth—almost 500 km across at the base. The great canyon, Coprates, stretches about one-sixth of the way around the Martian equator. Huge polar caps advance and retreat by as much as 35° latitude every season. Extremely violent dust storms occur on Mars, but its atmosphere is much thinner than the earth's. Dry channels are frequently found in the equatorial regions of Mars, yet there is no evidence of any fluid on the surface now. Extensive systems of faults indicate tensions in the planet's crust, but Mars does not seem to have tectonic activity like the earth.

In spite of the differences between Mars and the earth, it is becoming increasingly clear that a common framework of ideas is useful to describe the two planets, as well as the moon. All three seem to have been formed at the same time, more than 4 billion years ago, and to have undergone considerable geological activity, which might be understood in terms of the thickness of their lithospheres. The moon appears to have a very thick lithosphere, 300 to 1000 km deep, and therefore no tectonic activity. The plates in the earth's lithosphere are about 50 km thick, so they can be easily heated and moved by forces underneath.

At the recent colloquium, several speakers suggested that the activity of Mars is intermediate between that of the earth and the moon, and that Mars has a lithosphere about 200 km thick. In the Tharsis region, delineated by a ridge of three volcanoes, there is a fracture system which is characteristic of uplift or doming of the lithosphere, according to Michael Carr, of the United States Geological Survey (USGS) in Menlo Park, California. But if the lithosphere were spreading in the Tharsis region, it would have to undergo subduction at some other place on the planet, in effect burying itself there. There is no evidence of subduction, so apparently no plate movement is taking place. But evidence of less dramatic activity is abundant. Areas of fretted terrain, which are examples of tectonic fractures, are quite prevalent.

While geological activity varies immensely among planets and planetesimals with different sizes and compositions, the record of crater formation on Mars and the moon seems to suggest that many planets share the same history of bombardment by interplanetary debris. One of the most exciting developments of the recent colloquium was that two researchers, who had previously made different estimates of the cratering history of Mars, now seem to agree that the Martian cratering history is almost identical to that of the moon. Cratering records are compiled by counting the number of craters of a given size on parts of the surface with different ages. Ages on the moon have been precisely determined from samples retrieved by Apollo missions, and a curve that indicates extremely heavy bombardment of the moon about 4 billion years ago is well established (Fig. 1). No samples are available from Mars, but by assuming that the geological evolution of Mars has been fairly uniform, a cratering history for Mars can be derived. The curve for Mars indicated in Fig. 1 was calculated by Laurence Soderblom, at USGS in Flagstaff, Arizona, from counts of craters 4 to 10 km in diameter in various regions indicated. William Hartmann, of Science Applications Inc. in Tucson, Arizona, has derived a similar curve, but, because he estimates the history of erosion on Mars differently from Soderblom, Hartmann calculates somewhat higher cratering rates in the early part of the history.

From *Science*, vol. 183, pp. 187–189, 18 January 1974. Copyright 1974 by the American Association for the Advancement of Science. Reprinted with permission.

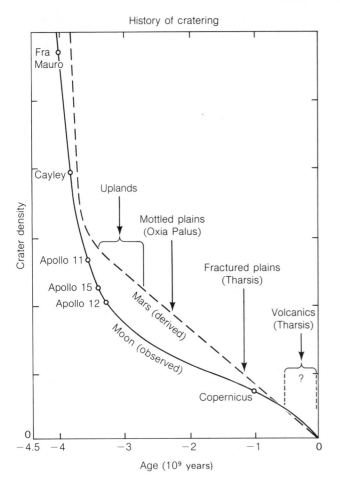

History of cratering

**Figure 1.** The number of 4- to 10-km craters per unit area versus the age of the cratered surfaces. Both bodies appear to have been heavily cratered 4 billion years ago, perhaps from the same source of debris.

The idea that Mars and the moon have similar flux histories appears to be novel and may unify a great deal of information about the early history of the solar system. As recently as 6 months ago, many investigators thought that the causes of cratering on Mars and the moon were quite different. But at the Pasadena colloquium it was suggested that the source of cratering debris was material in the asteroid belt that was ejected by perturbations from Jupiter, and the idea appeared to be rapidly accepted. The time required for the cratering rate on Mars to decrease by one-half, during the period of heavy bombardment, agrees well with estimate of the time required for Jupiter to sweep out half the material believed to have once filled the asteroid belt, and also agrees with the data from the moon. The specific rates of cratering need not be identical, since Mars is closer to the asteroid belt, but the ages of heavy bombardment should be the same for both Mars and the moon, as Fig. 1 indicates. (Whereas Mars was once thought to have a very high cratering rate even in recent times, with the result that many features were thought to be quite young, the theory of the "Jovian generator" suggests that the cratering rate on Mars is much lower, probably not more than twice the rate on the moon.) Eugene Shoemaker, of California Institute of Technology and USGS, who proposed the idea that Jupiter once sprayed a great flux of material across the solar system, pointed out that the same steep cratering curve found on Mars and the moon should be found on Mercury.

Whether the cratering on Mercury is similar to that on Mars and the moon should be known soon, because a Mariner spacecraft will fly by Mercury in March. On the other hand, the Galilean satellites of Jupiter, which are outside the asteroid belt, should have a very different cratering history, according to Soderblom. In 1977, two Mariner Jupiter-Saturn missions should photograph the Galilean satellites.

Apart from the growing consensus on a cratering history, the Mars colloquium was characterized by articulate and enthusiastic debate about past and present conditions of the planet. The major unanswered question, of course, is whether there was ever any liquid—particularly water—on the Martian surface. It seemed that those who hoped to find life were biased toward water, while those who didn't, were not. The principal benefit of the colloquiuim was probably that all the critical questions about Mars were raised and expressed clearly for a large audience of scientists. As Bruce Murray of Caltech repeatedly suggested, there are probably not enough hypotheses proposed yet to explain all the features of Mars, and the procedure of selecting the least unsatisfactory hypothesis inevitably leads to the wrong answer. Four major puzzles about Mars seem to be the history of the crust, the channels, the polar caps, and the volcanic activity of the planet.

Even from a quick look at the map of Mars, it is clear that one half of the planet is almost completely different from the other half. Although the equator is not the exact dividing line, the hemisphere that approximately corresponds to the northern hemisphere is only slightly cratered, while the southern hemisphere is very heavily cratered. The fact that Mars is apparently a bimodal planet, with two halves that have had completely different histories, is perhaps the greatest puzzle of all, suggests Harold Masursky, of USGS, Flagstaff. He thinks that there was an early episode on Mars when the lightweight crust was removed from half the planet and piled on the other half. What mechanism could do this is an extremely interesting question intrinsic to Mars. But Masursky suggests that the problem may be even more interesting because the history of the earth's surface has been traced back to the one big continent, Gondwanaland, and he points out that nobody tries to imagine what the earth looked like before that epoch. So Masursky suggests that, before the onset of the latest episode of continental drift on earth, perhaps Gondwanaland looked like Mars does now.

The channels on Mars have been heralded as evidence for aqueous fluids, and indeed many scientists think they were formed by running water—either condensing from the atmosphere or melting on the surface. But neither water nor carbon dioxide, the two principal gases of the atmosphere, could be stable as liquids under the present Martian conditions. Between 5 and 50 times the current atmospheric pressure is required for water to have a liquid phase, and far more pressure is required for liquid carbon dioxide. Liquid water clearly seems the more likely agent for fluvial processes. But if the channels were cut by water, the atmosphere must have changed drastically, and furthermore, one must ask where did water come from and where did it go?

Water could be stored as surface ice in the north polar cap, as subsurface ice or permafrost, as gas adsorbed to particles of finely ground material called the regolith, which covers much of the Martian surface or even in certain clays, which are possibly found on Mars. Evidence for polar ice, permafrost, and water in clays is somewhat sketchy. Infrared observations from the earth suggest that the regolith may contain 1 percent water, but no one knows how deep the broken soil extends. If the channels could be formed without running water, one argument for a large deposit of water is eliminated, and if some form of life were not expected on Mars, another reason to hope for water would disappear. But two more reasons to expect water remain. Several theorists think that Mars must have had a high concentration of volatile substances when it was formed, and that those volatiles—water among them—would not have escaped. Robert Sharp, of Caltech, thinks that some liquid is necessary over part of the history of Mars to account for the erosion of the surface and features such as etched pits, presumably caused by sapping of ground ice.

Although the majority of investigators seem to favor channels formed by water, others consider the recent symposium valuable because different channel formation processes were discussed. "At least we got the discussion off dead center," said one scientist. Stanley Schumm, of Colorado State University, Fort Collins, suggests that many of the channels, possibly even the braided channels, could have been formed from tectonic fractures that were modified. Erosion by blowing sand and dust is one possible modifying process. Schumm, who specializes in fluvial geomorphology, thinks that none of the channel patterns were necessarily produced by water, although they could be fluvial. Carr proposes that some Martian channels, especially the smaller sinuous channels, were caused by lava flow, but not the larger channels.

Next to the channels, the features of Mars that seem to arouse the most contention are the polar caps. Since the major constituent of the atmosphere, carbon dioxide ($CO_2$), can freeze onto the polar caps, which are at the temperature of 145°K, the atmosphere and the polar caps are closely coupled. The seasonal variation of $CO_2$ in the atmosphere seemed small, however, so Robert Leighton, of Caltech, and Murray have proposed that, in addition to the $CO_2$ caps that obviously advance and retreat each season, there is also a permanent $CO_2$ cap which acts as a buffer to moderate the atmospheric $CO_2$ variations. But there are problems with a permanent $CO_2$ cap. The water vapor in the atmosphere also varies seasonally, and Andrew Ingersoll, of Caltech, thinks that these pressure variations cannot be reconciled with the existence of a permanent $CO_2$ coldtrap.

The poles could also be the sites of permanent deposits of water ice. Infrared radiometry can distinguish water and $CO_2$ ices, but the north polar cap was covered with a cloud system, called a hood, when Mariner 9 observed it (Fig. 2), and observations of the south polar cap were severely limited because of the low inclination of the spacecraft's orbit.

Most of those at the conference seemed to agree that the polar regions are likely to be the key to understanding Mars. Besides changing ice caps, the poles have peculiar layered terrain, looking something like stacked poker chips, that almost everyone suspects were laid down during geological periods of alternating climates. The north polar region has the larger cap, and also a more pronounced topography. According to Conway Leovy, of the University of Washington, Seattle, perhaps the question of the existence of a permanent $CO_2$ cap can be answered by studying the rate at which $CO_2$ is deposited in the seasonal cap at the north pole of Mars. The layered terrain is particularly interesting because it may establish the dates of climatic variations, such as ice ages on Mars. If variations in solar activity have controlled the glacial ages of Mars, the layered terrain may provide a clue to the intermittent occurrence of glacial ages on earth.

As a result of the conference, planners for the Viking mission have decided to observe the north polar cap, which will be free of clouds when Viking arrives in 1976, from a high inclination orbit. The infrared radiometer on the 1976 Viking will have a small enough field of view to analyze the polar ice without any contribution from nearby soil, and it will also have two high-resolution cameras capable of convergent stereographic mapping that should be able to determine the thickness of the ice and the heights of the layered terrain.

A final puzzle, besides bimodality, channels, and polar caps, is the variety of volcanic features that are observed. When the great shield volcano, Nix Olympica, was observed in a sparsely cratered area, the theory was proposed that volcanism was very young and Mars had just "turned on." Indeed, Nix Olympica is recent, probably not more than 100 million years old, but recently close study has shown that there is also evidence of volcanism more than 2 billion years old. In addition to the shield volcanoes, there are complex volcanic centers, which may be similar to the sort of silicic volcanism on earth that produces large amounts of hot water as well as lava. On Mars, such volcanic centers are found in Alba, Hesperia, and north of Hellas, and are spread widely in Martian time, according to Masursky. Volcanism is generally assumed to have produced the Martian atmosphere, and Masursky stresses that the correlation between the ages of volcanism and the ages of channels, some of which appear to be 2 billion years old, indicates that atmosphere appeared early in Martian history and interacted with the surface for at least half of Martian time. As a limitation on atmospheric activity, Murray emphasizes that if water erosion had been active throughout much of Martian history all the ancient cratered terrain would be gone. But the idea that the atmosphere acted episodically rather than continuously would seem to circumvent this constraint, says Masursky. The atmosphere of Mars seems directly and intricately related to so many phenomena that it will probably be extremely difficult to develop a consistent atmospheric history. But the idea that Mars just turned on seems to have been laid to rest.

The detailed questions raised by the data from Mariner 9 almost all require more physical data about the planet. In a summary of the meeting, Richard Goody, of Harvard University, Cambridge, Massachusetts, suggested some further tests. Samples of surface materials are needed to learn compositions and ages of different terrain. Seismographic measurements should indicate the thickness of the lithosphere. Heat sensors on the surface could tell something about the forces interacting with the lithosphere from below. Measurements of such things as the ratio of argon to xenon in the atmosphere could

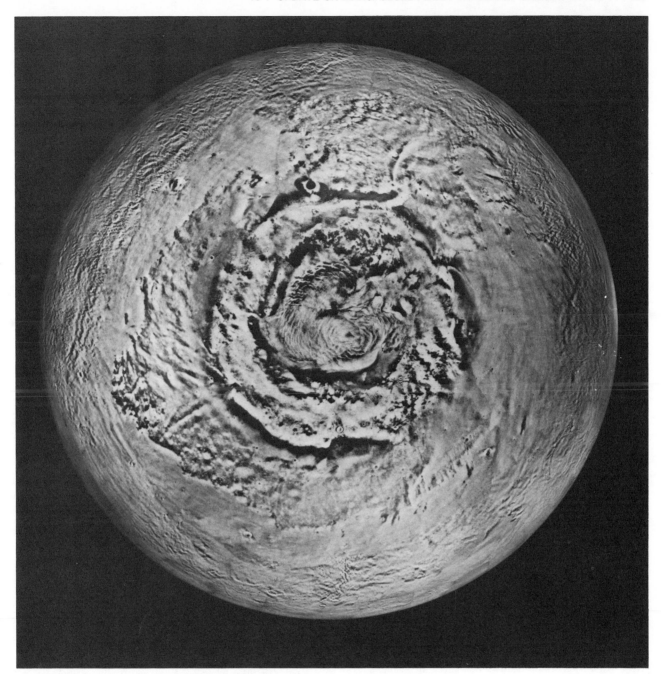

**Figure 2.** The north pole of Mars, as observed by the spacecraft Mariner 9. Layered terrain can be seen at the pole. The collar around the pole is cloud cover. [Source: Jet Propulsion Laboratory.]

indicate whether or not the origin of the atmosphere is really volcanism. Electromagnetic sounding could possibly detect ground ice.

Within the U.S. space program, most of these measurements will have to be made some time in the future. Viking is much more heavily instrumented for finding evidence of life than atmospheric or geological data. The two sites chosen for the Viking landers in 1976 are the mouth of a large channel, Chryse, and the region of highest water vapor, Cydonia. But Soviet scientists who were unable to attend the recent colloquium, do not expect evidence of life on Mars. Four Soviet spacecraft, presumably two orbiters and two landers, are en route to Mars now. They are expected to have sophisticated instruments for analyzing the atmosphere and the surface of the planet, and the first will arrive in early February.*

## Notes

1. The International Colloquium on Mars was sponsored by the campus and the Jet Propulsion Laboratory of the California Institute of Technology, 28 November to 1 December 1973.

*The Soviet spacecraft were not very successful; no major findings have been reported thus far. As we go to press, the Viking experiments are in operation on Mars.—Ed.

# 16

# Moons of Jupiter: Io Seems to Play an Important Role

William D. Metz

Jupiter and its Galilean moons were observed from the ground long before the space program began, and few people expected startling discoveries from terrestrial telescopes. But a recent observaton of the innermost Galilean moon has motivated a theory that seems to correlate many older observations from the earth as well as some of the new discoveries by Pioneer 10. Last year Bob Brown, at Harvard University, Cambridge, Massachusetts, discovered evidence of sodium on the moon Io. The initial observation was so straightforward that it could have been done 50 years ago. Furthermore, the existence of atomic sodium on a planetesimal seemed so unlikely that few people believed the initial observation until it was confirmed by high-resolution studies at the Smithsonian Astrophysical Observatory, Mount Hopkins, Arizona, by Brown and Fred Chaffee. A new model for sodium emission that was prepared just before Pioneer 10 reached Jupiter predicted several of Pioneer's discoveries, most notably the existence of an ionosphere on Io.

The innermost of the four large Galilean moons of Jupiter, Io has a 42-hour orbit around Jupiter and a mass comparable to that of the earth's moon, and for some time ammonia ice has been suspected on Io. To explain the observations of sodium emission, Michael McElroy, Yuk Ling Yung, and Bob Brown at Harvard propose that the surface of Io is covered with a layer of ammonia ice containing traces of sodium, and perhaps potassium and calcium. Even at the low surface temperature of Io, some ammonia would evaporate. Ultraviolet sunlight would then dissociate the ammonia to produce molecular nitrogen, which could excite sodium atoms to produce the emissions that have been observed. (Nitrogen would do this most efficiently if it were excited to a metastable state, perhaps by a mechanism similar to the auroral process on the earth.) The dissociation of ammonia would also produce atomic hydrogen, which is so light it would escape from the gravitational attraction of Io.

McElroy and his associates predict that there should be an extensive cloud of hydrogen, nitrogen, and sodium around Jupiter in the vicinity of Io's orbit. Pioneer 10 was not designed to find nitrogen or sodium, but a torus of hydrogen with a diameter of about the diameter of Io's orbit has been observed. The sodium emission model also requires hydrogen in the atmosphere of Io, as possibly detected by Pioneer 10, and the existence of an ionosphere formed by photoionization of sodium. The discovery of an ionosphere for Io was, of course, one of the most definitive early results from Pioneer.

Two other phenomena associated with Io can be incorporated within the sodium emission model. After Io emerges from Jupiter's shadow, it is unusually bright for about 15 minutes and then gets dimmer. Several people have suggested that ammonia from the atmosphere falls as snow onto Io while it is on the night side of Jupiter, then the brilliant snow evaporates as Io emerges into sunlight.

Another striking observation known for several years is that the bursts of radio noise from Jupiter are strongly correlated with the position of Io in its orbit. To explain this, Peter Goldreich of California Institute of Technology, Pasadena, and D. Lynden-Bell University of Cambridge, England, proposed that if Io had a conducting ionosphere, it could generate a large voltage as it cut through the magnetic field lines of Jupiter. The voltage could drive a plasma process, perhaps like the aurora, which would produce the observed radio noise. The sodium emission model provides the needed ionosphere, of course. In return, the generator model provides a method of exciting nitrogen and also heating the atmosphere to high temperatures needed for sodium to escape. No evidence for a torus of sodium around Jupiter has been found, but Lawrence Trafton at the University of Texas, Austin, and associates have made Earth-based observations that suggest a cloud of sodium emission around Io extending to about 20 times the diameter of the moon.

From *Science*, vol. 183, p. 293, 25 January 1974. Copyright 1974 by the American Association for the Advancement of Science. Reprinted with permission.

Other possible implications of the new model are that the gases from Io could be the source of trapped radiation belts, and that high-temperature neutral gases could move out far beyond Io's orbit, become ionized, and perhaps cause the stretching of the magnetic field that is observed.

The new observations suggest that Io affects the environment of Jupiter in many ways. Jupiter is unique, but the moons of Jupiter, and perhaps Saturn and Neptune, may be much more significant than their sizes alone suggest.

# 17

# Pioneer 10 Observations of the Ultraviolet Glow in the Vicinity of Jupiter

Darrell L. Judge and Robert W. Carlson

Abstract. *A two-channel ultraviolet photometer aboard Pioneer 10 has made several observations of the ultraviolet glow in the wavelength range from 170 to 1400 angstroms in the vicinity of Jupiter. Preliminary results indicate a Jovian hydrogen (1216 angstrom) glow with a brightness of about 1000 rayleighs and a helium (584 angstrom) glow with a brightness of about 10 to 20 rayleighs. In addition, Jupiter appears to have an extensive hydrogen torus surrounding it in the orbital plane of Io. The mean diameter of the torus is about equal to the diameter of the orbit of Io. Several observations of the Galilean satellites have also occurred but only a rather striking Io observation has been analyzed to date. If the observed Io glow is predominantly that of Lyman-α, the surface brightness is about 10,000 rayleighs.*

The ultraviolet instrument (*1*) on Pioneer 10 is a two-channel photometer designed to observe the resonance emissions from atomic hydrogen and helium at 1216 Å (the H Lyman-α line) and 584 Å, respectively. The instrument uses a filter and photocathodes to isolate these two emission features.

Detailed calculations (*2*) show that the hydrogen and helium resonance lines are the strongest features to be expected from the outer atmosphere of Jupiter, and arise from resonance scattering of the incident solar hydrogen and helium lines; thus, only broadband isolation of these lines is required. The He channel uses a thin aluminum film as a filter and LiF as a photocathode, resulting in a spectral band pass of approximately 200 to 800 Å. The hydrogen channel uses the front surface of the Al film as a photocathode for which the photoelectric response extends up to about 1400 Å. This channel is sensitive to both hydrogen and helium emissions, but the former is more intense by more than an order of magnitude, so the recorded signals accurately represent the Lyman-α intensity.

The optical axis of the photometer is oriented at 20° to the spacecraft spin axis while the instantaneous field of view is approximately 1° by 20° with the longer dimension tangential to the 20° arc swept out by the spacecraft rotation.

From *Science*, vol. 183, pp. 317-318, 25 January 1974. Copyright 1974 by the American Association for the Advancement of Science. Reprinted with permission.

This orientation was chosen to give two views of Jupiter, the first occurring at approximately 50 Jovian radii ($R_J$), and outside the predicted radiation belts, while the second observation period occurred at about 10 $R_J$. In addition, several observations of the Galilean satellites were possible during the 5 days before closest approach.

During the first Jupiter viewing period emissions from the planet were observed in both the hydrogen and helium channels but the data obtained during the second viewing period suffered degradation due to the energetic radiation belt particles. The preliminary estimate of the hydrogen Lyman-α intensity is somewhat less than 1000 rayleighs (*3*) while that of the He emission is approximately 10 to 20 rayleighs. Helium emissions have not previously been observed from Jupiter although the presence of He has been speculated on for many years.

The present Lyman-α measurements are lower than the sounding rocket observations of Rottman *et al.* (*4*). This discrepancy may be partially due to variations in the solar Lyman-α flux, differences in atmospheric properties which depend on solar activity, and calibration uncertainties.

Ultraviolet emissions were also observed in the hydrogen channel from the innermost Galilean satellite Io (JI). It is reasonable to assume that these emissions correspond to Lyman-α since the surface of Io is thought to consist of hydrogen-bearing ices of $NH_3$ and $H_2O$ (*5*). The atomic hydrogen could result from photolysis and particle bombardment of the atmosphere and surface. If H Lyman-$\alpha$ is the main contributor, the source brightness is approximately 10,000 rayleighs. Such an intensity would seem to require an excitation mechanism in addition to resonance scattering. Aurora-like activity, as produced by energetic electrons accelerated toward Io by its motional electromotive force (6), seems an attractive possibility.

Finally, hydrogen channel signals were observed from the equatorial plane of Jupiter during periods when neither the planet nor its satellites were in the field of view. These emissions of several hundred rayleighs in intensity are tentatively interpreted as due to a toroidal cloud of neutral hydrogen in orbit around Jupiter, similar to the hydrogen torus proposed by McDonough and Brice (7) for Saturn and Jupiter (8). Preliminary analysis indicates that this gas cloud occurs at approximately the orbit of Io, suggesting that this satellite is the source.

## References and Notes

1. The instrument was built at the Analog Technology Corp., Pasadena, Calif., under the supervision of D. Willingham. The liaison with Ames Research Center and the spacecraft interface were handled by T. Wong and R. Tworowski.
2. R. W. Carlson and D. L. Judge, *Planet. Space Sci.* 19, 327 (1971).
3. One rayleigh corresponds to a column excitation rate of $10^6$ photons $cm^{-2}$ $sec^{-1}$.
4. G. J. Rottman, H. W. Moos, C. S. Freer. *Astrophys. J.* 184, L89 (1973).
5. T. V. Johnson and T. B. McCord, *Icarus* 13, 37 (1972).
6. P. Goldreich and D. Lynden-Bell, *Astrophys. J.* 156, 59 (1969).
7. T. R. McDonough and N. M. Brice, *Icarus* 20, 136 (1973).
8. ————, private communication, July 1973.

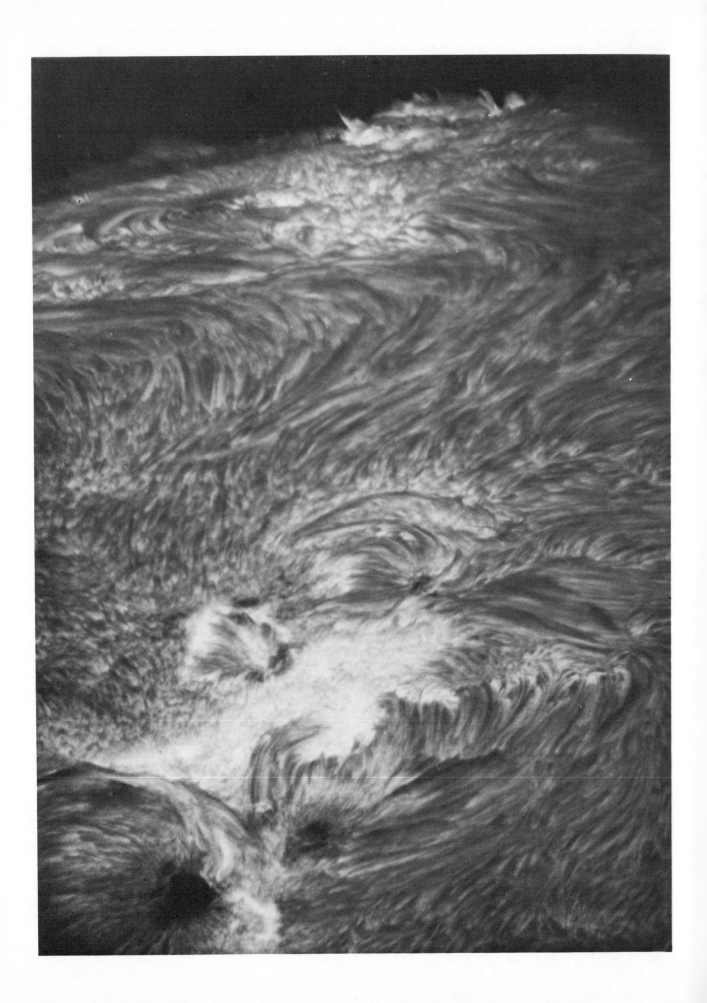

# PART V

# OUR SUN

"We are observing here a raw demonstration of the force or effect which causes sunspot groups to assume the 'correct' orientation." Solar physicist Spencer R. Weart, in "The Birth and Growth of Sunspot Regions," reports on an original investigation of the way in which sunspot groups become tilted. This article, taken from the elite and largely abstruse periodical of professional astronomers, the *Astrophysical Journal,* shows how, on occasion, a scientist can make a fundamental contribution to knowledge of an already much-studied phenomenon without resorting to complex equations or elaborate computer calculations.

It is known that sunspot groups usually exhibit a preferred orientation: the *preceding* spots (i.e., those at the leading edge of a group as it "moves" across the solar surface because of the rotation of the sun) are located slightly closer to the solar equator than the *following* spots of the same group. The preceding spots of all groups in a given solar hemisphere (northern or southern) usually have the same magnetic polarity during one eleven-year sunspot cycle, whereas the following spots have the opposite polarity. It is therefore logical to conclude that the preferred orientation, or tilt, of sunspot groups is also related to magnetism.

The most common method of solar observation is the H$\alpha$ patrol film, a time-lapse motion picture taken through a filter that isolates light at the wavelength of the prominent red spectral line of hydrogen (refer to the diagram on page 179 in Part VII of this book). Observations of the solar magnetic field, by means of *magnetograms,* must be obtained by more complex and time-consuming methods and are therefore less commonly available. Weart and his colleagues, by comparing selected H$\alpha$ pictures with magnetograms taken almost simultaneously, concluded that they could identify small magnetized regions, called arch filament systems (AFSs), directly on the H$\alpha$ films.

It seems that the AFSs bring magnetic fields, which are measured in terms of their *magnetic flux,* to the surface of the sun from deeper, unseen layers and

A high-resolution photograph of the sun taken in the hydrogen alpha line using the vacuum tower telescope of the Sacramento Peak Observatory. The solar limb is at the top. [Sacramento Peak Observatory, Air Force Cambridge Research Laboratories.]

that the AFSs develop into sunspots. At least, that is one theory. Weart used Hα patrol films as the raw material of his investigation. He made simple measurements of the tilts of structures on the sun with respect to the solar equator and developed a new theory to explain how the preferred orientation of a sunspot group arises, as described in "The Birth and Growth of Sunspot Regions." Another technique, called the *Doppler movie,* is also referred to in the article. A Doppler movie is a film that combines pictures taken in a series of adjacent wavelengths repeatedly for a given period; it can be analyzed through an understanding of the Doppler effect to determine the motions of solar structures along the line of sight.

In his next article, "What Makes Active Regions Grow?", Weart describes a follow-up study of AFSs and their relation to the growth of sunspots. In particular, he investigates the origin of the *asymmetry* of sunspot groups—that is, the tendency of the preceding part of a group to be larger than the following part. He proposes a simple theoretical model to explain this effect in terms of the properties of AFSs and finds evidence in the Hα films to support his model. His research includes studying groups with *reversed asymmetry* (i.e., the following part is larger than the preceding part). More research on this subject is needed; the author notes that, "We have seen the birth of only two sunspot groups with reversed asymmetry, but the reader is invited to search for others . . . ." Weart also mentions *supergranulation,* a pattern of large, slowly changing convection on the solar surface.

The sun's influence on its surroundings is exerted by more than just its output of light. Solar particles and magnetic fields pervade interplanetary space and envelop the earth. One of the editors explores this topic in "The Solar Wind Blows Some Good for Astronomy."

# The Birth and Growth of Sunspot Regions

Spencer R. Weart

Abstract. *We have observed in Hα some 100 active regions during their first few days. The first sign is almost always a bright spot with dark arches. The arches for different new regions are oriented differently with respect to lines of latitude, and there is no striking predominance of any particular tilt. This suggests that the magnetic-field direction below the surface of the Sun has a random component. Most new regions may be classified as having either "correct" tilt (NE in the northern hemisphere, SE in the southern) or "contrary" tilt. All regions which survive the first few days take up a correct tilt; however, regions correctly tilted at the start are more likely to survive and grow than contrary ones. This indicates that the effect which causes sunspot groups to be oriented with the preceding spots nearer the equator operates after the emergence of magnetic flux. We also find that interactions between separate flux tubes are important, for when two systems of arches arise in proximity, they are more likely to grow into a large active region.*

## I. Introduction

Study of the development of active regions is of major interest to solar physicists. According to widely accepted theory, activity occurs where magnetic flux emerges from within the photosphere. However, the nature of the subsurface field, and the way flux tubes emerge and interact, remain mysterious, and so the development of a given active region is more or less unpredictable.

As we reported previously (Weart and Zirin 1969), activity is often seen first in hydrogen Hα photoheliograms. A region of unusual Hα brightness, crossed with parallel dark arches, usually precedes any indication of activity in magnetogram data or sunspot drawings. Examples are shown in Figure 1. Probably the first to comment on this phenomenon was Waldmeier (1937); Ellison (1944) and Bruzek (1967, 1969) have studied the dynamics of these regions when they are well developed; Roberts (1970) has analyzed the motions of the

From *Astrophysical Journal,* vol. 162, pp. 987–992, 1970, The University of Chicago Press. Reprinted with permission.

arches as seen in Doppler movies. Bruzek calls them arch filament systems (AFSs); we will use this term throughout, but we emphasize that our studies concern new AFSs during their first day or so whereas Bruzek has studied a later stage of development. He and Roberts have shown that each arch rises slowly, as expected for the emergence of tubes of flux from below the surface. Also as expected, the arches span the dividing line between opposite magnetic polarities. An AFS is thus a relatively simple and comprehensible thing, particularly when it appears by itself in a quiet part of the Sun.

We expect that study of the new AFSs can cast some light on several questions. Is the subsurface magnetic field strictly parallel and latitudinal, as Babcock's (1961) theory suggests, or is it complex, with random radial and longitudinal components, as Leighton's (1969) computational model requires? The first appearance of an AFS gives us as close a view of the subsurface field as present techniques allow. Does magnetic flux rise solely because of the magnetic buoyancy described by Parker (1955), or do the twisting of flux tubes and other effects enter? What causes active regions to take up a preferred tilt, with the preceding polarity closer to the equator than the following polarity? These forces appear in their most elementary form when they affect the development of an isolated arch filament system.

## II. The Observations

We have looked at Hα time-lapse movies of the Sun which covered much of the period 1967–1969, and have noted about 100 young AFSs. In some cases we followed a developed active region back in time to its origin, and almost always found a typical AFS as the first sign. We learned to detect most AFSs within a few hours of their appearance. They are easily observed in a large-scale solar movie, but only with practice can one see them on a patrol film which puts the whole Sun into a 35-mm format.

Since material is streaming down both ends of the arches and thus produces Doppler-shifted light, AFSs are particularly visible in the wings of Hα, especially when they are near the

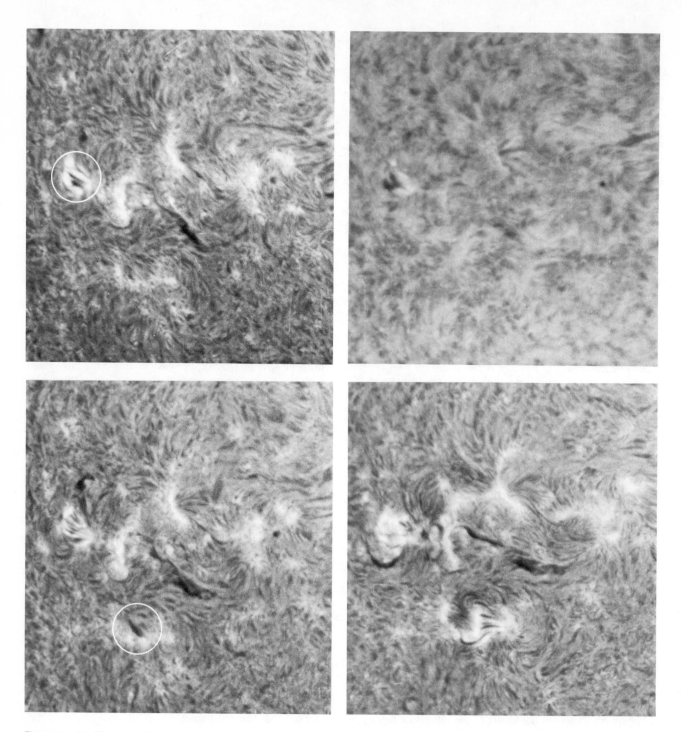

**Figure 1.** Hα filtergram showing typical features in the development of new AFSs. Upper left: 1968 September 23, 20:59:45 U.T. The area circled, which had been somewhat active for several days, now showed the extreme brightness and the dark arches characteristic of a new AFS. Upper right: 22:14:36 U.T. The new AFS was particularly visible in the line wings (here, Hα − ½ Å). Lower left: September 24, 00:01:05 U.T. The AFS was well developed. Meanwhile, the first large arch of another new AFS (circled) appeared within a cell of the chromospheric network. Lower right: 15:29:15 U.T. The first AFS became a normal sunspot group. The newer AFS faded out (although occasional arches appeared as late as 23:00), but meanwhile a third new AFS appeared immediately to the west of it and grew rapidly. This last area eventually developed into an important active region. Photographs taken at Robinson Laboratory, California Institute of Technology. North is at top, west at right.

limb. The extreme brightness of the underlying material is the most characteristic sign of a new AFS; during the first day or two of its existence a new AFS is the brightest Hα feature on the solar disk except flares. Another feature of AFSs which helps one locate them in movies is the vigorous activity of the arches. Often, too, there are small surges, and in a few cases there are no arches but many surges.

In the majority of our cases an initial orientation could be assigned to the AFS, taken from the roughly parallel arches which usually appear during the first few hours. This orientation shows the direction of the emerging magnetic field: we expect arches to lie along field lines, and comparison with magnetograms and sunspot drawings, when available, confirms this. We measured the angles of orientation of arches for ten AFSs and estimated the rest by eye.

Before statistics were compiled, a number of AFSs were eliminated for various reasons—too near the limb for good observation; later development unobservable due to poor weather or limb passage; proximity of other activity which might perturb the AFS; unusual complexity which made it impossible to assign a unique orientation to the arches; poor seeing; etc. This still left us many examples of relatively simple and well-observed AFSs.

## III. Orientation and Growth of New Arch Filament Systems

A diagram of the initial orientations of forty-one new AFSs is given as Figure 2. The accuracy of measurement varies, but in the worst cases the angle is defined within about ±15°. Several facts are at once obvious from this figure. First, AFSs can appear tilted at any angle to the solar equator, although angles greater than 70° are rare. Second, there is no striking tendency for AFSs in a given hemisphere to be oriented more toward, say, the east than the west.

The latter fact is unexpected, for well-developed active regions do show a nearly invariable orientation: the preceding polarity is closer to the equator than the following polarity. Thus, regions in the northern hemisphere have a northeast tilt, southern ones have a southeast tilt. No such strong tendency is found during the first day in the life of an AFS. Table 1 shows statistics for 61 AFSs whose orientation was sufficiently clear to be categorized as "northeast" (i.e., arches running NE-SW) or "southeast." We see that there is a tendency in the northern hemisphere for AFSs to have the expected tilt, but there are twelve cases to the contrary; in the southern hemisphere (where we have fewer cases) we see no statistically significant preference. We will show later that there is a possible source of bias in our measurements, for out of a number of AFSs which appear on the Sun, we are more likely to

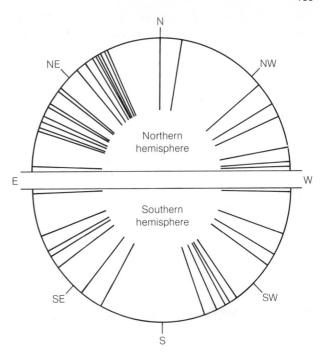

**Figure 2.** Orientations of new AFSs. Each line shows the average direction of the arches in a single AFS on the first day of its appearance. Except for avoidance of tilts greater than ±70°, the orientations are random.

observe those which have the "correct" tilt (preceding polarity nearer the equator). Thus, part of the mild preference shown in our data for correct tilts may be spurious.

Most of these AFSs also appeared as new sunspot groups on the Mount Wilson sunspot drawings; where the polarity of the new spots was recorded, we invariably found that the preceding spots had preceding polarity, regardless of whether they were closer to the equator than the following spots.

Eventually nearly all sunspot groups take up a correct tilt of small magnitude. The tilt of most AFSs must therefore change with time. And as we reported earlier (Weart and Zirin 1969), most AFSs do rotate during their first day or two of life; that is, the orientation seen at later times differs from the initial tilt. In virtually all cases the rotation is toward the equator. Thus if an AFS in the northern hemisphere initially tilts 40° to the northeast (clockwise from the equator), the next day it will be, say, 20° to the northeast. A northern AFS with an initial tilt of 40° to the northwest will later be 20° to the northwest and still later, will have the correct NE orientation.

We are observing here a raw demonstration of the force or effect which causes sunspot groups to assume the "correct" orientation. We note that this force operates *after* the flux has emerged, and that it must do more work on an AFS with contrary tilt than on one with a correct tilt. We next consider whether this affects the later development of the active region.

The majority of AFSs, like their accompanying sunspots, disappear within a day or two of their appearance. Only a few of the rest develop into active regions of major importance. We have estimated the importance of fifty active regions, using as a parameter the number of sunspots averaged from ESSA and Mount Wilson drawings. Seven of these regions become

TABLE 1
*Initial orientation of arches in new AFSs.*

|  | NE | SE |
| --- | --- | --- |
| Northern hemisphere | 23 | 12 |
| Southern hemisphere | 12 | 14 |

unusually large (at least twenty-five spots on the fifth day after appearance); these have been excluded from the statistics and will be discussed in more detail later.

The average number of spots on the day after initial appearance was $3.6 \pm 3.5$ for correctly tilted AFSs and $3.1 \pm 3.0$ for contrarily tilted ones. (We give standard deviations as an indication of the spread of the data, though the distribution was more nearly Poisson than Gaussian.) Once again we see no significant difference between correctly and contrarily tilted AFSs at their first appearance. However, correctly tilted AFSs are definitely more likely to survive and grow. Thirty percent (9 of 30) of the correctly tilted AFSs grew between the first and fifth days after their appearance; none (0 of 13) of the contrarily tilted ones grew.

As mentioned earlier, this fact may have introduced some bias into our study. Since some AFSs were found by tracking back developed groups to their origin, and since a preponderance of well-developed groups stem from correctly tilted AFSs, we have probably missed some contrarily tilted short-lived AFSs that otherwise would have been included.

We estimated the magnitude of AFS angular tilt measured from the equator regardless of whether it was eastward or westward, and found no significant correlation with the later size of the regions. We also found no relation between tilt and latitude of the AFSs.

## IV. Growth of Major Active Regions

We now turn to the seven regions we observed from birth until they became major centers of activity (at least twenty-five spots on $D + 5$). The AFSs which evolved into these regions were, as a set, distinctly abnormal. Their tilts reveal little (three predominantly correct, two predominantly contrary, two uncertain), partly because most of these regions were so complex from the start that the assignment of orientation was difficult. Magnetograms, when available, confirmed that these AFSs were never simple bipolar magnetic regions; either there was a neck of preceding polarity intruding deeply into following polarity or there were more than two concentrations of polarity. We also observed a number of complex AFSs which failed to develop into major active regions, but these were few in comparison with the many simple AFSs. Thus, on the average, AFSs which appear complicated during their first day or two of existence are much more likely than simple AFSs to become major regions. This fact can be deduced directly from magnetograms.

In at least four, and possibly all, of our seven major regions, the initial appearance was less that of a single AFS than of two AFSs next to one another. In two cases the Mount Wilson sunspot drawings (with polarity measurements) confirmed that we were indeed observing an interacting pair of bipolar magnetic regions. We have observed several other cases of paired AFSs which did not become very large regions, but the majority of these did develop more than the average AFS. Hence the appearance of two interacting AFSs is a strong signal that a large active region is being born. In some cases such regions develop very rapidly, growing to a full-sized active region in less than 6 hours. In other cases (we have observed three)

a normal AFS is seen on one day and is joined the next day by a second AFS, which then becomes the main center of activity.

## V. Discussion

We are now in a position to reach some tentative conclusions about the solar subsurface magnetic field and the generation of active regions.

First, our data show that the initial tilts of AFSs are highly randomized (within $\pm 70°$ of the equator). Then either the subsurface magnetic field is highly randomized or else the field lines are subject to arbitrary twisting as they emerge. We believe the latter explanation unlikely, for AFSs emerge too rapidly (in a few hours) for reasonable forces to operate; moreover, the arches commonly emerge spanning a cell of the supergranular network, suggesting that convective motions have helped to lift the field lines straight up from below the surface.

Our data also shed light on the force which causes sunspots to take up correct tilt (preceding sunspot nearer the equator). This acts after the magnetic flux emerges, for the initial tilt is random. Schmidt (1968) has suggested that this is a Coriolis force, but in some cases our AFSs rotate so rapidly (e.g., $20°$ in less than an hour) that only a shift of magnetic-field lines or the eruption of new field could explain the observations. The effect could be geometrical: perhaps the lines of magnetic field which rise at the outset are more steeply slanted away from the equator then those which rise later. Such would be the case, for example, if the entire flux tube were twisted (right-handed twist for a northern-hemisphere AFS correctly tilted; left-handed twist for one contrarily tilted). More study of the dynamics of arches in new AFSs might clear up this question.

The way flux initially emerges affects its later history. Flux which has correct tilt from the start is much more likely to form a persistent sunspot region than is flux of contrary tilt. There are two possible explanations of this fact: either correctly and contrarily tilted flux tubes are of different types (e.g., twisted with opposite handedness), or else they are treated differently by the force which rotates sunspot groups. We note that the paucity of tilts greater than $70°$ does tend to split AFSs into two distinct groups.

Finally, AFSs interact with one another, as would be expected of flux tubes. The majority of very large active regions grow from such interactions. A case of this was noted recently by Fortini and Torelli (1968), who found two bipolar sunspot groups interacting to form a region that later produced a class III flare. We have also noted many cases of new AFSs interacting with preexisting magnetic regions; indeed, at least 10 percent of all new AFSs emerge within preexisting sunspot groups, although we have chosen not to study these very complex cases at present. The fact that two adjacent AFSs are likely to produce a major sunspot group indicates that the interactions between flux tubes play an important role in the dynamics of active regions. Unfortunately, we know of no theoretical studies on this matter. But it seems clear that new flux tubes rise to the surface not only because of magnetic buoyancy, but sometimes partly because of convection (since new AFSs may appear spanning a network cell, but never spanning a cell boundary), and sometimes because of an inter-action with other flux tubes.

The many movies on which this study is based were taken at Big Bear Solar Observatory and Robinson Laboratory, California Institute of Technology, by a group under Dr. H. Zirin of Hale Observatories, and at the Rye Canyon Obser-vatory of Lockheed Corporation by a group under H. Ramsey. We are deeply indebted to all those involved in this difficult work of photoheliography. Sunspot drawings provided by Dr. R. Howard of Hale Observatories helped us greatly. Mr. Roger Chevalier's work in compiling data was invaluable. We also thank Dr. Zirin, Dr. Howard, and Dr. R. B. Leighton of California Institute of Technology for their many useful comments and suggestions. This work was funded in part by NASA grant 05 002 034 and NSF grant 1472.

## References

Babcock, H. W. 1961, *Ap. J.*, 133, 572.
Bruzek, A. 1967, *Solar Phys.*, 2, 451.
————. 1969, *ibid.*, 8, 129.
Ellison, M. A. 1944, *M.N.R.A.S.*, 104, 22.
Fortini, T., and Torelli, M. 1968, in *I.A.U. Symposium 35* (New York: Springer-Verlag).
Leighton, R. B. 1969, *Ap. J.*, 156, 1.
Parker, E. W. 1955, *Ap. J.*, 121, 491.
Richardson, R. S. 1948, *Ap. J.*, 107, 78.
Roberts, P. H. 1970, unpublished doctoral thesis, California Institute of Technology.
Schmidt, H. O. 1968, in *I.A.U. Symposium 35* (New York: Springer-Verlag).
Waldmeier, M. 1937, *Zs. f. Ap.*, 14, 91.
Weart, S., and Zirin, H. 1969, *Pub. A.S.P.*, 81, 480.

# 19

# What Makes Active Regions Grow?

Spencer R. Weart

Abstract. *We have studied the growth, or failure to grow, of well over 100 active regions. Most growth is connected with the emergence of a large batch of flux in the shape of a new Arch Filament System (AFS). During the recent sunspot maximum, new AFSs appeared at a rate of nearly one per day over the entire Sun. We see them popping up at random points, not only in quiet parts of the Sun, but also within old active regions; and more often than chance would predict, they come up right in the middle, on the site of preexisting AFSs. These AFS appearances and rejuvenations account for most of the growth of active regions.*

*We present evidence for the hypotheses that (1) a twist in the flux tubes of new AFSs is a key factor in determining which new AFSs will grow, and (2) this twist is related to the well-known asymmetry of sunspot groups.*

## I. Introduction

If we watch magnetic flux as it first emerges on the solar surface, we may learn something about how active regions are formed. Like psychologists, we study the infant to know the man. In recent years we have learned to identify flux in Hα pictures within an hour or so of its emergence, and some remarkable regularities have been discovered.

Usually a region of new flux looks like the object in figure 1. It is characterized by (1) brightness in the center of Hα, typically in two patches with a darker lane separating them; (2) two or more roughly parallel dark arch filaments; and (3) a size of around 20,000–40,000 km. Generally there are sunspots, but not always. Bruzek (1967, 1969) calls these things Arch Filament Systems or AFSs. The dynamics of the arch filaments have been studied by Bruzek, Roberts (1970), Vrabec and Weart (1972), and others. They are absorbing features a few thousand kilometers above the photosphere, rising in the center, with material streaming down both ends like water streaming off a whale's back when it surfaces. In a small minority of cases, however, new flux appears without any arch

From *Astrophysical Journal,* vol. 177, pp. 271–276, 1972, The University of Chicago Press. Reprinted with permission.

filaments, although there may be small surges that look like filaments (fig. 2). Moreover, regions which do have arch filaments but which are several days old and no longer changing are commonly called AFSs. Zirin (1972) has therefore proposed the term Emerging Flux Regions or EFRs for all compact areas of new bipolar magnetic flux. Clearly, AFS is a phenomenological term and EFR implies a judgment as to what is going on.

We will add to the nomenclature by defining another term, *new AFS:* an AFS not more than two days old. Even when the history of the region is unknown, new AFSs as a group can be distinguished from old ones, according to our observations of several dozen examples. The criteria are (1) new AFSs are extremely bright in the wings of Hα; (2) new AFSs are nearly all between 20,000 and 30,000 km across—distinctly smaller, on the average, than old AFSs; and (3) new AFSs are accompanied by the development of pores, i.e., infant sunspots. We regard new AFSs as a large subset of the set of EFRs. This paper is about new AFSs. Using the above observational criteria, we have identified over 150 examples, using films from the Caltech, Big Bear, Lockheed, and Aerospace observatories.

The study of new AFSs has yielded a coherent picture. Flux rises from below the surface, aided by convection, and changes after emergence. To some extent, we can use characteristics of new AFSs to predict the size of the regions that later develop (see Weart 1970a, and references therein).

But serious questions remain. What precisely is going on in new AFSs, and why do some of them develop into huge active regions while most others peter out in a few days? We would expect none of them to last more than a few days, if we hold strictly to the model of the sunspot cycle devised by Babcock (1961) and Leighton (1969). In this model, flux rises to the surface when the subsurface field is stronger than a certain critical value $B_c$; once it has emerged, it disperses. The critical value is supposedly set by the Parker (1955) mechanism: when a flux tube has a field of a few hundred gauss over a diameter of 1000 km or so, it will be buoyant, although not overwhelmingly so compared with the strength of supergranular convective motions. Helped by convection, however, the

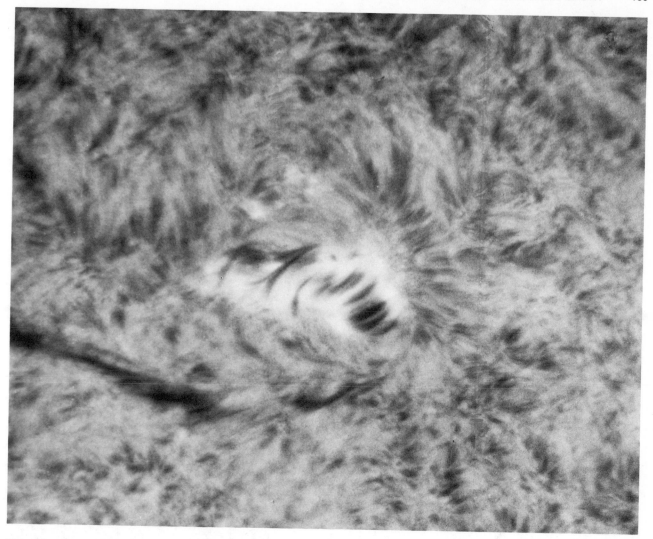

**Figure 1.** A new AFS a few hours old. The right-hand part shows the typical size, brightness in the center of the Hα line, and parallel arches; the additional complexity of the field on the left contributed to the growth of this AFS into a major active region. [Big Bear Solar Observatory, 1968 September 1.]

tube will bob to the surface—and indeed we do see such tubes, the arch filaments, appearing in the middle of supergranular cells. So the Babcock-Leighton model seems good up to this point. But surprisingly enough there are cases, relatively few but important, in which the flux does not simply emerge and disperse. New flux keeps appearing near the same area until fields of $5B_c$ and more are present—a large active region forms. How can this be explained?

In this paper we present evidence which may hint at the solution: (1) sometimes new AFSs appear by chance in the vicinity of an old active region and add to it; (2) sometimes an old AFS is rejuvenated, adding a new batch of flux; (3) an unknown mechanism tends to aid the growth of only those AFSs whose field lines have a certain twist.

## II. New Flux in Old Active Regions

Many but not all new AFSs appear in quiet parts of the Sun. We find that on the average one AFS appears on the visible side of the Sun every $2\frac{1}{2}$ to 3 days. These statistics come from counts made from films for 1967 February–May and from movies and ESSA prints for 1969 July–September, so they represent the sunspot maximum. We do not know what happens at sunspot minimum. If we take into account AFSs on the invisible side of the Sun, those overlooked because they were too near the limb, etc., then the true number of new AFSs on the entire Sun was probably nearly one per day during this period. (The new AFSs were almost always present on Mount Wilson magnetic maps, as bipolar magnetic regions with fields ≥40 gauss over a region some 20,000 km or more across for each polarity.) The new AFSs appear at random locations, so far as we can tell. They are widely scattered in latitude, and they neither favor nor avoid the vicinity of old active regions, prominences, etc. (This is not to deny the results of Bumba and Howard 1965, for we could not detect the large areas of weak polarity which they correlate with new flux emergence.)

Occasionally a new AFS is found very near an old active region, or even right in its midst. These can only be seen on

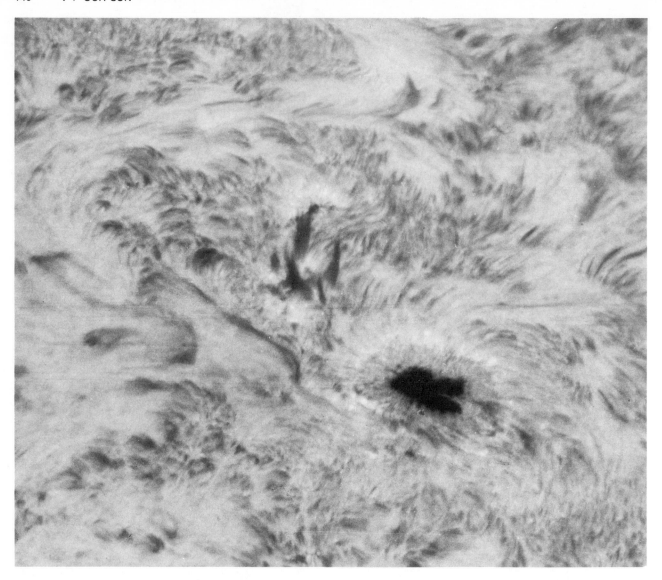

**Figure 2.** Above and to the left of the sunspot we see an EFR which, unlike most, has few or no arch filaments. Instead, there are a great many small surges. [Big Bear Solar Observatory, 1970 July 7, blue wing of Hα.]

films with excellent resolution, since otherwise the arch filaments and brightness are lost in the murky tangles of the region. Nevertheless we have identified over 25 cases.

In eight of these cases (and in a number of other cases we have seen on films of lower resolution), a new AFS appears right on the site of a previous AFS. That is, the AFS is rejuvenated, and shows all the characteristics of a new one: a compact set of parallel arch filaments, brightness in the center and wings of Hα, and new pores between the old sunspots. Figure 3 shows such a rejuvenated AFS. This region started as a new AFS some three days earlier, on 1970 May 24; on May 26 a new AFS appeared in the middle and eventually produced a large spot (second from the right). Here we see a second rebirth (May 27), with more new spots.

AFS rejuvenation happens several times more often than if the new AFSs were just coming up at chance locations. It is an important, hitherto overlooked process in the growth

of active regions. Over a much longer period (5 months), Prata (1971) has reported a case of four consecutive regions appearing as new AFSs on the same site. It would be rash to speculate on the mechanism that causes AFS rejuvenation, though it must have something to do with the subsurface pattern of magnetic and convective forces that accompanies normal bipolar sunspot groups—the hidden part of the iceberg.

In the rest of our 25 cases new AFSs appear at random locations in and near old active regions. They appear in about equal numbers preceding, following, or in the midst of the spot group, polewards or equatorwards of it, and near or far from the neutral line between opposite polarities. In the same films, we see as usual many new AFSs arising in quiet parts of the Sun; only about 15 percent of new AFSs other than rejuvenations appeared "in or very near" old active regions. We also estimate that roughly 16 percent of the active zones of the Sun were "in or very near" old regions during this same period,

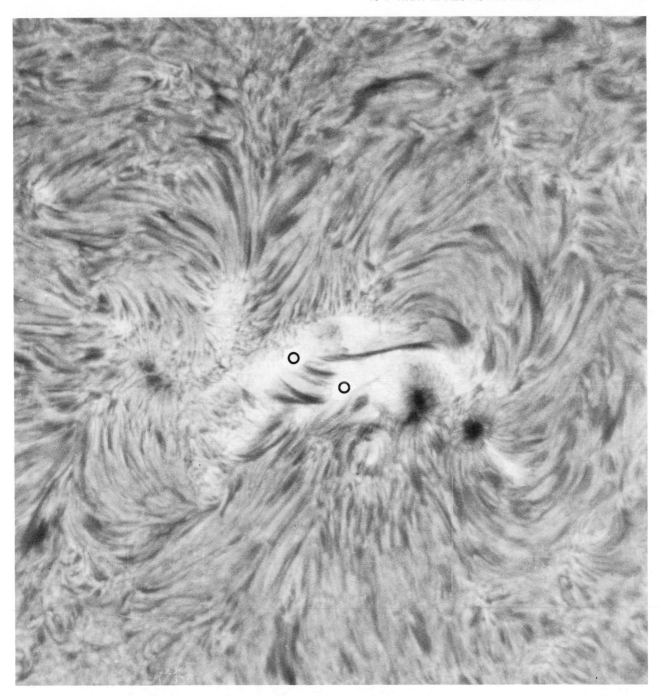

**Figure 3.** This new Arch Filament System is emerging at the same spot where two AFSs appeared and grew in succession over the previous three days. Such rejuvenations are an important way active regions keep growing. The circles show the location of two new sunspots which appeared on white-light pictures in the course of the day. [Big Bear Solar Observatory, 1970 May 27; north at top, east at left.]

1967–1969. So once again we find that the emergence of new (unrejuvenated) AFSs is random: they neither choose nor avoid preexisting active regions.

Of course, the active region shows new life when a new AFS pops up by chance nearby, as Zirin (1970, 1972) shows for some examples. We have seen no evidence for any significant growth of an active region except near a new or rejuvenated AFS or other compact Emerging Flux Region.

## III. Twisted Flux and Asymmetric Sunspot Groups

A long-standing puzzle of solar physics is the asymmetry of sunspot groups. Why is there often one large preceding spot and a cluster of smaller following ones? Why are nearly all groups slightly tilted, with the preceding spots nearer the equator? Several people have speculated that these and other features of sunspot groups can be explained by twisted flux tubes (see Leighton 1969 and references therein).

There is some evidence that the field lines in new AFSs are twisted. Most new AFSs appear to rotate; that is, the orientation of the arch filaments changes as new flux emerges. The tilt may change by 20° in under an hour, so it is too fast to explain with Coriolis or convective forces; it can only be a geometric effect. The simplest field configuration which would give this effect is a twisted flux tube. In such a tube the orientation of the field lines does change with depth in the right way; the deeper lines, emerging later, are oriented closer to the equator (see fig. 4).

We have previously (Weart 1970a) given evidence that most new AFSs fall into one of two distinct groups: those whose initial tilt relative to the equator is "correct" (preceding polarity nearer the equator), and those whose initial tilt is "incorrect." Both types rotate during the first day or so of their existence but in opposite directions, as if they stemmed from flux tubes twisted with opposite handedness. An "incorrectly" tilted AFS never rotates so far as to take up a "correct" tilt, nor vice versa. (We found one apparent exception to this rule, a new AFS which was very steeply tilted "incorrectly" and which later took up a "correct" tilt [Mount Wilson group 16430], but it rotated in the opposite direction from all other "incorrectly" tilted AFSs we have seen. Evidently it was twisted with "correct" handedness but, unlike most, first appeared so steeply tilted that the following magnetic polarity was slightly westward of preceding polarity. This chain of events is verified by Hα pictures and Mount Wilson sunspot drawings and magnetograms.)

As we reported earlier, new AFSs whose initial tilt is "correct" are much more likely to grow over the next five days than are "incorrect" ones. We will now assume that all these facts are related to twists in the magnetic field lines, give a heuristic model, and discuss a test of the model.

We assume the following: (1) the subsurface solar field is partly in the form of twisted flux tubes. (2) There are roughly equal numbers of flux tubes of each handedness. Emerging, they give "correctly" and "incorrectly" tilted new AFSs in roughly equal numbers, as observed. But (3) only those new AFSs that are twisted with the proper handedness survive, and this explains the observed asymmetry of sunspot groups. That is, some unknown mechanism (perhaps Coriolis forces or differential rotation) encourages tubes of one handedness to form large sunspot groups, and inhibits those of opposite handedness.

To test this theory, we inspected a uniform sample of some 60 new AFSs seen on movies for 1967 February–August. Eliminating the small number whose initial field pattern was too complex to be determined, we find only one AFS of initially "incorrect" tilt which still had sunspots five days after it appeared. This is a small fraction—but should not this single case be enough to disprove our contention?

The group that appeared 1967 June 3 (Mount Wilson group 16391) had initially "incorrect" tilt and lasted about a week. On drawings for June 6 and 7 (fig. 5) we see that its tilt is still "incorrect," though it has rotated somewhat. Moreover, it has a large *following* spot and a cluster of smaller preceding spots: it is a mirror image of a normal group. So we can still say that asymmetry of sunspot groups is explained by twisted flux tubes; in this case, a tube of "incorrect" handedness managed for once to survive.

Naturally we wondered if further examples could be found. A search turned up only one other relatively long-lived "incor-

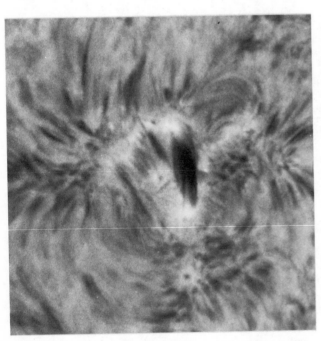
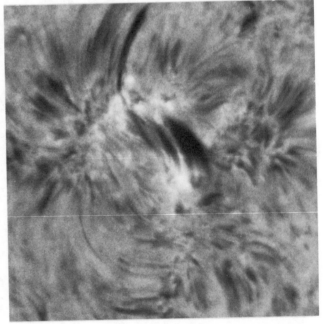

16:13:25

17:44:20

**Figure 4.** Typical "rotation" of a new AFS. In a short time, one set of arches is replaced by another set less steeply inclined. [Big Bear Solar Observatory, 1971 June 17, blue wing of Hα; north at top, east at left.]

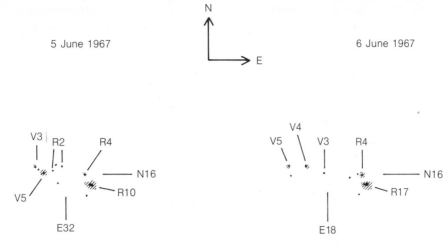

**Figure 5.** These drawings show an exceptional sunspot group: it emerged with a flux tube twisted with "incorrect" handedness (as inferred from AFS tilt and "rotation") but nevertheless survived for over 5 days. This northern-hemisphere group is the mirror image of a normal group, having larger following (*R*) spots that are closer to the equator than the preceding (*V*) spots. Latitude and longitude are indicated. [From Mount Wilson Observatory drawings, group 16341.]

rect" new AFS. The group that appeared late on 1971 June 16 (Mount Wilson group 18459) had a clear new AFS associated with it on June 17. The arches were inclined very steeply with "incorrect" tilt, and rotated as nearly all such AFSs do. The change in tilt (fig. 4) was most noticeable in the interval 16:15–17:45, when one set of arches was replaced by another set that was some 20° closer to a line of latitude. (The rotation continued, more slowly, for several days, as may be seen in fig. 6.) This, then, is a pure case of a new AFS with "incorrect" handedness. It survived four days. We predicted that it would retain its "incorrect" tilt, and the sunspot drawings of figure 6 show that it did. Moreover, no large preceding spot appeared—on the contrary, in the last two days it was given a Mount Wilson classification of βf ("bipolar, following polarity dominant"). The ESSA drawing for June 19 shows a large following spot.

We have given preliminary evidence for the following hypotheses: (1) the asymmetry of active regions stems from twisted flux tubes; (2) these emerge as new AFSs and then interact with other forces to grow, depending on the handedness of the twist; (3) the new AFSs emerge at random locations in the zones of solar activity; (4) they may revive, adding new flux to the center of an older bipolar sunspot group.

There are a number of ways these hypotheses can be tested and extended. We have seen the birth of only two sunspot groups with reversed asymmetry, but the reader is invited to search for others; we predict that he will find that most or all such groups stem from new AFSs of "incorrect" tilt and rotation. If the reader studies the evolution of active regions, we predict he will find that revivals of AFSs (and occasional new AFSs nearby) are the key to all regions which grow to major size, except those whose field configuration was confused

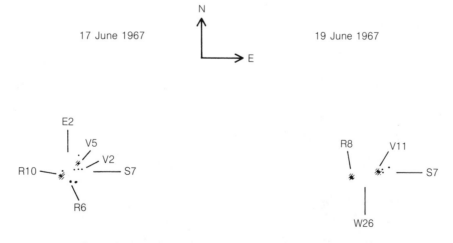

**Figure 6.** A second example of a reversed group, this time in the southern hemisphere. We predicted from the "incorrect" handedness of the AFS on June 17 that the group would retain its normal tilt and would not develop a dominant preceding spot. [From Mount Wilson Observatory drawings.]

from the start. Much of this work could be done from magneto-grams or from sunspot drawings with known polarities. We suggest to the theoretician that he study the effect of Coriolis and differential rotation forces on twisted flux tubes or that he seek the mechanism of AFS rejuvenation in the interactions of magnetic flux and convection beneath bipolar sunspot groups.

We do not know how the flux tubes get twisted in the first place. Supergranular convection currents can twist the indi-vidual arch filaments—the long upper filament in figure 3, for instance, looks rather helical—but convection may have too little energy to wind up the whole many-stranded "rope" (Weart 1970b). Deeper currents may be involved, perhaps the meridional circulation proposed by Durney and Roxburgh (1971).

Many people contributed to taking the Hα movies used in this study, at Big Bear Solar Observatory under H. Zirin and at Lockheed Observatory under H. Ramsey, and I owe thanks to all of them. Dr. Zirin in particular contributed many ideas. I am indebted to Dale Vrabec and others at Aerospace Observa-tory for use of their films and for fruitful discussions. This work was funded in part by NASA grant 05-002-034 and NSF grant GA-13642.

## References

Babcock, H. W. 1961, *Ap. J.*, 133, 572.
Bruzek, A. 1967, *Solar Phys.*, 2, 451.
———. 1969, *ibid.*, 8, 129.
Bumba, V., and Howard, R. 1965, *Ap. J.*, 141, 1492.
Durney, B. R., and Roxburgh, I. W. 1971, *Solar Phys.*, 16, 3.
Leighton, R. B. 1969, *Ap. J.*, 156, 1.
Parker, E. N. 1955, *Ap. J.*, 121, 491.
Prata, S. 1971, *Solar Phys.*, 19, 92.
Roberts, P. H. 1970, unpublished doctoral thesis, California Institute of Technology.
Vrabec, D., and Weart, S. R. 1972, in preparation.
Weart, S. R. 1970a, *Ap. J.*, 162, 987.
———. 1970b, *Solar Phys.*, 14, 274.
Zirin, H. 1970, *Solar Phys.*, 14, 328.
———. 1972, *ibid.*, 22, 34.

# The Solar Wind Blows Some Good for Astronomy

20

John C. Brandt

From the time of the high priests of Central America's ancient sun kingdoms to the bikini-clad bodies strewn across today's resort beaches, sun worshipers have been awed by the sun's regular appearance, its light and its heat. Most of us today are also aware that the prodigious energy of the sun somehow powers all of life on this planet and, assured by astronomers that the sun will continue to shine beneficently for a long time, we are satisfied and pass on to other things.

Nonetheless, reports of solar flares, sunspots and other solar phenomena periodically impinge on the public consciousness—reminders that things are not so simple. Only recently people began to hear of something quite new—the "solar wind." Virtually unknown until 1958 and extensively measured for the first time in 1962, the solar wind—a steady emission from the sun's corona of subatomic particles, mostly protons and electrons—has proved to be a significant component of the entire solar system. Scientists are just beginning to understand its effects on such permanent residents of interplanetary space as meteoric dust, the moon and the planets themselves, including our own.

But the widespread actions of the solar wind—and indeed its very existence—were first perceived through the study of those infrequent visitors to the solar system, comets—specifically comet tails. There are two major types of comet tails and both can be seen in the photograph of the comet Mrkos. A homogeneous tail curves back toward the path of the comet's orbit opposite to its motion; it consists chiefly of dust particles which scatter sunlight and produce a yellow glow. The other tail streams out almost directly away from the sun and is visible as a blue glow when seen in color photographs. Spectrographic analysis has shown that this kind of tail is a gas composed primarily of ionized molecules, the most prominent being carbon monoxide—that is, CO molecules that have lost one electron and therefore carry a positive electrical charge. Physicists call such an ionized gas a *plasma*. (See "Physicists probe the ultimate source of energy," SMITHSONIAN, December 1972.)

Astronomers long thought that the weak pressure of light from the sun was sufficient to force the small dust particles and gas from a comet's head back into its tails. But here and there along the ion tail one can see denser clouds of expelled gas or knots, as they are called. These knots have been seen to move down the tail and away from the sun at far greater accelerations than can be accounted for by the pressure of sunlight.

In the early 1950s an eminent German astrophysicist, Ludwig Biermann, suggested that the force needed to move the knots at such speeds could arise from the effect of the corpuscular radiation—an ionized gas or plasma streaming out from the sun at velocities of hundreds of miles per second. Since comets showed the effects of the corpuscular radiation regardless of the time or location of the comet, Biermann suggested that the solar corpuscular radiation is emitted from the sun in essentially all directions and at all times.

That the sun gives off some of its mass (its "corpus"), as well as energy, was not an altogether new or surprising idea. The sun's visible disk, called the photosphere, is surrounded by a tenuous, extremely hot plasma called the corona. Various independent methods had shown that the corona's temperature is in the vicinity of four million degrees Fahrenheit. Typical protons in a gas that hot have a velocity of 110 miles per second; some are much slower, some are much faster. A small fraction of the protons in the corona must, it was reasoned, exceed 270 miles per second, which is the velocity protons need to escape the sun's gravity.

By the late 1950s, astrophysicists, notably Sydney Chapman of the Geophysical Institute, University of Alaska, and Eugene N. Parker of the University of Chicago, had produced mathematical models for the extended solar corona. In Parker's model there was an outflow of solar particles dense enough, it turned out, to account for the effects of Biermann's corpuscular radiation on comet tails.

Not every scientist accepted this as a fact, but it was a sufficiently interesting hypothesis that detecting devices were placed on board the earliest satellites to seek any trace of what Parker had by then christened the "solar wind." The first actual measurements of this phenomenon were made by the

From *Smithsonian*, vol. 3, no. 10, pp. 30-35, 1973. Reprinted with permission.

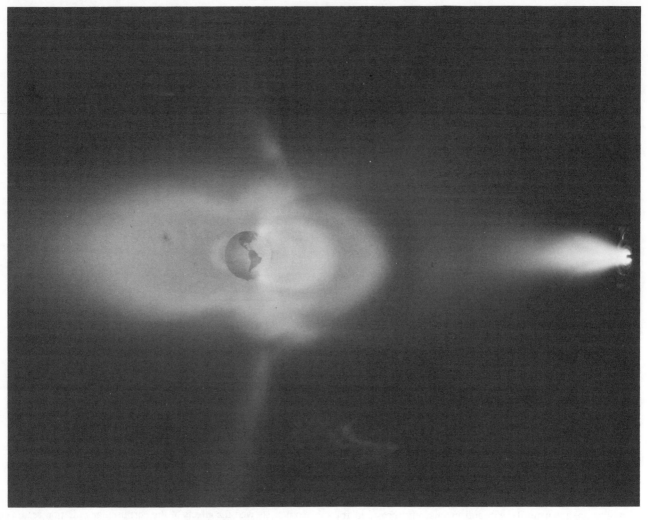

In a demonstration of solar wind, NASA scientists beam an
ionized gas into a magnetic field (caused by a magnet in the
model Earth). The result is an "auroral" glow around the
globe. [NASA Lewis Research Center.]

Russian satellite Lunik III in 1959, and any remaining doubts
about its existence were removed in 1962 by the extensive
measurements made by the American Venus probe, Mariner 2.
Since that time most of what we have learned about the solar
wind has derived from instruments on board space probes.

Approximately one decade of space measurements has
established the properties of the solar wind as it blows into
the vicinity of the Earth. It is a completely ionized gas consisting chiefly of some 80 protons and electrons per cubic
inch on the average, though the density varies between one-
tenth to ten times the average. The temperatures of the protons
and electrons are roughly 100,000 degrees Fahrenheit and
400,000 degrees Fahrenheit respectively. In addition to protons
and electrons, heavier nuclei have been discovered in the solar
wind, such as helium, carbon and oxygen.

Near the Earth, this constant stream of the sun's corpus
flows at an average speed of about 300 miles per second, though
the speed, too, varies—from one-half to twice the average.

One might expect that the solar wind, like sunlight, would
flow radially from the sun, that is, as if on a straight line from
the center of the sun outward into space. But measurements
made near the Earth have shown that, as these particles stream

from the sun's corona, their paths bend slightly in the direction
the sun rotates. This slight aberration is of profound significance: It means the sun's rotation is slowing down.

The slowdown is the result of both the solar wind and the
sun's magnetic field. Everyone is familiar with the action of the
Earth's magnetic field on a small magnet called a compass: A
force is exerted which lines up the compass needle with the
Earth's magnetic field lines (see drawing). A magnetic field
also tends to deflect protons and electrons—such particles
spiral around magnetic field lines and move along their length
rather than moving across them.

The sun also has a magnetic field and as solar wind particles
stream outward, they cannot cross the magnetic field lines.
Thus, the solar wind flow draws out the sun's magnetic field
and carries it along. The magnetic field lines act like "elastic
strings" which connect the sun and the solar wind. As the sun
rotates, these "strings" exert a force that pulls the solar wind
particles in the direction of the sun's rotation.

If the sun's magnetic field were strong enough, the entire
solar wind would rotate once every 27 days, as the sun itself
does, but the actual magnetic field is only strong enough to
cause a small deflection.

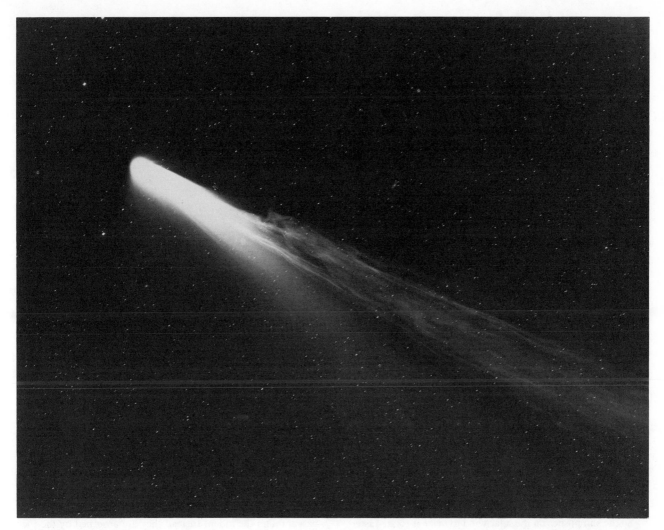

Above the smooth dust tail of the comet Mrkos, seen in 1957, is a kinky tail of ionized gas, or plasma, that streams out more directly away from the sun, the result of the solar wind. [Hale Observatories photograph.]

Ludwig Biermann, a German astrophysicist, first noticed the solar wind's actions in the tails of comets.

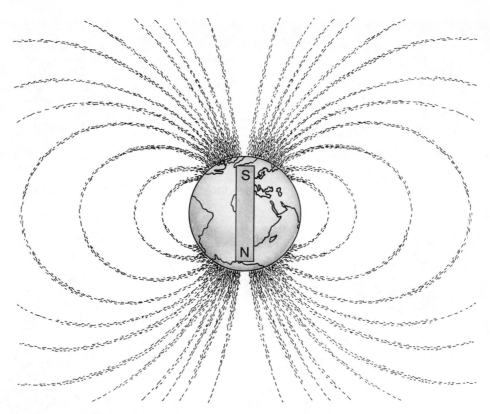

Earth's magnetic field: It is as though a bar magnet was embedded inside the planet and, as in the classic game with magnets, iron filings are shown pulled into positions along the field lines.

So the sun pushes on the solar wind to make it rotate faster. But the crucial point is that, according to one of Newton's laws, every action produces its own reaction. The solar wind is also pushing on the sun to make it rotate more slowly.

Measurements made near the orbit of Earth have shown that the solar wind motion in the direction of the sun's rotation is about four miles per second, and from that figure we can calculate the rate at which the sun is being braked. At present rates, the sun's rotation would be slowed to 40 percent of what it is today in some four billion years.

There are, of course, many stars in the universe with the same mass and photospheric temperature as the sun, and, of these, astronomers have found, the younger stars rotate faster than the older ones. This is now easy to explain—if these stars have magnetic fields and stellar winds similar to the solar wind. But closer to home, in our solar system, we know that the solar wind interacts in some way with almost everything. The moon is a case in point.

Satellites placed in lunar orbit have determined that the magnetic field in the solar wind apparently passes through the interior of the moon and comes out the other side relatively undisturbed. If the moon had a metallic core, as does the Earth, this could not occur. Thus, the solar wind permits us literally to probe the lunar interior, and the evidence, backed up by other observations (but not without some controversy), points to a cool, rocklike interior.

Of course, the protons and electrons themselves do not pass through the moon. Instead they are absorbed by the rock and dust on the moon's surface. This may explain why the lunar surface material is rather dark and reflects only a small fraction of the sunlight. Laboratory experiments in which terrestrial rock powder has been bombarded with a homemade solar wind have indicated that this is the case, and now that genuine lunar samples are available for study, it is only a matter of time before this question is answered. Space probes are also beginning to return more precise insights into the solar wind's effects on Mars and Venus.

The most thoroughly studied effects of the solar wind so far are its interactions with the Earth, and here again we are dealing with a magnetic field. While currents deep in the Earth's interior are believed to cause the Earth's magnetic field, it is easier to think of a powerful bar magnet embedded in the planet with its axis pointing to a spot near Thule, Greenland.

The Earth's magnetic field lines, if left to their own devices, would not be unlike the pattern formed when you place a strong magnet beneath a piece of paper on which iron filings have been sprinkled (again, see drawing). But the particles in the solar wind cannot move across the Earth's magnetic field lines any more than they can cross the field lines of the sun. Thus, some 40,000 miles in space in the direction of the sun, the solar wind collides with the Earth's field lines. The field lines, unable to sustain themselves against the impact of the solar wind, have been blown in an opposite direction, away from the sun, forming the Earth's geomagnetic tail, which is some 800 Earth radii (or three million miles) long and perhaps even longer.

It is very much as if the Earth, surrounded by its invisible magnetic armor that we call the magnetosphere, were a projectile in a wind tunnel. Where the solar wind separates and flows around the projectile Earth is where the Earth's magnetosphere terminates. This is called the magnetopause. And just as in those split-instant photographs one sees of bullets or of prototype planes in a wind tunnel, a bow wave forms about 16,000 miles beyond the magnetopause. This comes about because the relative speed of the medium (air or solar wind plasma) with respect to the obstacle (the bullet or the Earth's magnetic field) is faster than the speed of sound in the medium.

The region between the bow wave, or shock, and the magnetopause is called the magnetosheath, an area inhabited by the solar wind after passing through the shock and by irregular magnetic fields, tattered fragments of the sun's "elastic strings."

In a diagram, this complex interaction of solar wind and magnetic field may seem like a relatively stable affair. Yet highly sensitive compasses have shown that the Earth's magnetic field is subject to almost continuous aberrations. And changes in the density and velocity of the solar wind occur frequently. Occasionally solar flares erupt from the corona; vast dense streams of protons and electrons collide with the Earth's magnetic field, distorting it further and causing a wide range of geomagnetic storms and other activities, ranging from the magnificent sight of an aurora to a teletype machine typing out nonsense all by itself.

The solar wind was discovered so recently that it is barely mentioned in most undergraduate astronomy textbooks, and its role in the solar system and beyond has only been glimpsed. (The answer is not known, among other questions, as to how far out from the sun the solar wind blows.) Not only will more be learned about this prominent presence in our corner of the universe but the solar wind itself will serve as an increasingly valuable tool for understanding other astronomical phenomena. For example, far beyond the Earth the solar wind collides with interstellar gas and observing its effects may shed light on the interstellar gas unobtainable by other methods. The solar wind, indeed, may become a large laboratory for investigating the nature of plasmas, the most common material of the universe.

The sun's disk [here eclipsed by the moon] is surrounded by the corona, a tenuous glowing gas at four million degrees Fahrenheit. That is hot enough for some particles to escape, forming the solar wind. [High Altitude Observatory]

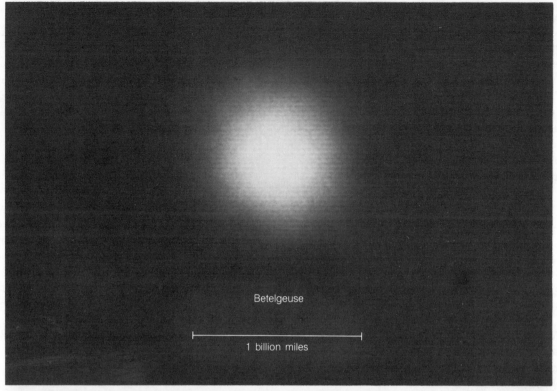

Betelgeuse

|———————— 1 billion miles ————————|

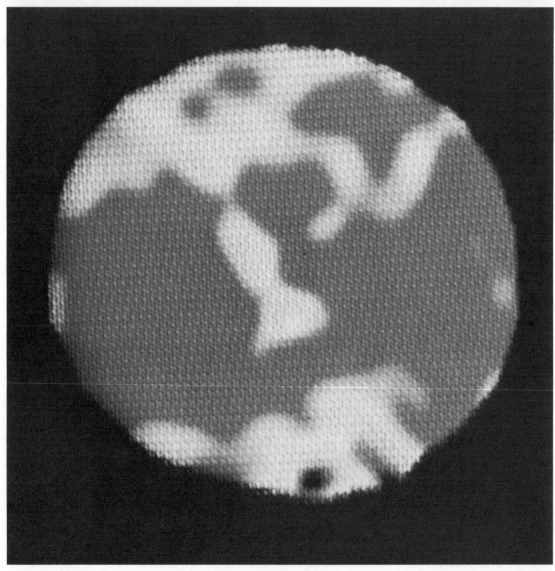

# VI

# THE STARS

The observation of cepheid variable stars has been perhaps the most important tool in determining the scale of distances in the universe. This topic is reviewed by Lick Observatory astronomer Robert P. Kraft in his article "Pulsating Stars and Cosmic Distances."

"We spent another seven years verifying the data before we were certain enough of our evidence to publish the results: an object, which could only be called a planet, was orbiting another star." Peter van de Kamp tells the story of his famous discovery in "Barnard's Star: The Search for Other Solar Systems." Since this article was written, other astronomers have attempted unsuccessfully to confirm his findings, while Professor van de Kamp has reanalyzed his data and defended them. Until confirmation is accomplished, however, we will lack generally accepted proof of the existence of planetary systems outside our own.

"While most known contact binaries are just barely in contact, in that the stars are only slightly larger than their lobes, a few are known to be substantially larger, and have a thick, connecting neck." Robert E. Wilson used an electronic computer to help illustrate his account of the strange effects that occur when two stars are so close together in "Binary Stars: A Look at Some Interesting Developments."

Photograph made with Mayall telescope seems to show structure on the supergiant star Betelgeuse (diameter about five-hundredths of an arc second). If real, the "starspots" may correspond to gigantic convection cells, larger than our sun. Lower picture has been enhanced by computer to bring out detail. [Kitt Peak National Observatory.]

# 21

# Pulsating Stars and Cosmic Distances

Robert P. Kraft

Our present picture of the universe—its structure, size and age—rests to a large extent upon observations of a few pulsating stars. Each of these stars waxes and wanes as much as one full magnitude (2.5 times) in brightness according to a fixed rhythm ranging in period from less than a day to more than 50 days. In general, the longer the period, the greater the luminosity of the star. Such stars are called cepheid variables after their prototype, star delta in the constellation Cepheus; the most familiar of them is Polaris (the pole star), which brightens and fades in a period of 3.97 days. We do not know what causes the pulsation of cepheid variables, nor what it signifies in the biography of a star. Some 40 years ago, however, by a bold stroke of invention, the variable luminosity of these stars was made to furnish a distance scale that gives astronomy its reach into the cosmos beyond the immediate neighborhood of the solar system.

The new distance-scale at once made it possible to locate the center and to measure the dimensions of our galaxy. A few years later the presence of cepheid variables in celestial objects such as the Great Nebula in Andromeda helped establish that these "nebulae" are themselves galaxies—island universes as large as our own located at immense distances out in space. But in recent years the profound usefulness of the cepheid distance-scale has been almost overshadowed by its defects. Corrections in the scale have made it necessary for the dimensions of the observable universe outside our galaxy to be doubled, and still further revisions may be required. Be-

cause the age of an expanding and evolving universe can be deduced from its distance scale, cosmologists have concurrently had to revise the age of the universe upward, from two billion to perhaps 10 billion years. These corrections and further refinements still in progress derive from closer study of the cepheids themselves. It now seems safe to say that the cosmic distance-scale will not again expand so radically, and that it is at last ready for secure calibration.

In all likelihood we shall achieve this objective still without understanding why the cepheids pulsate. Among the 15,000 stars listed in the monumental new Soviet *Variable Star Catalogue*, edited by B. V. Kukarkin, P. P. Parenago, Y. Efremov and P. Kholopov, about 3,000 exhibit the regular pulsation of the cepheids. Spectroscopic observation shows that the surface temperature of these stars varies upward and downward in phase with their light. Apparently they also expand as they brighten and contract as they fade. In the 1920's Sir Arthur Eddington was able to show theoretically that the rate of pulsation must be related to the mean density of the cepheid (its mass divided by its volume), much as the period of a pendulum on earth is governed by its length. But we have no mechanism to explain this behavior, and we cannot say why a star becomes a cepheid.

The most important advance in our knowledge of the cepheids—and the most drastic revision of the distance scale—came a decade ago with Walter Baade's discovery that the stars of the

universe may be divided into two major populations. To Population I, made up of young, hot, short-lived stars, he assigned the brighter and longer-period cepheids that appear in the arms of spiral galaxies. The fainter and shorter-period cepheids associated with the globular clusters that swarm around the centers of galaxies Baade placed among the older and longer-lived stars of Population II. While astronomers now believe that Baade's two populations represent an oversimplification and that stars are more continuously graded in age, the cepheids seem mostly to belong to the extreme ends of the population spread.

At present we imagine that the young Population I cepheids represent a phase in the life of any star. If we plot the color (that is, the temperature) of stars against their absolute luminosity (their intrinsic brightness corrected for distance), most of them occupy a rather well-defined "main sequence" [*illustration on page 158*]. To the right of the main sequence is a scattering of other stars, most of them "red giants." Between the main sequence and the red giants is an "instability strip" containing the cepheids. We presently conceive that a star starts out bright and hot, after a very rapid stage of gravitational contraction; then, after the star has consumed a certain amount of its hydrogen fuel, it begins to cool. Thus in terms of the color-luminosity diagram a star spends most of its life on or near the main sequence, but eventually evolves to the right. When it reaches the instability strip, it begins to pulsate. As the star passes through this strip, in the course of a few million years, its pulsation slows and lengthens in period. Upon reaching the end of the strip it ceases to pulsate and becomes a red giant. Ul-

From *Scientific American*, vol. 201, pp. 48–55, July 1959. Copyright © 1959 by Scientific American, Inc. All rights reserved. Reprinted with permission.

timately it dims into the graveyard of the white-dwarf stars [see "Dying Stars," by Jesse L. Greenstein; SCIENTIFIC AMERICAN, January, 1959].°

This hypothetical account does not, however, cover the evolution of the old Population II cepheids found in globular clusters. Perhaps these enter the instability strip by evolving "backward" from the red-giant phase instead of from the main sequence. Most of the globular-cluster cepheids have very short periods of less than a day, but even those having longer periods can be clearly distinguished from Population I stars of similar period. Long-period globular-cluster cepheids are on the average 1.5 magnitudes fainter than the younger long-period cepheids, exhibit quite different spectra and have masses only about a fourth as large.

The cepheids are highly luminous stars. Polaris, the nearest of them, is not a particularly bright cepheid, but it is about 600 times brighter than the sun. The brightest Population I cepheids are more than 10,000 times more luminous than the sun! This is a fortunate circumstance so far as the measurement of extragalactic distances is concerned, because it means that such stars make themselves visible at very long range.

In order to understand how pulsating stars can furnish a distance scale, we must go back 50 years to the work of Solon I. Bailey and Henrietta S. Leavitt of the Harvard College Observatory. Bailey carried out an extensive investigation of the cepheids in globular clusters within our own galaxy. He found that almost all had periods of less than a day, except for a few that had periods in the range of 12 to 20 days. Miss Leavitt later studied the cepheid variables that appeared in great numbers in photographs of the Clouds of Magellan, the two small galaxies that are companions of our own; she found that most of these cepheids had periods of more than a day. Even more remarkable was Miss Leavitt's discovery that the average apparent brightness of the Magellanic Cloud cepheids is directly correlated with the length of their respective periods of pulsation. Bailey had found no such dependence of luminosity on period in the globular-cluster cepheids, at least those with a period of less than a day.

Astronomers soon recognized the promise of Miss Leavitt's finding. It was known even then that the Magellanic

LONG-PERIOD CEPHEIDS are found among young Population I stars such as these in the Andromeda Nebula. A small section of one arm of the Nebula was photographed at two different times with the 200-inch telescope on Palomar Mountain. The marked star at upper left is a cepheid variable; the other star (*invisible in bottom picture*) is a nova.

°See also Articles 26 and 28, by Ben Bova and Louis C. Green, respectively, in this reader.

**SHORT-PERIOD CEPHEIDS** are found in globular clusters of old Population II stars. These photographs of M 3, one of the globular clusters of the Milky Way, were taken 18 hours and 43 minutes apart with the 100-inch telescope on Mount Wilson. Four of the

Clouds are distant congregations of stars. Thus the cepheids in the Clouds are all at virtually the same distance from the solar system, and the light of all is attenuated to the same extent by its journey to the earth. Miss Leavitt's measurements of the varied apparent brightness of these stars could therefore be taken as indications of their relative absolute brightness. Here was a potential yardstick for measuring really long distances in the universe!

It was obvious that, if the distance of the Magellanic Clouds could be ascertained, one could determine the absolute brightness of the cepheids. Miss Leavitt's period-luminosity scale could then be used to find the distance to any stellar system or subsystem containing cepheids by turning the problem around: Measure the period of the cepheid, read off its absolute luminosity from the period-luminosity scale, compare this with the observed apparent luminosity of the cepheid and find the star's distance by applying the law that the intensity of light varies inversely with the square of the distance. Of course the accuracy of such a measuring rod depends on the assumption that cepheids in all parts of the universe obey the same period-luminosity law Miss Leavitt had derived from the cepheids in the Magellanic Clouds. This turned out to be a pivotal assumption.

At the time of Miss Leavitt's discovery there was unhappily no way to ascertain the distance of the Magellanic Clouds. Stellar-distance measurement still depended on direct trigonometric parallax, which is effective only for nearby stars. Against the background of stars distributed in the depth of space at all distances from the sun, a nearby star appears to shift its position as the earth travels from one side of the sun to the other. It is thus possible to measure the distances of such stars by simple trigonometry [see illustration on page 157]. Even these distances are so large that it is convenient to describe them with a unit called the parsec. We say that a star is at a distance of one parsec if its parallax, that is, half its shift of position, equals one second of arc. But the nearest star has a parallax of slightly less than .8 second of arc. This corresponds to a distance of slightly more than 1.3 parsecs, or 25,000 billion miles. Sirius, the brightest star in the sky, is 2.7 parsecs away, and the parallax of a star at a distance of 100 parsecs is only .01 second. Such small angles cannot be determined very precisely; a distance of about 30 or 40 parsecs is the practical limit for determination by direct trigonometric means.

The cepheids are so rare in space that the nearest of them—Polaris—is 90 parsecs away. It is clear, therefore, that trigonometry could not be used to de-

termine the distance of a single cepheid, and could yield no information on the absolute brightness of even the nearby cepheids.

How, then, could the distance to any cepheid be obtained? Before Miss Leavitt had made her discovery, astronomers had devised a method for measuring what might be called the "middle distances" of our galaxy. With so many stars on our photographic plates we may assume that many stars in any given group have the same absolute brightness. We may also assume that the motions of these stars, either radially in the line of sight or transversely across the sky, will be at random. Now with the spectrograph we can determine the actual radial velocity of any observable star, independent of its distance from us. The spectrum is shifted toward the violet if the star is approaching and toward the red if it is receding, and the extent of shift gives us the velocity of its motion. On the other hand, the apparent transverse motion across the line of sight (called the proper motion) does depend on distance. If the stars of our given group move, on the average, with the same actual velocities independent of distance, then the proper motions of these stars will appear to get smaller with distance. Of course relatively few stars are near enough to the sun to have exhibited any proper motion during the

more than 200 cepheids in the cluster are marked by pairs of horizontal lines. On comparing the marked stars in the two photo-graphs, one can see a small but perceptible change in their luminosity. The star at top right, for example, becomes brighter.

first century of photographic astronomy. But when we have determined the statistical spread of the radial velocities, it is reasonable to suppose that the proper motions vary in the same range. Since the distribution of proper motions does decrease with distance, the identification of the spread in radial velocities with the spread in proper motions indicates the average distance to the group of stars under consideration.

With the mean distance obtained in this way, one can correct the mean apparent magnitude of the stars for the effect of distance and get the average absolute magnitude. From studies of this sort in 1913 Ejnar Hertzsprung of Denmark found an average absolute magnitude of −2.3 for a cepheid with a period of 6.6 days. (On the magnitude scale the lower number refers to the brighter star; stars brighter than the first magnitude have negative magnitudes.) Hertzsprung's result was based on only 13 nearby cepheids for which the proper motions were known. But astronomers now had the absolute luminosity value needed to convert the apparent luminosity of any cepheid to absolute luminosity by reference to Miss Leavitt's period-luminosity scale.

In 1918 Harlow Shapley of the Mt. Wilson Observatory saw how the scale could be applied to determine the dis-

tances of the globular clusters in our galaxy. He fitted the long-period cepheids (periods of 12 to 20 days) of the globular clusters to the period-luminosity scale for the cepheids of the Magellanic Clouds. From this he determined the absolute luminosity and hence the distance of the long-period cluster stars. Using this determination of the distance to the clusters, he deduced that the mean absolute magnitude of the numerous fainter cepheids in the clusters with periods of less than a day was a little brighter than zero (i.e., some 100 times brighter than the sun). Shapley then had a scale to measure the distance to the clusters that contain only faint, short-period cepheids. From the globular-cluster distances thus derived, he deduced that the globular-cluster system was centered on a point about 16,000 parsecs from the sun in the direction of the constellation of Sagittarius. It seemed reasonable to identify this point with the center of our galaxy. Shapley had obtained the first good estimate of the size of any galaxy. Later determination of the luminosities of these shorter-period cluster cepheids, obtained by proper-motion and radial-velocity studies, have verified Shapley's deduction and shown his estimate to be of the right order.

The period-luminosity scale could also be used to estimate the distances to any nearby galaxy that contains ceph-

eids. Edwin P. Hubble and his associates at the Mount Wilson Observatory soon ruled off the distance to the Magellanic Clouds and to the Great Nebula in Andromeda. By the comparison of apparent to absolute magnitude thus effected for these and other more distant galaxies, the cepheid distance-scale made it possible to calibrate the spectrographic shift toward the red for the measurement of distances to the throngs of even more distant galaxies so faint and tiny that the cepheids and other stars in their populations cannot be resolved. Cosmologists working from these data were able to estimate the size of the universe and its age from the time of its initial expansion. All this extrapolated from the observation of the peculiar process of cepheid pulsation that we do not yet fully understand!

In the next 25 years, however, astronomers and cosmologists encountered numerous difficulties that cast increasing suspicion on the period-luminosity relationship upon which the whole edifice was built. All other galaxies, as measured by the cepheid distance-scale, were smaller in size than our own, a peculiarly self-aggrandizing result. As nuclear physicists succeeded in calibrating the rate at which uranium and thorium have been decaying to lead in the rocks of the earth, their "clocks" made the earth ap-

156

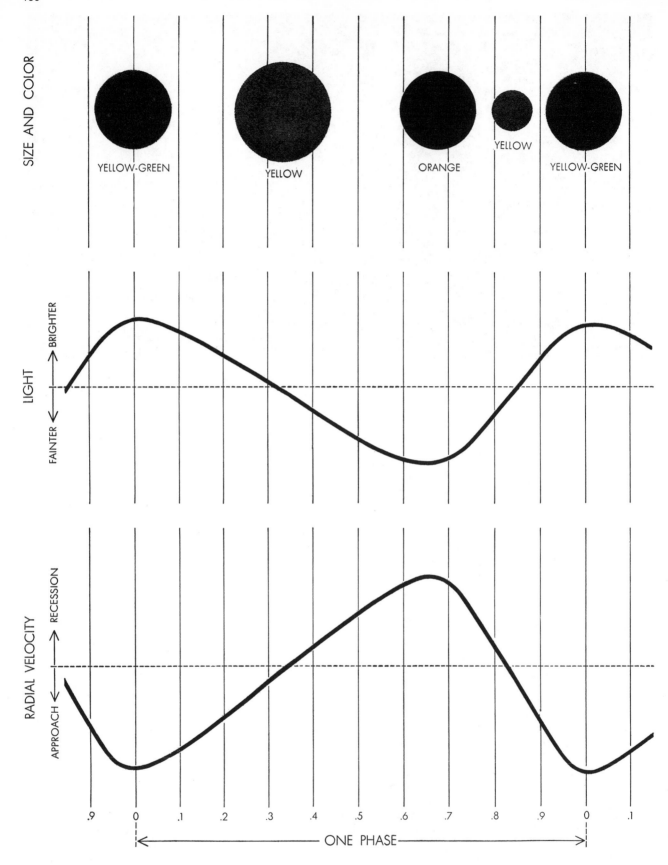

CYCLE OF A TYPICAL CEPHEID includes changes in color (*top*), light (*middle*) and radial velocity (*bottom*). Maximum light coincides with the bluest color (*yellow-green*), that is, with the highest surface temperature. The fluctuations in radial velocity are probably the result of changes in the size of the star such that its radius is largest (*large yellow disk*) midway between maximum and minimum light, and smallest (*small yellow disk*) in the opposite part of the cycle when the light is increasing. The relative sizes have been exaggerated in the drawing for purposes of clarity; the change in radius is never more than 20 per cent.

pear considerably more ancient than the universe. There was difficulty also in reconciling Eddington's calculation of the mean density of the cepheids with density estimates derived from the relationship of the observed luminosity of these stars to their rate of pulsation.

An observation by Hubble and Baade finally opened the way to a test of these suspicions. They pointed out that, if the distance to the Andromeda Nebula had been correctly measured, then the brightest stars of the globular clusters surrounding its central region appeared to be too faint compared to the brightest stars in the globular clusters of our own galaxy. If these bright stars in the Andromeda Nebula were assigned the same absolute brightness as the corresponding stars in our galaxy, then the cepheids visible in the Andromeda Nebula and many of the longer-period cepheids in our own system would also have to be assigned a higher absolute magnitude with respect to the shorter-period cepheids of the globular clusters that had formed the basis of Shapley's scale. Could it be that the globular-cluster cepheids obeyed a period-luminosity law different from that observed for other cepheids?

Such a possibility was foreshadowed in 1940 by an observation made by Alfred H. Joy at the Mount Wilson Observatory. He found a marked difference between the spectrum of a 15-day cepheid in the vicinity of the solar system and a 15-day cepheid in a globular cluster. Then, during the war years, Baade was able to devote the 100-inch telescope on Mount Wilson almost full time to his study of the stellar populations in the Andromeda Nebula. In dividing all stars into two populations he also found a basis for classifying the cepheids into two species.

With the 200-inch telescope in operation on Palomar Mountain shortly after the end of the war, Baade set out to observe the two types of stars "side by side," that is, at the same distance. Unfortunately not even the 200-inch telescope can resolve the faint short-period cepheids in the globular clusters of the Andromeda Nebula. But Baade was able to measure the Population I cepheids of that galaxy with great accuracy against the brightest globular-cluster stars, for which absolute magnitude had been established with the help of the Population II cepheids in our galaxy. Shapley had set the absolute magnitude of these stars at −1.5, based upon his determination that the shorter-period cluster cepheids have an absolute magnitude of zero. The

distance to the Andromeda Nebula, calculated from its Population I cepheids in accord with the established period-luminosity scale, predicted that the bright globular-cluster stars should have an apparent magnitude of 20.9. Baade found that these stars were actually magnitude 22.4. In other words, they were 1.5 magnitudes fainter.

This demonstrated that the estimate of the distance to Andromeda was too small by a factor of about two. It also showed that the absolute brightness of the Population I cepheids in the Andromeda Nebula was 1.5 magnitudes brighter than had been indicated by the period-luminosity scale. They have a lower apparent magnitude because they are farther away than had been supposed. With distance to the Andromeda Nebula doubled, its size also doubled, bringing it into line with the size of our own galaxy. These results were dramatically confirmed when A. D. Thackeray and A. J. Wesselink of the Radcliffe Observatory in South Africa discovered short-period cepheids in the Large Magellanic Cloud at exactly the magnitude predicted by Baade.

Hindsight now fully explains the discrepancy in the period-luminosity scale. With the Population I cepheids advanced 1.5 magnitudes in luminosity, there is a discontinuity in the scale that clearly divides the cepheids into two types. We also understand why this distinction was missed in the early part of this century. The young Population I stars in the arms of our spiral galaxy lie close to its central plane; the brighter light of these stars is accordingly dimmed by the clouds of dust and gas in which stars are formed. The older Population II stars, which resemble the stars in globular clusters, have had time to drift above and below the galactic plane, so their dimmer light reaches us without obscuration. By a remarkable coincidence the interstellar absorption of the light from the Population I cepheids almost exactly equals the difference in the actual brightness of Population I and Population II cepheids, that is, 1.5 magnitudes. No such obscuration dims the light of Population I cepheids in the Andromeda Nebula or the Magellanic Clouds; their lower apparent magnitude is now correctly attributed to their greater distance. Thanks to this combination of circumstances Shapley was able to fit the long-period cepheids in globular clusters to Miss Leavitt's period-luminosity curve for the cepheids in the Magellanic Clouds. He could not

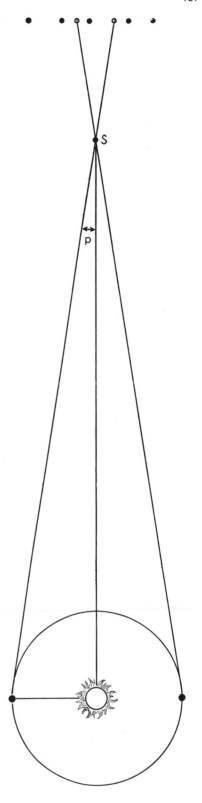

PARALLAX is used to measure the distance of nearby stars. As the earth moves around the sun (*bottom*) a nearby star (S) appears to change its position (*open circles*) in relation to stars much farther away. When the parallactic angle *p* is one second of arc, the star's distance from the sun is 19,000 billion miles, or 3.26 light-years, or one parsec. Here the change in the apparent position is much exaggerated.

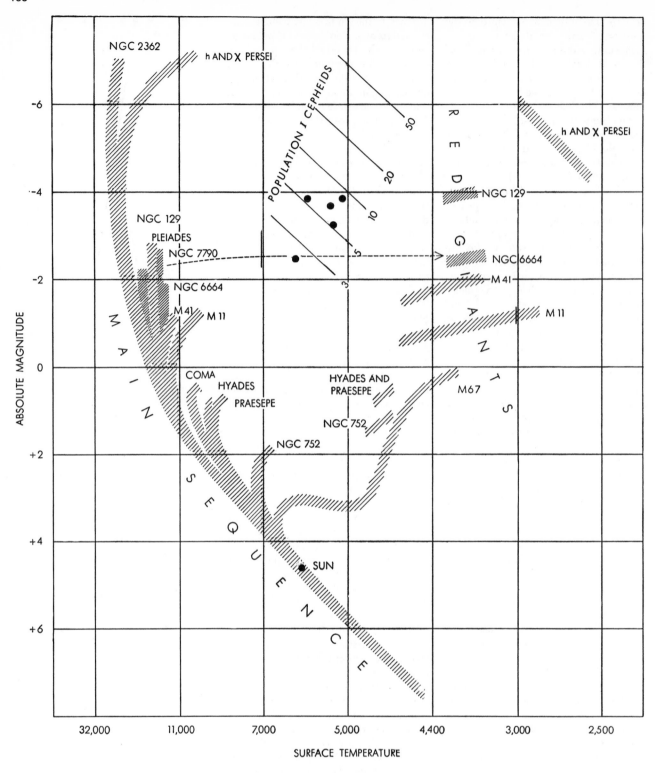

GALACTIC CLUSTERS OF THE MILKY WAY are plotted by determining the absolute magnitude and the intrinsic color of their stars. In this diagram the colors have been converted to approximately equivalent surface temperatures in degrees Kelvin. Galactic clusters are identified by their names or numbers; fine hatching marks those that contain long-period Population I cepheids. Five cepheids of the galactic clusters are shown in the "cepheid domain" which is marked off according to the length of the period in days. The broken line is the evolutionary track of a cepheid of NGC 6664. Originally the star was located among NGC 6664 stars on the main sequence, but about 100 million years ago it increased in luminosity about one magnitude and began to decrease in surface temperature, so that it moved horizontally across the diagram toward the cepheid "instability strip." Later it will become a red giant with a very large radius and a surface temperature of about 4,000 degrees K., like the present NGC 6664 red-giant stars.

have known that the two types of stars are quite different objects.

From the time of Miss Leavitt's first observations the distinction between the two species of cepheids had also been obscured by a scatter of about one magnitude in the positions of the stars along the mean line of the period-luminosity curve. For many years this was attributed to observational error and possibly to internal absorption within the Magellanic Clouds. But the scatter could also result from a bona fide physical departure of a given star from the mean line. This is a point of more than academic interest; such uncertainty in the magnitude of a particular star corresponds to a factor of 50 per cent in the computation of its distance. The range of error is too great if the objective is to measure the distance to a galaxy in which only one or two cepheids are available. Accurate determination of distances to individual cepheids has also assumed new importance in the study of our own galaxy. Population I cepheids might be expected to outline the spiral arms of our galaxy and, being very luminous, to carry our knowledge of the spiral structure to considerable distances from the sun.

We are now certain that the scatter is real. Highly accurate photoelectric measurements of cepheids in the Small Magellanic Cloud by Halton C. Arp of the Mount Wilson and Palomar Observatories have established that the scatter is very much larger than the errors of observation. Allan R. Sandage of the same observatories has offered an explanation. Sandage predicts from the theoretical period-density relation that the period-luminosity law must be amended to take account of a third variable. This variable is the surface temperature of the star.

Observations of certain cepheids for which highly accurate surface temperatures and absolute magnitudes can be derived seem to confirm Sandage's theory. These stars are members of loose clusterings of very young Population I stars in our galaxy called open or galactic clusters. The first two were found by John B. Irwin in 1955 at the Radcliffe Observatory. Others were located by Sydney van den Bergh and myself, and the number of such cepheids is now about 10. Their colors (hence surface temperatures) and absolute luminosities are obtained by yet another method for determining distances to stars. We may expect stars that are close together on the color-luminosity dia-

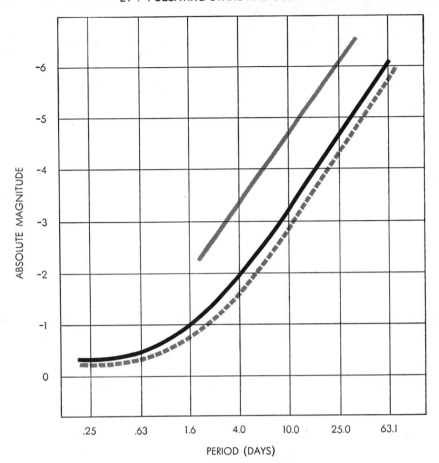

PERIOD-LUMINOSITY RELATION used by Harlow Shapley fitted all cepheids into one curve (*black*), with the short-period cluster variables at the lower end. The period-luminosity relation of Walter Baade divides the cepheids into Population I (gray) and Population II (*broken line*). The latter stars are fainter than Population I stars of the same period. On magnitude scale, brightness increases by a factor of 2.5 from −1 to −2, and so on.

gram, and thus are similar in color and spectral characteristics, to have the same absolute brightness. By matching some of the stars of a cluster to similar stars for which the distance is known, we can derive the distance to the cluster. We can then determine the luminosity of the other stars in the cluster. Unfortunately most of the galactic clusters are obscured by interstellar material. This material not only absorbs light, but also reddens it, making the surface temperature of a star seem lower. By observing these stars in several colors, however, it is possible to derive intrinsic colors and surface temperatures.

With Sandage's period-luminosity-surface temperature relationship apparently well sustained, we can now determine the distance of a single cepheid if we know its surface temperature and period. The procedure may be demonstrated by reference to the color-luminosity diagram on the opposite page. On this diagram the cepheid variables occupy a band that, at a given position, reaches horizontally across a temperature range of about 1,000 degrees absolute and vertically over a factor of about six in absolute magnitude. Sandage has computed the lines of constant period, which slope down diagonally to the right on the diagram. To use the diagram, we locate a given cepheid (not necessarily from a galactic cluster) on its appropriate period line. We then draw a vertical line from the base of the diagram corresponding to the temperature of the star. The intersection of this line with the period line gives us the luminosity with high precision. We can then obtain the distance to the star with what is hoped to be an error of less than 10 per cent.

The final result of these studies of cepheids in galactic clusters should be a useful and accurate period-luminosity-surface temperature chart for the cepheid variables. Astronomers may expect soon to have a much more reliable scale for measurement of long distances inside our own galaxy and beyond.

# 22 Barnard's Star: The Search for Other Solar Systems

Peter van de Kamp

For eighteen long years—during which the earth was convulsed by World War II, the reign of nuclear terror started, and new nations began to proliferate around the globe—my staff and I patiently made thousands of photographs of a single star; then measured differences in position of 1/25,000 of an inch. In 1956 we found what we were looking for. We spent another seven years verifying the data before we were certain enough of our evidence to publish the results: an object, which could only be called a planet, was orbiting another star. We had found a planet of Jovian dimensions, the first such planet ever found outside our own solar system.

To make the photographs we used a relatively modest 24-inch telescope. We worked in eastern Pennsylvania, a part of the country not renowned for clear skies. The real search came when we bent over the measuring machine, plotted the results, and then made the mathematical analyses that revealed what no earthbound telescope had ever shown, a planet orbiting another star.

The story actually began in June, 1916, when the astronomical community was surprised by the discovery that a faint red dwarf star was moving across the sky far more rapidly than any other known star. The motion was real, of the sort that astronomers call "proper motion." It takes about 170 years for the star to cross an apparent distance in the sky equal to the moon's diameter; this may sound a bit tortoiselike, but it is far faster than the motion of any star discovered before or since.

The discoverer, Edwin Emerson Barnard, was an amateur astronomer then on the staff of the Yerkes Observatory in Chicago. It had been known for nearly 200 years that the stars were not fixed on the celestial sphere but move relative to each other. Edmund Halley, whose namesake comet will return in 1986, first recorded these motions. How much a star will appear to move relative to other stars depends greatly on how close it is and in which direction it is moving relative to our line of sight. A star moving straight toward or away from us will not appear to be moving at all relative to the stars around it.

Barnard's Star, a 9.5-magnitude object in the constellation Ophiuchus, has a proper motion of 10.31 seconds of arc annually. It is 5.9 light-years away, the second nearest star. It appears to be moving across our line of sight at about 55 miles a second and to be coming toward us at 67 miles a second. The star's intrinsic luminosity is about 1/2300 that of the sun.

Because of its proximity, Barnard's Star is more than an oddity; it offers the best chance of detecting planets orbiting another star by observing their gravitational effects on the motion of their parent body. It is not now possible to detect such planets by direct means, for reasons that become clear if we reverse the situation. From the nearest star, Alpha Centauri, the largest planet in our system, Jupiter, would at best appear as a 23-magnitude object separated from the sun, now a first magnitude star, by no more than four seconds of arc. A still fainter earth would be totally lost in the glare of the sun.

Gravitational effects are observable at such distances, however; as long ago as 1844 F. W. Bessel discovered the unseen stellar companions of Sirius and Procyon by this method. The technique works because even if an object has no light of its own, its mass affects the path of a nearby object. Newton's laws of motion tell us that a single star, or the center of mass of a multiple system, will move at uniform speed in a straight line unless affected by some outside force. Any deviation reveals the presence of an unseen companion.

Barnard's Star thus offers a unique opportunity to detect unseen companions small enough to be planets. The task is simply stated: Plot the deviations in the path of Barnard's Star across the sky, and then calculate how large—or small—a body would be required to account for them. Performing the task was something else. The deviations, called perturbations, amounted to about 0.03 seconds of arc. Measuring them was like detecting a one-inch movement at a distance of one hundred miles.

The basic instrument is a long-focus telescope adapted for photography, really just a long-focus camera. The long-focus characteristic yields large-scale portrayal of small portions of

From *Natural History* Magazine, April 1970. Copyright © The American Museum of Natural History, 1970. Reprinted with permission.

Looking like automobile headlights, these double dots are actually twin images of faint stars caught in two photographs taken 11 months apart. The images that appear tilted in the center are of Barnard's Star. To make this picture the negatives were shifted so that the later image of all the stars fell to the right of the earlier. The images of Barnard's Star appear tilted, however, because of its perceptible movement toward the north (top of the picture) in the 11-month interval.

the sky, which in turn permits precise measurement of small, angular displacements. The precision of the instrument, the photographic plate, and the measuring machine are such that very high positional accuracy can be obtained. A single photographic exposure furnishes a position relative to a reference background of "fixed" background stars with an accuracy of about 0.04 seconds of arc. Multiple exposures, up to four a night, can improve the accuracy to 0.02 seconds of arc. By further combining several nights' observations into one "normal point," positions with an accuracy of 0.01 seconds of arc or better are obtained.

Measurements must not only be accurate; they must be compatible. As observations are made over a period of decades, the telescope must be kept optically constant, filters and photographic emulsions should not be changed, and not only should the same measuring engine be used but ideally the same person should use it. Use of more impersonal measuring machines in the future may yield further improvements in accuracy over the measurements made so far, which rely on the visual bisection of star images by individual measurers.

At the Sproul Observatory we began a systematic program of measurement in 1937. By the end of 1968 we had 3,036 plates, containing 10,452 exposures taken on 766 nights; these included 25 plates taken during the interval 1916-19 following the star's discovery. The hardest part of the work is not making the photographs, but determining thousands of accurate positions.

The position of Barnard's Star is measured against a reference background of faint stars (page 161). Positions are measured in two coordinates, parallel and perpendicular to the celestial equator, known as right ascension and declination. The reference stars are moving too, of course, but because they are much farther away their proper motions are much smaller and are known with sufficient accuracy to be taken into account.

Once the positions are plotted, they have to be corrected for the effects of the earth's revolution around the sun, for the motion of the reference stars, for the motion of Barnard's Star itself, and for the changing perspective as Barnard's Star comes closer to us. After eighteen years of photographing, plotting, and analyzing, we found the evidence of perturbations for which we were hoping. When our corrections had all been made, we were left with a systematic deviation, mostly in right ascension, with a cycle approximately one-fourth of a century. The subsequent decade has confirmed this interpretation. There appears to be no other way to explain this deviation than by interpreting it as a perturbation caused by an unseen companion.

If there is just one companion, our calculations show it has a mass one and a half times that of Jupiter and revolves around its sun every 25 years, at a distance 1.5 times that at which the earth circles our sun. Instead of a nearly circular orbit, such as all planets in our solar system exhibit, however, this companion would travel in a highly elongated ellipse. My first reaction to this unexpected feature was, why not? Should we demand nearly circular orbits just because we were born on a planet that happens to have one? My next thought was that perhaps we should be impressed by the near-circular orbits of the planets in our solar system. Immediately an alternate interpretation of the perturbations became possible. Within the uncertainties of the data, the asymmetry ascribed to a highly elliptical orbit could easily be explained by the presence of two planetary companions with different periods of revolution in near-circular orbits. The idea of two companions had been contemplated ever since dips in right ascension turned up in the data for both 1955 and 1956. One dip could have been bad luck; two in consecutive years could not be disregarded so easily. I felt a little as though I were retrograding from the Keplerian viewpoint of elliptical orbits to the Copernican scheme of circular orbits; I felt better after reminding myself that highly elliptical orbits are contrary to the only experience we have.

Simple trial and error (I wasn't going to let a computer do this; I wanted to savor for myself the pleasures of playing with various orbits, for a slide rule and simple desk calculator are all that is necessary in these ultimate studies) showed that two circular, corevolving, nearly coplanar orbits with periods of 26 and 12 years and radii of 4.7 and 2.8 times the earth–sun distance, respectively, represented the observations very well. (The corresponding figures for Saturn are 29.5 years and 9.5 times the earth–sun distance; those for Jupiter are 12 years and 5.2 times the earth–sun distance.) The masses of both perturbing objects worked out to be still closer to that of Jupiter than had the mass of the hypothesized single companion.

Because of the feeble luminosity of Barnard's Star, the two planets would have surface temperatures so low that any discussion of life on them is out of order.

It seems clear that these two companions, which I call simply B1 and B2, are true planets and not just very small stars in a complex multiple system. Both stars and planets are spheres of matter. Stars result from spheres of matter having so much mass that contracting under the force of gravity, they heat up enough to start nuclear reactions and glow with their own light. Their mass classifies them as stars even if at the end of their life they collapse to planetlike dimensions. Planets are considered to be the end product of smaller contracting spheres that lack sufficient mass for nuclear reactions, or possibly the result of the accretion of scattered material. Planets never shine by their own light.

The present study does not permit any interpretation that would involve more than two planets. It would be totally impossible to discover a planet the size of the earth. The mass of the earth is only 1/319 that of Jupiter, not enough to cause noticeable perturbation of the central star.

This gravitational technique is so sensitive, however, that if Barnard's Star had no unseen companions, it would show deviations caused by the perturbations of Jupiter and Saturn on our sun. The wobble of the sun in its path caused by Jupiter and Saturn, and to a lesser extent by Uranus and Neptune, is large enough to produce wobbles in the paths of nearby stars. The wobble produced in Barnard's Star is not enough for Jupiter and Saturn to be discovered if for some reason they had been previously overlooked, but it is large enough to be allowed for in analyses of the path of Barnard's Star.

The size of the perturbation observed in Barnard's Star makes it certain that something is there; the amplitude is ten times the margin of error. Whether the one- or two-planet hypothesis is to be preferred may be decided in the next several years, when we can test our predictions against new observations.

To date, Barnard's Star is the only star other than our sun to show clear evidence of having planets. Other nearby stars are under intense study, and there appears to be some tentative evidence for planets circling still other stars, but any definite statement is several years away. Barnard's Star is the second closest to us. The farther away a star is, of course, the more difficult it is to detect the very slight wobble any planets would cause.

The search for extrasolar planets has been going on for 32 years, slowly, painstakingly, with none of the excitement of

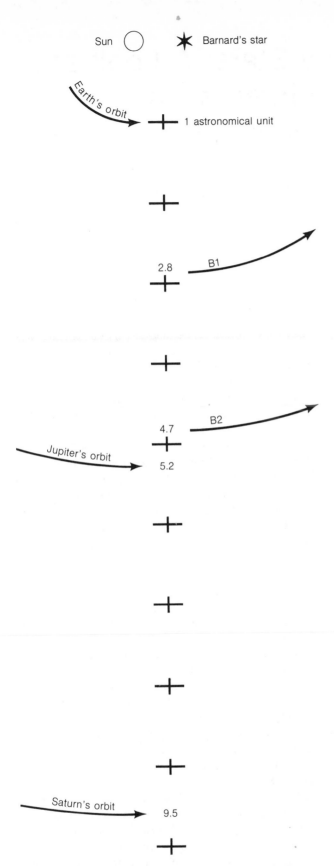

Sun ◯    ✹ Barnard's star

Earth's orbit ↘ ┿ 1 astronomical unit

┿

2.8  ┿ B1 ➚

┿

B2 ➚
4.7  ┿
Jupiter's orbit ➘ 5.2

┿

┿

┿

┿

Saturn's orbit ➙ 9.5

┿

Scale drawing compares orbits in our solar system with those of planets orbiting Barnard's Star. One astronomical unit is the mean distance from the earth to the sun, about 93 million miles.

# 23 Binary Stars: A Look at Some Interesting Developments

Robert E. Wilson

It has been estimated that more than half of the stars in our galaxy are multiple systems, in which two or more stars orbit one another. Among these the binary stars are particularly interesting for several reasons: From principles of celestial mechanics, we expect only the binaries to have orbits which repeat at regular intervals. Furthermore, only the binaries should form "close" systems, in which the separation of the components is not very large compared to their sizes. Although such close binaries are relatively uncommon in space, we shall see that they are of great interest. Their binary nature can be discovered even at great distances because as they move around their orbits, their velocity relative to us changes, leading to corresponding changes in the Doppler shifts of their spectral lines. In many cases, the stars can also be seen to eclipse each other, making detection considerably easier. All in all thousands of such systems have been catalogued and studied. While it is generally very difficult to determine the physical characteristics (such as mass and radius) of single stars directly, the laws governing binary star systems can be used to measure these quantities. Indeed, we obtain virtually our only direct data on stellar masses, and most of the accurate determinations of stellar radii and luminosities from the work on binary systems.

Some well known stars are found to be binaries. For example, Sirius, the brightest star in our sky has a companion which is a white dwarf. And the nearest star system, Alpha Centauri, consists of two stars, with a third revolving a much larger distance away. However the components of these binaries are so widely separated that virtually no physical interaction occurs between them. We shall be interested here in binaries in which the separations are relatively small and the physical interactions relatively important.

## The Main Areas of Research

The study of binary stars begins with the same problems and principles which are encountered for single stars but presents

From *Mercury,* the Journal of the Astronomical Society of the Pacific, September/October 1974. Reprinted with permission.

many new ones as well. Some of the main areas of research on these objects include: 1) *Decoding the information in the observational record:* None of the close binaries can be separated visually, even in a large telescope. For such systems we analyze changes in the brightness and the radial velocity (the motion toward or away from us) to determine the characteristics of the two component stars. 2) *Developing a theory of the direct physical interaction of binary stars:* Stars which are very close to each other are expected to interact, both by tides and by radiation. Many are found to be exchanging material between components, and all do so eventually. 3) *Predicting the evolution of such systems:* The proximity of the other star will alter the life cycle through which each star goes. The results of (1) are compared with predictions based on (2) and on single star evolution theory to discover the various stages which occur.

## Basic Assumptions

As a good starting point for thinking about such problems, consider the following simple question. What figures (i.e. shapes) do stars in binary systems assume as a result of their own rotation and tidal interaction with the other stars? This question is a direct part of (2) and, as we shall see, has important consequences with regard to (1) and (3). When the stars are sufficiently well separated so that tidal effects can be ignored it is easy to compute their shapes, but the effects of rotation, tides, and noncircular orbits in combination result in a very difficult problem. Fortunately, even for the closest binaries we find a special case which has a fairly simple solution and which is quite common in nature. This is the case in which the components rotate as our moon does, with the same period as the orbital motion. (Thus the moon always keeps essentially the same face toward the Earth.) This kind of rotation is called *synchronous* and is much easier to treat mathematically than is nonsynchronous rotation. Synchronous rotation and circular orbits are the rule for very close binaries because tidal drag has the effect of producing just these conditions. The synchronous case is relatively simple because there are no relative motions of any part of the system with respect to any other. Although the binary system revolves in space as a whole, in effect "it has no moving parts". To make things even simpler, we can make one further simplifying assumption which, it can be shown, introduces extremely little error. This simplification is that, although the stars may be relatively large and considerably distorted, they *attract one another* nearly as if their entire masses were concentrated into mass-points at their centers.[1] With these assumptions, the mathematics by which this problem is then solved need account only for the gravitational attractions of these two mass points, according to Newton's law of gravitation, and for the force due to the rotation of the entire binary system about its center of mass.

Before examining the specific results with regard to the figures of rotating binary stars, let us consider a simpler example, the rotating earth. What principle governs the particular shape assumed by the earth? As is customary, we adopt mean sea level as defining the figure of the earth. The ocean surface unerringly forms a smooth surface (apart from waves, of course) without hills or valleys, and we wish to consider just how the water arranges itself to do this. In common language, we say that "water seeks its own level." Should any irregularities temporarily be created over the ocean surface, they are removed by flows until no further flows are necessary. Where is the water surface when a steady condition has been reached or, to put it another way, what does the water "understand" by "its own level"?

What we mean by a level surface in this sense is one on which the potential energy is the same at all places. This will include rotational as well as gravitational potential energy. If the former were zero (no rotation) the earth would be spherical, since surfaces of constant gravitational potential about a mass point or spherical mass are spherical. Rotational forces are, in fact, very small on the earth's surface compared to gravitational forces, so the earth is not far from being spherical.

Suppose now a significant amount of water were added to or taken from the earth's oceans so that mean sea level became slightly higher or lower. Of course, this new ocean surface would still coincide with a surface of constant potential energy (equipotential surface), but a slightly larger or smaller one. We see, therefore, that there are an infinite number of such surfaces around the earth, a particular one of which happens to mark present-day sea level.

and more tidally elongated and rotationally flattened. Just as the actual shape of the sea level on Earth is given by such a surface, so too the shapes of the stars in a binary system will be determined by its equipotential surfaces. Stars which are very small compared to their separation will be virtually spherical, while those which are larger will be increasingly egg-shaped (with the inward-facing end smaller than the other, just like an egg) and also rotationally flattened.

What should we expect to find at the place where the family of surfaces around $M_1$ merges with that around $M_2$? First, consider an intuitively obvious idea. It is certain that there must exist a *balance point*—including the attractions of $M_1$ and $M_2$ and also rotational force—somewhere on the line of centers between the components. At that point, any matter which is somehow forced to rotate with the system "will not know which way to fall"—it will be in balance. Material slightly closer to $M_1$ than this point will fall toward $M_1$ (or if it is part of star 1, will remain so), and conversely for material which is slightly closer to $M_2$. This balance point is called the inner Lagrangian point ($L_1$ point) after J. L. Lagrange, who studied the celestial mechanics of the problem, and has unique significance for our problem. Obviously, if we draw successively larger equipotential surfaces around $M_1$ we shall eventually draw one which includes the $L_1$ point, and similarly for $M_2$. Now the mathematical analysis shows that the largest equipotential surfaces which *completely enclose* one star or the other are those which include the $L_1$ point. This is not surprising, for otherwise we could find part of the surface of one star on the "wrong" side of the balance point—that is, in a region where it should be gravitationally dominated by the other component. The analysis also gives the detailed shapes of these largest closed equipotential surfaces, and we see that each comes to a point on the inner facing side. The volume enclosed by this largest equipotential surface is called the *Roche lobe* of the star; it sets the largest dimensions the component can have before starting to spill its material onto the other component. We now see that stars which are successively larger will experience increasing distortion of their shape until they reach the Roche lobe surface when, in effect, a hole opens up at the $L_1$ point and further size increase is prevented by

## Roche Lobes

The case for stars is similar to that of the earth in some respects. The "surface" is a fluid, this time a gas, and we expect the gas to become arranged so as to have constant density along the surfaces of constant potential energy. The main complications for *binary* stars are that now we have *two* sources of gravitational attraction and the center of rotation (center of mass) is not at the center of either mass but in between them. However the problem of locating the equipotential surfaces in this case was solved about a century ago by E. Roche. Figure 1 shows some examples of these surfaces for two idealized point masses which we will call $M_1$ and $M_2$. Those surfaces which are close to the point masses are nearly spherical, while those which are successively larger are more

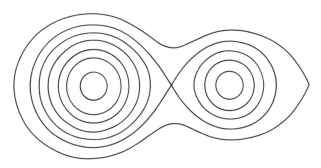

Figure 1.   Equatorial cross-section of the Roche surfaces of constant potential energy for a binary star system. As explained in the text, the shapes of binary stars are defined by such surfaces.

simple loss of material to the other star. Two interesting points immediately become evident:

1) A binary component which is undergoing a steady expansion (something which occurs at various stages in the normal evolution of stars) will rather accurately assume the dimensions of its Roche lobe since rapid loss of material through the "hole" at the $L_1$ point prevents achieving a larger size, while the continuing expansion rules out any smaller size. The situation is very much like the fixed level reached by the water in a tub which has an overflow port.

2) The Roche lobe dimensions can be significantly exceeded only if *both* stars have filled their respective lobes, so that neither can serve as a sink for the material of the other. We have in this case the well-known situation of a *contact binary*. While most known contact binaries are just barely in contact, in that the stars are only slightly larger than their lobes, a few are known to be substantially larger, and have a thick, connecting neck.

To summarize, we find that in binary star systems the components are virtually spherical when they are small compared to their Roche lobes, become progressively more egg-shaped as they approach the size of their Roche lobes, and should actually develop a point on one end when they exactly fill their lobes. Thus the degree of tidal distortion for each component depends on how large it is compared to its own Roche lobe. Except for contact binaries, a star cannot be significantly larger than its Roche lobe[2] because it will, almost immediately, lose any matter which is outside the lobe to the other component. Thus stars undergoing evolutionary expansion in a binary system will assume the size and shape of their Roche lobes with near-exactness.

We can now ask which Roche lobe in a given system is the larger? The location of the $L_1$ point relative to the binary component stars can be computed quite accurately if, and only if, the ratio of their masses is known. A familiar example is provided by the balance point in the earth-moon system, which is rather close to the moon (although in this case the relative position is not fixed because the moon's orbit is eccentric.) If the $L_1$ point is closer to the less massive body, then obviously (cf. Fig. 1) the less massive body must have the smaller Roche lobe.

### Specific Examples

For a specific example, consider the prototype of the eclipsing binaries, Algol, or $\beta$ Persei. Algol is actually a triple system, but we are interested now only in the eclipsing pair—a nearly spherical B8 main sequence star[3] whose eclipse provides the main brightness variation, and a tidally distorted sub-giant star of Type K or G, which causes the main eclipse and is in turn covered by the B8 star in a shallow secondary eclipse. The upper part of Figure 2 shows a series of computer-generated "pictures" of the Algol system. (The lower part of the diagram will be discussed presently.) In this and subsequent figures, the two stars are shown in four different positions (phases) of their orbits. If we take a complete orbit to go from a designa-

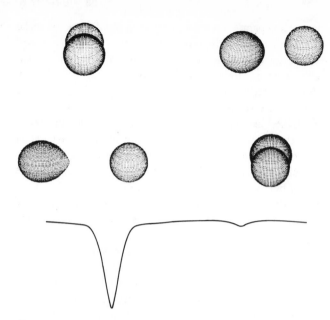

**Figure 2.**   Computer-generated pictures of the semi-detached system of Algol (the demon star) at various phases in its orbit. A light curve for one orbit is also shown.

tion of 0.0 to 1.0, the upper left picture shows the system at phase 0.0, the upper right at 0.125, the lower left at 0.25 and the lower right at 0.50. We see that, despite the fact that one is called a sub-giant, the components have nearly equal sizes, but that one is far more distorted than the other. The reason is easy to state—the spherical B8 star is about 5 times more massive than its cooler companion. We understand this state of affairs because the B8 star, being much the more massive star, has a much larger Roche lobe than the sub-giant. It, therefore, is small compared to its lobe whereas the sub-giant is large compared to its lobe and in fact fills it entirely. Of course, even if one knew nothing about the Roche model the situation could easily be rationalized just by saying that the more massive star should produce the larger tides in its companion. However, although this seems reasonable for Algol, imagine a binary in which the massive primary is considerably larger than the Algol primary and the light secondary considerably smaller than the Algol secondary, but with the same 5 to 1 mass ratio. Then the primary could more nearly fill its Roche lobe than might the secondary and we would have a case in which the *more massive* star is the *more distorted*. Therefore it is neither the mass ratio alone nor the size ratio alone which determines relative distortion, but rather the sizes of the components compared to the sizes of their Roche lobes which, in turn, are determined by the mass ratio. The larger the star compared to its Roche lobe, the greater the distortion.

Viewed another way, we can say that tides (whether dynamic, as ocean tides, or static, as we are now discussing) are due to differences of gravitational forces. The moon attracts the near side of the earth more strongly than the far side and thus "stretches" the earth (or ocean) along the moon-earth line. We can imagine this tide being increased either by having the moon be more massive, or by having the earth larger in diameter. In the first case, we increase all forces and thus also their

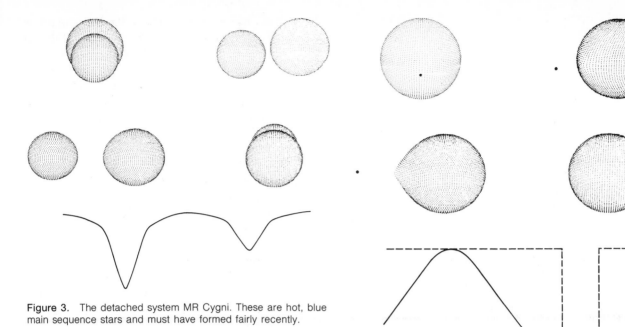

Figure 3. The detached system MR Cygni. These are hot, blue main sequence stars and must have formed fairly recently.

Figure 5. The x-ray binary Hercules X-1, or HZ Herculis. The large tide is raised by the small orbiting dot, which is thought to be a neutron star, and is the source of x-radiation. If drawn to scale, the dot would be invisible. The dashed line indicates the observed x-ray emission for one cycle.

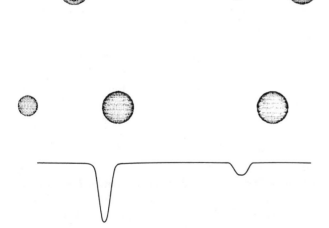

Figure 4. The well-detached system EE Pegasi. These main sequence stars are so far inside their Roche lobes as to be virtually spherical.

differences, while in the second case we increase only the differences, but in either case the tidal effect is increased.

Further examples of binaries in various stages of filling their lobes are given in the upper portions of the next five figures. Figure 3 shows the binary MR Cygni, which is known as a *detached* system because neither component fills its Roche lobe. Note that both components show strong tidal distortion, but not such great distortion as the Algol secondary. Algol (Fig. 2), in which one component fills its lobe while the other does not, is an example of a *semi-detached* system. In contrast, Figure 4 shows the system of EE Peg, in which both components are small compared to their lobes, and therefore are close to being spherical.

**Unusual Binary Systems**

Let us now turn to some of the more unusual binary systems which have been discovered. Figure 5 shows one model for the appearance of a rather complicated system called HZ Herculis (or Her X-1). The binary nature of this object was discovered in 1971 by a group headed by R. Giacconi, who observed repeating eclipses of its x-radiation with the UHURU satellite. Only one component of this system is visible to optical telescopes. The diagram shows this star of middle spectral class which is filling its Roche lobe, although it is not certain at this time that the star actually does so—it could be a little smaller.

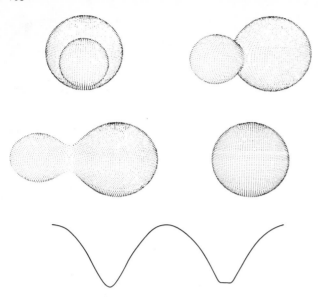

**Figure 6.** The contact binary RZ Tauri. The components exchange both energy and material through the connecting neck.

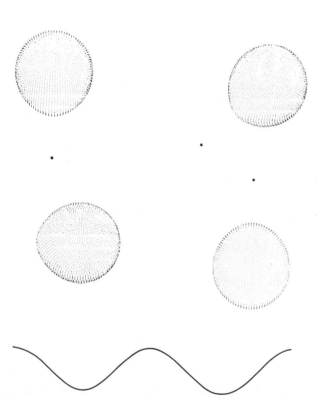

**Figure 7.** The x-ray binary Cygnus X-1, or HDE 226868. The orbiting dot may be a gravitationally collapsed object, or black hole. The orbit is inclined by only about 30° to the plane of the sky, so essentially we are "looking down" on the system. The vertical scale on the light curve has been stretched by a factor of ten relative to the scales on Figures 2–6, because the variation is very small in this case.

The small dot represents the source of the x-rays which, at present, is generally believed to be a neutron star.[4] If the dot were printed to scale, it would be far too small to be seen, since the radius of a neutron star is of the order of 10 kilometers. There is some uncertainty in the masses of the components of HZ Her, but reasonable values would be 1.7 solar masses for the optical star and 0.8 solar masses for the x-ray star. Here we find an example (a rather extreme one) in which it is the *more* massive component which has the greater tidal distortion. Indeed, since neutron stars are so extremely small, the x-ray component could afford to be larger by a considerable factor and still remain essentially spherical, while its more massive companion, being about the size of its Roche lobe, has very large permanent tides.

Figure 6 shows the contact binary RZ Tauri. Here both components exceed their lobes and are connected, as we have discussed. In such cases there must be one smooth equipotential surface to act as a boundary for both components, as shown. If the contrary were the case (i.e. if the "surface" of one component were at a higher potential level than that of the other) flows would occur between components until such differences were eliminated.

The last figure shows Cygnus X-1 (also known as HDE 226868), another binary system which is a strong source of x-rays. In this system there appears to be a good chance that the small x-ray component is a black hole—a star whose final collapse is still proceeding, but has been brought to a virtual halt for outside observers by the gravitational slowing of time.[5] The oscillations of a radiating atom on such an object would be similarly halted, insofar as we could observe, so that the only practical means for detection of black holes would be through their gravitation. At present, the main and perhaps only hope for finding black holes seems to lie in their possible occurrence in binary star systems. Here the gravitational field would be detectable through its effect on the motion of the other binary component and perhaps through its role in producing x-rays. Transfer of material between the two components would be important not only in the evolution of the system but also in generating x-rays as material comes near the black hole. Transferred gas would become heated by compression to perhaps 50 million degrees in falling toward the black hole. Gas at such a temperature radiates primarily x-radiation, which would be observable except when emitted from a region extremely close to the black hole. Observational and theoretical work on this interesting system is proceeding all over the world. We should also note that the same process can lead to x-ray emission from neutron stars and possibly white dwarfs in binary systems.

Each of the Figures (2-7) includes a theoretical light curve showing one cycle of the repeating variation of brightness for each system as seen from the earth. The effects of tidal distortion are evident for MR Cygni, RZ Tauri, and Cyg X-1 in that these systems show brightness variation even between the eclipses, being brightest when the elongated stars are seen broadside and faintest when the narrow ends are viewed. Thus, while the more sudden changes in brightness are due to one star moving in front of the other, the general curvature in the light curves between the eclipses is due to tidal distortion (called the ellipticity effect). In fact, as can be seen from Figure 7, Cyg X-1 has no eclipses because of the direction from which we view the system, and its entire variation is due to the ellipticity effect.

Five of the binaries exhibit variation due to a process called the *reflection effect*, but it is most easily noticed for HZ Her, Algol, and MR Cyg. It appears mainly as a brightening near the time of the secondary eclipse. The dip in brightness signifying the secondary eclipse, therefore, appears "at the top of a little hill" on the light curve. This is because the inner-facing side of the cooler component is heated by the strong radiation from the hotter component and glows more brightly than the outward-facing side. Just before and after the time of secondary eclipse we are looking directly at this heated side, so the combined output from the binary system appears especially bright. Of course, the cooler star also heats the hotter one somewhat, but this is a smaller effect unless the components have equal temperatures. For RZ Tau (Figure 6) the two reflection effects virtually cancel one another because the temperatures are nearly equal.

The brightness variation of HZ Her (Figure 5) is particularly interesting because it is due almost entirely to the reflection effect. Here the source of heating is the very strong x-radiation from the x-ray star. We see no eclipse in the computed optical (visible) light curve (solid line) because the tiny x-ray star covers only an insignificant fraction of the disk of the normal star. Naturally, the eclipse of the x-ray star by the normal star is detectable only in the x-ray observations. Notice the rectangular profile of the x-ray eclipse in the schematic x-ray "light curve" (dashed line).[6] Here there is no evidence of the normal star except for its blocking of the x-radiation. The absence of any gradual transition (partial eclipse) regions (the sharp corners in the dashed line) indicates that the x-ray star is very much smaller than the normal star and that the edge of the normal star is sharply defined.

Among the other binary systems we can find examples of quite different types of eclipses. In particular, we can distinguish among *total* eclipses (where the star is completely covered), *annular* eclipses (where a ring of the eclipsed star remains visible), and *partial* eclipses (where a total or annular stage is not reached). In the first case, the light curve appears flat (unchanging) while one star is out of view. Those of Algol and MR Cygni are partial and in each case the light curve shows a rapid decline to minimum followed by an equally rapid recovery. The secondary (shallower) eclipse of EE Peg is total like the x-ray eclipse of HZ Her, but in this case the two components have comparable dimensions so that partial phases precede and follow the (flat) total section. Another such total eclipse is shown at the secondary minimum of RZ Tau.

The primary eclipse of EE Peg is annular; that is, for an interval near mid-eclipse, the disk of the smaller star is contained (projected) entirely within that of the larger star, just as is the moon during an annular eclipse of the sun, leaving a ring of unobscured surface of the larger star in view. As one can see from the pictures of the system, the bottom of the eclipse light curve *would* be flat in this case if the large, eclipsed

star had a uniformly bright surface, but in real stars a phenomenon known as *limb darkening* (darkening toward the edge of the visible disk) provides a rounded bottom, as shown in Figure 4. This effect makes the eclipse noticeably different from the (pointed) partial eclipses of Algol and MR Cyg and, of course, from the total secondary eclipses of EE Peg and RZ Tau. Limb darkening occurs because the light emitted in the observer's direction from the limb (edge) of a star comes, for the most part, from relatively high layers in the star's semi-transparent photosphere. Since these high layers are cooler than the deeper layers which are seen when viewing the center of the disk, they radiate less strongly so that the limb appears dark relative to the center of the disk.

## Binary Star Evolution

Having seen how a study of the gravitational interaction of double stars can lead to an understanding of the circumstances under which we expect transfer of material from one component to the other, we now turn to the consequences of this mass transfer. This is a complex issue, involving many problems which remain to be explored, but if we limit the discussion to a few basic principles, we can gain some idea of the important progress which has been made in understanding binary star evolution, mostly within the last decade.

One of the best established rules governing the evolution of single stars is that all phases of evolution proceed faster for stars of greater mass. For any star we expect a relatively long quiescent existence during which the radius and luminosity change only very slowly, followed by a relatively brief interval of rapid expansion, during which the star becomes a red giant, just before the effective "death" of the star. A massive star begins its expansion much sooner than a low-mass star. This makes it quite puzzling to consider the case of Algol, for example, in which the primary component of perhaps 5 solar masses appears to be in the early stages of its evolution, while the secondary, of about 1 solar mass, is already undergoing its evolutionary expansion. That is, the secondary completely fills its Roche lobe and we know from spectroscopic observations that it is spilling matter onto the large star through the $L_1$ point. The solution to this paradox is now understood, as a result of the work of many astronomers, among whom J. Crawford, R. Kippenhahn, D. Morton, B. Paczynski, and M. Plavec are particularly noteworthy. We now know that in systems such as Algol, the mass transfer occurs on so large a scale as to *reverse* the mass ratio. The star which was originally more massive did indeed begin its evolutionary expansion first; but, in the process, it transferred so much material to the other star that it has become the less massive component of the system.

Several questions arise immediately: Why is the mass transfer on so large a scale, with most of the system's mass being involved, and so rapid, with the mass ratio being reversed in only 10,000 to 100,000 years? If *most* of the mass of the original primary is lost to the secondary, why is all of it not lost in this way, thus converting the binary to a single star? That is, it may seem curious at first sight that the standard mass exchange process is so spectacular as to dump, say, 80% of the mass of one star onto the other, yet stops short of transferring 100%.

To see why this is so, let us consider what happens to the Roche lobe of the original, more massive primary component at the beginning of mass exchange. We suppose that this star has been expanding and has just become as large as its Roche lobe, so that it spills a small amount of mass through the balance point onto the other star. We know from the dynamical laws governing such systems that we can expect this event to alter not only the mass ratio but also the orbital period and separation of the two stars. If the mass ratio changed without corresponding changes in period and separation, the system's total angular momentum would change (which is not permitted, of course, unless some material leaves the entire system). Now it turns out that the distance between the stars must *decrease* when the flow is from the massive to the low-mass component. This means that all the orbital dimensions in the system, including the size of the primary star's Roche lobe, shrink to a smaller scale than before. Furthermore, we have already noted earlier that the *relative* size of the Roche lobe depends (only) on the mass ratio, with the star of larger mass having the larger lobe. These two effects, taken together, cause the primary Roche lobe to shrink significantly with the transfer of a fairly small mass. However, this in turn leads inevitably to further mass loss because the star now finds itself again slightly overspilling its now smaller lobe. We therefore have what is usually called a positive feedback process in that a small initial transfer of mass leads to conditions which encourage further transfer, and so on until the flow becomes quite large. Such a binary system is said to be in the rapid phase of mass transfer.

We now ask what stops this runaway process before all the mass is transferred to the secondary? Recall that the orbital separation shrinks *when the transfer is from the more massive to the less massive star*. At some stage in the procedure the masses become equal; after this the components *must separate* in order to conserve angular momentum because flow will then be from the less to the more massive star. Thus one of the effects causing the shrinking of the primary star's Roche lobe will be reversed. Eventually the primary will find itself in a situation in which it no longer overspills its Roche lobe because the lobe no longer shrinks as matter is transferred.[7] However, by the time this happens, most of the star's mass will have been transferred to the secondary, the mass ratio will have been reversed, and we shall have a system perhaps like that of Algol.

The binary has now reached the end of the rapid phase of mass transfer. Further mass exchange is discouraged because it now tends to make the original primary star *smaller* than its lobe. However the evolutionary expansion, which started the process, will not have stopped, and will now continue to produce a relatively leisurely mass flow from the original primary to the secondary.[8] This is the slow phase of mass transfer, which we see today in Algol and many other semi-

detached binaries. Subsequent developments in the system will depend to a considerable extent on the particular masses and even on the original chemical composition and internal evolutionary state of the individual stars, but in one way or another they must account for most of the very strange and unusual binaries we see, including those with white dwarf, neutron star or black hole components.

## For Further Reading

Abell, George: *Exploration of the Universe* (1969: Holt, Rinehart and Winston) is a good introductory text with a chapter (15) on binaries.

O. Eggen: "Stars in Contact" in *Scientific American* June 1968, also concerns close binaries and stellar evolution.

D. Thomsen: "Where Space is Sharply Curved" in *Science News* Feb. 23, 1974 contains some recent observations of binary systems which may contain neutron stars or black holes.

## Notes

1. For a spherical star this rule is exact. That is, the gravitational field *outside* a spherical star is identical to that which would be produced by an equal mass squeezed down to very small radius. For a star with tidal or rotational distortion, the rule is not far from correct because real stars have most of their mass concentrated into a fairly small high-density core, with the outer, distorted regions having much lower density.

2. It has recently been argued that a significant excess over Roche lobe dimensions is possible in certain brief phases of binary star evolution, but this is not established at present.

3. A main sequence star has not yet begun its evolutionary expansion, or is expanding only very slowly because the hydrogen fuel in the core is not seriously depleted. The letters O, B, A, F, G, K, M refer to a classification scheme of stars by their temperature (by means of features in their spectra).

4. One can think of a neutron star as an object of extremely high density, which is composed almost entirely of neutrons. A typical neutron star would have a mass equal to that of the sun compressed into a sphere of 10-kilometer radius. Neutron stars should only be formed as a result of the supernova process, and it is now generally accepted that the observed pulsars are neutron stars. Some pulsars are found where supernovae are known to have exploded. Neutron stars have been discussed in several semi-popular articles (e.g., M. A. Ruderman, *Scientific American*, February, 1971, p. 24.) [See also Articles 28 and 42 by Louis C. Green and Stephen P. Maran, respectively, in this reader.]

5. See the May/June, 1974, issue of *Mercury* for a thorough discussion of these objects. [See also Articles 26 and 27 by Ben Bova and S. Chandrasekhar, respectively, in this reader.]

6. The actual observed x-ray variation is far more complicated than the diagram indicates, and has stimulated much work on physical processes associated with

Her X-1. Even the observed optical light curve differs in important ways from the illustrated computed curve, and several ideas have been advanced regarding the cause of these departures from a simple reflection effect model. However, there is no doubt that the main variation of about 1.5 magnitudes is due to the reflection effect. Thus, for the sake of clarity, we have ignored these further complications.

7.  In quantitative work it is necessary to account also for the changing equilibrium radius of the star which is losing mass. Only the most essential features of the process are described here.

8.  Notice that observers would now call the original primary the secondary, and vice versa, because of the mass ratio reversal.

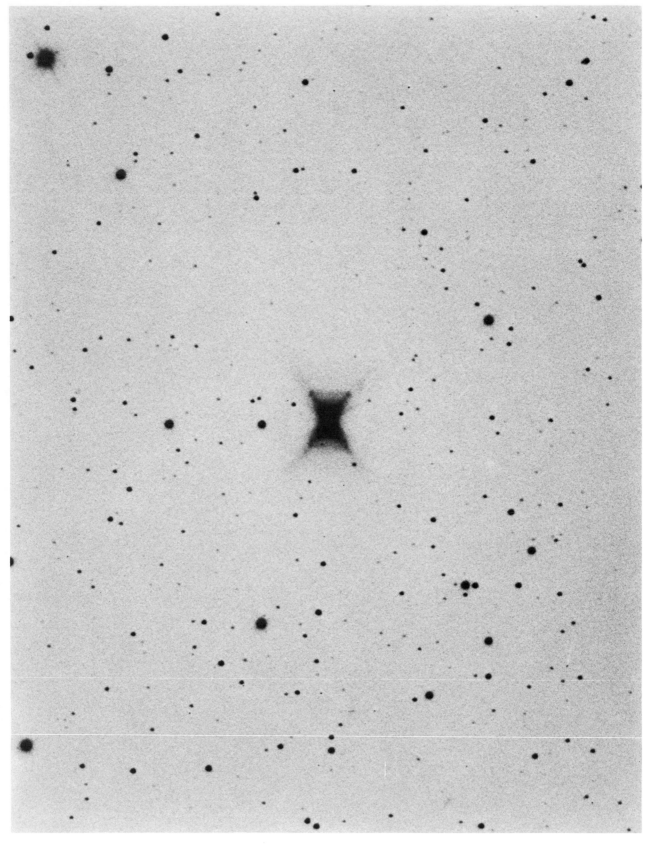

The Red Rectangle, photographed by T. R. Gull with the Mayall telescope (negative print). According to one interpretation, a new solar system may be forming in the waist of the hour-glass-shaped nebula, where intense infrared radiation is observed. The less-detailed photograph in which the nebula was first seen showed it to be roughly rectangular, leading to its present name. [Kitt Peak National Observatory.]

PART **VII**

# THE BIRTH AND DEATH OF STARS

"If open clusters evolve physically, each of them should pass through a rather brief period of glory and then subside and become a less spectacular grouping of old stars." Astronomer Bart J. Bok, an authority on our Milky Way galaxy, reports how star clusters, dark globules, and interstellar gas and dust yield clues to "The Birth of Stars."

". . . studies of cluster stars provide evidence that the universe has expanded from a tremendously hot primeval atom that was formed in one unique event 12 or 13 billion years ago." Estimates of the age of the universe change as new observations and theories are published. One technique for investigating this question is to study the evolution and chemical composition of the very old objects described by theoretical astrophysicist Icko Iben, Jr., in "Globular-Cluster Stars."

Three possible last stages in the life cycle of a star are white dwarfs, neutron stars, and black holes. They are examined in different degrees of technical detail by science writer Ben Bova ("Obituary of Stars: Tale of Red Giants, White Dwarfs and Black Holes"), mathematical astronomer S. Chandrasekhar ("The 'Black Hole' in Astrophysics"), and astronomer Louis C. Green ("Ordinary Stars, White Dwarfs, and Neutron Stars"). White dwarfs are now familiar objects to astronomers, although their physical properties have been studied only since 1914. (Bova is inaccurate in stating that the first white dwarf was discovered in 1915. The extremely compact nature of this star, called Sirius B, was first recognized at about that time, but its existence was actually suspected in 1844 and it was first seen in 1862. The Sirius story can be found in an article by editor Maran, which appeared in the August/September, 1975 issue of *Natural History* magazine.) The existence of neutron stars was predicted in the 1930s, although confirmation of their presence had to await the discovery of pulsars in the late 1960s. Green describes the physical properties of neutron stars, whereas Maran (in Part XI of this reader) relates the history of the pulsar in the Crab Nebula. It now seems certain that the recently discovered pulsars are indeed the once hypothetical neutron stars and that most of the binary systems that produce x rays also contain neutron stars. However, it seems likely that in one of these x-ray binaries, called Cygnus X-1, the small component is not a neutron star; rather it is a black hole. Since the articles included in this part of the reader were written, new data from x-ray satellites have also suggested that black holes may be present in some globular clusters. However, this conclusion is both tentative and controversial.

The short, rather technical selection on "Dynamic Relaxation of Planetary Systems and Bode's Law" is included to stimulate the interest of those students who are using computers in their undergraduate studies, even though they are unlikely to be familiar with the particular *fourth-order Adams-Moulton routine* that author Jack G. Hills employed as a computational technique. Astronomer Hills, who wrote this paper as a graduate student at the University of Michigan, proposes in it a test of the theory of *dynamic relaxation*. According to this theory, planets will eventually follow orbits that obey numerical relationships similar to that of Bode's Law in our own solar system, regardless of their original orbits. The reasoning is that the gravitational pull of one or more planets on another planet will gradually move it into a final orbit that is not significantly affected by further gravitational effects. Hills explains that "To test this hypothesis eleven planetary systems each with a central star of 1 $M_\odot$ [solar mass] . . . were evolved from random initial orbits on an IBM 360/67 computer." The planetary systems had different mass functions, that is, different sets of individual planetary masses. Hills found that regardless of the mass function and the original orbits, the planetary systems all evolved into stable arrangements in which the orbits obeyed laws of Bode's type. Two problems to try are (A) to verify that equation (1) from Hills's paper indeed agrees with Bode's Law for the solar system; and (B) to obtain data on the inner moons of Jupiter, Saturn, and Uranus from tables in a textbook and check on Hills's statement that these moons also fit Bode's-type laws. If the theory of dynamic relaxation indeed also applies to the orbits of planetary satellites such as these, why do you suppose that the outer moons don't follow the Bode's-type laws of the inner moons? Why do some moons follow retrograde orbits, whereas all the planets have prograde orbits?

# The Birth of Stars

Bart J. Bok

It is clear that the stars in our galaxy are of many different ages. The great majority of them, such as the sun, are old by cosmic standards and will live for some billions of years. At the end of their life many of them will subside to being white dwarfs: tiny stars about the size of the earth that will eventually cool to a dark cinder. There are other stars, however, that are consuming their nuclear fuel so rapidly that they cannot have existed as stars for more than a few million years; they include the hot blue-white giants such as Rigel in the constellation Orion. They must evolve rapidly through the gradual exhaustion of their internal energy sources and become less conspicuous. If stars have finite lifetimes and can grow old and die, it would seem that they must also be born. What is the evidence for the birth of stars in the galaxy?

The visible band of the Milky Way marks an almost perfect great circle in the sky, indicating that the sun and its planets are close to the central plane of the galaxy. Counts of the stars along the Milky Way show that the fainter (and therefore the more distant) stars are much more concentrated toward the galactic equator than the brighter ones, which suggests that the Milky Way system is a large galaxy that is highly flattened. The uneven distribution of faint stars along the Milky Way shows that the sun is not located at the center of the galaxy; modern estimates place the sun at a distance of 30,000 to 33,000 light-years from the center. The direction toward the center is marked by the great star clouds in the constellation Sagittarius.

The main body of the galaxy is a disk with an overall diameter of some 100,-000 light-years. The flatness of the disk suggests that it rotates at a fairly rapid rate, and that is the case. It does not, however, rotate like a solid wheel. The inner parts complete a circuit around the center much more quickly than the outer parts do. The observational evidence for galactic rotation is that stars in the neighborhood of the sun generally move in almost circular orbits around the galactic center at an average rate close to 250 kilometers per second. The distance from the sun to the galactic center is so great that it takes the sun 250 million years to complete a single circuit; we may call that interval a cosmic year. The principal gravitational force that controls this motion is produced by the central star clouds of the galaxy, which have a total mass of perhaps 50 billion suns. The mass of the entire Milky Way system is estimated to be a little more than 100 billion solar masses.

The appropriate conditions for the formation of stars exist only in the central plane of the galaxy or close to it. The processes of star birth have apparently exhausted themselves in the spheroidal "halo" surrounding the galaxy, where old stars prevail. These processes, however, are very much under way in the central disk, where interstellar gas and dust intermingle with young stars. It is believed that the interstellar gas and dust are the material from which stars are now being formed. Dense concentrations of gas and dust that are on the way to becoming real stars are termed protostars. It has been suggested that protostars are formed when some of the gas and dust associated with the spiral arms of the galaxy piles up into clouds. A more specific possibility is that observed small dark clouds of interstellar dust grains and even smaller dark "globules" of dust collapse into stars or clusters of stars. A third suggestion is that protostars originate with gas ejected in the explosions of supernovas; a few such clouds are visible as luminous filaments.

We shall be considering each of these processes in turn.

The stars in the galaxy are being continually reshuffled. In the course of one million years—only a two-hundred-and-fiftieth of a cosmic year—two stars now close together but having a difference in velocity of one kilometer per second will have separated by three light-years. Therefore in less than one cosmic year some groups of stars may be dissipated and others may be formed afresh; the appearance of the galaxy should show profound changes. The physical makeup of the Milky Way system will also change over intervals of the order of one cosmic year. As we have seen, the rate at which some kinds of stars use up their supplies of energy is quite high. The blue-white Type O and Type B giant stars are a good case in point. They consume their supplies of nuclear fuel at such a prodigious rate that they cannot keep it up for more than a fraction of a cosmic year. The majority of these stars have probably existed 10 million years since their formation, which is only a twenty-fifth of a cosmic year.

The loosely bound "open" star clusters are the best indicators that star formation is still a continuous process. Within a couple of thousand light-years of the sun there are a good many open clusters with luminous Type O and Type B stars. Open clusters are groups of several hundred stars within a volume roughly 10 light-years in diameter. The presence of Type O and Type B stars implies that many such clusters must be quite young on the cosmic time scale. If open clusters evolve physically, each of them should pass through a rather brief period of glory and then subside and become a less spectacular grouping of old stars.

In the first stage of the cluster's development the protostars will evolve and move toward the "main sequence" of

From *Scientific American*, vol. 227, pp. 49–61, August 1972. Copyright © 1972 by Scientific American, Inc. All rights reserved. Reprinted with permission.

the Hertzsprung-Russell diagram [*see illustration on page 178*]. Most of the stars will spend their early life on the main sequence, where they will use up their principal supplies of nuclear energy. These supplies are exhausted at the highest rate by the intrinsically bright stars, which are the first to evolve away from the main sequence. Most of them will shed much of the gas in their atmosphere and evolve toward the white-dwarf stage. Stars that have an intrinsic brightness less than or equal to the sun's will spend a few billion years on the main sequence. An evolved star cluster will have very few highly luminous stars and a number of white dwarfs.

If, as seems likely, the formation of loose clusters with Type O and Type B stars has been taking place at a reasonably steady rate over the past 10 to 20 cosmic years, one should expect to find 100 old loose clusters for every spectacular young one. This is not the case. Where have the clusters gone? The scarcity of old open clusters probably in-

dicates that they evolve rapidly and dynamically. Relatively straightforward calculations show that clusters with a few hundred members and with diameters of the order of 10 light-years lead a precarious existence in the galaxy. The nucleus of the galaxy exerts tidal forces on them that are highly disruptive. The individual stars are gravitationally so loosely bound to the cluster system that quite a few of them, particularly the less massive ones (which move the fastest), will escape from the cluster in a cosmic year or so. In addition, encounters and near-passages between star clusters and clouds of interstellar dust and gas will tend to loosen up most open clusters and disrupt them in less than a cosmic year.

What are the processes that allow the supply of young open clusters to be maintained? There are several places where the birth of clusters and stars may be in progress. The Pleiades and Hyades clusters in the constellation Taurus will almost surely not be there a cosmic year in the future, but replacements seem to

be on the way.

To understand the processes of star formation more fully, it is necessary to ask first how the galaxy was formed. The oldest star clusters and individual stars are found at large distances from the central plane of the Milky Way. This fact would seem to imply that shortly after the universe was formed the Milky Way system became a separate unit, a large, nearly spherical blob of gas. That may have happened 40 or 50 cosmic years ago. When the first condensation began to form in the original gas and dust, stars and star clusters were probably born all through the large blob of gas. Globular star clusters, which are much more symmetrical and much richer in stars than open clusters, were apparently formed between 20 and 40 cosmic years ago, suggesting that conditions in the original gas cloud were relatively quiescent. As time progressed the gas began to be concentrated more toward the central plane of the galaxy, where it somehow achieved its present rotational

**LANE OF DUST AND GAS** is plainly visible in the plane of the spiral galaxy Messier 104 in Virgo, which was photographed with the 200-inch Palomar telescope. The galaxy is surrounded by a spherical "halo" of mostly old red stars, and star birth is confined to the galactic plane. The Milky Way galaxy, if it were viewed from the outside, would probably look quite similar to Messier 104.

properties. Younger stars and clusters were formed in the gas cloud as the cloud became increasingly more flattened.

The galaxy is currently in a stage of development where its central gas and dust layer is remarkably thin: only 1,000 light-years thick in the vicinity of the sun. At present star birth seems entirely confined to this thin layer of gas and dust. It is fortunate for astronomers interested in the birth of stars that the evolutionary processes are continuing near the central plane of the galaxy and even more fortunate that the sun and the earth occupy a position well suited for the observation of these processes.

## The Interstellar Medium

In order to get a clear picture of star formation and protostars one must consider the physical conditions in the interstellar medium. It is in this medium that the concentrations that give rise to protostars are formed. The composition and physics of the interstellar medium are now quite well understood. The principal constituent is hydrogen. The ionized hydrogen atom can be detected in a great variety of ways. In the visible portion of the spectrum it announces its presence by the Balmer recombination lines observed after a free electron is captured by a positively charged hydrogen nucleus (a proton). The Balmer lines are emitted as the electron cascades to the second level of the neutral (un-ionized) atom [*see illustration on page 179*]. Radio astronomers catch the hydrogen atom in transitions between very high energy levels, or they observe the continuous spectrum of radiation generated as an electron whizzes past a proton. Neutral atomic hydrogen emits and absorbs radiation at the radio wavelength of 21 centimeters and can be readily observed by radio telescopes. Molecular hydrogen has been detected by means of ultraviolet observations from rockets. It seems only a matter of time before its distribution will be charted in detail through ultraviolet observations, possibly combined with infrared and radio observations.

Also present in interstellar space are helium, nitrogen, carbon, oxygen and many other elements. The spectral lines emitted by these atoms and their ions are best observed in the brightly glowing emission nebulas. When any gas is heated, it will emit radiation at certain wavelengths, depending on its composition. If the gas is cooler than a star behind it, it will absorb the star's radiation at those same wavelengths. The elements present in the interstellar medium are checked whenever possible by the interstellar absorption lines observable in the spectra of distant stars. For every 10,000 hydrogen atoms there are on the average 1,200 helium atoms, two carbon atoms, one or two nitrogen atoms, three or four oxygen atoms, one neon atom, one sulfur atom and traces of some of the heavier atoms such as iron and chlorine. If a laboratory chemist were able to analyze a sample of the interstellar gas, he

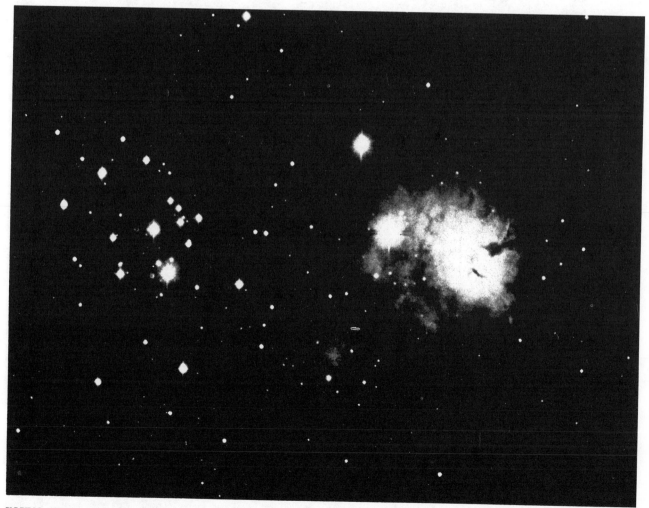

**YOUNG STARS AND A LARGE NEBULA** are found side by side in the region of Messier 8, photographed by the author with the 90-inch reflecting telescope at the Steward Observatory of the University of Arizona. It seems as though the processes of star birth have been more or less completed in the cluster on the left, and dust and gas are apparently condensing in the nebula on the right.

would conclude that it was a mixture of hydrogen and helium with impurities!

Not only atoms and ions but also molecules have been detected in interstellar space. In the 1930's some of the telltale absorption lines found in the spectra of distant stars were attributed to simple diatomic molecules such as CH, CH$^+$ and CN. The past 10 years have witnessed an amazing increase in the number of molecules detected. In 1963 lines in the radio spectrum were discovered that were attributed to the hydroxyl radical (OH). Three major discoveries followed in 1968 and 1969: molecules of ammonia (NH$_3$), water vapor (H$_2$O) and formaldehyde (H$_2$CO). In 1970 a number of additional lines were discovered, notably one for carbon monoxide (CO).

It is striking that some very complex molecules, for example methyl alcohol (CH$_3$OH), are present. The list of interstellar molecules has now been extended to some 30. Most of the molecules are found in regions of space where cosmic dust prevails. The Great Nebula in Orion, the gaseous clouds near the center of the galaxy and the quiescent large globules and larger clouds of cosmic dust within a few hundred light-years of the sun have proved to be the favorite hunting grounds, but molecules continue to pop up in surprising concentrations in many unexpected spots. They are often found in regions near sources emitting strong infrared radiation.

There has been much discussion in recent years about the composition and physics of cosmic dust grains. Their presence in interstellar space is conclusively demonstrated by the existence of dark nebulas and globules of all kinds and by the reddening of starlight as the dust grains scatter the bluer light. Stars close to the galactic equator are reddened, showing that the central plane of the galaxy is rich in dust grains. The infrared radiation observed from many objects in the galaxy seems to indicate that stars or protostars are embedded in thick clouds of dust grains. The grains are tiny particles with a diameter of the order of .0005 millimeter. At first it seemed that they were specks of "dirty ice" built of simple molecules of carbon, nitrogen and oxygen combined with hydrogen and possibly contaminated with iron and oth-

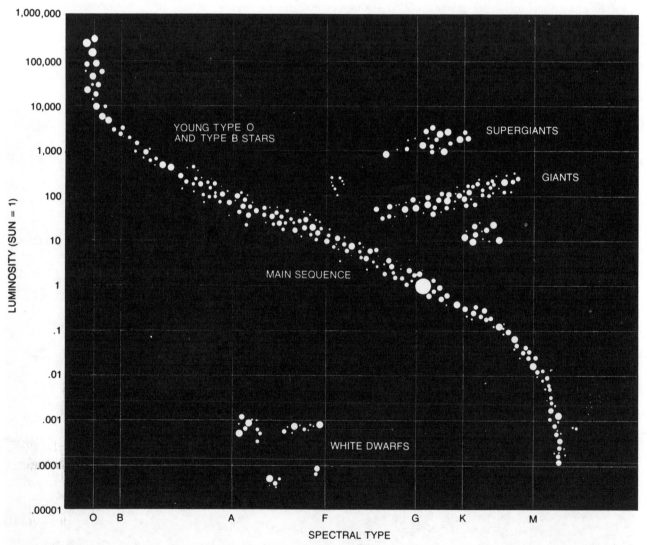

**HERTZSPRUNG-RUSSELL DIAGRAM** relates the luminosities and the spectral classes of stars. Luminosity in terms of the sun is plotted along the vertical axis and spectral classes along the horizontal axis. The spectral type corresponds to temperature and color. Type O and Type B stars such as Spica in Virgo are bluish white and have a temperature of about 20,000 degrees Kelvin or higher; Type M stars such as Antares in Scorpio are red and have a temperature of about 3,000 degrees K. The hot young Type O and Type B stars typical of the spiral arms of galaxies occupy the top left corner of the diagram. They are extraordinarily luminous and are consuming their nuclear fuel at such a prodigious rate that they cannot have existed in their present state for more than 10 million years. Most older and more "normal" stars such as the sun (*large dot*) are grouped along a band called the main sequence that stretches from the top left corner to the lower right corner of the diagram. These main-sequence stars have a lifetime of billions of years.

er substances. That interpretation was abandoned when observations in the infrared failed to yield evidence for frozen water. Cosmic dust grains were also found in considerable abundance in regions hot enough for "dirty ice" to have evaporated. Grains of graphite with an icy mantle were suggested, and for a while the hypothesis held much favor. Next it was found from studies in the infrared that silicate particles are probably abundant in the atmospheres of very cool stars rich in oxygen; it seemed not unlikely that many of these particles would be lost to the interstellar medium. Present conceptions are that the interstellar grains are most likely a mixture of graphite, silicate and iron grains. They probably originate mostly in the extended atmospheres of infrared stars and starlike objects. It is not out of the question that some grains have been ejected into interstellar space as a by-product of supernova explosions. The trend is definitely away from supposing that cosmic dust grains were formed to any great extent in the cool interstellar clouds.

The neutral atomic hydrogen in the spiral arms of the galaxy is probably at fairly low temperatures: about 100 degrees Kelvin (degrees Celsius above absolute zero). The gas in the regions between the arms is in all likelihood at a far higher temperature, possibly as high as 10,000 degrees K. It is more rarefied than the gas in the spiral arms, but because of its higher temperature there may be a pressure equilibrium between the gas in the spiral arms and the gas in the regions between the spiral arms.

What are the conditions inside the clouds of cosmic dust grains: the dark nebulas? Radio studies of formaldehyde and other molecules found in dark clouds have shown that the temperatures inside these clouds may be as low as five degrees K., and in some places they are probably even lower. There seems to be little doubt that conditions deep inside large dark nebulas and probably in globules are right for the formation of many kinds of molecules. Gaseous atoms should often stick to the very cold solid interstellar grains inside these clouds. The small solid grains therefore provide surfaces where atoms can combine to form molecules. They also serve to filter out most of the ultraviolet radiation that might penetrate inside the dark cloud or globule. The key to the formation of interstellar molecules seems to lie in low temperature and in the absence of ultraviolet radiation, which would inhibit the formation of molecules and would also destroy the molecules that had formed.

Many kinds of dark nebulas have all

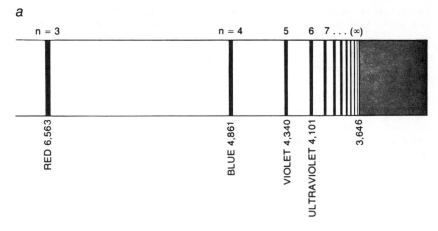

a

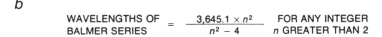

b

$$\text{WAVELENGTHS OF BALMER SERIES} = \frac{3,645.1 \times n^2}{n^2 - 4} \quad \text{FOR ANY INTEGER } n \text{ GREATER THAN 2}$$

c

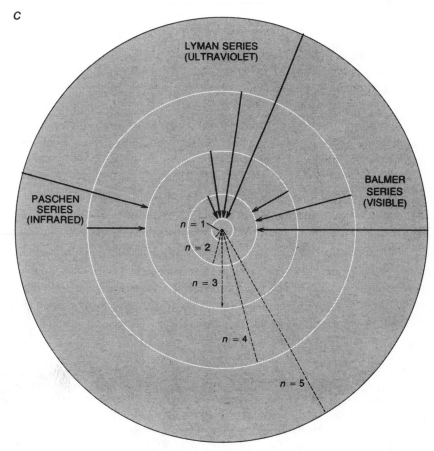

BALMER LINES in the visible portion of the hydrogen spectrum (a) indicate that the principal constituent of the interstellar medium is hydrogen. The lines are produced when the single electron of the hydrogen atom cascades down to the second energy level of the atom. The exact wavelength of the line is specified by a simple formula (b). If the electron drops from the third energy level ($n = 3$) to the second ($n = 2$), it emits light at the wavelength of 6,563 angstroms (where one angstrom is $10^{-8}$ centimeter). If a free electron is captured from infinity ($n = \infty$) by a positively charged hydrogen nucleus and falls to the energy level $n = 2$, the wavelength of light emitted is 3,646 angstroms. Similar series of spectral lines are produced when the electron drops to other energy levels of the atom (c).

the necessary properties for becoming protostars. In a given nebula counts of the number of stars seen through it and studies of their colors yield information about the cloud's absorption of light and its probable distance from the sun. The cloud's linear diameter and its mass in terms of the cosmic dust can then be estimated. If the spectral lines of molecules are found within the cloud, they tell of the composition and mass of the gas associated with the dust. The relative intensities of certain molecular spectral lines and the presence or absence of others indicate the order of magnitude of the temperatures inside these prestellar concentrations of dust. Turbulence within the clouds widens the spectral lines, giving information about turbulent velocities inside the clouds. The turbulent velocities are quite low, often only a few kilometers per second. All these properties must be known before a theoretician can try to construct a hypothetical model of a protostar.

### Star Formation and Spiral Arms

Spiral galaxies rank among the grandest of celestial spectacles. They have been studied intensively in recent years with both optical and radio techniques. The spiral arms and their related features, which show up magnificently in photographs of neighboring galaxies, have been subjected to intense scrutiny, and their composition and dynamics are beginning to be understood. The Milky Way has the same kind of spiral structure. Work on it and neighboring galaxies has shown that the spiral arms are concentrations of interstellar gas and cosmic dust of more than average density associated with hot Type O and Type B stars. It is significant that the young Type O and Type B stars are found, singly or in clusters, along the spiral arms; they show up in photographs as brilliant beads lighting up the string of the spiral arm. These young stars generate the ultraviolet radiation that produces the beautiful bright emission nebulas characteristic of spiral features. Radio studies show that the interstellar hydrogen and other gases are distributed more densely along the spiral features than between them. Dust is found throughout the arms, and as one would expect it is associated with the interstellar gas. The highest dust concentrations are found mainly along the inner parts of the arms closest to the central nucleus of the spiral galaxy.

The stars commonly found in the spiral features are without exception very young. They are between 10 and 25 million years old, ages between a quarter and a half of 1 percent of the age of the sun and the earth, or at most a tenth of a cosmic year. It is therefore only natural to look for evidence of continuing star birth as a phenomenon of spiral structure. Are the physical conditions in spiral features such that the formation of protostars will naturally take place? How will the protostars, once they are formed, collapse into the dense nebulous objects that are observed by their infrared emission? How will these objects in turn develop into the hot Type O and Type B stars and other young stars that are found preferentially in the spiral features?

Theoretical work on the spiral structure of both the Milky Way and external galaxies provides good evidence that phenomena of shock and compression are responsible for such features. The theory most widely accepted for the production and maintenance of large-scale spiral structure is the density-wave theory developed by C. C. Lin of the Massachusetts Institute of Technology and Frank H. Shu of the State University of New York at Stony Brook. The theory predicts that a gravitational potential wave of spiral form rotates within each spiral galaxy, moving like a boomerang in the central galactic plane. When the density wave passes through the interstellar medium, dense concentrations of dust and gas are produced, which become detectable as spiral arms. W. W. Roberts of the University of Virginia has shown that a high-pressure shock wave accompanies the density wave. If clouds of cosmic dust and gas of more than average density are present in the interstellar material, they are compressed to five or 10 times their original density—possibly beyond the critical stage necessary to form protostars. The shock wave may serve as a trigger mechanism that sets off the process of star birth along a spiral line. Conditions for the piling up of dust and gas are most favorable along the inside of the spiral arms. Here is where the early beginnings of star formation are found. The entire process takes place within a short time: an interval of the order of one million to 10 million years. One would expect to find the newborn stars along the inner edge of a spiral arm. Roberts points out that the conditions for the compression of the available clouds cease to exist as the shock wave passes out of the gas. The spiral line of compression is quite narrow.

When the initially cool clouds of dust and gas collapse, they heat up. They should first become visible as murky clouds with a star deep inside, and these may be observable only in the infrared. Ultimately many of these young stars should become blue-white supergiants rich in the ultraviolet radiation capable of exciting the bright emission nebulas.

As the shock wave moves through the interstellar medium it will inevitably pass through some regions of below-average density where the compression of the shock wave is insufficient to produce conditions of collapse followed by the formation of protostars. It is therefore quite understandable that along a spiral arm we do not find a smooth distribution of young clusters and associations of stars. There should be voids between the lines of concentration of protostars and young objects. This picture of star formation seems to have very strong observational backing: the various stages are as plainly visible to the eye as plants in successive stages of development in a garden.

It would not be right to present the shock-wave theory as one that has received unquestioned acceptance. Some of the theoreticians working in the field

| OBJECT | AVERAGE RADIUS (LIGHT-YEARS) | ESTIMATED MASS (MASS OF SUN = 1) | ACCRETION OF MASS IN 100 MILLION YEARS (MASS OF SUN = 1) |
|---|---|---|---|
| LARGE CLOUD | 12 | 2,000 | 1,000 |
| LARGE GLOBULE | 3 | 60 | 30 |
| SMALL GLOBULE | .1 | .2 | .05 |

**DARK NEBULAS** of interstellar grains can be roughly grouped according to three categories: large cloud, large globule and small globule. Such objects are often associated with areas of strong infrared emission and are believed to be potential birthplaces of stars. Photographs on the next three pages are pictorial examples of objects in each of these categories.

COALSACK NEBULA in the southern Milky Way, photographed by the author with the Curtis-Schmidt telescope at the Cerro Tololo InterAmerican Observatory, is one example of a large cloud of interstellar grains. Although the nebula is visually very dark, it is actually quite transparent. Faint stars can be seen through the obscuring matter. Recent studies have shown that the most transpar-ent sections of the Coalsack dim background stars by only one magnitude (two and a half times) in brightness; some regions, however, are much denser and dim stars by some five magnitudes (100 times). Many nebulas provide the right conditions for the formation of interstellar molecules such as formaldehyde, and may also eventually collapse into protostars or a cluster of protostars.

are strongly opposed to the entire concept of the Lin-Shu density-wave theory of spiral formation. Their guess is that large-scale magnetic fields play a major role in the formation of spiral arms and protostars. All parties seem agreed, however, that spiral structure and protostar formation go together, whatever the mechanism that started it all may be.

The dark nebulas are another class of objects that seem to suggest that star birth is taking place in the galaxy. They often appear as dark holes or clouds against a rich stellar background. Small interstellar grains are the chief known constituent of these nebulas. The reason the nebulas are dark is that the light from the stars beyond them is absorbed and scattered by the tiny grains. Interstellar gas is probably associated with the dust in such nebulas.

## Dark Nebulas and Globules

The distance to a typical large cloud of interstellar grains can be estimated rather closely because the cloud reddens the light from stars behind it. It is therefore not difficult to distinguish between foreground and background stars, and the cloud can be assigned a rough distance. If the distance is known, the cloud's dimensions can be deduced from its apparent diameter in the sky.

A typical large cloud has a radius of some 12 light-years; the dust in it alone has a mass about 20 times the mass of the sun. Various studies have suggested that there is much gas associated with these objects; the amount of gas is generally estimated to exceed the mass of the cosmic dust by a factor of 50 or 100. In the course of time these clouds must sweep up a considerable amount of matter from the surrounding interstellar medium. It looks as though the amount of interstellar gas and dust swept up in an interval of 100 million years is equal to about half the estimated mass of the cloud. Most of the dark clouds should roughly double their mass in time intervals of the order of one cosmic year.

Globules of dust are objects of special interest. On photographs large globules often look like "holes in heaven." In a region with a rich and smooth background distribution of stars one suddenly encounters a darkened spot that looks like an area of low sensitivity in the photographic emulsion. There can be little doubt that these dark holes are roundish clouds or large globules of dust floating by themselves in interstellar space. Some long-exposure photographs with large modern telescopes show the background stars faintly coming through the obscuring matter; some globules must represent very dense small clouds of cosmic grains. Small globules are most often seen as tiny dark specks projected against the luminous background of a bright nebula. It is important to note that globules do not always accompany luminous nebulas, which suggests that conditions for their formation differ from one luminous

TWO LARGE GLOBULES appear as regions of low sensitivity of the film in this photograph in red light made with the 48-inch Schmidt telescope on Palomar Mountain. The globule near the top is No. 134 in the list compiled by Edward Emerson Barnard and the bottom one is No. 133. Given enough time, large globules will probably collapse into protostars and ultimately into actual stars.

nebula to another. No background light shines through the smallest dark specks. The masses assigned to globules are only guesses at a minimum figure. Moreover, it has not been possible to measure the gas content of globules. They contain such small amounts of interstellar gas that observational evidence for its presence is difficult to obtain. Some radio astronomers have discovered small, dense concentrations of the hydroxyl radical (OH), but these are mostly near emission nebulas and not at the position of the small globules.

Within 1,000 light-years of the sun, a small distance compared with the diameter of the galaxy, there are approximately a dozen large clouds and 100 fair-sized globules. It is not at all certain how many small globules there are. Small globules can only be seen projected against the bright emission nebulas, since they cover too small an area of the sky to be distinguishable against the nor-

mal background of stars. It is not known at present whether the small globules are selectively associated with the periphery of emission nebulas or are distributed more or less regularly in the galaxy. On the whole I favor the first suggestion. If small globules were present throughout the central plane of the galaxy, one would expect to see them projected against the luminous background of every bright nebula. As I have noted, that is not the case. The small globules are probably clouds literally rolled up into little dust balls by the pressure exerted by the expanding gas at the periphery of the nebula. The pressure of the ultraviolet radiation emitted by the hot Type O and Type B stars in the heart of each emission nebula probably assists in the formation of the globules. Dark clouds and globules seem to be units that have no choice except to gradually collapse into protostars or break up into clusters of protostars. Although the pressure

waves emitted by a bright nebula may contribute to the formation of the smallest globules, the large clouds and large globules will probably collapse rather quietly on their own under the force of their own gravitation.

The Coalsack Nebula in the southern Milky Way is a fine example of a dark-nebula complex. It covers an area of the sky about five degrees square directly adjacent to the Southern Cross. Photographs show it to be a region of below-average transparency. The stars that lie beyond the Coalsack shine through it with their brightness diminished on the average by from one to three magnitudes (2½ to 15 times). There are some very black spots in the nebula, generally oval in shape, through which no stars can be seen at all. We can almost imagine that the Coalsack Nebula is a dark cloud that is being fragmented into smaller units of cosmic grains; each of these units may eventually become a protostar. The

SMALL GLOBULES appear as tiny black spots near the top right corner of this photograph of the nebula IC 2294 made with the Curtis-Schmidt telescope at Cerro Tololo. Small globules are most often seen projected against the fringes of such luminous nebulas. Smallest of these globules have diameters of the order of the size of the solar system. They also are likely to collapse into protostars.

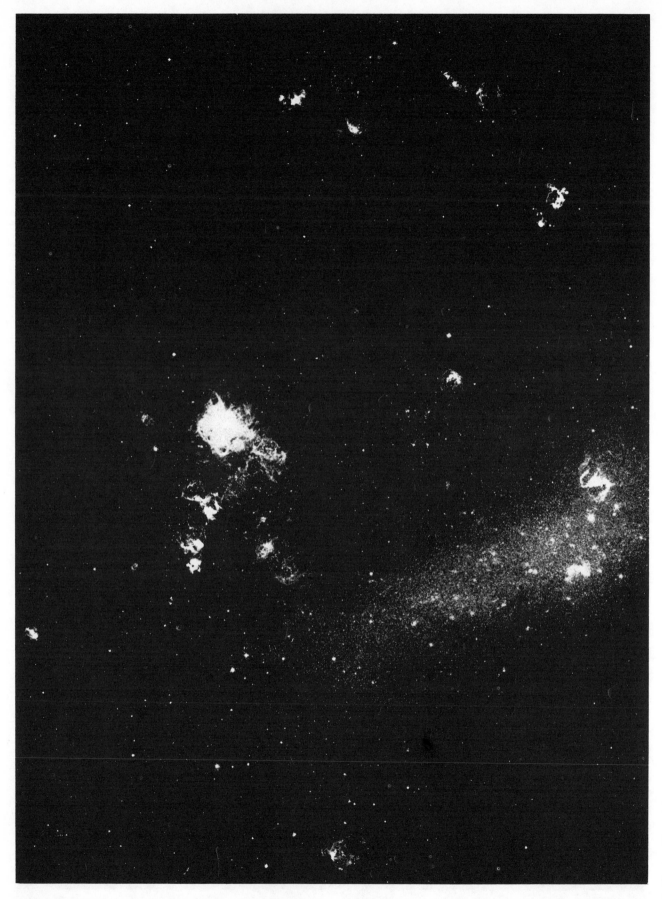

NEBULA 30 DORADUS (*left*) in the large Magellanic Cloud (*right*) is rich in bright nebulosity and young blue-white Type O and Type B stars, as shown in this photograph made by the author with the Curtis-Schmidt telescope at Cerro Tololo. This is one of the best examples of a region in which star birth should currently be taking place. The nebula is so large and luminous that if it were placed where the Great Nebula in Orion is now, it would fill the entire constellation of Orion and cast shadows on the earth at night.

Coalsack Nebula looks like the place where a star cluster is about to be born!

## Supernova Explosions

A supernova explosion, in which a star nearly obliterates itself, is among the most spectacular of celestial phenomena. In A.D. 1054 Chinese astronomers observed a gigantic supernova explosion in the constellation Taurus at the position where we now find the Crab Nebula. There is a stellar cinder left behind, which is observed as a radio and optical pulsar and which appears to be a neutron star that has totally collapsed and is rotating on its axis in the incredibly short period of a thirtieth of a second. The Crab Nebula is 7,000 light-years from the sun and belongs to the Milky Way system. In 1971 attention was drawn to another supernova pulsar located about 45 degrees south of the celestial equator in the constellation Vela in the middle of the enormous luminous cloud recently named the Gum Nebula. The nebula is extremely diffuse and extends to a distance of 30 degrees or more from the central pulsar. The pulsar is some 1,400 light-years away, five times closer than the Crab Nebula [see "The Gum Nebula," by Stephen P. Maran; SCIENTIFIC AMERICAN, December, 1971].

The sun is less than 500 light-years from the nearer rim of the Gum Nebula (which is named for its discoverer, the late Australian astronomer Colin S. Gum). There are no known historical records to suggest that the explosion was observed in ancient times, but that is hardly surprising. The event must have taken place at least 11,000 years ago and possibly as much as 30,000 years ago. The explosion must have had two immediate effects on the surrounding interstellar medium. First, large amounts of gas enriched with elements heavier than helium must have been added to the interstellar medium around the supernova. Second, tremendous amounts of energy must have been transmitted to the surrounding interstellar medium in the form of explosive shock waves. The neutral atomic hydrogen that must have been there before the supernova outburst would have been ionized by the outpouring of ultraviolet energy. The supply of fresh energy is by no means exhausted; the tiny, rapidly rotating pulsar continues to pour it into the interstellar medium. Pulsars and supernovas are potentially a productive source of cosmic ray particles with energies high enough to help maintain the ionization

of the surrounding medium. One of the most striking visible features of the Gum Nebula is its complex filamentary structure, the kind of effect one would expect from the passage of energetic shock waves through the interstellar medium.

It has been suggested that stars were formed or are still being formed in the region of the Gum Nebula. Several highly luminous and apparently young stars are close to the nebula and may have originated at the same time as the star that later became the supernova pulsar. Some of these stars are currently traveling at high velocity away from their apparent common point of origin.

The gaseous filaments visible in photographs of the Gum Nebula must represent highly condensed gas. Perhaps these filaments will ultimately break up into strings of protostars or young stars. Many workers in the field consider it likely that supernovas are conducive to the formation of protostars. The fact remains that no mass production of protostars has been observed in such regions. Moreover, there does not seem to be an abundance of young Type O and Type B stars near supernova remnants.

It may be naïve, however, to look for protostars or young stars in the regions close to recent supernovas. The collapse of a gas cloud into recognizable protostars or young stars is a process requiring at least 100,000 years, and in most cases as much as 10 million years. The Crab Nebula supernova explosion was observed less than 1,000 years ago; the supernova explosion at the heart of the Gum Nebula cannot have taken place much more than about 30,000 years ago. Even the youngest hot stars observed in the region near the Gum Nebula must have predated the supernova explosion by hundreds of thousands of years. The condensations from which these hot stars were born must have originated long before the recent supernova outburst that produced the pulsar. Supernovas may have much to do with the triggering of star birth, but so far we do not know precisely what it is.

Interstellar clouds have been studied in several quite different regions of the electromagnetic spectrum: X-ray, ultraviolet, visible, infrared and radio wavelengths. Such studies have yielded basic information on the properties of clouds, some of which may be on their way to becoming protostars. A variety of dark clouds and infrared objects have been discovered and classified. On the basis of such information we can ask: What kinds of mechanism are responsible for the development of a protostar and its collapse into a star?

## The Formation of Stars

In a recent survey of theories of the formation of stars, Derek McNally of the University of London lists several processes that may be at work; in the end he favors star formation by collapse with gravity as the major cause. His conclusions are generally confirmed by the studies of other astrophysicists, notably C. C. Hayashi and his colleagues at Kyoto University. R. B. Larson of Yale University has drawn special attention to a process by which one of two things could happen. Either a cloud will collapse into many different units and form a cluster of stars or it will collapse much faster near the center than in its outer parts. The second alternative means that a star would be formed mostly from the material near the center of the cloud, and that a young star would be embedded in a large envelope of dust and gas. Such a star would have a truly murky atmosphere! Much attention is being given to how the extended atmosphere might collapse. It seems likely that the protostar would rotate and that this rotation would play an important part in holding it up for some time. In one way or another the protostar must get rid of some of the angular momentum that is stored in the rotating cloud of gas and dust. It can do so most readily by forming dusty shells around itself that in turn break up into planets. The theory for the formation of a protostar from a dust cloud seems to lead almost naturally to the formation of a planetary system.

Recent investigations of infrared objects strongly support the theory that stars are formed by collapsing clouds. Eric E. Becklin and G. Neugebauer of the California Institute of Technology have discovered an infrared point source near the heart of the Great Nebula in Orion that is almost surely a very young star. F. J. Low and D. E. Kleinmann of the University of Arizona have found a second object close to the same region; it seems to be a compact dust nebula, probably with a newborn star or cluster of stars near its center.

Evidence for nebulas enclosed in dust shells has been forthcoming from radioastronomical observations as well. Peter Mezger and his colleagues at the Max Planck Institute in Bonn have found a number of emission nebulas that emit strongly in radio wavelengths but are not detectable at visual wavelengths. The hypothesis is that these are "cocoon nebulas": brilliant nebulas embedded in clouds of interstellar grains. Their radiation in the radio region can pass through the surrounding dust clouds but their radiation in the visible region cannot. They might be observable in the infrared.

A class of intrinsically faint stars, the T Tauri stars, almost certainly represents a very early stage of stellar evolution. T Tauri stars vary irregularly in their energy output and show strong emission lines that are presumably produced in their extended outer atmosphere. The spectra of such stars also exhibit absorption lines that are formed deeper in the atmosphere. The lines are broad and fuzzy, indicating that the stars are either rotating rapidly or continuously ejecting mass. There is much evidence to support the hypothesis that gases are continuously flowing out of the atmosphere of a T Tauri star. Such stars are most often found in groups, generally near or within dark nebulas. The fact that they cluster together is so marked that V. A. Ambartsumian of the Byurakan Astrophysical Observatory in the U.S.S.R. gave them the name "T associations." E. Mendoza of the National Autonomous University of Mexico has found that T Tauri stars are strong emitters of infrared radiation.

George Herbig of the Lick Observatory has apparently observed the formation of one truly new star, FU Orionis, which suddenly appeared in 1936. Quite recently Guillermo Haro of the Tonantzintla National Astrophysical Observatory in Mexico has drawn attention to a star that behaves much like FU Orionis: the faint variable star V 1057 in Cygnus. V 1057 Cygni has recently flared up and is now very bright in the infrared. Haro expresses the opinion that FU Orionis and V 1057 Cygni were originally T Tauri stars that have now advanced to the next evolutionary stage, which is represented by one or more characteristic long-term flare-ups. Herbig and Haro have discovered a number of small bright nebulas (now named Herbig-Haro objects) that are interspersed with the edges of dark nebulas, mostly in regions where T Tauri stars are abundant. T Tauri itself, the prototype for which the class is named, is embedded in such a nebula.

The general scheme I have described here is one favoring the formation of protostars through the process of gravitational collapse in clouds of interstellar gas and dust. I should point out that not every astronomer and astrophysicist favors this scheme. David Layzer of the Harvard College Observatory has developed the following theory of the related formation of stars and galaxies. In the beginning the mass of the universe was distributed quite irregularly. Fragmentation and clustering took place on a large scale, and there were some blobs of hot, dense plasma in which the gravitational field was much stronger than average and in which there were also large electric fields. Layzor believes conditions in these blobs would have been conducive to star formation. Another position has been taken by Ambartsumian and is strongly supported by Halton C. Arp of the Hale Observatories. They consider it likely that violent explosions in the nuclei of galaxies (including the nucleus of our own) may have much to do with the origin and maintenance of spiral structure. They suggest that associations of young stars along with interstellar gas and dust would be a direct result of the ejection of material by such explosions. So far neither the Ambartsumian-Arp theory nor Layzer's has been developed to the point where it can be checked in detail by observations.

It seems quite natural that star birth should be occurring now in the spiral arms of the Milky Way and of neighboring galaxies. Many dark nebulas and globules composed of interstellar gas and dust are seen almost in the act of collapsing into protostars or their close relatives. Objects that are either protostars or very young stars have also been observed. Infrared objects provide a natural link between small dark clouds and relatively normal stars; they may be cool, dense dust clouds with a star or a cluster of stars near the center. The cocoon nebulas may also supply newborn stars. The T Tauri stars seem to be the next stage and help to bridge the gap between the protostars and the young stars.

Comprehensive research is continuing on the problems of change in the galaxy and the related questions of the birth of stars and their early evolution. It is a good thing to describe the galaxy in all its majesty and to study the properties of its many components. The final aim, however, goes further. We want to know how the galaxy came into being, how the stars were formed and what the history and the future of the Milky Way system is.

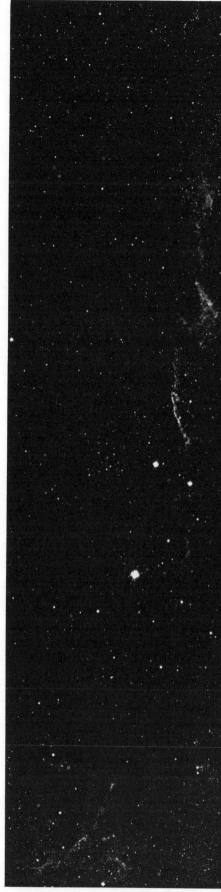

GUM NEBULA, photographed by the author with the Curtis-Schmidt telescope at Cerro Tololo, has an overall diameter of

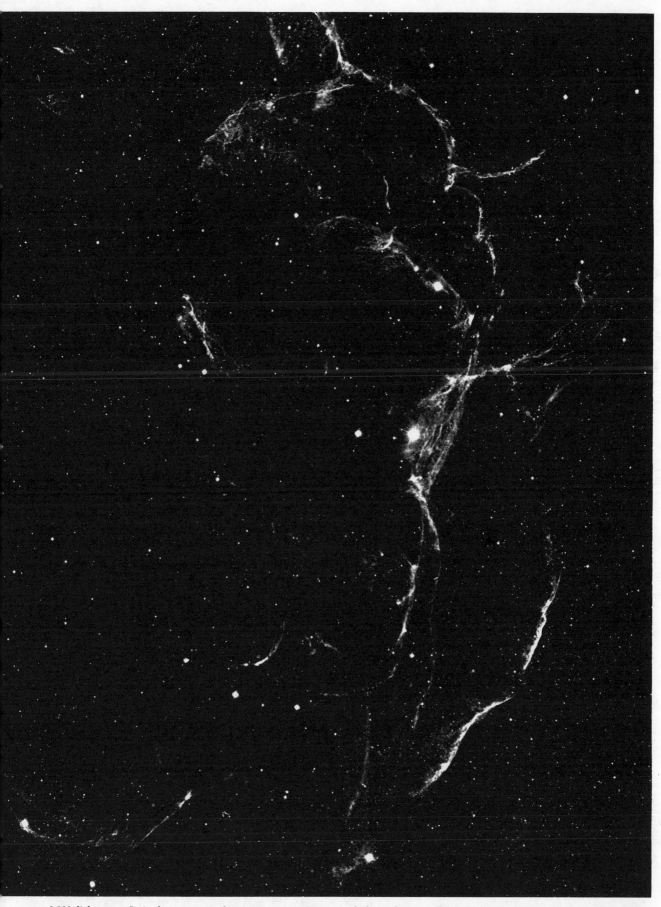

some 2,300 light-years. It is the remnant of a supernova explosion that may have occurred as long as 30,000 years ago. A pulsar, the stellar cinder of the supernova, is located near the center of the nebula in the constellation Vela. The delicate filaments must be gas that is highly condensed and enriched by metals from the supernova; they might ultimately break up into strings of young stars.

# Globular-Cluster Stars

Icko Iben, Jr.

If we could view our galaxy from outside, we would see a giant pinwheel made up of billions of stars rotating slowly around a compact, brilliant nucleus. Looking more closely, we would detect that the spherical volume of space above and below the pinwheel was not empty but was filled with billions of much fainter stars, and scattered about in this "halo" we would also see some 200 fuzzy but brightly glowing globules made up of stars. Close inspection would reveal that each globule consists of from 100,000 to a million stars, most of which are of low luminosity. These are the globular clusters. If one could collapse some hundreds of millions of years into a few minutes, one would see that the 200 clusters are traveling in giant elliptical orbits around the nucleus of the galaxy, closely resembling the old-fashioned picture of electrons whirling around the nucleus of an atom [see illustration on page 190].

The globular clusters are fascinating astronomical objects in their own right, but what has pushed them to the forefront of theoretical interest has been the recognition that the stars in globular clusters are exceedingly ancient and contain clues to the early history of the universe. Indeed, some of the first estimates of their age were so high (20 to 25 billion years) that they could not be reconciled with the apparently much younger age of the universe, as inferred from the recession velocity of distant galaxies. The

velocity measurements imply that all the galaxies emerged from a primordial fireball no more than 12 or 13 billion years ago.

My own interest has been in making theoretical models of stars of various initial masses and compositions to see exactly how they would evolve and how rapidly they would evolve as their nuclear fuel was consumed. From such studies one can hope to refine estimates of the age of globular-cluster stars, to establish if the ages are really inconsistent with the estimated age of the universe and to draw some inferences about the cosmological conditions surrounding their origin. The conclusion, in brief, is that the stars in globular clusters are about as old as the universe is inferred to be, and that the initial abundance of helium in cluster stars is remarkably close to that formed in "big bang" models of the universe. Hence studies of cluster stars provide evidence that the universe has expanded from a tremendously hot primeval atom that was formed in one unique event 12 or 13 billion years ago.

Direct observation of individual cluster stars provides just enough information to make comparison with theoretical models worthwhile. Although one can determine none of the bulk properties (mass, radius or intrinsic brightness), one can estimate the surface temperature from the star's color and judge whether it is more luminous or less luminous than its neighbors. From the star's spectrum one can also estimate the abundance of elements heavier than helium. Unfortunately it is the helium content that most directly determines the course of a star's evolution. Like our own sun, cluster stars are so cool that helium atoms remain in their lowest states of excitation and thus yield little spectroscopic information. Elements

heavier than helium, on the other hand, are readily excited, which makes their abundance easier to estimate. It turns out that cluster stars are strikingly deficient in heavy elements: they contain only from a tenth to a hundredth as many heavy atoms as stars of equal mass in the disk of the galaxy do. Moreover, the farther a cluster star (or other halo star) is from the center of the galaxy, the more deficient it seems to be in heavy elements.

Knowing the surface temperature of stars in a given cluster and their relative brightness, one can plot the distribution of cluster stars in a two-coordinate scheme known as a Hertzsprung-Russell, or H-R, diagram [see illustration on page 191]. The horizontal axis in an H-R diagram represents surface temperature (as inferred from its color); the vertical axis represents luminosity. After plotting a few thousand stars one sees that the great majority fall in a simple linear pattern (the "cluster locus") that rather resembles the head and beak of a bird. The head is thrown back so that the beak points to the upper right corner of the diagram. The tip of the beak is the tip of the "red giant" branch of the cluster locus: the branch containing the coolest (reddest) and brightest stars in the cluster. The shoulder and back of the bird consist of stars 100 to 10,000 times less luminous, whose surfaces range up to 2,000 degrees hotter on the Kelvin, or absolute, scale. Since the great majority of cluster stars (as well as galactic-disk stars) fall in this range, they are known as main-sequence stars. (The sun falls in the middle of the main sequence.) The breast of the bird, which forms an almost horizontal line in the diagram (the "horizontal branch"), is made up of stars approximately 50 times brighter than the sun, with temperatures ranging from the sun's temperature (5,800 degrees K.) to

GLOBULAR CLUSTER M 3 in the constellation Canes Venatici (opposite page) is one of the brighter clusters visible in the Northern Hemisphere. This cluster contains some 500,000 stars and is thus a typical member of its class. Photograph was taken with the 200-inch Hale telescope on Palomar Mountain.

From Scientific American, vol. 223, pp. 27–39, July 1970. Copyright © 1970 by
Scientific American, Inc. All rights reserved. Reprinted with permission.

more than 11,000 degrees K. In this branch stars with surface temperatures between 6,500 and 7,500 degrees fall within the "instability strip." Known as RR Lyrae stars, they dim and brighten rhythmically at precise intervals ranging from six hours to about a day.

If one assumes, as seems reasonable, that all the stars in a cluster are nearly the same age, how can one account for their distribution along the cluster locus? How do they differ from one another? Since all seem to have about the same content of heavy elements, it is reasonable to assume they all started with the same relative abundance of hydrogen and helium. This leaves only mass as the probable variable.

This assumption, however, does not tell us whether mass simply correlates with a specific position on the locus, which would imply that a star is born to its "station" in life, or whether mass determines a rate of evolution through a succession of stations on the locus. Since stars must "burn" their substance in order to shine at all, the latter proposition seems more likely. We still do not know, however, in which direction a star may travel along the locus or whether stars at birth can appear at many different points or at only a few. To answer such questions one must construct model stars and see what kinds of answer emerge. Finally, model studies should be able to tell us if stars of the same age but of different initial mass would indeed distribute themselves as actual stars in an H-R diagram do.

In these idealized models we suppose that a star is a perfect sphere composed of a sequence of concentric shells. At any given instant, as a consequence of simple physical laws, the temperature, pressure and composition will vary from one shell to the next. As the star evolves and exhausts a given nuclear fuel, the shells change in physical structure and composition. During a star's lifetime the temperature in the core can vary from 10 million to a billion degrees K. Densities in the core can reach millions of grams per cubic centimeter. Matter inside the star is predominantly in the form of a plasma, consisting of free atomic nuclei and free electrons. At several stages in a star's life the electrons in the core become "degenerate," a quantum-mechanical state in which the frequency of electron collisions depends less on temperature than on electron density.

Under these conditions the pressure in the core can be maintained not by an increase in temperature but solely by an increase in density. The significance of this will become clear as we proceed.

The design of a stellar model begins with a specification of composition and total mass. The symbols $X$, $Y$ and $Z$ are respectively used to indicate initial abundance of hydrogen, helium and heavier elements. For a typical globular-cluster star we can let $X = .699$, $Y = .3$ and $Z = .001$. (The choice of .3 for $Y$ is actually the consequence of many such computations, as the reader will discover.) Let us also assume that the model cluster star has eight-tenths the mass of the sun. When it begins to burn nuclear fuel, it has .7 the sun's radius and luminosity [see illustration on page 193]. As in the sun, the principal nuclear reaction is the fusion of two protons (hydrogen nuclei) to form a deuteron, the nucleus of deuterium (heavy hydrogen). Subsequently a deuteron and a proton combine to create a light helium nucleus. Two light helium nuclei then combine to form a heavy helium nucleus, releasing two protons in the process. The net result is that for every heavy helium nucleus formed, four protons disappear.

In our model star significant amounts of energy are released throughout a fairly large central region. Near the center of this region and in much of the stellar envelope energy is transported by convection, by matter in turbulent motion. Throughout the main volume of the star, however, energy is carried outward by photons, which are scattered, absorbed and reemitted. This process is called radiative diffusion. As photons approach the surface they are scattered less and less until they finally emerge from the star in straight lines. Although we do not yet have a theory to predict how fast material particles boil off the surface, there is spectroscopic evidence that red giants may boil off as much as a tenth of a solar mass in 100 million years, or perhaps 10,000 times the rate at which particles leave the sun in the "solar wind."

What happens as time passes and the supply of hydrogen in the center of the star is depleted? The key to the answer rests on the simple fact that the fusion of four protons to create one helium nucleus reduces the number of free particles in the star's interior. At a given temperature the pressure in the interior is directly proportional to the number of colliding particles. With fewer particles in the center the star begins to contract until the compressive force of gravitation is exactly balanced by the resisting force,

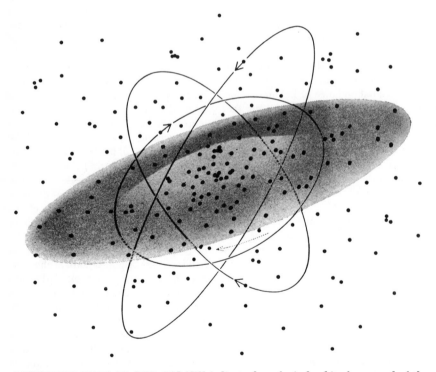

**SCHEMATIC VIEW OF OUR GALAXY** indicates hypothetical orbits for several of the globular clusters that evidently had enough mass to condense into gravitationally bound groups before the main body of the galactic disk was formed. The disk consists of about 100 billion stars, has a diameter of about 100,000 light-years and makes one revolution in about 200 million years. The globular clusters and several billion individual stars of low luminosity form a "halo" around the disk. Cluster stars and individual halo stars, unlike disk stars, are notably deficient in elements heavier than helium. This characteristic and others lead to the conclusion, based on theoretical models, that cluster stars (and presumably the individual halo stars as well) are older than the brightest stars that are found in the galactic disk.

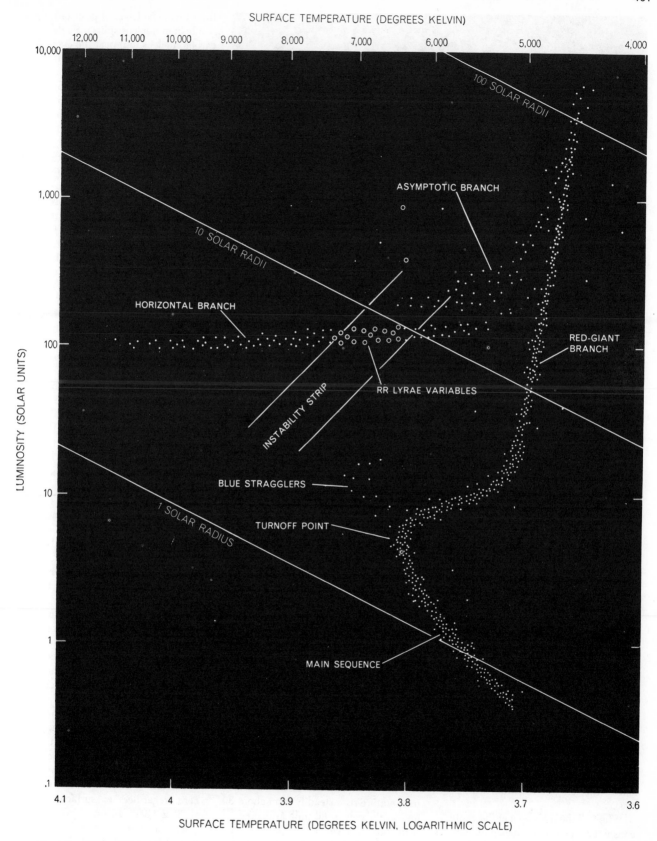

SURFACE TEMPERATURE (DEGREES KELVIN)

SURFACE TEMPERATURE (DEGREES KELVIN, LOGARITHMIC SCALE)

**TEMPERATURE-LUMINOSITY DISTRIBUTION** of the stars in a typical globular cluster is plotted in a Hertzsprung-Russell (H-R) diagram. The surface temperature of the star is derived from a formula based on color: the redder the cooler, the bluer the hotter. Since the absolute luminosity of a star cannot be determined directly, the vertical scale simply indicates relative brightness. The mean curve defined by the distribution is the "cluster locus." A star begins to shine when the temperature in its core be-comes high enough to fuse hydrogen nuclei into helium nuclei. At this stage it makes its appearance on the main sequence. As the star evolves it sometimes expands and sometimes contracts. On various grounds it is believed all the stars in a globular cluster are nearly the same age. They are distributed throughout the cluster locus, as we see them today, because some stars (the most massive ones) evolve much faster than others. The author's studies have helped to establish the rate of evolution in the globular-cluster stars.

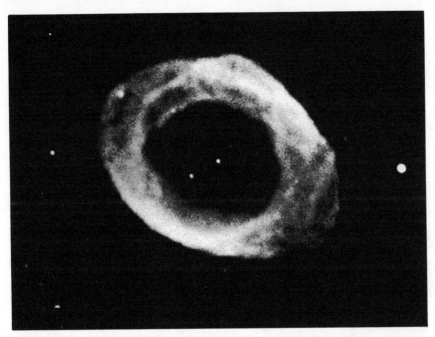

RING NEBULA in the constellation Lyra is one of the most famous planetary nebulas. It consists of a small, hot star surrounded by a shell of glowing gas. The shell may represent material that was ejected during an unstable stage toward the end of the star's evolution.

which is a product of the particle density and temperature. As particles fall toward the center of the star they acquire kinetic energy from the gravitational field. This energy, in turn, is converted by collisions into the energy of thermal motion. Thus the temperature rises along with the particle density.

As a result of the increased temperature in the contracting core the fusion of the remaining hydrogen proceeds more rapidly, releasing more energy that must find its way to the surface. To accommodate the increased flow of energy the outer shell of the star is forced to expand. Hence the apparent paradox: as the interior of the star contracts, the surface volume increases.

This sequence of events, which can be traced in a stellar model, has its counterpart in the H-R diagram. While hydrogen still burns at the center the luminosity of the star increases more rapidly than its radius does. The surface temperature therefore rises, and the track of the star in the H-R diagram is upward and to the left [see illustration on page 194].

When all the hydrogen at the center is consumed, the outer envelope begins to expand more rapidly than the luminosity increases. When this happens, the star has arrived at the "turnoff point" where it leaves the main sequence and starts on its way toward the red-giant branch.

The rising pressure in the growing core, where hydrogen is exhausted, causes the electrons in the core to be-

come increasingly degenerate. Two new phenomena become important. First, energy can flow efficiently through a "sea" of degenerate electrons by conduction, just as heat flows through a metal bar. Thus the temperature rises at about the same rate everywhere in the core. The second consequence is that within the core neutrinos and antineutrinos start being produced in pairs in huge numbers. Their source is a plasma process that cannot readily be duplicated in terrestrial laboratories. Because neutrinos can pass through ordinary matter almost as if it were not there, they escape unimpeded from the core of the red giant, carrying the energy that produced them. The curious net effect of the escape of neutrinos is that energy leaves the center so rapidly that temperatures do not rise as fast as they do somewhat away from the center, where the degeneracy is not so great.

If one makes a computer model of a typical red giant 70 percent as massive as the sun, one finds that the core temperature rises steadily from about 30 million degrees to 80 million degrees, a temperature at which the fusion of helium is kindled. Until that time comes hydrogen remains the nuclear fuel, but at the higher temperatures now prevailing the fusion process proceeds predominantly by the carbon-nitrogen cycle. In this cycle a carbon nucleus absorbs four protons in sequence; along the way two of the four protons are converted to neu-

trons by the emission of positrons. Finally a helium nucleus splits off, releasing a carbon nucleus to repeat the cycle. (The cycle owes half of its name to the transient appearance of an isotope of nitrogen.) Since there is no more hydrogen left in the core, the carbon-nitrogen cycle proceeds just outside the core in an exceedingly thin shell (about 2,000 miles thick): the radius of the shell is only about three times the radius of the earth [see illustrations on pages 196 and 197].

Because a young red giant consumes hydrogen at a high rate it evolves rapidly up the giant branch of the H-R diagram, growing steadily brighter. Throughout this period the size of the core remains constant while the envelope of the star expands by a factor of 10 or more as it grows slightly cooler. One can think of the hydrogen-burning shell in a red giant as a furnace fixed in space that draws in fuel from above and deposits the ashes (helium) in a bin (the core) that steadily grows denser and hotter.

We find in our studies that a star seven-tenths as massive as the sun requires about three billion years to move from the turnoff point on the H-R diagram to the tip of the red-giant branch [see illustration on page 194]. Thus we have evidence, based on physical principles, that the direction of stellar evolution indeed proceeds from the main-sequence portion of the H-R diagram to the red-giant portion. We also discover that if one considers a model only slightly more massive, say a star of .8 solar mass, the time needed to evolve from the turnoff point to the base of the red-giant branch is much shorter. A star of .7 solar mass traverses this portion of the H-R diagram in about 2.5 billion years; a star of .8 solar mass takes only half as long. Both stars, however, spend very nearly the same amount of time in the red-giant stage: about 300 million years.

We have now reached the stage in a star's evolution where changes occur exceedingly rapidly. When this brief period ends, a star that has reached the tip of the red-giant branch reappears on the horizontal branch of the diagram, where, regardless of surface temperature, it is about 40 times brighter than the sun and about 25 times brighter than it was at the turnoff point. The pathway to this new location involves violent changes in the star's structure.

When the core of the red giant, which has been steadily growing in mass and getting hotter, approaches half the sun's mass, helium fusion is kindled. Pairs of helium nuclei fuse into carbon nuclei and some of the carbon nuclei take up another helium nucleus to form oxygen. With further slight increases in temperature and density more energy is produced than can be carried away even by degenerate electrons. The core becomes hotter and fusion proceeds still faster.

At first the pressure that supports the core is primarily due to degenerate electrons and is therefore nearly independent of temperature. Eventually, however, the portion of the electron pressure that increases with temperature begins to dominate the temperature-independent portion. At that point the core expands rapidly until the electrons are no longer degenerate.

The entire process is described as the "helium flash." At its peak the core temperature exceeds 300 million degrees; when the flash ends, the core temperature falls back to about 100 million degrees. The star, no longer a red giant, now embarks on a quiet period in which helium burns at the center of the core and hydrogen continues to burn at the edge of the core [see illustrations on

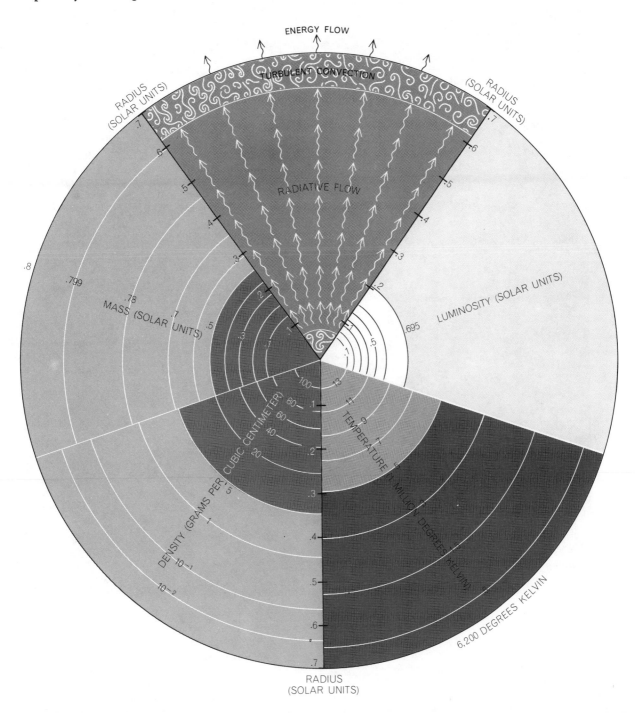

MAIN-SEQUENCE STAR is represented in a pie chart that shows how different properties vary with distance from the center. This model represents a typical globular-cluster star containing 69.9 percent hydrogen, 30 percent helium and .1 percent of elements heavier than helium. Its surface temperature is about 500 degrees Kelvin higher than the surface of the sun. Since it is only eight-tenths as massive as the sun it has only seven-tenths the sun's radius and luminosity. Luminosity is a direct function of energy production, 99 percent of which takes place in a burning hydrogen core that contains half of the star's total mass. In the very center of the core matter is mixed by turbulent convection. Throughout most of the star's volume energy is carried by radiative flow, a process involving the repeated absorption, reemission and scattering of photons. Turbulent convection reappears in the outer shell of the star.

194

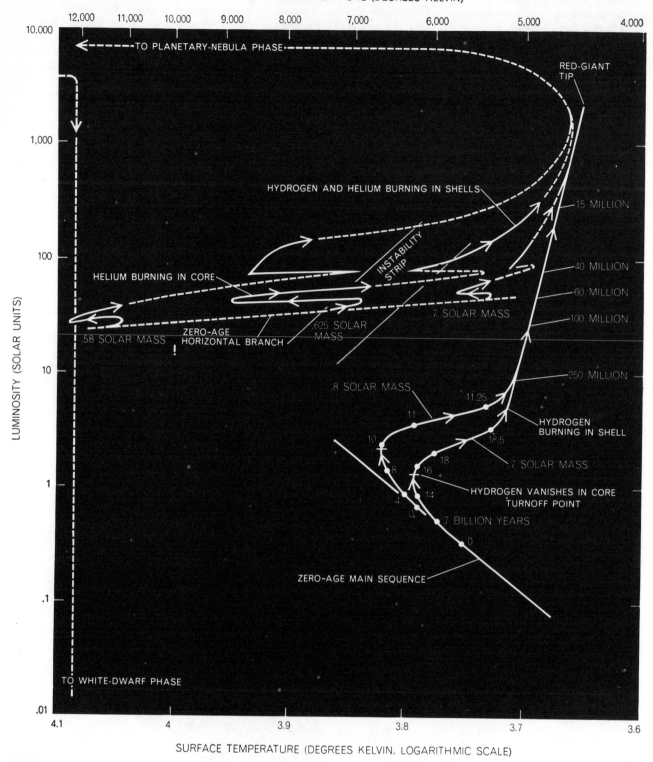

TRACKS OF MODEL STARS in an H-R diagram closely follow the pattern of actual stars in a cluster locus (*see illustration on page 191*). "Zero-age main sequence" defines the location of all homogeneous models of a given composition that are burning hydrogen in their center. A model of .7 solar mass exhausts its central hydrogen and reaches the "turnoff point" (*horizontal bar*) in slightly less than 16 billion years. A model of .8 solar mass reaches the turnoff much more rapidly: in about 10 billion years. The two evolutionary tracks merge as both stars begin moving up the red-giant branch of the diagram. In this stage hydrogen burns in a thin shell surrounding a core of inert helium. The time required by either model to reach the tip of the red-giant branch from various lower points is shown in millions of years. At the tip of the red-giant branch the helium in the core of the model star becomes hot enough to burn, creating a violent "helium flash," after which the model turns up on the "zero-age horizontal branch," defined as the position of all stars of the same composition that have begun to burn helium quietly. The location of stars on this branch is very sensitive to mass. The models traverse the initial solid portion of the tracks in about 60 million years; they traverse the succeeding broken-line portion in only about five million years. At the end of that time the star's central helium is gone and energy is provided by helium and hydrogen burning in separate shells. The final broken portion of the track on the diagram ("To planetary-nebula phase") is speculative.

*pages 198 and 199*]. The hydrogen-burning shell is much thicker but also much weaker than it was during the final years of its red-giant period. The helium-burning core and the hydrogen-burning shell make roughly equal contributions to the star's luminosity. For a typical cluster star of .625 solar mass the phase in which helium burns in the core lasts about 60 million years.

At some time in this period a star of about .6 to .625 solar mass will pass through the instability strip, where, as an RR Lyrae star, it will dim and brighten periodically. The rhythmic variation can be traced to instability in a zone near the surface where helium and hydrogen are only partially ionized. The degree of ionization depends sensitively on pressure and temperature. Any instability that increases ionization subtracts energy from the flow of energy reaching the surface and the star dims. Ionization, however, increases the number of free particles in a given volume, so that the shell simultaneously starts to expand. As the shell expands it cools, and electrons recombine with nuclei. The recombination releases energy, and the star again brightens. As the star keeps "hunting" unsuccessfully for a stable balance its luminosity fluctuates.

A star of .625 solar mass will enter and leave the instability strip more than once as it moves back and forth across the horizontal branch, swinging first to the cooler red end, then back to the hotter blue end and finally back to the red. During this last swing the star develops a thick helium-burning shell in addition to its hydrogen-burning shell. Because of numerical complexities that arise, our understanding of what happens thereafter is not complete.

Evidently instabilities may become so severe that the outer hydrogen shell can collapse while the inner helium shell is expanding. A little later the two shells reverse their direction. One can see that if an oscillation develops and becomes strong enough, the material lying outside the helium shell may be thrown violently out into space. This is perhaps how a planetary nebula is formed [*see illustration on page 192*]. The remnant star may then survive as a brilliant blue star several thousand times brighter than the sun and with a surface temperature of more than 12,000 degrees. Thus it will lie outside the limits of the H-R diagram as depicted in the accompanying illustrations. We cannot predict whether or not all stars of low mass will eject an expanding nebular shell.

Even if no mass is ejected, a star as light as .625 solar mass cannot become hot enough to initiate the fusion of the carbon and oxygen nuclei that remain after all the hydrogen and helium are consumed. The star has no choice but to contract under the influence of gravitation until most of its interior electrons are degenerate. During contraction the star's luminosity is due almost entirely to the release of gravitational energy. The collapse is finally halted by the pressure of degenerate electrons. The only source of energy then remaining is the heat stored in carbon and oxygen nuclei. The star now dims rapidly along a track of constant radius. (The luminosity along the track is too low to be shown in the illustration on the opposite page.) The star has turned into a white dwarf with a radius scarcely larger than the radius of the earth. In this condition it emits so little radiation, even though its surface is hotter than the sun, that it can shine for billions of years.

From such model studies we can now say with some confidence that the way stars in a globular cluster distribute themselves in an H-R diagram is a consequence not of differences in age or of initial composition but solely of differences in initial mass. We have also followed the changes in surface temperature and luminosity of our models as they evolve, step by step, through the H-R diagram. We can now interpret the important features of the cluster locus and identify the evolutionary stage of a star at any particular point.

Let us imagine a cluster made up of model stars of many different initial masses but all the same age. For concreteness we shall assume that this common age is about 12.5 billion years [*see illustration on page 200*]. The main sequence is still populated by stars lighter than .7 solar mass. Stars that are only slightly heavier, about .75 solar mass, have reached the turnoff point, having exhausted the hydrogen in their core. Stars between .75 and .78 solar mass initially have rounded the turnoff and are moving rapidly up the red-giant branch. Stars of .8 solar mass initially have already been through a helium flash and are no longer red giants.

The horizontal branch is populated by stars that began life with at least eight-tenths the mass of the sun. Because they have lost mass along the way their present masses lie between about .6 and .8 solar mass. In general the red stars on the horizontal branch are more massive than the blue ones.

Finally we see that some of our stellar models are burning both helium and hydrogen in thick shells. In the H-R diagram these stars are spread in a broad band, the "asymptotic branch," that slants upward from the red end of the horizontal branch toward the red-giant branch. The area of this band and the distribution of stars within it depend on the amount of mass lost by these stars during their red-giant phase.

Hence we are able to account for the characteristics of most of the stars in a real cluster. Perhaps the most puzzling stars to explain are the "blue stragglers," a small number of stars that seemingly refuse to turn off the main sequence. Actually they may represent a brief transitional stage between the red-giant and horizontal-branch phases. Another suggestion is that they may be rare double stars so close together that their evolution is radically affected.

With this background we are now in a position to estimate how much helium was present in cluster stars at the time of their formation. Once the helium abundance is estimated we can use that value to make model stars coincide with the relative positions of actual stars in H-R diagrams of clusters and from this derive an estimate of the age of real clusters to see whether it conflicts with or supports estimates of the age of the universe derived from other evidence.

Of particular importance in making helium-abundance estimates is the upper section of the red-giant branch, beginning at a point directly opposite the horizontal branch. This point can serve as a luminosity reference point in comparing model stars with actual ones.

When astronomers sample the stars in a cluster to plot a cluster locus, they find that the great majority fall within the main sequence. Fewer than .1 percent fall on the segment of the red-giant branch that is of particular interest; a comparable number occupy the horizontal branch. The reason for the sparsity of stars in these regions is simply that as a star becomes older, the rate at which it moves along its evolutionary track accelerates.

It is a fair assumption that the number of stars in a given segment of the H-R diagram is proportional to the time a single star spends there. Moreover, the rate at which stars leave a given segment must be roughly equal to the rate at which stars arrive from a preceding segment. Direct observation shows that the ratio of the number of horizontal-branch stars to the number of red-giant-branch stars lies between .8 and 1. These, at least, are the ratios for the only two clusters that have been adequately studied. Partial data for 10 other clusters fall in the same range.

The trick now is to adjust the composition of theoretical models so that their rate of evolution provides a steady-state cluster in which the ratio of horizontal-branch stars to giant-branch stars matches the observed ratio: between .8 and 1. The important composition variables in these models are Y, the fractional content of helium, and Z, the fractional content of heavier elements. (Hydrogen, X, then constitutes the remainder.) Fortunately spectroscopic observations provide direct information about Z. The observed abundances of heavy elements lie between .001 and .0001 (between .1 percent and .01 percent). Using these values as limits in our models, we can obtain the desired ratios of horizontal-branch stars to red-giant-branch stars by adjusting the helium content, Y, between .26 and .32 (26 and 32 percent). The mean value, .29, represents the best guess for the initial helium abundance for stars in globular clusters [see upper illustration on page 201].

Now we are ready to estimate age. Here the critical ratio is between the luminosity of stars on the horizontal

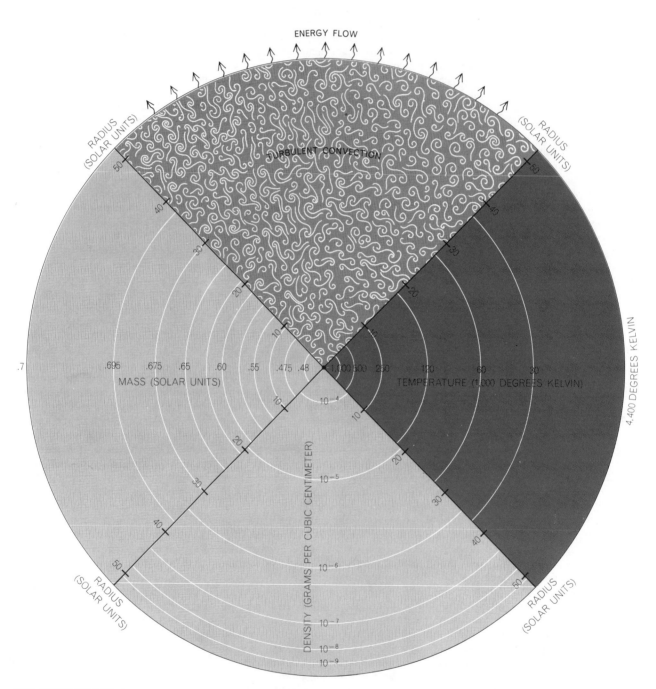

**RED-GIANT MODEL** shows the dramatic changes that occur in the stellar envelope when a star of .7 solar mass exhausts the hydrogen in its center and turns off the main sequence. Now burning hydrogen in a thin shell, it enters a new stage of life on the red-giant branch of the H-R diagram. Whereas its radius had been smaller than the radius of the sun, it now expands rapidly to more than 50 solar radii and its luminosity jumps more than a thousandfold. The young giant is now 975 times brighter than the sun. The surface temperature, however, drops nearly 1,500 degrees, so that the color of the star is now red rather than yellow. Most of the star's mass and its entire energy source are concentrated within the small dot in the center of this illustration (see illustration on opposite page).

branch and the luminosity of more massive stars of the same age that have just exhausted the hydrogen at their centers and thus are just reaching the turn-off point. In actual clusters horizontal-branch stars are 25 times brighter than stars at the turnoff. In our model clusters this ratio of brightness varies with the age of the model: the older the cluster, the greater the ratio. When we choose

values of Z, as before, between .001 and .0001, and introduce values of $Y$ between .26 and .32, we obtain the desired luminosity ratio when the age of the model cluster lies between 10 and 15 billion years [*see lower illustration on page 201*]. The most probable value for the age of the cluster is 12.5 billion years with an uncertainty of ±3 billion years.

Even allowing for the uncertainty in these age estimates it seems clear that globular clusters, and probably other halo stars as well, are among the oldest objects in our galaxy. We know that metal-rich stars in the galactic disk are much younger; some of the brightest among our near neighbors are only a few hundred million years old. The sun itself was born no more than five billion years

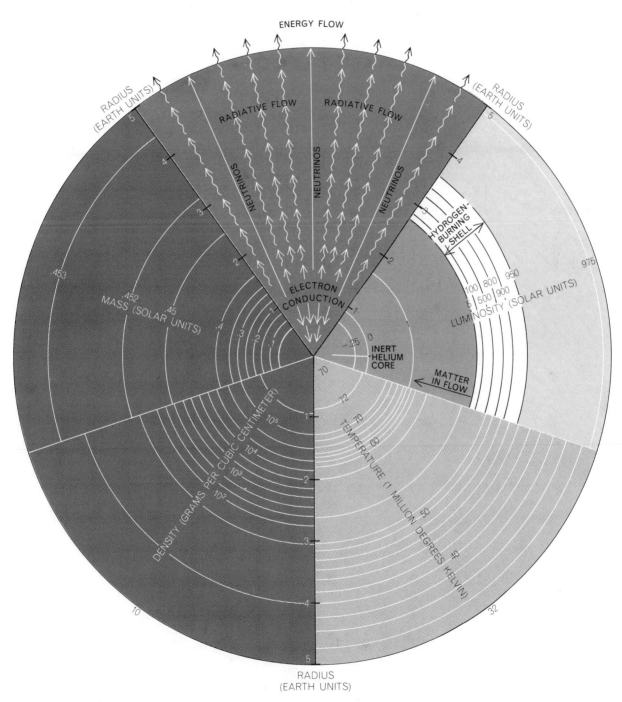

**CORE OF RED GIANT**, whose envelope is diagrammed on the opposite page, is only five times the radius of the earth. All the star's energy is produced in a thin shell of hydrogen, barely 2,000 miles thick, whose mass is only about 300 times that of the earth. The maximum temperature in the inert core of helium is not reached at the center. This comes about because neutrinos, produced copiously by a plasma process near the center, escape through the giant's envelope almost as if it were transparent, removing energy at an enormous rate. The deficit is made good by energy conducted inward by "degenerate" electrons. The helium core finally becomes hot enough to ignite a thermonuclear reaction, producing the helium flash that ends the star's days as a red giant.

ago. It seems safe to say that in the galactic disk the overwhelming majority of stars are no more than about 10 billion years old and that they have been forming continuously throughout this period from clouds of gas comparatively rich in elements heavier than helium.

Hence there appears to be a sharp break in the history of our galaxy that coincides with copious production of

heavy elements within a fairly short period of time. The concentration of metal-rich stars near the galactic center and in the galactic disk suggests that most of the production of heavy elements may have occurred in these regions *after* the gaseous spheres that became the globular clusters were already formed. An attractive hypothesis is that the galaxy collapsed from an original cloud of gas com-

posed almost entirely of hydrogen and helium. The masses that became globular clusters separated out before the final stages of collapse, between 12 and 13 billion years ago. Soon thereafter densities near the center of the collapsing protogalaxy became favorable for star formation. Presumably some of the star masses that coalesced were so large they consumed fuel at an extremely high rate

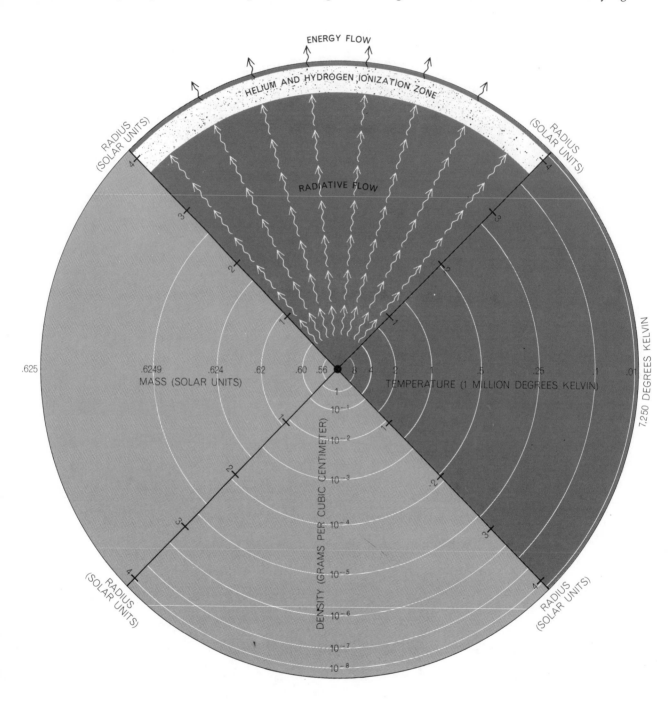

"HORIZONTAL BRANCH" STAR, once a red giant, has shrunk back to four solar radii, assuming a model whose mass is .625 times that of the sun. It is 43 times brighter than the sun and its surface is some 1,500 degrees hotter. The star's energy is produced in a volume represented by the central dot. This tiny volume encloses a central core of burning helium and a shell of burning hydrogen

(see *illustration on opposite page*). Helium fusion contributes about a third, and hydrogen fusion the balance, of the star's energy. In the outer mantle hydrogen and helium are only partially ionized. When this zone becomes deep enough, it induces the periodic fluctuations in brightness characteristic of stars called RR Lyrae variables, which populate "instability strip" in the H-R diagram.

and thus burned themselves out in less than 100 million years. Such stars would have been hot enough to form elements heavier than carbon and oxygen. Some of them undoubtedly exploded, showering heavy elements out into space. The great majority of stars we see today in the disk of our galaxy were formed after these explosive events and so were able to incorporate sizable amounts (a few percent) of the elements heavier than helium.

According to this hypothesis, the sub-condensations of gas remaining outside the galactic disk in the final stages of collapse remained isolated, each contracting under its own weight. At the center of some of these "protoglobular" clusters conditions may have become favorable for star formation even earlier than at the center of the galaxy. In others star formation may have begun later, in which case clusters might vary in age by as much as five billion years. We shall have to wait for more observations and further refinements in theory to decide whether or not this is so. For the present, however, it seems reasonable to assume that globular clusters are all nearly the same age and that this age—12.5 ± 3 bil-

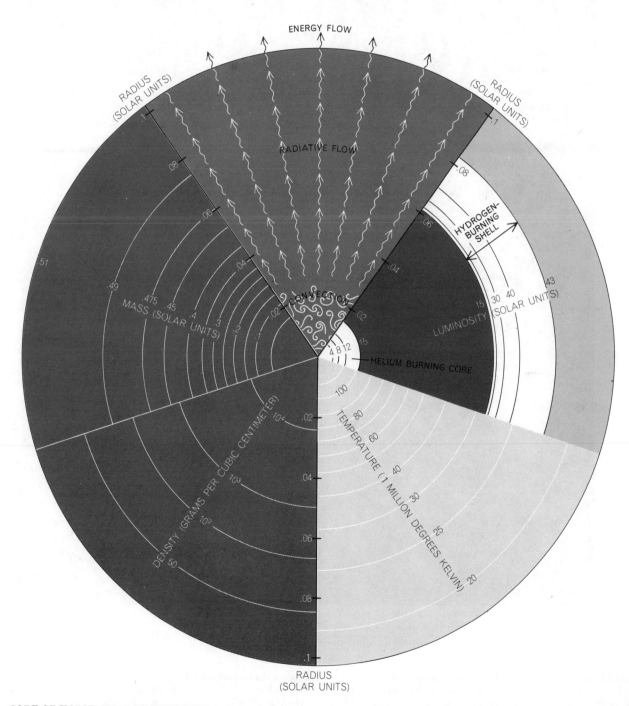

CORE OF HORIZONTAL-BRANCH STAR is the core of the star of .625 solar mass depicted on the opposite page. More than 80 percent of the star's total mass is contained within a volume whose radius is only a tenth that of the sun. In this tiny core helium is converted to carbon by nuclear fusion, and some of the carbon nuclei capture helium nuclei to form oxygen nuclei. The hydrogen-burning shell is much thicker but less energetic than it was in the red-giant stage. Helium is exhausted about 100 times more rapidly than hydrogen was exhausted when the star was starting life on the main sequence. Thus a star spends only about a hundredth as much time (the length of time is typically 100 million years) on the horizontal branch of the H-R diagram as it does on the main sequence.

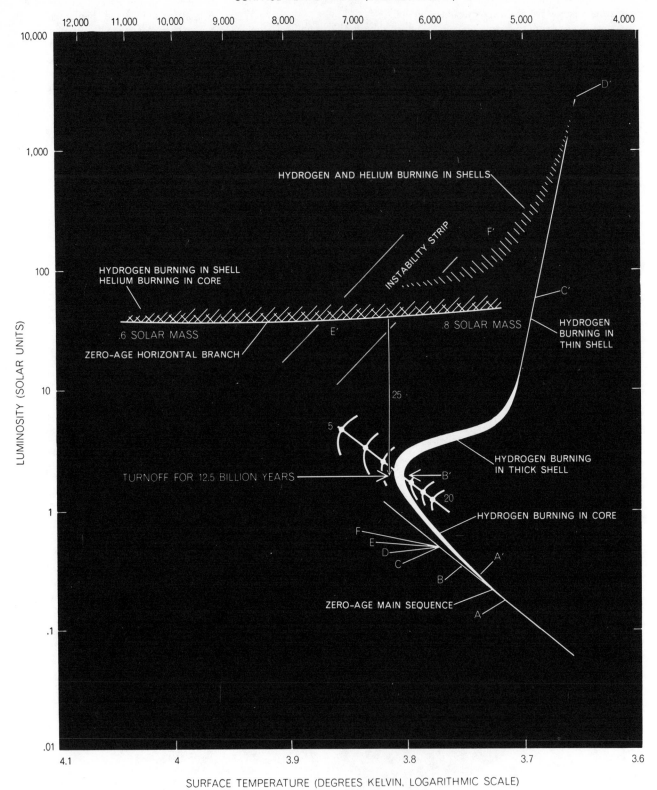

SURFACE TEMPERATURE (DEGREES KELVIN)

SURFACE TEMPERATURE (DEGREES KELVIN, LOGARITHMIC SCALE)

LUMINOSITY (SOLAR UNITS)

HYDROGEN AND HELIUM BURNING IN SHELLS

INSTABILITY STRIP

HYDROGEN BURNING IN SHELL
HELIUM BURNING IN CORE

HYDROGEN BURNING IN THIN SHELL

.6 SOLAR MASS

.8 SOLAR MASS

ZERO-AGE HORIZONTAL BRANCH

HYDROGEN BURNING IN THICK SHELL

TURNOFF FOR 12.5 BILLION YEARS

HYDROGEN BURNING IN CORE

ZERO-AGE MAIN SEQUENCE

**CLUSTER LOCUS OF MODEL STARS** looks like this at the end of 12.5 billion years. The stars that are spread along the locus from A' to D' are all of the same age and initial composition but they differ in mass. Stars at A' began at point A 12.5 billion years earlier; stars at B' began at B, and so on. The turnoff points for theoretical clusters ranging in age from five billion to 20 billion years are also shown by dots placed at intervals of 2.5 billion years. The 12.5-billion-year age appears to be the most probable age for globular-cluster stars because model horizontal-branch stars are then brighter than model stars at the turnoff point by the factor actually observed in real clusters: a factor of roughly 25. The horizontal branch in the diagram is constructed from tracks for models with masses between .6 and .785 solar mass that are burning helium at the center. The masses chosen for these models are arbitrarily lower than the masses of red-giant models in order to compensate for the evidence that stars in the red-giant stage throw off significant amounts of mass. Thus stars lying between D' and F' originated as more massive stars between D and F on the main sequence.

lion years—marks the time when our galaxy collapsed to its present dimensions.

How does this interact with estimates of the age of the universe based on the observed rate of recession of galaxies? If we interpret the observed recession rate according to the simplest big-bang model, the universe began with the explosion of a primordial atom between seven and 13 billion years ago, the precise time depending on the average density one assigns to matter in the universe now. Current knowledge of the average density does not exclude either end of the age range. The lower the present average density, or the more "open" the universe, the closer the big-bang estimate approaches 13 billion years. In view of our estimate of cluster age, it seems that the universe may be very open indeed and that our galaxy and the cluster stars within it may have been formed very shortly after the "beginning."

The fact that galaxies are receding from one another does not by itself constitute proof that we are living in a big-bang universe; the fact of recession can also be accounted for by steady-state models, which assume that the average density of matter in the universe is maintained constant by the continuous creation of matter. An additional fact, however, would seem to make the big-bang model the more appropriate description. This fact is the abundance of helium in the primitive gas, when computed for models of the primordial fireball. The figure obtained agrees very closely with the values disclosed by our studies of the primitive gas from which the oldest stars were formed.

In big-bang models temperatures in the primordial fireball are at one stage so high (above 10 billion degrees) that matter is almost entirely dissociated into protons, neutrons and electrons. As the fireball rapidly expands and cools, protons and neutrons can begin to fuse into deuterium nuclei. Later deuterium nuclei can fuse into helium nuclei and nuclei of still heavier elements. For the simplest big-bang models, however, the matter emerging from the fireball contains between 23 and 28 percent helium; virtually all the rest is hydrogen. The fraction of elements heavier than helium is more than 100 times smaller than can be detected in the most metal-deficient globular-cluster stars.

The effective absence of heavy elements in the matter emerging from the fireball phase strengthens our inference from purely galactic evidence that most of the heavy elements were absent in the

protogalaxy and were made in massive stars during a phase of galactic collapse.

The abundance of helium in matter emerging from the fireball phase is strikingly close to the initial helium abundance we have estimated for the oldest metal-poor stars. This agreement provides perhaps the strongest evidence yet uncovered for the hypothesis that the

matter in the universe was once crammed together at extremely high densities and at temperatures in excess of 10 billion degrees. The fact that under such conditions most of the energy in the universe was in the form of electromagnetic radiation (photons) gives added meaning to the phrase "And God said, Let there be light."

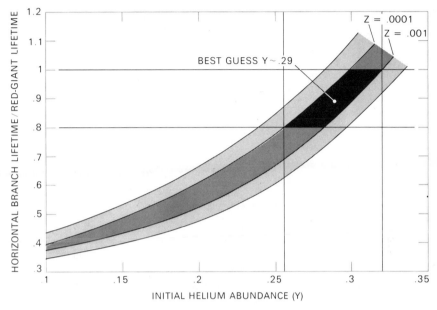

HELIUM CONTENT OF CLUSTER STARS can be derived from model studies by specifying the abundance of elements heavier than helium (labeled $Z$) and by specifying the ratio between the time models spend on the horizontal branch and the time spent on the red-giant branch. In actual clusters this ratio seems to lie between .8 and 1. The most probable value of $Z$ is between .001 and .0001 (.1 and .01 percent). The best-fitting models show an initial helium abundance ($Y$) between .26 and .32, with a mean of about .29.

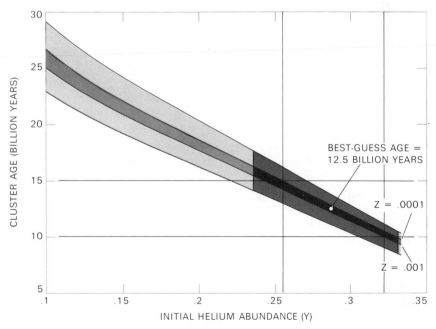

AGE OF CLUSTER STARS can be computed once the initial abundance of helium and heavier elements ($Z$) is known. The ages of models are adjusted until stars in the horizontal branch of the cluster locus show a luminosity some 25 times higher than models that are just leaving the main sequence. The most probable age is 12.5 ± 3 billion years.

# 26 Obituary of Stars: Tale of Red Giants, White Dwarfs and Black Holes

Ben Bova

In the predawn darkness of September 2 last year, Robin Conway, a British radio astronomer, received a frenzied phone call from America. An x-ray-emitting star named Cygnus X-3 was exploding.

Radio astronomers in Canada and the United States had observed titanic flares of energy coming from Cygnus X-3, and they wanted Conway to get the huge Jodrell Bank radio telescope on it in time to catch this unexpected and dramatic event. Within hours, observatories all over the world were aiming their instruments at Cygnus X-3.

For about ten years astronomers had been observing sites in the sky that are powerful emitters of x rays. Theoretical considerations led them to the conclusion that these x-ray *sources,* as astronomers refer to them, are dying stars—or at least they were dying [hundreds or] thousands of years ago when they emitted the x rays we now receive on Earth.

In the month following Cygnus X-3's first flare-up, four separate bursts of energy were detected and [some] astronomers believe they witnessed the death throes of a star. A few go even further, claiming that Cygnus X-3 is actually a star that has collapsed and disappeared from our universe: a black hole. In other words, they have "seen" something absolutely invisible. A black hole is the ultimate in stellar deaths. The dying star literally performs the old joke of "digging a hole, jumping in, and pulling the hole in after it." No light, no heat, no particle of matter can escape from a black hole; it is a cavity in space.

Of considerable interest to us on Earth is the question: Will the sun one day disappear in a black hole, dragging its planets, including Earth, down with it? Not likely, according to current theories. But, like all the stars, the sun will die. How? When? To answer that, consider another question: What keeps the sun basically at its present size and shape?

The sun is a star, as nearly perfect a sphere as you'll find anywhere in nature. Its diameter is 865,400 miles, more than 100 times that of the Earth—but the sun is only medium-sized

as stars go. If you could see a time-lapse movie of the sun's formative years, you would see a huge distended cloud of gas and dust falling inward on itself—a process called gravitational collapse. This event took place nearly five billion years ago, but it happened in an astronomical eye-blink: a scant 50 million years.

Then the gravitational collapse stopped, leaving the sun as a sphere of glowing gas. The gravitational force is still there, still trying to squeeze the sun's two billion billion billion tons of matter (that's $2 \times 10^{27}$ tons) into a still smaller sphere.

But balancing the inward-pulling force of gravity are the outward-pushing forces of gas and radiation pressure. The gas pressure at the sun's core is at least 1.3 billion times the pressure of air at sea level here on Earth. At the core of the sun, temperatures reach some 15 million degrees Kelvin (which is about 27 million degrees Fahrenheit), hot enough to allow thermonuclear reactions in which hydrogen is converted into helium and releases energy, adding to the outward pressure of the gas.

For nearly five billion years the outward and inward forces have been neatly balanced, keeping the sun at about the same size and shape, although its mass is changing imperceptibly. But in another five to ten billion years, the sun's supply of hydrogen fuel will begin to run low. Most of its core will be denser and hotter than it is now, some 100 million degrees Kelvin at its center. The helium itself will then begin to undergo fusion reactions, forming carbon, oxygen and neon.

Gravity will yield somewhat to increased gas and radiation pressures, and the sun's outer layers will expand. The shining surface of the sun, called the photosphere, will become distended and cooler, and will turn redder in color. The sun will become what astronomers call a *red giant*. The sun's outer envelope will become hugely bloated and reach out to engulf the inner planets, including Earth.

As the fusion processes in the sun's core produce constantly heavier elements, the core temperature will keep rising. The higher the core temperature, the more easily new fusion reactions start with the heavier elements, leading to the creation of still heavier elements and still higher temperatures.

From *Smithsonian,* vol. 4, no. 4, pp. 54–63, 1973. Reprinted with permission.

A character in a Hemingway novel was once asked how he went bankrupt. "Two ways," he said. "Gradually and then suddenly." The same thing happens to stars, and gravity is the waiting banker who eventually forecloses all the loans.

Were we to be present, we would see that in the aging star each new cycle of element-building goes faster than the previous one. Each cycle brings the sun closer to disaster. Through it all, gravity is constantly being outfought by rising gas and radiation pressures.

Ultimately, fusion reactions at the sun's core produce iron. Further fusion reactions with iron do not release energy, they require it. The game is over. The temperature of the star's core slows its upward spiral, energy production at the core shuts down, the outward pressures of gas and radiation die away. And gravity, which has been waiting all this time, becomes the victor. The sun will now collapse.

Into a black hole? Apparently not; the sun is too small a star for that. Instead it will become what is called a *white dwarf.*

Astronomers discovered the first white dwarf in 1915—the tiny companion of the bright star Sirius. (Since Sirius is known as the Dog Star, its white dwarf companion was inevitably dubbed the Pup.) White dwarf stars were then thought to be the end point in stellar evolution. They are just as massive as an ordinary star, but they are squeezed down to a size more like that of the Earth—some 100 times smaller than even a medium-sized star.

The question then became: How does a star go from its distended red giant phase down to a collapsed white dwarf? The problem sounds more like something out of a fairy tale than a problem in astrophysics, but it was—and is—a critical problem for the understanding of the death of stars.

Stars have been known to explode. Sometimes they explode rather mildly, in cosmic burps called *novas.* A nova outburst does not damage the star much—some novas may repeat their flare-up every few thousand years. Of course, anyone living on a nearby planet would be incinerated by even a mild nova outburst. There are also more catastrophic stellar explosions, called *supernovas,* that release as much energy in 24 hours as the sun emits in more than a billion years.

After the first white dwarf was discovered in 1915, many astronomers believed that red giants eventually exploded and the white dwarfs were the remainders left after the cataclysms. Now we know this is not *always* the case. Stars that are of the sun's mass may end up as white dwarfs without any traumatic explosions at all.

Computer analyses show that as gravity begins to compress the sun, after its red giant phase has ended, the density and temperature inside rise. As its inner temperature soars, any unused fusible materials—from hydrogen to iron—will eventually reach their [nuclear] ignition temperature and explode like cosmic firecrackers. Eventually the increasing density and temperature produce a braking action that halts the collapse. The sun's collapse may take place over a span of a few million years, gradually sinking from its grossly distended red giant diameter down to a size more like our own Earth's—about 8,000 miles across.

When all the material in the sun has been squeezed down to this small a size, it is so dense that if you had a teaspoonful of it, it would sink to the center of the Earth instantly, plowing through rock and metal as if they were quicksilver.

Again, why does the gravitational collapse stop? Why won't the sun get even smaller and denser? The answer lies in what happens inside the sun.

Like all stars, the sun is composed of plasma, which consists of ions (atomic nuclei that have been stripped of their orbital electrons) and free electrons. As the density of the plasma increases, these ions and electrons collide more frequently. They are squeezed more and more, until at a density of about a thousand tons per cubic inch the electrons resist further compression. This produces the counterforce that finally balances out against gravity and stops the sun from collapsing any further.

So now we have the sun as a white dwarf, just about the size of Earth, though still with about 330,000 times more mass. Henceforth, it simply cools off, as the heat generated from its collapse dissipates slowly into space. This process may take billions of years, but eventually the sun will be nothing more than a cold, dark cinder, circled by frozen planets.

So much for the sun; a fairly ordinary star. But there are many stars with greater mass than the sun and if you thirst for yet more drama in the skies, consider the death agonies of these more massive stars. Stars that are at least half again as massive as the sun don't stop collapsing at the white dwarf stage—and they reach that stage differently.

If there is enough unburned fusible material in the star's outer shell after its career as a red giant and an unstable period as a pulsating star, the rapidly rising heat of the core may trigger those superficial explosions called novas.* These blow off some of the star's outer envelope of plasma and may destroy any planets nearby. But the core of the star survives. It keeps on shrinking, passing quickly through the white dwarf stage. The electrons' resistance to compression cannot counteract the pull of gravity because the gravitational force is much stronger, being proportional to the star's mass. The collapse goes on. The electrons are squeezed into the protons (particles of the atomic nuclei) in the plasma, turning all the protons into neutrons.

We now have a mass at least equal to the sun's, the core remainder of an originally more massive star, that consists entirely of neutrons packed side by side into a sphere no more than 100 miles wide—probably more like ten miles wide. It contains some $10^{57}$ neutrons, squeezed together at a density of about a billion tons per cubic inch.

We have a *neutron star.*

If the star at this stage is not more than about three solar masses, then the tremendous repulsive forces that the neutrons exert on each other will resist any further gravitational crunch. The brakes are on—neutron brakes this time—and the collapse stops.

But there are still the outer layers of the star, even though much of this material might have blown off in one or more

---

*By current thinking, such events might resemble planetary nebula ejections rather than nova explosions; the latter events probably occur after mass transfer onto a white dwarf star in a binary system.—Ed.

explosions. What's left of this outer shell falls in on the tiny neutron core, since gravity is always at work. The impact creates enough heat to drive the core's surface temperature up to billions of degrees for a fraction of a second. Under these circumstances most of the heat energy is converted into the strangest particles in all of physics: neutrinos.

Not to be confused with neutrons, neutrinos are aloof little particles which, under ordinary circumstances, can zip through a wall of lead *50 light-years thick* without being stopped at all.

But the conditions around a neutron star are far from ordinary. The densities and temperatures of the plasma are so high that even the evasive neutrinos can travel only a few feet before they are deflected or absorbed. Most of their tremendous energy is imparted to the plasma clouds, heating them to tens of billions of degrees. This creates a supernova explosion that blows away everything except the tiny core of the neutron star.

All this—the collapse of the neutron core, the infall of the shell of plasma, the heating that forms the neutrinos, the core supernova explosion—all this happens in a few seconds.

The result? Look at the Crab Nebula, that wildly twisted cloud of plasma a few light-years across, still expanding at several hundred miles per second more than 900 years after the core supernova that created it (SMITHSONIAN, June 1970).* The Crab Nebula glows with a full spectrum of electromagnetic energy: radio waves, infrared, visible light, ultraviolet, x rays and even gamma radiation. And in the center of the Crab Nebula, beautifully verifying all the theories and calculations, is a pulsar!

---

*Article 42 in this reader.

## No Little Green Men

For the pulsars, most astronomers now believe, are actually neutron stars that are emitting sharply timed bursts of radio energy.

The first pulsar was discovered in 1967 by Jocelyn Bell and Antony Hewish of Cambridge University. Within a few months a half dozen more were found, including the pulsar in the Crab Nebula, which is designated NP 0532 (see figure at the bottom of this page).

When the pulsars were first discovered, their uncannily precise radio bursts led some astronomers to wonder if these might not be radio signals from intelligent beings. For a few weeks they were informally called LGM signals—for "little green men." By the end of 1967, Thomas Gold of Cornell University and several other theoreticians had proposed natural models that seemed to explain the pulsar phenomenon without resorting to interstellar civilizations.

Gold's theory is the most widely accepted at present. He pictures the pulsar as a neutron star that is spinning rapidly while surrounded by fairly dense plasma clouds, the remnants of the supernova explosion. The observed timing of the pulsars' radio bursts, all grouped around the 1 to 100 times-a-second mark, fit in well with the expected spin rate of a 10- to 100-mile-wide neutron star.

The discovery that the pulse rate of the Crab Nebula's pulsar was lengthening supported Gold's explanation. Early in 1969 the Crab Nebula pulsar was detected visually and photographed. The optical pulsations are in synchronization with the radio pulses, as theory predicted they would be. At least one pulsar shows a rate of pulsation that occasionally speeds up, only to gradually slow down again. This most likely means that this neutron star is jerkily shrinking, still being crushed to a smaller size by gravity.

How far can the crush go? According to theoretical physicists, such as Kip S. Thorne of the California Institute of Technology, under the right circumstances a star can literally go straight out of this universe. Into a black hole.

Rotating like a searchlight, the pulsar in the Crab Nebula spins, beaming light and x rays toward the Earth about 30 times a second, blinking off, on (center), and off again. The pulsar is a neutron star, the remains of a more massive star that suffered a huge explosion, creating the Crab Nebula. [J. S. Miller and E. J. Wampler, Lick Observatory.]

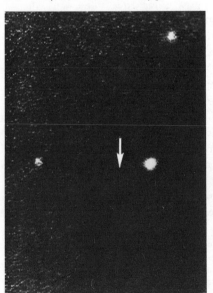
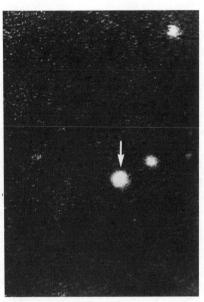
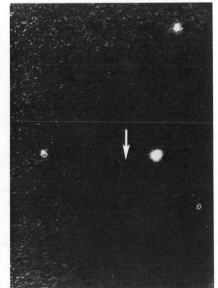

Let's take another look at a neutron star, using the eyes of imagination to peer through the swirling carnage of plasma clouds that surrounds it.

The neutron star has already suffered a gravitational collapse and supernova explosion. If the core's mass is not more than about three times the sun's mass, its collapse stops at this phase. For several thousand years the neutron star will radiate more x rays than any other type of electromagnetic energy.

The neutron star can also cause radio and optical pulses, although these are likely to be coming from the plasma cloud surrounding the star, rather than from the neutron star itself. After a hundred million years or so, the neutron star's temperature cools down and it will become a quiet, dark chunk of extremely dense matter measuring some 10 to 100 miles across.

But if the neutron star is more than about three solar masses, the gravitational infall doesn't stop. The star keeps on shrinking, and the gravitational force gets mightier with each microsecond.

As the interior density goes past the ten-billion-tons-per-cubic-inch mark, the neutrons themselves are squeezed down into smaller subatomic particles called hyperons. At this point, no possible braking force can stop the collapse. The gravitational force becomes all-consuming. For a star of about three solar masses, once its radius gets down under two miles, it winks out. (This is called the *gravitational radius,* the radius at which light can no longer escape from the star.) When the star collapses past that size,* it has dug itself a black hole in space-time and disappeared into it.

But the gravitational collapse does not stop simply because the outside world can no longer see the star. According to an early theory, the star keeps on collapsing until it reaches the almost unimaginable state of zero volume and infinite density. Such a point is called a *Schwarzschild singularity,* after the German physicist Karl Schwarzschild (1873-1916).

What would it look like if you could watch the final disappearance of a star as it collapsed into a black hole? You would have to be able to see in x-ray wavelengths, because that is what the collapsing star is radiating. And you would have to be able to see through the plasma clouds surrounding the scene. Most of all, you would have to look very fast, because the whole affair happens in less than a second. (According to computer calculations, that is; under the influence of such titanic gravitational fields, perhaps time itself becomes warped out of all recognition.)

One moment the dying star is hanging there in the midst of the plasma cloud, then suddenly it shrinks like a pricked balloon, getting smaller and smaller.

The photons of x-ray energy must now work harder to get away from the fast-increasing gravitational field. This shifts them to longer wavelengths. Perhaps they are shifted even far enough to turn into light waves.

At last—still within the space of a second, remember—the collapse will seem to slow down. A few x rays are struggling

*Actually, the star vanishes when it collapses within the Schwarzs-child radius (as defined by S. Chandrasekhar on page 209). This equals twice the gravitational radius and is thus four miles in the present case.—Ed.

up from the rim of the black hole. The star finally disappears, but there's a dim halo left, where those last few x rays are taking tortuously spiral paths to work out of the gravitational pit that the star has dug for itself.

If we can envision an interstellar spacecraft, it could approach a black hole quite closely without being sucked in. If the spacecraft were going fast enough, it could orbit around the hole's gravitational radius, just as a satellite can orbit around Earth if it has the proper velocity. But should the spacecraft wander inside the gravitational radius—goodbye! It can never get out again, no matter how mighty its engines.

Gravitational collapse into a black hole can also happen to objects larger than individual stars—an entire galaxy could theoretically collapse. Interstellar space might be strewn with "potholes" ranging from a few miles to a few million miles across, black holes where individual stars or whole galaxies have collapsed in on themselves.

We can now see why there was such worldwide excitement over Cygnus X-3 when it flared up. Here was an x-ray source that must be a neutron star and possibly might be a black hole.

Cygnus X-3 is located in the region of the sky mapped out by the constellation Cygnus, the Swan. In 1966, a group of scientists led by Riccardo Giacconi of American Science and Engineering found that there were several sources of intense x-radiation in Cygnus. X-3 was the third such source discovered in that area of the sky. It is apparently far out on the fringes of the Milky Way galaxy, roughly 30,000 light-years away from us.

Cygnus X-3 was soon found to be emitting radio waves as well as x-rays, although it was a rather weak radio source. But on September 2, 1972, a team led by P. C. Gregory of the University of Toronto, using the 150-foot radio dish antenna of the Algonquin Radio Observatory in Ontario, was surprised when they discovered that Cygnus X-3 was suddenly 1,000 times stronger in radio emission than it had been only two days earlier.

X-ray telescopes aboard the satellites *Uhuru* and *Copernicus* showed no increase in the star's x-ray output. (Since x rays do not penetrate the Earth's atmosphere, x-ray detectors must be placed in high-altitude balloons or rockets, or aboard satellites.) The radio emission was not coming from the star itself but from the plasma cloud surrounding it.

Over the next month, four different outbursts were observed, some of them with "flares" that lasted only about an hour. Generally, though, the active periods were about 24 hours long, indicating that the cloud producing them is one light-day in diameter—that is, about two-and-a-half times the diameter of our solar system as we know it.

## A Star with a Dead Partner

Remo Ruffini of Princeton University believes that Cygnus X-3 is a black hole which began its life as a very massive star and now has exploded and collapsed. It is part of a binary, or two-star system, he claims, citing evidence that indicates Cygnus X-3 is rotating around a partner star every 4.8 hours. The x-ray and radio output may be caused by matter from the partner star being sucked into the black hole. Ruffini also believes that at least two other x-ray sources—Cygnus X-1 and Small Magellanic Cloud X-1—are also black holes that were once members of binary star pairs. Other astronomers and physicists disagree strongly. They feel that it is much too early to declare these objects black holes. The evidence is at best hazy and uncertain, and even the black hole theory is questioned (see box, page 207).

If astronomers have detected black holes, they are literally studying things that no longer exist in our observable universe. Or do they?

In that strange world inside a black hole, where an entire star is crushed down toward zero volume and infinite density, what physical rules are in effect? In this realm, theoreticians rely on mathematics and—in a sense—on poetry. Even the relativistic physics of Einstein can't explain what happens at such densities. No one knows what happens, physically, down at the bottom of a black hole—except that it may be bottomless. Or even open-ended.

Several theoreticians have pointed out that the mathematics of gravitational collapse and Schwarzschild singularities apply only to perfectly spherical bodies. Stars are not exactly perfect spheres; galaxies are even less so. As they are gravitationally crushed, it is likely that any deformations in their shapes will become exaggerated, not smoothed out. Roger Penrose, a British mathematical physicist, has shown that a nonspherical body would not be completely crushed to zero volume. For reasons known best to the mathematicians, such a body can escape dwindling down to a singularity, but it can't stay in the same physical location where it collapsed. In effect, the black hole turns into a tunnel.

### Space Warps and Time Warps

We can visualize this by drawing on an analogy that relativistic cosmologists have often used. The universe consists of four dimensions: three spatial dimensions, and one dimension of time. Picture space-time as being represented by a thin, very flexible sheet of rubber. It is curved, perhaps quite intricately convoluted.

Massive bodies such as stars can be thought of as tiny ball bearings resting on this rubber sheet. The bigger and more massive the star, the deeper the dimple it makes in this otherwise smooth sheet. For a star or galaxy that is collapsing into a black hole, the dimple starts to look more like a tunnel—a long, thin tube stretched in the fabric of space-time by the gravitational collapse of the massive body.

If the body does not dwindle to a singularity, then the tube-tunnel might emerge somewhere else in space-time. The star or galaxy might have dug its way out of one place in the universe and reappeared somewhere, and perhaps *sometime*, else. Several cosmologists have suggested that the enigmatic quasars, which appear to be incredibly distant and more powerful than a hundred ordinary galaxies, might be the explosive reemergence of a collapsed galaxy. In this view, a quasar represents a "white hole" at the end of a tunnel through space-time that began with a black hole.

No one has seriously proposed explaining the physics of this phenomenon. Where even relativistic physics breaks down, you can't expect more than a shrug of the shoulders when you ask questions. At the densities and gravitational-field strengths involved, it seems clear to the theoreticians that the entire fabric of space would get badly bent. Dare one use the science-fiction term, "space warp?" Just as space is warped and stressed under the titanic gravitational forces, time itself might be stretched, warped, changed. Certainly we shouldn't expect time to flow at the same rate inside a black hole as it does elsewhere.

What would such tunnels be like? Would they stay intact, forming a sort of underground railway system crisscrossing the fabric of space-time? What happens to time inside such a tunnel?

Could it be that, if we ever do build star-faring spacecraft, far from avoiding black holes, we will seek them out, looking for a "Northwest Passage" between here and the most distant regions in space—or, perhaps, beyond space?

## Who, theoretically speaking, needs black holes?

The idea of black holes began as just that—a theoretical necessity in the now dominant concept that the universe is expanding, the result of a single, massive event known as the big bang. Recent observations, such as the discovery of objects called quasars that appear to be moving away at fantastic rates, support this theory and have made trouble for another school of thought which holds that the universe is in a steady state.

Indeed, steady-state theorists are in something of the same position as later Ptolemaic astronomers whose model of the heavens had to become ever more complicated to explain each new observation. Nevertheless, steady-state theory has its adherents and one of them, according to the *New Scientist,* under a headline saying: ". . . are Black Holes really necessary?", suggests that black holes do not exist. In the upper realms of mathematics in which cosmologists regularly deal, J. V. Narlikar argues that whatever your cosmological preference, it is possible to explain the universe without resorting to black holes.

Be that as it may, astronomers are scouring the skies, seeking them. Not only has Remo Ruffini of Princeton University claimed to have found them (see page 205), but so have five observers at Mt. Stromlo Observatory in Australia.

Optically observing an x-ray star (designated 2U 0900-40) in the southern hemisphere, they believe they found a binary star system consisting of one very massive, visible star being orbited by a much smaller, invisible and therefore dead star—a black hole. As matter from the large star is ripped away by the intense gravity of the orbiting black hole and disappears into it, energy is emitted in the form of x rays.

Even better than finding a black hole would be to watch a pulsar collapse into one, for that would be to see a model of the ultimate collapse of the universe—two situations in which the law of physics are transcended. Such are the thoughts, at least, of John A. Wheeler of Princeton, a chief black hole theorist who spoke at the Smithsonian Institution, National Academy of Sciences symposium on the nature of scientific discovery last April (see SMITHSONIAN, April 1973). And what, in the way of scientific discovery, lies beyond black holes? Said Wheeler: "the greatest mystery of all—why there is something rather than nothing."

# 27

# The "Black Hole" in Astrophysics

S. Chandrasekhar

When a stone is thrown upwards, it will normally ascend to a certain height and will then fall back. The height to which the stone ascends, before it falls back, will be the higher, the higher the initial speed with which it is projected. And if it is projected with a sufficient speed, it will escape from the earth's gravity. Quite generally, the stronger the initial gravitational pull of the parent body, the faster a stone on it must be projected if it is to escape from the gravity of the larger body.

It is one of the consequences of the general theory of relativity that light will also be similarly influenced by gravity. Normally, however, the effect is very small. But it is measurable: for example, the bending of light from a distant star, as it grazes the sun during an eclipse, has been measured. For this reason, one can ask whether or not a body like the sun could become sufficiently compressed that the force of gravity on its surface would become so strong that even light could not escape from it. If such circumstances could be realized, then the sun would not be visible from the outside: it would become a "black hole."

It can be calculated that if the sun, with its present mass, could be shrunk from its present radius of 700,000 kilometers to a radius of 3.75 kilometers, then light emitted tangentially from its surface would go around and around the sun. If the sun were to shrink still further—to a radius of, say, two kilometers—then no light it emitted could emerge. It would, in fact, recapture everything that might be thrown or emitted from it; and no one from the outside could see it.

The possibility of such a phenomenon happening is a clear prediction of very general ideas and was indeed contemplated by Laplace as long ago as 1798; the phenomenon depends only on the present observationally verified fact that the propagation of light is influenced by gravity.

The question which remains is whether objects as massive as the sun can ever shrink to such small dimensions. Can they, in fact, become so small that they will disappear from view and become "black holes" in the sky? Current developments relating to the evolution of stars answer both these questions

in the affirmative; and it appears that black holes of the kind predicted by theory do in fact occur in nature. The theoretical and observational grounds for these expectations are derived from considerations relating to the final stages in the evolution of stars.

The radiation of a star is derived from nuclear processes taking place in its interior. In the case of the sun, the process consists in the burning of hydrogen into helium; and at its present rate of radiation, the consumption of hydrogen is so small that the sun can continue to radiate for a time of the order of $10^{11}$ years.*

On the other hand, a star which is ten times as massive as the sun—and there are many such stars—is known to radiate energy at a rate which is about 10,000 times greater. Consequently, it will exhaust its internal sources of energy in a relatively short time, of the order of $10^7$ years.

It is estimated that the galaxy has existed for a period of something like twenty billion years. Therefore, if a star in the galaxy loses all of its energy sources in ten million years, the question as to what happens to it after it has exhausted its source of energy becomes an important one.

As a star exhausts its source of energy it will begin to contract. The density increases as the contraction proceeds. When the star has reached a density of about $10^6$ grams per cc, there is a well defined state into which it can settle down if its mass is less than 1.5 times that of the sun. This is the state which one observes in the so-called white dwarfs (the companion of Sirius is an example). White dwarfs are stars which are very faint and not very massive but which do have densities of the order of $10^6$ grams per cc. The important proviso here is that in order that a star may have the possibility of settling into such a state, its mass must be less than the limit stated.

If the star should have a mass in excess of the limit, there is no way in which its collapse could be arrested at the white dwarf stage, *i.e.* when its radius is a few thousand km. Accordingly the more massive stars must contract further.

The next stage where the contraction could be arrested

From *The University of Chicago Magazine*, Summer 1974. Copyright 1974 by The University of Chicago. Reprinted with permission.

*When it is $10^{10}$ years old and has consumed 10 percent of its hydrogen, the sun will leave the main sequence.—Ed.

would be when the atomic nuclei are so tightly packed that the density of matter becomes comparable to that which exists inside the atomic nucleus. That density is not $10^6$ grams per cc, but a million million grams per cc. The question arises: Can a star settle down when it reaches nuclear density of $10^{12}$ to $10^{13}$ grams/cm$^3$—the density represented in the so-called neutron stars? That depends on the star's mass; stable neutron stars can exist only for a restricted range of masses.

Calculations show that, while the permissible range of masses for stable neutron stars is subject to uncertainties, the range *is* narrow: the current estimate is between 0.3 to 1.0 times the solar mass.* The principal conclusions that follow from the foregoing considerations can be summarized as follows.

Massive stars in the course of their evolution must collapse to dimensions of the order of 10 to 20 km once they have exhausted their nuclear source of energy. In this process of collapse, a substantial fraction of the mass may be ejected. If the mass ejected is such that what remains is in the permissible range of masses for stable neutron stars, then a neutron star will be formed. While the formation of a stable neutron star could be expected in some cases, it is clear that their formation is subject to vicissitudes.

It is not in fact an *a priori* likely event that a star initially having a mass of, say, ten solar masses, ejects, during an explosion, subject to violent fluctuations, an amount of mass just sufficient to leave behind a residue in a specified narrow range of masses. It is more likely that the star ejects an amount of mass that is either too large or too little. In such cases the residue will not be able to settle into a finite state; and the process of collapse must continue indefinitely until the gravitational force becomes strong enough to hold the radiation. In other words, a black hole must form.

The radius that a star of mass $M$ must have, in order that it may not be visible from the outside, is called the Schwarzschild radius; it is given by $2GM/c^2$ where $G$ is the constant of gravitation and $c$ is the velocity of light. For a mass equal to that of the sun, this radius is 2.5 km.

The phenomena attendant upon stars going into the "black hole"—*i.e.* going inside their Schwarzschild radii—will be described in different ways—by one who is observing it from the outside, and by one who is moving with the surface of the collapsing star.

*Some physicists give estimates as high as three solar masses. —Ed.

Imagine that the observer on the surface of the collapsing star transmits time signals at equal intervals (by *his* clock) at some prescribed wavelength (by his standard). So long as the surface of the collapsing star has a radius that is large compared to the Schwarzschild radius, these signals will be received by the distant observer at intervals that he will judge as (very nearly) equally spaced. But as the collapse proceeds, the distant observer will judge that the signals are arriving at intervals that are gradually lengthening, and that the wavelength of his reception is also lengthening.

As the stellar surface approaches the Schwarzschild limit the lengthening of the intervals, as well as the lengthening of the wavelength of his reception, will become exponential by his time. The distant observer will receive no signal after the collapsing surface has crossed the Schwarzschild surface; and there is no way for him to learn what happens to the collapsing star after it has receded inside the Schwarzschild surface.

For the distant observer, the collapse to the Schwarzschild radius takes, strictly, an infinite time (by his clock) though the time scale in which he loses contact in the end is of the order of milliseconds.

The story is quite different for the observer on the surface of the collapsing star. For him nothing unusual happens as he crosses the Schwarzschild surface: he will cross it smoothly and at a finite time by his clock. But once he is inside the Schwarzschild surface, he will be propelled inexorably towards the singularity: there is no way in which he can avoid being crushed to zero volume at the singularity, and no way at all to retrace his steps.

It is currently believed that the X-ray star Cyg X-1 is a binary system in which one of the components (the X-ray component) is a black hole. The reason for this belief is derived from the following two circumstances. First, the star exhibits a rapidly fluctuating intensity on a time scale of 50 milliseconds or less. From this fact we can conclude that the X-ray emitting regions must be "compact," with a linear dimension less than $10^4$ km. Second, from the observations relating to the binary nature of the system, it has been deduced that the minimum mass of the compact star is six solar masses. Since this minimum of six solar masses is already considerably in excess of the limiting mass for white dwarfs, and of all estimates of the upper limit for stable neutron stars, the conclusion is inevitable that we are here dealing with a black hole.

# Ordinary Stars, White Dwarfs, and Neutron Stars

Louis C. Green

As was indicated in the writer's article in last November's issue [*Sky and Telescope*, vol. 40, p. 260], physical theory tells us that there are four basic types of stars in the universe: ordinary stars, white dwarfs, neutron stars, and the still-unobserved collapsed stars.

Ordinary stars are those we see when glancing at the heavens on any clear night. White dwarfs were first recognized as such in the 1920's, while pulsars have been discovered and identified as neutron stars in the last three years.

The reasons for believing pulsars to be neutron stars were given in the April, 1969, issue [of *Sky and Telescope*], page 216. Briefly, the argument is this: The exact periodicity with which each pulsar radiates its radio energy implies a relation to a fundamental property such as oscillations of the source, or orbital motion in a binary system, or rotation.

The first possibility can be discounted, for the oscillation period of a gravitating mass decreases as its density increases, and even white dwarfs are not dense enough to oscillate with periods shorter than about a second, whereas, for example, the period of the Crab pulsar (NP 0532) is only 0.033 second.

As for the second possibility, the period of a close binary consisting of two extremely dense stars could be short enough to fit pulsar observations. However, such a system would lose energy very rapidly through gravitational radiation, and although the periods of a number of pulsars are known to be lengthening, the rate of change is much slower than it would be for such a very close binary.

For an ordinary star or white dwarf to rotate with a typical pulsar period, the equatorial velocity at such a sizable star's surface would be greater than the speed of light. Since [matter cannot attain the speed of light] this possibility is eliminated.

We are left with only two alternatives, oscillation or rotation of neutron stars. However, their oscillation periods would be several orders of magnitude shorter than the shortest observed pulsar period. But no argument has been found that eliminates rotating neutron stars, and the model of a pulsar as a rotating neutron star has been elaborated by various workers with results that agree in significant ways with data from pulsar observations.

## Hydrostatic Equilibrium

To understand the remarkable properties of neutron stars requires review of modern ideas about stellar structure. Stars—whether ordinary, white dwarf, or neutron—are in hydrostatic equilibrium. The only exceptions are the obviously changing objects: supernovae, novae, and other intrinsic variables.

If a star is in hydrostatic equilibrium, at any point inside it the weight of the overlying, downward-pressing layers exactly balances the total outward pressure of both gas and radiation. For the star to remain stable, this balance between contractive and expansive forces must be very precise indeed.

But in Cepheid and Mira-type variables, for example, the balance has in some way been slightly disturbed. Suppose the outward pressure has become a little too great, forcing the star to expand, with its outward-moving layers gradually gaining momentum. As the volume increases, the internal pressure drops (cooling the interior) until it balances gravitational attraction. However, because of the momentum the motion does not stop, and the star expands beyond the equilibrium size. This further reduces the pressure, allowing gravitation to slow the outward motion, which in time halts.

At this stage, however, the internal pressure is not enough to maintain the star in its extended form. It therefore begins to contract, accelerating as it goes. When the point is reached at which pressure and gravitation are equal, the inward motion carries the star to smaller than the equilibrium size. Finally, when the inward motion ceases, the pressure is so great the whole cycle begins again.

It might be thought that such oscillations would die out quickly, but observations and theory agree that the viscosity of stellar material is low. Once radial pulsations are begun, they continue for a very long time.

Strictly speaking, every star changes with time. Main-

From *Sky and Telescope*, vol. 41, pp. 18-20, 1971. Reprinted with permission.

sequence stars use up the hydrogen in their cores and move upward and to the right in the Hertzsprung-Russell diagram. But such evolutionary changes are extremely slow, requiring millions or billions of years, depending upon the star's mass. Hence we may regard these objects at any one moment as being in equilibrium. Even for the pulsating Cepheid and Mira variable stars, the deviations from hydrostatic equilibrium are very small.

Physical conditions are very different in novae or in supernovae, where changes in the interiors may take place in a matter of minutes, or even more rapidly. In such cases hydrostatic equilibrium is no longer a satisfactory approximation.

It is easy to compute what would happen to a star if by some means we could turn off the internal pressure completely. A star like the sun would collapse at a remarkable rate, its whole physical structure and energy output being drastically altered in roughly 25 minutes. A star of greater mean density would collapse even faster. Such a turning off of the internal pressure is very close to what we believe happens when a star reaches the supernova stage.

At that point, its structure and interior temperature are such that a large fraction of the internal energy is suddenly transformed into neutrinos and antineutrinos. These tiny chargeless, massless particles stream rapidly out of the star, carrying with them the energy required to support the overlying layers. The layers collapse on the interior, gaining kinetic energy at the expense of gravitational potential energy as they fall. This implosion results in an enormous energy release in the interior, leading to an explosion which spreads outward into the region of unburned nuclear species and is accelerated by the consumption of these fuels.

## Energy Transport in a Star

In addition to hydrostatic equilibrium, a second equilibrium condition applies to any stable star: the total energy liberated in the stellar interior must be balanced by the energy radiated from the surface. Small, temporary deviations from this requirement can be allowed without any drastic change in the general structure of the star, but any long-continued under- or overproduction will lead to temperature and pressure changes to which the star must adjust.

The energy generated in the interior can reach the surface in three ways: by radiation, convection, and conduction. In the first case, photons diffuse outward toward the surface, under the guidance of the outward temperature gradient, by repeated absorptions and emissions.

In convection, masses of material move upward from the hotter, lower layers to regions of less pressure, where they expand and cool. If, after the expansion, the material has a greater density than its surroundings, it will move downward to its place of origin. This situation is called stability against convection, and when it applies in an ordinary star, radiation is the principal means of energy transport. But if the mass element continues to move upward, bringing hotter gas to the outer layers, energy transport is chiefly by convection.

These two modes of energy transfer are the important ones for ordinary stars, but for white dwarfs and neutron stars conduction plays the dominant role. Here collisions transfer energy from one ion or electron to another, so the internal heat is brought to the surface by diffusion, again being guided by the temperature gradient.

## The State of Matter

For a quantitative theory of stellar evolution, we need not only the two equilibrium conditions and the energy-transport equations, but also three gas-characteristic equations.

One of these gives the pressure of the gas as a function of its density, temperature, and chemical composition. The second tells the rate at which energy is generated at each point in the star's interior; this rate also depends upon density, temperature, and chemical composition. Finally, we must know how the opacity of the stellar material to radiation changes with these same variables.

For radiative transfer, calculations or laboratory experiments must therefore be made to ascertain to what extent the various atoms and ions absorb or scatter radiation of all wavelengths, from the infrared to X-rays. If conduction is an important mode of energy transport, it is necessary to know how conductivity depends on the composition and physical state of the material. In white dwarfs, the conduction is primarily by electrons, while in the early supernova stage the conductivity becomes virtually infinite, since neutrinos and antineutrinos reach the surface of the star from its deep interior almost without any collisons.

The properties of gases at low temperatures and pressures are familiar to us in studying the earth's atmosphere. As the temperature increases, gas molecules tend to dissociate into their component atoms, and the atoms become increasingly ionized. In this range of density and temperature, a gas may be said to obey the *perfect gas* law. This statement means that interactions between the particles of a gas are not very important in describing its behavior as a whole.

But if the density is greatly increased, regardless of temperature, the material reaches the state of *degenerate matter*. The electrons become degenerate first, and under much more extreme conditions the nuclei do too. In everyday life we have some contact with matter in a closely related state, for the electrons in a metal behave in this fashion. The characteristics are those which come to the fore when the wave nature of matter is prominent.

Like photons of radiation, electrons and protons have the properties of both particles and waves. As we approach atomic dimensions, the wave characteristics of matter become increasingly noticeable. In a metal, the wavelengths of the valence electrons (those in the outermost shells) become of the same order as the separations of the nuclei. Under these circumstances an electron can no longer be assigned to any one specific nucleus, but rather it is free to move from the neighborhood of one to that of another. When an electric field is applied, the electrons move among the ions, making the metal a good electrical conductor. The Pauli exclusion principle, which, as often stated, says that no two electrons in an atom can possess the same set of quantum numbers, now applies to the electrons shared over many nuclei.

The English astrophysicist Sir Ralph H. Fowler, who in 1926 was the first to realize that the white dwarf stars are made of degenerate matter. [Photograph by Ramsey and Muspratt, Ltd.]

In degenerate matter, the nuclei are so close together that the inner as well as the outer electrons are free to move, and the pressure is largely that of the electrons. If the degeneracy is not complete, the pressure is still temperature-dependent to a greater or lesser extent. But if degeneracy is complete, the pressure depends primarily on the number of electrons per unit volume and not upon the temperature. If the electrons are also relativistic (that is, with kinetic energies of several times their rest-mass energies, $m_0c^2$), the dependence of pressure on electron density is somewhat less strong, but still independent of temperature.

In ordinary stars, conduction is usually much less effective than radiation as a means of energy transport. On the other hand, degenerate matter is a good conductor of heat, for much the same reason metals are. Thus, in white dwarfs and neutron stars energy is transported mainly by conduction.

In the interior of a white dwarf, the ions are so closely spaced that the potential energy of the electrostatic interactions is much larger than the kinetic energy associated with the ions heat motions. This state of affairs corresponds exactly to that in crystals at ordinary temperatures and pressures. Therefore, it is reasonable to speak of the degenerate matter in a white dwarf [interior] as *crystalline*, although its temperature may be around 15,000,000° Kelvin. This conclusion tells us that each

white dwarf should be largely of uniform temperature, and also that its cooling rate may be very much faster than we had formerly supposed.

## Neutron Star Properties

Stellar evolution from the main-sequence stage onward is a story of the exhaustion of successive nuclear fuels in the core and surrounding layers. Each exhaustion is followed by a contraction of the star, which increases the internal temperature and brings the next nuclear fuel to the ignition point. However, in time the nuclear fuels are all exhausted, and the star either continues its contraction to become a collapsed object or it changes into a white dwarf or a neutron star, and shrinkage stops.

As the contraction proceeds, the electrons become degenerate. We have already noted that the pressure then no longer depends upon temperature. The star may cool, but it remains stable. Such white dwarfs have densities of perhaps $10^5$ to $10^8$ grams per cubic centimeter. But if the star is sufficiently massive (above the Chandrasekhar limit of 1.4 suns), densities greater than $10^8$ can be attained, and the electrons become relativistic. The pressure no longer rises so steeply with increasing density, and a range of further contraction is opened.

At densities above $10^{10}$, electrons begin to be forced onto the protons to form neutrons. Said differently, the equilibrium between neutrons, protons, and electrons shifts in favor of the neutrons. The pressure ceases to climb rapidly with rising density. Contraction proceeds, and when the density reaches $5 \times 10^{13}$ about 90 percent of the electrons and protons have disappeared.

However, short-range but very strong repulsive forces between neutrons now become increasingly noticeable. These forces and the growing neutron degeneracy cause a rapid rise in pressure, which halts further contraction (provided the mass is less than the Oppenheimer-Volkoff limit of about 2.0 suns). The result is a stable neutron star.

Even before the shrinkage stops, densities of greater than $10^{15}$ can be reached, and reactions between neutrons and the less abundant protons and electrons yield a number of unfamiliar elementary particles, mu-mesons, hyperons, and those heavy particles called resonances. The properties of such a mixture are unknown at present.

Current views with regard to neutron stars were summarized by Malvin Ruderman of Columbia University, at a recent international conference on thermodynamics held in Great Britain.

At the moment of the creation of a neutron star in a supernova explosion, the temperature is about $10^{12}$ degrees K., he and others believe. Thereafter, due to neutrino and antineutrino emission, the star cools exceedingly rapidly to $10^{10}$ in less than a second and to $10^8$ in the following 1,000 years.

The acceleration of gravity at the surface of a neutron star is around $10^8$ times the solar value, so that its atmosphere is only a few centimeters thick. Because of the electron degeneracy, the temperature is uniform from a few meters below the surface to the center. In the first few kilometers there are still some nuclei present. Their separations are far less than in the

The white dwarf Sirius B is seen here to the lower right of the much brighter and six-pointed image of Sirius A. The latter is symmetrically flanked by diffraction images produced by a coarse wire grating for measurement purposes. This photograph was taken with a 26-inch refractor and a hexagonal diaphragm by I. W. Lindenblad. [U.S. Naval Observatory photo.]

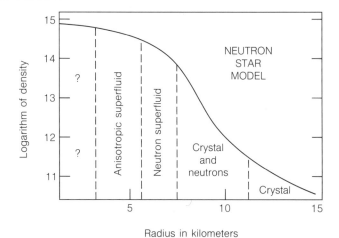

Malvin Ruderman, of the physics department, Columbia University, New York City, made this diagram to illustrate his model of the internal layers in a neutron star of 30 kilometers diameter. The center of the star is at the left, and the surface is at the right.

white dwarfs, and their electrostatic interactions are correspondingly higher, much exceeding the energies of their thermal motions. This region is therefore crystalline in character, and is called the *crust* of the neutron star.

In the next five kilometers or so inward, the neutrons probably behave as a superfluid, first with properties like liquid helium, at greater depths with a curious anisotropic character as well.

Two kinds of superfluid are known in the laboratory: the superfluid phase of $He^4$, and the superconducting state of various metals. Both appear only at extremely low temperatures, near absolute zero. The similarities of these two states have become increasingly apparent to physicists in the last decade. In both cases, the explanation of the observed behavior in terms of interacting particles is much the same, except that the electrons tend to repel one another.

The conditions under which superfluid helium can exist are described by the accompanying *phase diagram*. Temperature is plotted along one axis and pressure along the other. The boundaries are indicated between the domains in which the material behaves as a gas, liquid, or solid. In this chart for $He^4$, we see that at low pressures and at temperatures higher than about 1° K. helium is a gas. At higher pressures there are

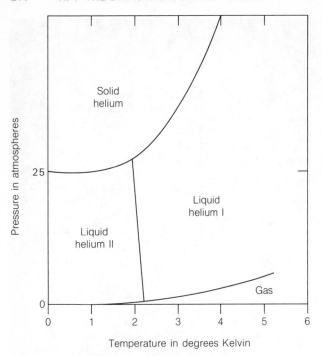

Phase diagram of helium at low temperatures. [Adapted from R. D. Parks, *Superconductivity*, New York, 1969, courtesy Marcel Dekker, Inc.]

two varieties of liquid helium. Helium I acts like an ordinary liquid, but helium II is a superfluid, showing no friction either internally or with its container's walls. Helium II remains a superfluid from about 2° K. down to the lowest attainable temperatures.

Insight is gained from the fact that only the ordinary helium isotope of mass 4 shows superfluidity; the mass-3 isotope does not. Nuclei of He[4] have zero nuclear spin; such particles are called *bosons*. Nuclei of He[3] have a spin of $\frac{1}{2}$, and are *fermions*. The basic units of any superfluid are bosons.

All electrons possess a spin, and therefore for the electrons in a superconductor to become bosons they must form pairs, with the spin axes of the two electrons in opposite directions. Since the electrons are similarly charged, they repel one another, and the formation of a boson would be difficult were it not for the charge compensation that results from the ions acting as intermediaries. Similarly, each neutron possesses a spin, and for the neutrons inside a neutron star to form a superfluid, they must also combine in pairs with oppositely directed spins to make up the necessary bosons. Since neutrons carry no charge, the process is somewhat simpler for a neutron superfluid than for a superconductor.

Physicists find it convenient to describe the superfluid properties of helium II in terms of a two-fluid model, consisting of a normal and a superfluid component. (The actual presence of two fluids is not implied.) The fraction that is superfluid increases from 0.0 at the interface between helium I and helium II to 1.0 at a temperature of absolute zero. The normal component behaves like an ordinary liquid of very low viscosity, whereas the superfluid acts like a classical ideal fluid without viscosity and with no drag at a boundary surface. But the superfluid portion has an additional property: the product of the tangential component of its velocity times the path length summed for each element of the path around a closed circuit is always quantized, in much the same way as the discrete energy levels in an atom. Furthermore, this "circulation" around any two circuits must be the same if they can be transformed into one another.

These properties yield many remarkable results which have been verified in laboratory experiments. For example, an electric current set up in a ring of superconducting metal will flow for months or years without any battery or other device to drive it. Similarly, a superfluid liquid set in motion in a circular path continues to move around that path for very long periods of time.

As was mentioned, in the more massive neutron stars central densities of at least $10^{15}$ grams per cubic centimeter occur, and we do not know the properties of matter that will result. But for the less massive neutron stars, the superfluid state extends from the crust all the way to the center.

Of course, this description of the interior of a neutron star is derived from theory, since the range of densities far exceeds anything attainable in the laboratory. However, this is also true of the interiors of the planets and the sun, for which prediction agrees remarkably well with observation at many points. Though the extrapolation is greater for neutron stars, the overall characteristics of these strange objects are fairly well understood, with details of the picture still to change as knowledge increases.

# Dynamic Relaxation of Planetary Systems and Bode's Law

**29**

J. G. Hills

Bode's law, which states that the semi-major axis $r_i$ of the $i$th planet from the Sun is given by

$$r_i = r_1 a^{(i-1)} \qquad (1)$$

where $a$ is a constant, is obeyed by the inner satellites of Jupiter, Saturn and Uranus as well as the planets.[1] Because the initial conditions in these satellite systems are not likely to have been the same as in the planetary system, this suggests to me that Bode's law may have resulted from a process of dynamical relaxation. That an epoch characterized by strong dynamical encounters occurred before the planets relaxed into stable orbits is also suggested by the surprisingly large differences between the inclinations of their angular momentum vectors of rotation and those of orbital revolution.

To test this hypothesis eleven planetary systems each with a central star of 1 $M_\odot$ but with widely different planetary mass functions were evolved from random initial orbits on an IBM 360/67 computer by numerically integrating the equations of motion. The integrations were done in two dimensions to save computer time. They were performed in double precision using a fourth-order Adams–Moulton routine with the length of the time steps being controlled by the truncation errors. Special precautions were taken to reduce these errors during close encounters. The median cumulative error, $\Delta E$, in the energy $E_0$ of a system was $\Delta E / E_0 \sim 10^{-5}$.

Despite large differences in planetary mass functions and initial conditions the various systems tended towards stationary states having a number of common characteristics, the principal one being a tendency for the periods of adjacent orbits to be small integer fractions of each other, that is, they show the commensurabilities of the mean motions of classical celestial mechanics (see ref. 2). This is illustrated by Figs. 1 and 2, which give the frequency of period ratios of planets in systems 5 and 11, the two systems which were evolved for the longest periods of time. Figs. 3 and 4 give the time variations of the semi-major axes of the planets in these two systems while

Fig. 5 shows these for system 9, the only one in which two planets were in stationary orbits from the beginning of the integrations. These two planets were accidentally placed into nearly commensurate orbits with a period ratio of (13/6). The tendency towards commensurate orbits is also evident in Figs. 1 and 2 despite the two systems having at best reached only quasi-stationary states.

A quantitative test of the correlation between the observed frequency pattern of the period ratios, $R_i$, shown in Figs. 1 and 2 and the commensurate points is difficult to construct because any given $R_i$ is, in general, not independent of the $R_i$s which proceeded it. If the distribution of $R_i$s tends to be Gaussian around the commensurabilities and if the commensurabilities are equally probable the function

$$F(\Delta) = \sum_{i=1}^{N} \sum_{j=1}^{8} \exp\{-[R_i - (C_j + \Delta)]^2/\sigma\} \qquad (2)$$

where the $C_j$s are the commensurabilities 3/2, 5/3, 2, . . . 3 will have a maximum near $\Delta = 0$. In Fig. 6 we have plotted this function for the $R_i$s of system 5 with $\sigma = 0.025$. This shows a maximum near $\Delta = -0.01$, which is less than $0.05$ of the average spacing between commensurabilities. This suggests a high correlation between the frequency pattern of the $R_i$s and the commensurabilities.

The three systems which have been discussed cover a wide spectrum of initial conditions and planetary mass functions. In system 5 each of the four planets has ten times the mass of one of the Jovian planets while in system 11 each of the six planets has a mass of $5 \times 10^{-4}\, M_\odot$. The initial semi-major axes and eccentricities of the planetary orbits in these two systems were chosen at random, although, due to an early misjudgment about conditions in the solar nebula, the initial semi-major axes in system 11 were only allowed to lie in a narrow range between 6 and 8 A.U. In system 9 the masses of the planets are $8 \times 10^{-4}\, M_\odot$, $3 \times 10^{-4}\, M_\odot$ and $1 \times 10^{-7}\, M_\odot$. As such widely different systems could evolve towards stationary states characterized by commensurate orbits, it seems likely that nearly all planetary systems will evolve towards these stationary states. The initial conditions and planetary

*From Nature, vol. 225, pp. 840–842, 1970. Reprinted with permission.*

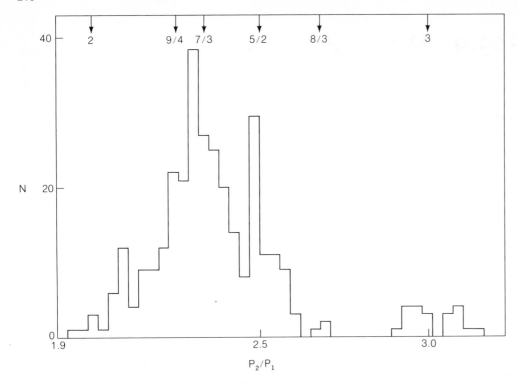

**Figure 1.** Frequency of the ratio of the periods of planets 2 and 4 in system 5. This was evaluated at intervals of ten years starting 1,000 yr after the beginning of the planetary evolution.

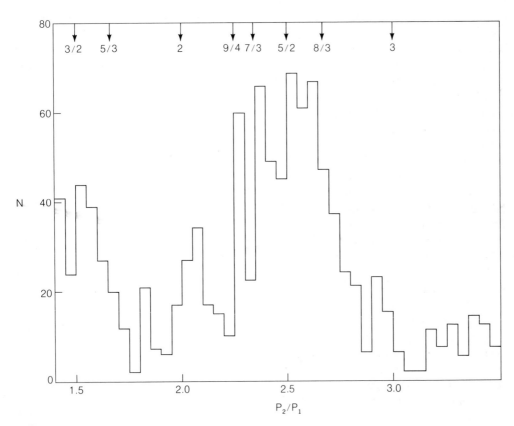

**Figure 2.** Frequency of the ratio of the periods of the two outermost planets in system 11 except when the ratio is less than 1.4, in which case the third outermost planet is used in place of the second. The ratio was evaluated at intervals of ten years starting 1,000 yr after the beginning of the planetary evolution.

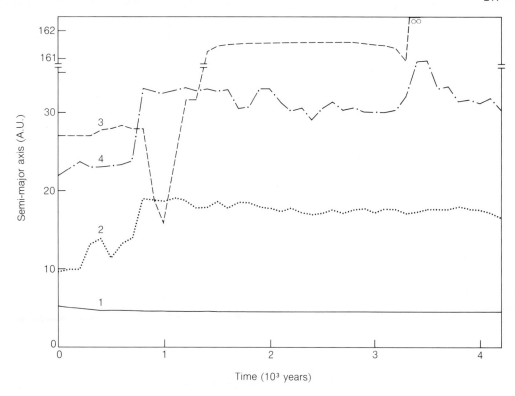

**Figure 3.** Time variation of the semi-major axes of the planets in system 5.

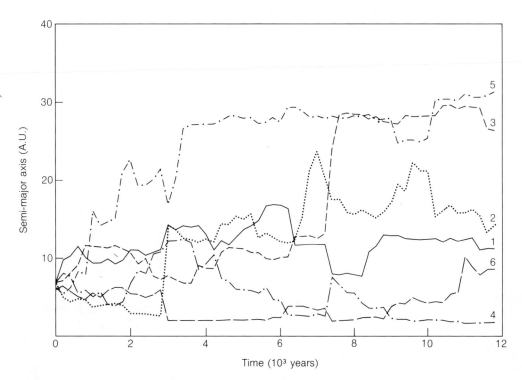

**Figure 4.** Time variation of the semi-major axes of the planets in system 11.

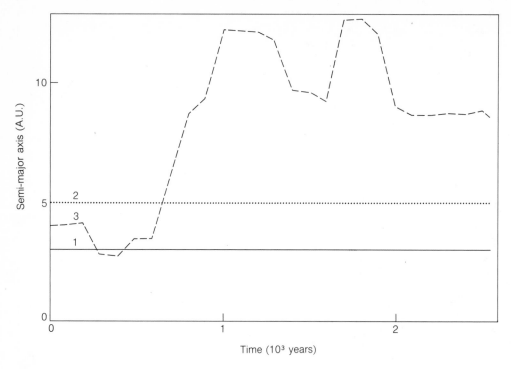

**Figure 5.** Time variation of the semi-major axes of the planets in system 9.

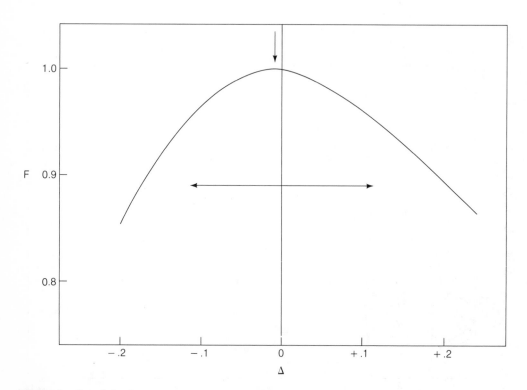

**Figure 6.** Correlation between the frequency pattern of the period ratios of system 11 shown in Fig. 2 and the commensurate points. The arrow indicates the position of the peak of the curve, and the horizontal bar shows the average spacing between commensurabilities.

mass functions seem only to determine the time necessary to reach such a state.

If the dynamical evolution of the semi-major axes of the planets in a system can be approximated by a succession of random walks for which the length of each step is directly proportional to the mass of the perturbing planet, the fact that system 5, which has planets ten times as massive as the Jovian planets, reached a quasi-stationary state in about $2 \times 10^3$ yr suggests that the Jovian planets reached such a state in about $2 \times 10^5$ yr. Considering the large uncertainty in this calculation, a reasonable estimate of the time necessary for the Jovian planets to evolve to their present orbits after their formation is $10^5$–$10^6$ yr.

Recent investigations[3–6] have shown that commensurate orbits are more frequent in the solar system, particularly among the Jovian planets and among their satellite systems, than is expected from chance alone. For those commensurate pairs of orbits which have been studied in detail it has been found that the commensurability condition leads to resonant perturbations of large amplitude, but these cause only periodic changes in the orbital elements.[7] It thus seems likely that commensurate orbits are the final signatures of a system of strongly interacting planets with commensurability being the stablest possible arrangement of their semi-major axes.

Bode's law results from the fact that some commensurabilities are more favoured than others. We see in Figs. 1 and 2 that most commensurabilities lie in a narrow range between (9/4) and (8/3); this is true for the other computer simulated systems as well. Because of this small spread in the period ratios, the period of the $i$th planet from the Sun is given closely by the relation

$$P_i = P_1 A^{(i-1)} \tag{3}$$

where $A$ is constant, but this is just Bode's law as it can be trivially converted to equation (1) by Kepler's third law. Pluto

is usually cited as being the most flagrant violator of Bode's law. This results from the approximate character of the law. The orbit of Pluto is actually highly commensurate with the orbit of Neptune, the period ratio being nearly (3/2). We see in Fig. 2 that the second largest clumping of commensurate points is near (3/2), but these commensurabilities are considerably less frequent than those in the range (9/4)–(8/3). The smaller commensurabilities are much more frequent in the satellite systems than in the solar planetary system and in the computer simulated planetary systems. The reason for this will be discussed in a later paper.

The implications of these results on theories of the origin of planetary and satellite systems as well as a more detailed report on the analysis of the numerical experiments will also be given in a later paper.

I thank Professor F. T. Haddock and Dr. R. L. Sears for helpful discussions.

## References

1. Brandt, J. C., and Hodge, P., *Solar System Astrophysics* (McGraw-Hill, New York, 1964).
2. Brouwer, D., and Clemence, G. M., in *Planets and Satellites* (edit. by Kuiper, G. P., and Middlehurst, M.) (Univ. Chicago Press, 1961).
3. Roy, A. E., and Ovenden, M. W., *Mon. Not. Roy. Astron. Soc.*, 114, 232 (1954).
4. Roy, A. E., and Ovenden, M. W., *Mon. Not. Roy. Astron. Soc.*, 115, 296 (1955).
5. Goldreich, P., *Mon. Not. Roy. Astron. Soc.*, 130, 159 (1965).
6. Dermott, S. F., *Mon. Not. Roy. Astron. Soc.*, 141, 349 (1968).
7. Hagihara, Y., in *Planets and Satellites* (edit. by Kuiper, G. P., and Middlehurst, M.) (Univ. Chicago Press, 1961).

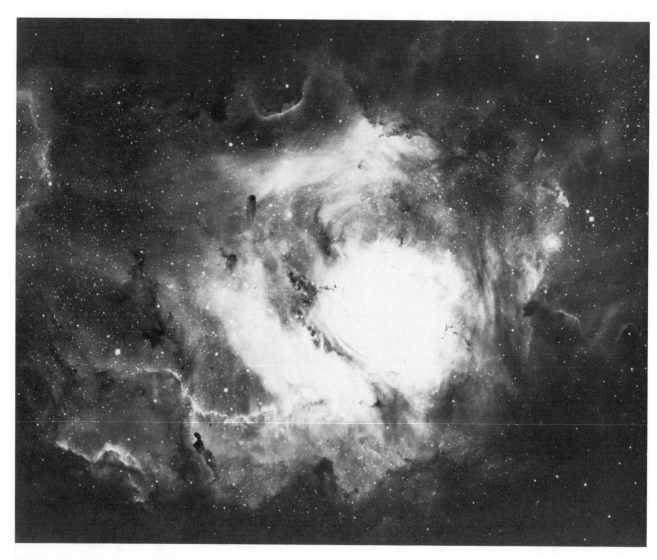

M8, the Lagoon Nebula, a bright region of
ionized hydrogen in the constellation
Sagittarius, photographed with the Mayall
telescope. [Kitt Peak National Observatory.]

# INTERSTELLAR MATTER AND THE GALAXY

An extensive collection of chemical compounds has been found to exist in space, as reported by radio astronomer Barry E. Turner in "Interstellar Molecules." Turner describes the relationship of these molecules to the cool, dark dust clouds in which stars are born and tells how the physical properties of the interstellar matter can be deduced from radio observations.

"The density wave theory predicts that the only stable and semipermanent spiral configurations will be those showing two trailing spiral arms emanating from opposite sides of the nucleus." Bart J. Bok discusses the leading theory for the origin of spiral arms in galaxies and explains how it compares with observations of stars and interstellar matter in the Milky Way and other galaxies in "Updating Galactic Spiral Structure."

Must we rely on remote telescopic studies to investigate interstellar matter? Perhaps not, according to A. G. W. Cameron, author of many theories concerning the origin and evolution of the sun, planets, stars, and even the galaxy as a whole. Cameron believes that certain meteorites (called *type I carbonaceous chondrites*) that have landed on Earth are actually agglomerations of interstellar dust particles. He explains his idea in "Interstellar Grains in Museums?". Chondrites are stone meteorites, as distinct from metallic ones, and often contain round mineral structures called chondrules that distinguish them from ordinary rocks. *Type I carbonaceous chondrite* is a technical expression that distinguishes a particular group of the meteorites from others on the basis of mineral composition; in particular, they appear to have been subjected to lower temperatures than the other chondrites. Cameron mentions *ferromagnetism*, a naturally occurring magnetic property of iron, nickel, cobalt, and certain rare earth elements and alloys, and *superparamagnetism*, a magnetic property displayed by groups of microscopic particles.

# Interstellar Molecules

Barry E. Turner

Between the stars of our galaxy there are vast clouds of gas and dust. These interstellar clouds were discovered some 200 years ago by William Herschel, who described them as "holes in the sky" because they obscured the light of the stars behind them. The obscuring property was long attributed to the dust: tiny particles of unknown composition that redden and polarize the light from many stars. Over the past 35 years it has gradually been discovered that the interstellar medium also comprises many different species of free molecules, including some moderately complex ones. The molecules are densest where the dust is dense. They are found in regions where stars appear to be forming and in the outer atmosphere of cool stars. Observations of molecules provide information on the physical conditions of such regions, which until recently have been largely inaccessible to astronomers.

The quantitative study of the interstellar medium began after the spectrograph made it possible to analyze the light from stars in detail. In 1904 Johannes Franz Hartmann of Germany suggested that an absorption line of ionized calcium seen in the spectrum of certain bright stars originated in interstellar space, in other words, that there were calcium ions (atoms stripped of one electron or more) between the earth and the stars that absorbed the light of the stars at certain wavelengths. Later neutral, or un-ionized, sodium was also found to be a constituent of the interstellar medium. By 1937 it was recognized that hydrogen is the most abundant element in the cosmos and must therefore make up the

bulk of the interstellar medium. Presumably the hydrogen was in the form of single atoms rather than the diatomic molecule ($H_2$). Today we know that only trace amounts of calcium and sodium are present in comparison with hydrogen.

In the visible region of the spectrum atomic hydrogen (H) can be observed only in the ionized state by its emission of what are known as the Balmer recombination lines. These spectral lines are produced when a free electron is captured by a hydrogen nucleus and cascades down the energy levels of the atom. Ionized hydrogen is found near very hot stars, where it is visible in the form of the glowing clouds known as emission nebulas or H II regions. The ionized hydrogen of these nebulas, which is associated with much smaller amounts of ionized atomic helium, oxygen, nitrogen, carbon and other trace elements, has been studied for many years. The ions are indicators of the temperature and density of the clouds of gas immediately surrounding the hottest stars—clouds out of which these stars are evidently born. The high temperatures (10,000 degrees Kelvin) and relatively high densities (100 atoms per cubic centimeter) that characterize the H II regions were long ago recognized as having no relation to the conditions that prevail in the cold, dark regions of space far from hot stars.

The first interstellar molecule was discovered in 1937. It was the free chemical radical of carbon and hydrogen (CH). The ionized radical ($CH^+$) and the cyanogen radical (CN) were identified during the next four years in the spectra of

a few bright blue-white Type O and Type B stars, often the same stars against which the clouds of calcium and sodium had been observed. These trace constituents are not, however, useful for study of the general interstellar medium. One reason is that they can be observed only immediately in front of the brightest stars. Another is that the clouds containing them must be dense enough to produce an observable absorption line but not so dense that the light from the background star is too attenuated. Very specific conditions are thus required. Thirdly, observations at visible wavelengths can penetrate the interstellar medium only 2,000 or 3,000 light-years, because these wavelengths are effectively absorbed by interstellar dust. Fourthly, observations of trace constituents such as calcium, sodium, CH, $CH^+$ and CN give no information about the total amount of gas present between the stars. In addition, it cannot be proved that these constituents are not in some way physically associated with the very few hot stars in whose direction they are seen.

This deadlocked situation might never have changed except for the rapid development of radio astronomy over the past 20 years. In fact, one of the first major triumphs of radio astronomy, which came in 1951, was the detection of interstellar atomic hydrogen by means of its spectral line at the radio wavelength of 21 centimeters. It was quickly realized that observations at radio wavelengths could penetrate completely through the galaxy, a distance of some 60,000 light-years, because radio waves are not appreciably absorbed by the interstellar

From *Scientific American*, vol. 228, pp. 51–69, March 1973. Copyright © 1973 by Scientific American, Inc. All rights reserved. Reprinted with permission.

dust. The 21-centimeter line of hydrogen was therefore utilized immediately in the years after 1951 as a tool for studying the gas in our galaxy from one edge to the other. This gas accounts for between 5 and 7 percent of the total mass of the galaxy. In the course of the 21-centimeter explorations the entire spiral structure of the galaxy was laid out; before that only faint nearby glimmerings of it had been revealed by the distribution of stars.

What physical and chemical conditions of the interstellar gas could be deduced by examining the 21-centimeter hydrogen line? The first pictures were crude, but they indicated that on the average the gas had a temperature of 100 degrees K. and a density of between one atom and 10 atoms per cubic centimeter in the spiral arms of the galaxy and .1 atom per cubic centimeter between the spiral arms. Certain observational refinements and theoretical considerations have led to a "two-component" model of the interstellar medium. One component consists of cool clouds of gas with a density of between 10 and 100 atoms per cubic centimeter and a temperature of 100 degrees K. or less. These clouds are in a pressure equilibrium with the second component, a hot gas with a density of about .1 atom per cubic centimeter and a temperature perhaps as high as 10,000 degrees K. The gas within the spiral arms consists of both components; the gas between the spiral arms consists mainly of the hot component.

That was more or less the picture of the interstellar medium in 1968, the year marking the birth of molecular astronomy as we now know it. In 1963 radio astronomers had detected the hydroxyl radical (OH). By 1968 OH had been observed in a few dozen directions in the galaxy, nearly all of them being toward H II regions and young stars detectable primarily at infrared wavelengths. Because the H II regions radiate at a temperature higher than the temperature of the interstellar medium and the OH gas embedded in it, the observers expected that the OH gas would absorb some of the radiation from the H II regions. Therefore it was expected that the spectral lines of the OH gas would be seen as an absorption feature in the radio spectrum of the H II regions.

## Emission from Hydroxyl

In many cases, however, the OH was observed not as an absorption feature but as intense, narrow emission features in the radio spectrum of the H II regions

[see illustration on following page]. If the OH molecules had been absorbing the radiation as had been expected, it would have indicated that most of them were in a low energy state. The opposite behavior indicates that the energy states of the OH molecules are quite different from what one would expect if the OH gas were in equilibrium with its cool surroundings. The OH gas is evidently being "pumped" by some mechanism into an excited condition in which it is able to amplify the background radiation. Such behavior describes an interstellar hydroxyl maser.

The maser is the predecessor of the laser; it amplifies radio waves instead of visible light. In an interstellar maser the OH molecules are excited so that most of them are in a high energy state instead of a low energy state. When one of the molecules drops to a low energy state, it emits a quantum of energy at a certain narrow line in the radio spectrum that stimulates other molecules to do the same. Some source of energy within the interstellar cloud keeps pumping the molecules back up to the excited state after they have fallen out of it, so that the process continues.

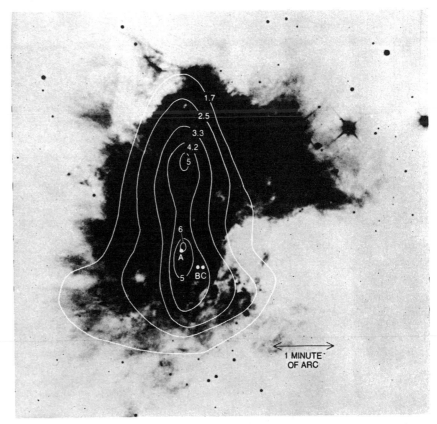

1 MINUTE OF ARC

CONTOUR MAP superimposed on a negative print of the Orion nebula shows the distribution of the interstellar molecule formaldehyde ($H_2CO$) at the wavelength of two millimeters. Numbers on contour lines are brightness temperatures: a measure of the intensity of the radiation in degrees Kelvin above the cosmic background radiation of three degrees K. The two regions of strongest molecular emission do not correspond to any objects seen at visible wavelengths; they do, however, coincide with areas of strong infrared emission. One such object is an infrared nebula found by D. E. Kleinmann and F. J. Low of the University of Arizona (A). Also shown is a tiny source of powerful maser emission (B) from the hydroxyl radical (OH) and water vapor ($H_2O$), the spectra of which are shown on the opposite page. This source lies very near a bright starlike source of infrared radiation (C) discovered by E. E. Becklin and G. Neugebauer of the California Institute of Technology. It is believed that these molecular clouds actually lie behind the bright Orion nebula in a dense region of gas not yet ionized by the central Trapezium stars. Carbon monosulfide (CS) is distributed in much the same way as the formaldehyde within this region. A cloud of hydrogen cyanide (HCN) in the same region is about twice the size of the clouds of carbon monosulfide and formaldehyde, and is similar to a cloud of the unidentified molecule "X-ogen." Carbon monoxide (CO) is observed in a gigantic cloud about one degree in diameter (more than 20 times the size of the carbon monosulfide cloud). It may exist throughout the ionized region of the nebula as well as surrounding it.

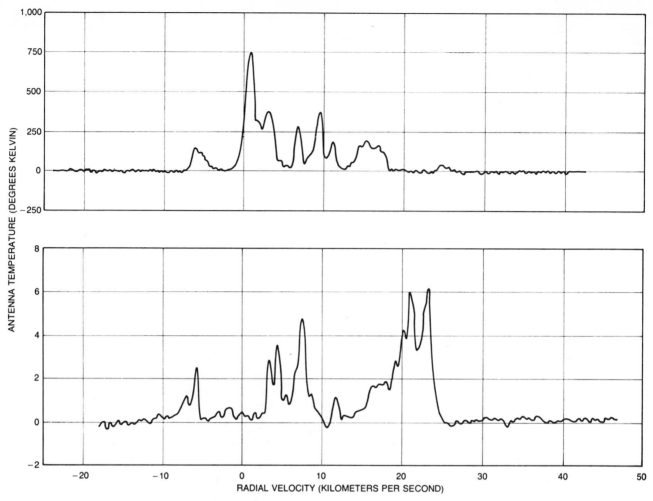

SPECTRA OF WATER (*top*) and of the hydroxyl radical (*bottom*) show that these interstellar molecules are emitting radiation rather than absorbing it. These emission spectra were taken in the direction of the infrared objects of the Orion nebula. They are calibrated in terms of the temperature of the antenna receiving the radiation (*vertical scale*); the brightness temperature is equal to the antenna temperature multiplied by the square of the ratio of the size of the telescope beam to the angular size of the source. Long-baseline interferometry has established that the angular size of these two emission sources is no more than .005 second of arc. This implies that the brightness temperature is in excess of $10^{13}$ degrees K. for both water and the hydroxyl radical. The emission from the water molecules is highly variable and changes on a time scale of a few days. Emission from the hydroxyl radical is strongly polarized. All these characteristics suggest that a maser is at work in interstellar space "pumping" the molecules into an excited state so that they amplify the weaker local or background radiation and produce the powerful emission signals that are observed.

Rather than contributing a fresh insight into the physical conditions of the interstellar medium, however, the observations of the peculiar OH emission presented many new questions about the excitation of the molecules of the medium. At the sites of strong OH emission the medium is very dense by interstellar standards: it consists of as many as $10^8$ particles per cubic centimeter. These sites seem to be confined to the atmosphere of cool young stars or to protostars: objects on their way to becoming stars.

### The More Complex Molecules

Interstellar molecules consisting of more than two atoms were first discov-ered late in 1968. They have profoundly altered our concept of interstellar chemistry and modified our views of physical conditions in space. In 1968 most astronomers believed the density of the interstellar medium was so low that it would be difficult to get more than two atoms to combine. They expected to find only diatomic molecules in interstellar space, and even those molecules would be short-lived because of the destructive effects of ultraviolet radiation and cosmic rays. Such ideas needed drastic revision when a group at the University of California at Berkeley (Charles H. Townes, William J. Welch, A. C. Cheung, David M. Rank and D. D. Thornton) found ammonia ($NH_3$) in several interstellar clouds in the region of the galactic cen-ter. Soon thereafter the Berkeley group detected emission signals from water vapor in several regions of the galaxy.

Although more complex molecules were subsequently discovered, ammonia and water had already established two of the most prominent trends observed in interstellar molecular clouds. First, the clouds· are quite dense compared with any other known interstellar region. Second, the physical conditions responsible for exciting the molecules to radiate or to absorb are quite different from terrestrial conditions. In some instances the conditions give rise to an interstellar maser. In others they create what might be called an interstellar refrigerator. At some wavelengths a molecule such as formaldehyde ($H_2CO$) absorbs more energy than the surrounding conditions would seem to allow if the conditions are such that the terrestrial laws of thermodynamics are applicable.

## Where the Molecules Are Found

Since 1968 the rate at which new interstellar molecules have been discovered has increased exponentially. The list now stands at 26 molecules. Where are these molecules found? How are they formed and destroyed? How can the astronomer utilize their signals to understand the physical and chemical processes that are operating in interstellar space?

Over the past two years the galaxy has been surveyed intensively at the wavelengths characteristic of the hydroxyl radical, formaldehyde and carbon monoxide (CO). The surveys have shown that these molecules, like atomic hydrogen, are strongly concentrated toward

the central plane of the galaxy in a layer that in the vicinity of the sun is less than 1,000 light-years thick. Within this disk the molecules appear to be distributed widely, reaching higher concentrations closer to the center of the galaxy. Carbon monoxide and formaldehyde are distributed in much the same way. Most of the other interstellar molecules, however, are observed only in a very few regions. Whether this distribution is because they do not exist in most regions, or because they are simply not excited in a way that would enable them to emit or absorb measurable signals, is not known.

It is now clear that the particular regions where molecules are found in interstellar space are just those regions that have a high density of particles. There are two reasons for this correlation. First, the higher the density of the interstellar medium, the greater the chance that molecules will be formed from the atoms present. The molecules

are formed either by a simple collision between the atoms of interstellar gas or by their coming together on the surface of a particle of interstellar dust. Second, many of the known interstellar molecules must be excited to certain energy states in order to emit detectable radiation. These states are best produced by constant collisions with the other atoms or molecules in a gas, which will happen more frequently in a denser gas. What kinds of interstellar region possess the high densities required?

## Dark Clouds and Protostars

Within 1,000 light-years of the sun there are about a dozen dark clouds that show up on photographs as holes in the background of stars. Perhaps 3,000 such clouds exist throughout the galaxy. They are typically 12 light-years in diameter, and the mass of the dust in them is about 20 times the mass of the sun. Some of

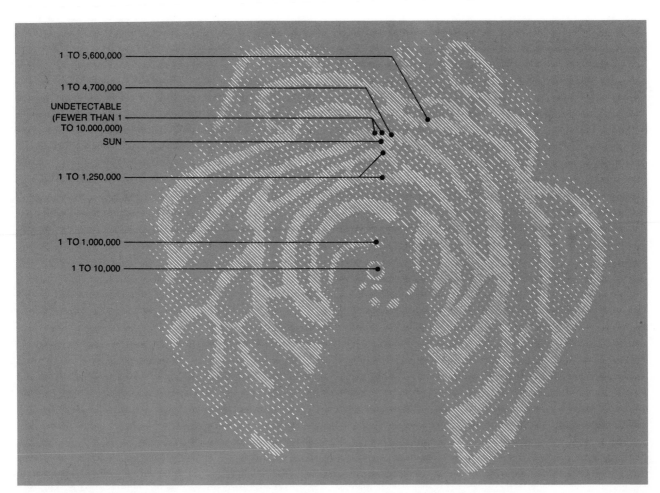

PLAN VIEW OF THE GALAXY shows the distribution of hydroxyl molecules throughout the interstellar medium. The galaxy was surveyed in the radio wavelength of the spectral line of atomic hydrogen; the map represents the way hydrogen is distributed throughout the spiral arms. The radio of the hydroxyl radical to hydrogen increases a thousandfold between the region near the solar system and the galactic center. There one hydroxyl radical is found for every 10,000 hydrogen atoms, but surveys in other directions show fewer than one hydroxyl radical per four million hydrogen atoms or are unable to detect hydroxyl-radical absorption at all.

the interstellar molecules—for example the hydroxyl radical, ammonia, formaldehyde and carbon monoxide—have been detected in these clouds, but most have not. From such observations it is inferred that the total mass of the gas in the clouds exceeds the mass of the dust by well over 100 times. The density of the gas is typically between a few hundred and 10,000 particles per cubic centimeter and the temperature is low: not more than four degrees K. in some clouds and perhaps as high as 25 degrees K. in the hottest ones. From the extent and mass of the dark clouds it can be calculated that they are gravitationally unstable and are collapsing to form stars.

It is believed the dark clouds slowly contract into what are called globules: small clouds of dust that show up on photographs of emission nebulas as completely opaque specks. Little is known about the gas content of these objects, and estimates of their mass are only guesses at a minimum figure. Recently, however, the hydroxyl radical and for-

maldehyde have been detected in a few globules. Preliminary analysis indicates that the temperature of the globules is even lower than that of the larger dust clouds and that the density is somewhat higher.

A globule is believed to be one of the final stages in the collapse of an interstellar cloud into a protostar: a denser concentration of dust and gas, perhaps about the size of the solar system, that is close to becoming an actual star. In the protostar stage of collapse the density and temperature near the center of the object rise sufficiently for it to radiate rather strongly in the infrared region of the spectrum. Theoretical work by R. B. Larson of Yale University shows that the protostar collapses much faster near the center than in the outer parts, so that newly born stars should be surrounded by a large envelope of dust and gas. That envelope in turn is surrounded by the remnants of the interstellar cloud. If the new star is a very luminous Type O or Type B star, it will emit enough ul-

traviolet radiation to ionize the hydrogen in the surrounding cloud, forming an H II region. The H II region will often be surrounded by a cloud of un-ionized gas that has not been reached by the ionizing radiation.

Where do molecules fit into this picture? It appears that they are associated with every one of the steps in the contraction of an interstellar cloud. We have seen that the hydroxyl radical, formaldehyde and carbon monoxide are observed in virtually all regions of the spiral arms and possibly between the arms. The larger dust clouds also contain ammonia. It is only when we turn to regions where it is suspected stars are being formed that we observe the remaining molecules. Here the molecules may be associated with the protostars themselves. Alternatively they may exist in the dense, un-ionized and mostly very massive clouds that surround the H II regions generated by newly formed stars.

A good example of the second kind of region is the Great Nebula in Orion. Em-

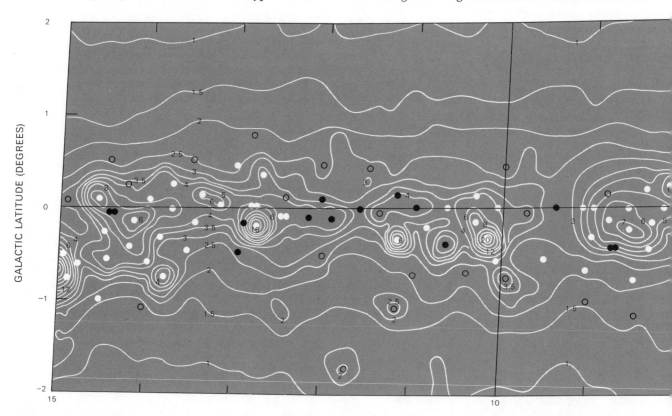

**EDGE-ON VIEW OF THE GALAXY** near the galactic center has been mapped at the radio wavelength of 11 centimeters by Wilhelm J. Altenhoff of the Max Planck Institute for Radio Astronomy in Bonn. The contour lines are measures of the intensity of the radiation from various radio sources such as H II regions (regions of ionized hydrogen near hot young stars) and the remnants of supernova explosions. The radiation is in the form of a continuous spectrum, as distinguished from the "line" spectrum that characterizes atoms or molecules in which emission or absorption occurs only in certain very narrow intervals of wavelength. The numbers on the contour lines are brightness temperatures. The dots represent regions searched by the author for emission or absorption signals from hydroxyl. The black dots indicate where the hydroxyl radical has been observed as a source of powerful emission as an interstellar maser. The white dots mark the locations where OH is seen to absorb some of the background radiation at its characteristic wavelength of 18 centimeters. Open circles are regions where no OH has been detected. Many of the regions of OH emis-

bedded in the Orion nebula is a cluster of very hot young stars known as the Trapezium cluster. These stars are the energy source of the nebula itself. Surrounding the nebula is a large cloud of neutral hydrogen; just behind this cloud is a dense dark cloud. In its center infrared astronomers have detected several objects that appear to be protostars.

Molecules are found throughout the region of the Orion nebula. A great cloud of carbon monoxide, occupying about one degree of arc in the sky and containing some 100 solar masses of the gas, extends outward from the nebula into the surrounding cloud of neutral hydrogen. The hydroxyl radical has a similar distribution. Smaller clouds of hydrogen cyanide (HCN) and of an unidentified molecule dubbed X-ogen are seen projected against the central portions of the nebula. In the immediate vicinity of the infrared objects are found strong concentrations of formaldehyde, methyl alcohol ($CH_3OH$), carbon monosulfide

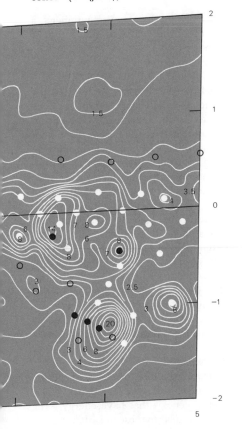

sion are also sources of strong maser emission from water vapor. As can be seen, OH is a widespread constituent of the galactic disk and is confined closely to the galactic plane (galactic latitude = 0 degrees). Carbon monoxide and formaldehyde are distributed in much the same way throughout the galaxy. Most other interstellar molecules have been observed only in very few sources.

(CS), the cyanogen radical (CN), ammonia and cyanoacetylene ($HC_3N$). There are also highly excited but very small clouds of the hydroxyl radical and water vapor that are emitting powerful radio signals of the maser type.

The distribution of these molecules within the Orion cloud is not accidental. Molecules exist in observable quantities only in regions that satisfy two conditions: (1) that the rate at which the molecules are formed exceed the rate at which they are destroyed, and (2) that the molecules be in a region that can be excited into the observed energy states. Regions of high density tend to satisfy both of these requirements. The hydroxyl radical (when it is observed as an absorption feature) and carbon monoxide are observed in states of low excitation and are also relatively immune to destruction by light. This explains why they are observed throughout the galaxy, and also why they are observed over a larger region in the Orion nebula than the other molecules. The molecules observed in the small central part of the Orion nebula are not observed throughout the galaxy but are seen only in a few apparently very dense areas in the vicinity of certain H II regions. In fact, it is not the Orion nebula but another H II region near the galactic center, known as Sagittarius B2, that is at present the only observed site of eight of the 26 known interstellar molecules. What makes the Sagittarius B2 cloud so unique is not its temperature (about 100 degrees K.) but its enormous densities (up to $10^8$ particles per cubic centimeter) and also its size (about 20 light-years), which together make its density in the line of sight 100 or more times greater than the known density in any other interstellar cloud.

## Interstellar Masers

One of the earliest indications that interstellar molecules were probably associated with protostars came from the maser emission of the hydroxyl radical and water vapor. Unlike the hydroxyl radical, which is often observed as a fairly normal absorption feature against sources that have a continuous spectrum in the radio region, water vapor is seen only in the form of powerful emission lines. Moreover, it is seen only in regions that can be smaller than the solar system, perhaps even smaller than the orbit of the earth. The emission lines of water vapor are so strong that if they were radiated from an ordinary heated body, the temperature of that body would have to be

$10^{13}$ degrees K. Such temperatures cannot be connected with the kinetic energy of the molecules or with any temperature of radiation unless the radiation has been amplified. It is these characteristics that imply that an interstellar maser is at work.

Soon after interstellar water masers were discovered in 1968 it was noticed that they are always found in regions where the hydroxyl radical is also emitting powerful maser radiation. The equivalent radiation temperature for the hydroxyl radical is about as high as that for the water. In some of these sources the rate at which energy is radiated at just the two narrow spectral lines of the hydroxyl radical and water is as much as the energy emitted by the sun at all wavelengths. The energy must come indirectly from the gravitational collapse of the protostar; indeed, the emission acts to cool the cloud and cause it to collapse even faster.

Theories explaining such a spectacular excitation of the hydroxyl radical and water, and why they are found together, have centered on various kinds of pumping and on the effects of strong infrared radiation. These processes work well only if the densities are as high as $10^8$ particles per cubic centimeter and the temperatures are several hundred to 1,000 degrees K. These conditions exist only in protostars such as the infrared sources in the Orion nebula. In fact, a rather large amount of energy (equivalent to 700 degrees K.) along with the high densities is needed to excite water molecules to radiate in the observed manner at all. This probably precludes interstellar water's ever being observed in any other type of object not having similar special conditions.

Conditions required to give rise to the maser action of the hydroxyl radical and water also exist in the outer regions of certain very cool stars. That brings us to the last category of celestial objects in which molecules are observed. A number of molecules—for example diatomic hydrogen ($H_2$), water, carbon monoxide, the cyanogen radical, diatomic carbon ($C_2$), methylidyne (CH), hydrogen cyanide (HCN) and acetylene ($C_2H_2$)—have long been known through their optical and infrared spectra in the atmosphere of cool red Type M stars and the stars known as Mira Ceti variables. It was not until 1968, however, that W. J. Wilson and Alan H. Barrett of the Massachusetts Institute of Technology detected powerful hydroxyl-radical signals at radio wavelengths from infrared stars not associated with H II regions. Many of these stars appear to be very young; several are Mira Ceti variables that G. Neugebauer and Robert B. Leighton of the California Institute of Technology had found earlier were strong sources of infrared radiation. Two years later maser emission from water vapor was also discovered in Mira Ceti variables; it is believed that these masers are being pumped by the strong infrared radiation.

Marvin M. Litvak of M.I.T. has shown that the hydroxyl radical would emit radiation at the frequency of 1,612 megahertz (million cycles per second) under the action of infrared pumping. In the infrared stars OH radiates predominantly at just that frequency. This is not the case for the hydroxyl radical in protostars, which radiates predominantly at 1,665 megahertz and may be pumped by chemical processes. The emission from the hydroxyl radical and water is thought to arise in the shell of gas and dust that surrounds these stars. Observations of the shell typically show features in the spectrum that are Doppler-shifted in a way suggesting that the shell is expanding or contracting at a velocity of between 10 and 30 kilometers per second.

Most or all of the stars with the hydroxyl radical and water vapor in their outer atmosphere appear to contain more oxygen than carbon. In other types of cool star the opposite is true. The hydroxyl radical and water vapor are not found in these "carbon" stars, but carbon-containing molecules such as carbon monoxide, carbon monosulfide (CS), HCN and CN have been observed in them at both radio and infrared wavelengths. The radio observations have proved to be particularly interesting because they indicate the presence in these molecules not only of carbon 12, the commonest isotope of carbon, but also of the rarer isotope carbon 13. In these stars the abundance ratio of carbon 12 to carbon 13 is only about 4 : 1; on the earth and in most of the interstellar medium the ratio is 89 : 1.

## Molecules as Interstellar Probes

Direct information about the constitution, location and velocity of interstellar clouds can be gained by detecting molecules and measuring their position and the amount by which their spectral features are Doppler-shifted. Combined with other knowledge, a somewhat deeper analysis of the intensity of the signals can yield information about the physical conditions and dynamics within the clouds. The particular spectral lines that are the signature of molecules arise when

KNOWN INTERSTELLAR MOLECULES number 26 at present.[*] They are listed (*on the opposite page*) in order of their discovery along with the wavelength and the telescope at which they were first detected. An asterisk by the wavelength indicates that the corresponding molecule has been observed at additional wavelengths since its discovery. Two molecules deserve special mention: the cyanogen radical (CN) was detected at the microwave wavelength of 2.6 millimeters in 1970, and carbon monoxide at the ultraviolet wavelength of about 1,400 angstroms in 1971. Only these two molecules have been observed in two entirely different regions of the spectrum. Eight of the molecules have so far been detected in just one source: the unusually dense and massive interstellar cloud Sagittarius B2, which is associated with an H II region not far from the galactic center. That hydrogen ($H_2$), the simplest of all interstellar molecules, should be observed in only two clouds so far is surprising because it ought to be the stable form of hydrogen in all interstellar clouds dense enough to yield other detectable molecules. The designation "100-inch, M.W.O." refers to the 100-inch reflecting telescope of the Mount Wilson Observatory; "84-foot, L.L.," to the 84-foot radio telescope at the Lincoln Laboratory of the Massachusetts Institute of Technology; "20-foot, H.C.O.," to the 20-foot radio telescope at the Hat Creek Observatory of the University of California; "140-foot, N.R.A.O.," to the 140-foot radio telescope at the National Radio Astronomy Observatory in Green Bank, W. Va.; "36-foot, N.R.A.O.," to the 36-foot radio telescope of the National Radio Astronomy Observatory located at the Kitt Peak National Observatory in Arizona; "N.R.L.," to an Aerobee rocket launched by Naval Research Laboratory; "210-foot, C.S.I.R.O.," to the 210-foot radio telescope of Commonwealth Scientific and Industrial Research Organization radio observatory located at Parkes in Australia.

---

[*]Since 1972, additional molecules and wavelengths have been observed, and some of the earlier molecules have been found in additional regions. As of early 1975, the identity of "x-ogen" was still a mystery, with $HCO^+$ as a leading possibility.—Ed.

the molecules or their constituent electrons change their state of motion. Each molecule tends to rotate around the axes of symmetry characteristic of it. Changes in rotation cause the molecule to radiate or absorb electromagnetic energy at wavelengths that typically lie in the microwave region of the spectrum: wavelengths between about one millimeter and six centimeters. The atoms within the molecule vibrate with respect to one another as well as rotate around one another. The vibrational motion can also change, causing the molecule typically to emit or absorb infrared radiation. Moreover, the electrons of the various atoms can jump from one orbit to another, and these transitions give rise to radiation or absorption in the visible or ultraviolet regions of the spectrum.

Carbon monosulfide is an example of the simplest possible molecule: a linear diatomic molecule. In its states of lowest electronic and vibrational energy, which are the only states populated in interstellar space, the possible motions of the molecule are simply movement in one direction and end-over-end rotation. The laws of quantum mechanics show that the corresponding energy states form a simple "ladder." A molecule can move up or down the ladder when it collides with another molecule or another kind of particle, or when it absorbs or emits a photon (a quantum of radiation). If left to itself, the molecule will spontaneously emit photons and cascade to the ground state, or lowest energy level, in a definitely specified time. The energy levels

are labeled by the quantum number $J$, a measure of the rotational momentum of the molecule.

For carbon monosulfide it takes an average of four hours for the molecule to drop from the first energy level to the ground state. The molecule spends even less time in the higher levels. The astronomer detects radiation from the higher levels, so that some mechanism must be acting to maintain the carbon monosulfide molecules in these levels, balancing their tendency to decay spontaneously. Fields of radiation could act in this way, but they would have to be fields that are much stronger than the ones that are directly observed in these clouds. A more likely mechanism of excitation is collisions with other particles.

For the kinds of particle we are considering and the temperatures involved we can estimate the density required to maintain molecules such as carbon monosulfide in excited states. Typically densities of one million particles per cubic centimeter result for the clouds of carbon monosulfide observed near H II regions. Other molecules that have energy-level schemes similar to those of carbon monosulfide include carbon monoxide, silicon monoxide (SiO), carbonyl sulfide (OCS), hydrogen cyanide and cyanoacetylene ($HC_3N$). Carbon monoxide has a much lower rate of spontaneous decay than the other linear molecules. Hence a lower rate of excitation by collision is required for its excited levels to be maintained. This is one reason carbon monoxide is observed over larger regions of space

| YEAR | MOLECULE | FORMULA | WAVELENGTH | NUMBER OF REGIONS | TELESCOPE |
|------|----------|---------|------------|-------------------|-----------|
| 1937 | METHYLIDYNE (IONIZED) | $CH^+$ | 3,958 ANGSTROMS* | 88 | 100-INCH, M.W.O. |
| 1937 | METHYLIDYNE | CH | 4,300 ANGSTROMS* | 60 | 100-INCH, M.W.O. |
| 1939 | CYANOGEN RADICAL | CN | 3,875 ANGSTROMS* | 14 | 100-INCH, M.W.O. |
| 1963 | HYDROXYL RADICAL | OH | 18.0 CENTIMETERS* | ~600 | 84-FOOT, L.L. |
| 1968 | AMMONIA | $NH_3$ | 1.3 CENTIMETERS* | 12 | 20-FOOT, H.C.O. |
| 1968 | WATER | $H_2O$ | 1.3 CENTIMETERS | 35 | 20-FOOT, H.C.O. |
| 1969 | FORMALDEHYDE | $H_2CO$ | 6.2 CENTIMETERS* | ~150 | 140-FOOT, N.R.A.O. |
| 1970 | CARBON MONOXIDE | CO | 2.6 MILLIMETERS | 60 | 36-FOOT, N.R.A.O. |
| 1970 | HYDROGEN CYANIDE | HCN | 3.4 MILLIMETERS | 10 | 36-FOOT, N.R.A.O. |
| 1970 | X-OGEN | ? | 3.4 MILLIMETERS | 8 | 36-FOOT, N.R.A.O. |
| 1970 | CYANOACETYLENE | $HC_3N$ | 3.3 CENTIMETERS* | 4 | 140-FOOT, N.R.A.O. |
| 1970 | HYDROGEN | $H_2$ | 1,060 ANGSTROMS | 2 | N.R.L. |
| 1970 | METHYL ALCOHOL | $CH_3OH$ | 35.9 CENTIMETERS* | 3 | 140-FOOT, N.R.A.O. |
| 1970 | FORMIC ACID | HCOOH | 18.3 CENTIMETERS | 1 | 140-FOOT, N.R.A.O. |
| 1971 | CARBON MONOSULFIDE | CS | 2.0 MILLIMETERS* | 20 | 36-FOOT, N.R.A.O. |
| 1971 | FORMAMIDE | $NH_2CHO$ | 6.5 CENTIMETERS* | 2 | 140-FOOT, N.R.A.O. |
| 1971 | CARBONYL SULFIDE | OCS | 2.5 MILLIMETERS* | 1 | 36-FOOT, N.R.A.O. |
| 1971 | SILICON MONOXIDE | SiO | 2.3 MILLIMETERS* | 2 | 36-FOOT, N.R.A.O. |
| 1971 | METHYL CYANIDE | $CH_3CN$ | 2.7 MILLIMETERS | 1 | 36-FOOT, N.R.A.O. |
| 1971 | ISOCYANIC ACID | HNCO | 3.4 MILLIMETERS* | 1 | 36-FOOT, N.R.A.O. |
| 1971 | HYDROGEN ISOCYANIDE? | HNC? | 3.3 MILLIMETERS | 3 | 36-FOOT, N.R.A.O. |
| 1971 | METHYLACETYLENE | $CH_3CCH$ | 3.5 MILLIMETERS | 1 | 36-FOOT, N.R.A.O. |
| 1971 | ACETALDEHYDE | $CH_3CHO$ | 28.1 CENTIMETERS | 1 | 140-FOOT, N.R.A.O. |
| 1972 | THIOFORMALDEHYDE | $H_2CS$ | 9.5 CENTIMETERS | 1 | 210-FOOT, C.S.I.R.O. |
| 1972 | HYDROGEN SULFIDE | $H_2S$ | 1.8 MILLIMETERS | 7 | 36-FOOT, N.R.A.O. |
| 1972 | METHANIMINE | $CH_2NH$ | 5.7 CENTIMETERS | 1 | 210-FOOT, C.S.I.R.O. |

than other molecules, and it makes it possible to trace interstellar clouds out to regions where the density is as low as about 100 particles per cubic centimeter.

By observing different transitions of the molecules within a cloud it is possible to monitor different regions of density. The higher energy levels require higher densities for the appropriate excitation. A given spectral line has a certain width that is determined partly by the temperature of the cloud (the higher the temperature, the broader the line) and partly by turbulence or large-scale motions within the cloud. In most sources of interstellar molecules turbulence is dominant over the temperature effect. In these cases the narrower the spectral line, the smaller the net turbulence and hence the smaller the region within the cloud that is emitting the spectral line. For example, the spectral line emitted by carbon monosulfide as it drops from the second energy level to the first ($J = 2$ to $J = 1$) is seen to be emitted from two distinct ranges of velocity [see bottom illustration on page 231]. The line resulting from a drop from the first energy level to the ground state ($J = 1$ to $J = 0$) is emitted over a much larger range of velocities. This is interpreted as indicating that within the large cloud of carbon monosulfide there are two smaller regions of higher density. Observations of more

highly excited molecules should further refine the density map.

How are the temperatures of a cloud determined from molecular observations? It turns out that molecules such as carbon monosulfide, which radiate their energy rapidly, are rather insensitive probes of temperature. Other molecules, such as carbon monoxide and ammonia, are more useful. If a molecule of carbon monoxide radiates its energy away much more slowly than it is excited by collisions, then it comes into thermal equilibrium with its surroundings. Under these conditions the intensity of the various transitions between energy levels can be exactly calculated as a function of the temperature only. Observations of the intensity of the transitions then make it possible to deduce the temperatures within the cloud. In this way carbon monoxide has yielded temperatures of between five degrees and 25 degrees K. in the large dust clouds, and typically of 100 degrees K. in the molecular clouds surrounding H II regions.

Ammonia is a unique probe of physical conditions in interstellar clouds because this molecule has several different types of energy state. The molecule is a tetrahedron with a nitrogen atom at an apex above the plane formed by three hydrogen atoms; the nitrogen atom can oscillate back and forth through the plane. As a result each level of the rotational energy, $J$, is split into two closely spaced levels by the oscillation. Transitions between these closely spaced levels are known as inversion transitions. They are observed at the microwave wavelength of 1.3 centimeters. For ammonia a second quantum number, $K$, refers to the component of angular momentum around the molecule's axis of symmetry. The quantum numbers are written together: $J,K$. Ammonia can change its value of $J$ by 1 by emitting or absorbing photons. Collisions may cause $J$ to change by more than 1. $K$, on the other hand, cannot change its value by the emission or absorption of photons. Collisions between particles in a gas allow transitions only between states whose $K$ is $3n + 1$ and $3n - 1$, or between states whose $K$ is $3n$ ($n$ is any integer). The species whose $K$ is $3n + 1$ and $3n - 1$ is called para-ammonia; the species whose $K$ is $3n$ is called ortho-ammonia.

The number of ammonia molecules in the various states of rotational energy can be directly estimated by observing the intensity of emission from the inversion transitions. Some rotational levels (such as the 2,1 level) decay in a time as short as 20 seconds to lower states (such

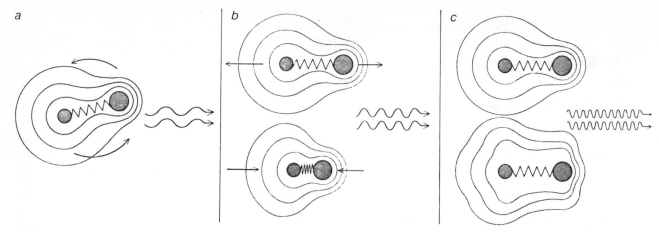

*a*    *b*    *c*

**MOLECULES OR THEIR ELECTRONS** can change their state of motion, giving rise to various lines in the electromagnetic spectrum. Each molecule tends to rotate around its characteristic axes of symmetry; in the case of carbon monosulfide, a simple diatomic molecule, the rotation is end over end (*a*). Changes in the rotation cause the molecule to radiate or absorb energy at wavelengths that typically lie in the microwave region of the spectrum. The atoms within the molecule vibrate with respect to each other as if they were joined by a spring (*b*). The vibrational motion can also change, causing the molecule to emit or absorb shorter-wavelength infrared radiation. Moreover, the electrons of the various atoms can jump from one energy state to another, changing the configuration of the electron orbits around the nuclei of the molecule (*c*). These transitions give rise to radiation or absorption in the visible or ultraviolet regions of the spectrum. In this case the carbon atom is to the left and the sulfur atom is to the right; the molecule is shown before the transition in the first excited state (*top*) and after the transition in the ground state (*bottom*). The contours represent the probability of finding an electron in various parts of the electron cloud. Inner contours are virtually the same for isolated atoms of carbon and sulfur as they are for the molecule; the probability of finding an electron decreases toward the outer contours.

as the 1,1 level) by spontaneously emitting a photon. To observe a spectral line from the 2,1 level as strong as some that have been detected requires a high rate of excitation by collisions to compete with the rate of deexcitation by radiation. Therefore the density must be very high. In fact, in the Sagittarius B2 ammonia source a density as high as $10^9$ particles per cubic centimeter may be required. Transitions between certain other states (such as 1,1 and 2,2) occur only as a result of collision, but they occur at a high rate at such high densities. These states are thus brought into thermal equilibrium with the surrounding particles on a time scale of about a year. The ratio of molecules populating the 1,1 state to molecules populating the 2,2 state (that is, the observed ratio of intensity of the spectral lines) should be controlled only by the temperature of the colliding particles. Arguments such as this one lead to temperatures of between 20 degrees and 80 degrees K. for various clouds in the region of the galactic center.

One other temperature can be derived from the observations of ammonia. It can be obtained from the presence of the two different species of ammonia: ortho-ammonia (to which the 3,3 state belongs) and para-ammonia (to which the 1,1 and 2,2 states belong). Transitions between these species can occur only if the spin of the hydrogen nuclei (protons) is changed. That process is likely to occur only when the molecules collide with particles of interstellar dust and are broken up and reassembled. The average time between such spin-changing collisions is probably at least a million years. Thus the ratio of the intensity of the spectral line for molecules in the 3,3 state to the intensity for molecules in the 1,1 state or the 2,2 state represents some temperature inside the cloud a million years or so in the past. Such temperatures are found to be higher than the present temperatures, suggesting that the clouds cool as they contract.

If observations of molecular lines can yield so much information about temperature and density in interstellar clouds, one would expect that they would also give an accurate measure of the abundance of the molecules. This is not the case for a number of reasons. One is that the intensity with which a cloud of molecules emits or absorbs electromagnetic energy of a given wavelength depends not only on the total number of molecules present but also on what fraction of them are in the energy states of the observed transitions. The fraction can be calculated if we know that the molecules are in thermal equilibrium with their surroundings. The majority of interstellar molecules are not in thermal equilibrium, however. The most spectacular examples of nonequilibrium are the hydroxyl-radical maser and the water maser. Another problem is that the receiving beam of a radio telescope is fairly wide, and such beam widths often fail to resolve small patches of molecular emission. The result is that the intensity of these patches is underestimated. Nevertheless, for most of the 26 molecules known it is thought the abundances that have been deduced are accurate to within a factor of 10 or so. On the other hand, the estimated amount of hydroxyl radical and of water in sources where they act as an interstellar maser may be in error by a factor of 100 or even 1,000. In spite of these uncertainties many conclusions important to the understanding of interstellar chemistry have emerged.

## Chemical Composition of the Clouds

Interstellar clouds that harbor detectable quantities of molecules consist almost entirely of dust particles and molecular hydrogen. The dust particles are observed directly, since they attenuate and redden the light of background stars. Because of severe technical difficulties molecular hydrogen has been directly observed in only two clouds: very tenuous regions in front of the stars Xi Persei and Delta Scorpio that dim the background stars by about one magnitude (roughly 2.5 times). These clouds are not known at present to contain any other molecules. Theoretical considerations, however, indicate that clouds whose extinction of background starlight exceeds about 1.5 magnitudes will contain hydrogen in molecules rather than atoms. The rate at which hydrogen molecules are formed on the dust particles will exceed the rate at which they are dissociated by ultraviolet radiation. It is a general fact that molecules observed in the ultraviolet or visible portions of the spectrum are never seen in the same regions as molecules observed in the radio portion. The reason is that the rate at which any molecule can emit or absorb photons at radio wavelengths is much smaller than the rate for photons at visible or ultraviolet wavelengths. Thus many more molecules are required in the line of sight to produce detectable signals at radio wavelengths.

Any cloud containing enough molecules to be detected at radio wavelengths also contains so much dust that the background stars against which the molecules are observed as absorption features at visible or ultraviolet wavelengths are completely obscured. The result is that the molecules detected at visible or ultraviolet wavelengths (such as molecular hydrogen and CH or CH$^+$) are observed only in low-density clouds where the extinction of starlight is low and where the density of the particles does not exceed some 100 per cubic centimeter. On the other hand, molecules observed at radio wavelengths are seen only in much denser, darker clouds, where the density of the particles ranges from 100 to $10^9$ per cubic centimeter and the extinction probably exceeds 50 magnitudes (a factor of more than $10^{20}$).

In these dense clouds of molecular hydrogen and dust, molecules are observed in only trace amounts. That is because the molecules contain atoms other than hydrogen, which on a cosmic scale are much less abundant than hydrogen. For every 10,000 hydrogen atoms there are three or four oxygen atoms, two carbon atoms, one or two nitrogen atoms and one sulfur atom. Here, however, an important fact emerges. If this cosmic abundance scale (determined largely from atomic spectra observed in stars near the sun) applies to interstellar clouds in general, then the observed molecules alone account for nearly all the carbon atoms that must be present. The observed molecules must also account for an appreciable fraction (perhaps 30 percent) of all the oxygen atoms. By way of contrast, perhaps no more than .0001 percent of the available nitrogen atoms appear in the observed molecules. Most of the nitrogen atoms are probably in the form of the unobserved diatomic molecule ($N_2$), which comprises most of the earth's atmosphere.

Two key conclusions follow from these facts. First, the formation of molecules per available atom in interstellar clouds must be a highly efficient process. Second, interstellar chemistry seems to favor the production of organic molecules, that

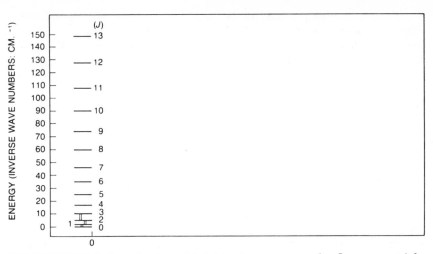

ENERGY LEVELS of a molecule are labeled by the quantum number $J$, a measure of the rotational momentum of the molecule. For carbon monosulfide it takes an average of four hours for the molecule to drop from the first energy level to the ground state. The molecule spends even less time at higher energy levels. The vertical bars between the energy levels $J = 3$, $J = 2$, $J = 1$ and $J = 0$ represent transitions of carbon monosulfide that have been observed in interstellar space by radio astronomers. The unit $cm.^{-1}$, known as an inverse wave number, is proportional to energy and refers to the number of wavelengths per centimeter. Transitions are measured in terms of changes in energy, and the inverse wave number is a direct indication of the amount of energy involved. For example, a difference in energy of five inverse wave numbers corresponds to a transition that has a wavelength of two millimeters; the corresponding spectral line would be in the microwave region.

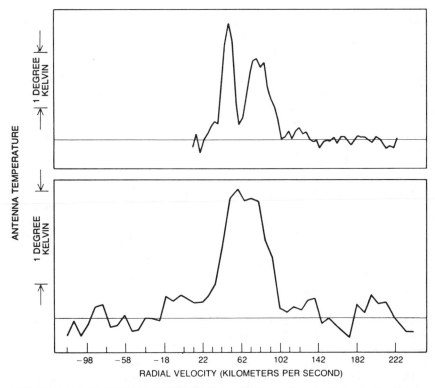

SPECTRA OF CARBON MONOSULFIDE in the interstellar cloud Sagittarius B2 are shown for two transitions of the molecule. A given spectral line has a certain width that is determined partly by turbulence; the narrower the spectral line, the smaller the net turbulence and hence the smaller the region within the cloud that is emitting the line. The line emitted by carbon monosulfide as it drops from the second energy level to the first (from $J = 2$ to $J = 1$) shows two distinct peaks, meaning that the line is emitted from two discrete regions each having a different radial velocity (top). On the other hand, the line resulting from a drop from the first energy level to the ground state ($J = 1$ to $J = 0$) is emitted over a much larger range of velocities (bottom). Collisions between particles account for the molecules in the higher energy state; these features indicate that within the large cloud of carbon monosulfide there are two smaller dense regions. Horizontal lines through spectra average zero-signal level around which there are some noise fluctuations.

is, molecules containing carbon. All the simplest organic molecules have been found in interstellar space, whereas many of the even simpler nonorganic species such as nitric oxide (NO), sulfur monoxide (SO) and the thiol radical (SH) have not been detected in spite of sensitive searches. No explanation for the fact that interstellar chemistry is predominantly organic has yet emerged.

## Molecules and Background Radiation

In 1965 Arno A. Penzias and R. W. Wilson of Bell Laboratories obtained evidence that the universe is bathed in a uniform field of radiation that has a temperature equivalent to three degrees K. at microwave wavelengths. If this field could be shown to have a "black body" spectrum, that is, to have a temperature equivalent to three degrees K. at all wavelengths, the most logical explanation for it would be that it was a remnant of the primordial fireball that marked the beginning of the universe's current phase of expansion. On the other hand, large deviations from a black-body spectrum would discount such an origin. Since 1969 observations of interstellar molecules have played an important role in deciding this fundamental question.

It has been known since 1969 that interstellar formaldehyde, like the hydroxyl radical and water, is excited in a very strange way. One of its transitions, at the wavelength of six centimeters, was observed by P. Palmer of the University of Chicago, B. M. Zuckerman of the University of Maryland, L. E. Snyder of the University of Virginia and David Buhl of the National Radio Astronomy Observatory. They saw the transition as an absorption feature in certain dust clouds against a region of sky that contained no known sources of radiation with a continuous spectrum. The radiation that was being absorbed must therefore be the three-degree cosmic background. Since the observation of formaldehyde is free from many of the technical difficulties of direct measurements of the background field, it is one of the best direct confirmations that the cosmic background exists.

Measurements of the excitation of CN, CH and CH$^+$ have recently provided estimates of the intensity of the three-degree radiation at radio wavelengths shorter than three millimeters. No other reliable information exists in this part of the spectrum. In the rarefied clouds where those three molecules are observed the rate at which they collide with other particles is so low that their relative populations in the lowest few rotational energy levels are governed largely by the rate at which they interact with photons from the cosmic background radiation. Measurements of the relative populations, deduced from observations of two or more transitions involving their energy levels, yield the intensity of the cosmic background radiation at the wavelengths of those transitions. CN verifies that the intensity is indeed about three degrees at a wavelength of 2.63 millimeters. CH and CH$^+$ provide similar, although less definitive, results at the wavelengths of .56 and .36 millimeter. It seems likely that further measurements of the excitation of the various molecules now becoming known will be of great help in determining the high-frequency part of the cosmic radiation spectrum.

## Isotope Abundances

On the earth and in the sun, comets and meteorites the different isotopes of the chemical elements are found in certain well-defined proportions. As I have indicated, the abundance ratio of carbon 12 to carbon 13 is 89 : 1. The ratio of oxygen 16 to oxygen 18 is 490 : 1, of nitrogen 14 to nitrogen 15, 270 : 1, and of sulfur 32 to sulfur 34, 22 : 1. In the kinds of stars that consume their nuclear fuels in processes involving carbon, nitrogen and oxygen the isotope ratios of these elements are altered. The ratio of carbon 12 to carbon 13 may be as low as 4 : 1, as we have seen, and the ratio of nitrogen 14 to nitrogen 15 may become very large. In stars that consume helium, carbon is processed in such a way that virtually all the carbon 13 is used up.

Stars continually recycle their material through the interstellar medium. They are created from the interstellar gas; they subject the gas to nuclear-burning processes, and they return the gas to interstellar space by ejecting matter or by exploding in a supernova. Therefore the isotope ratios in the solar system ought to be accidental ones resulting from nonequilibrium processes or from the mixing of gases from various sources.

The startling result provided by the study of interstellar molecules containing different isotopes is that the interstellar isotope ratios appear to be the same as they are on the earth. The one apparent exception is in the region of the galactic center, where the ratio of carbon 12 to carbon 13 seems to be about half as large as it is elsewhere. This fact might suggest that proportionately more interstellar material has passed through carbon-burning stars in the galactic center, ei-

ther because in that region there is less gas in relation to stars or because a larger fraction of the stars are of the massive carbon-burning type.

The fact that the bulk of the interstellar medium resembles the earth in isotope ratios would seem to indicate that interstellar chemistry has changed little in the five billion years since the earth was born. There are two possible explanations. Perhaps there are no regions (except for the galactic center) that are oversupplied with carbon-burning stars in relation to helium-burning stars. Alternatively, the interstellar gas must be well mixed over great distances and within a time that is comparable to the relatively short lifetime of the hottest stars.

## Molecular Life Cycles

As we have seen, the discovery of the simple molecules CN, CH and CH$^+$ came as a surprise to astronomers. They had predicted that because it was highly improbable that two atoms would come together to form a molecule, and because any such molecules would be rapidly destroyed by the harsh interstellar ultraviolet radiation, the abundance of the molecules ought to be vanishingly small. Given the fact that the rate of destruction of the complex organic molecules is even higher and the fact that they are observed in much larger quantities than CN, CH and CH$^+$, it is obvious that new theories for their existence are needed.

In the past few years several mechanisms for the formation of molecules have been proposed. They break down

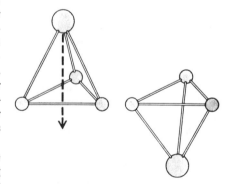

AMMONIA MOLECULE (NH$_3$) is a tetrahedron with a nitrogen atom at the apex above the plane formed by three hydrogen atoms. The nitrogen atom can oscillate from one side (*left*) to the other (*right*) through the plane. As a result each level of the rotational energy, J, is split into two closely spaced levels. Transitions between the two levels are inversion transitions and are observed at the wavelength of 1.3 centimeters.

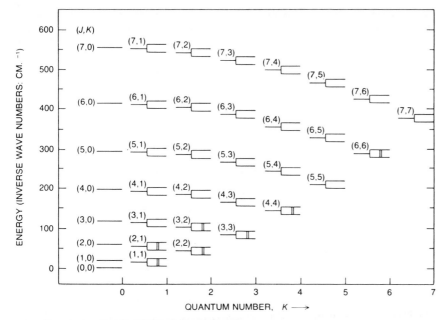

AMMONIA'S ENERGY LEVELS differ from those of carbon monosulfide because of the oscillation of the nitrogen atom. A second quantum number, $K$, refers to the component of angular momentum around the molecule's axis of symmetry. The two quantum numbers are written together: $J,K$. Ammonia can change its value of $J$ by 1 by emitting or absorbing a photon; collisions may cause $J$ to change by more than 1. $K$ changes only by collisions.

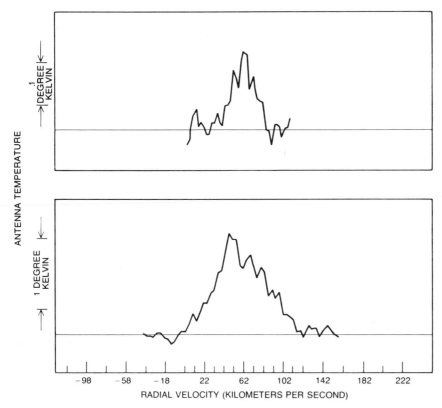

SPECTRA OF AMMONIA in Sagittarius B2 yield information about the density inside the source. The number of ammonia molecules in the various states of rotational energy can be estimated by observing the inversion transitions. Some rotational levels, such as the 2,1 level (top), decay in a time as short as 20 seconds to a lower state, such as the 1,1 level (bottom), by spontaneously emitting a photon. To observe a spectral line from the 2,1 level as strong as some that have been detected requires a high rate of excitation by collisions to compete with the spontaneous rate of decay. Therefore the density must be very high. In Sagittarius B2 a density as high as $10^9$ particles per cubic centimeter may be required.

into five major categories. First, the molecules might be built up by atoms colliding with and sticking to one another as they move around in the gaseous interstellar medium. Second, the molecules could form on the surface of dust particles out of atoms or other molecules that impinge on such surfaces. Third, the molecules might form from the evaporation or decomposition of dust particles when they are struck by energetic particles or photons or encounter shock waves or other heating phenomena. (This mechanism leaves open the question of how the dust particles themselves originated.) Fourth, the molecules might come into existence through collisions with atoms in the dense atmosphere of stars and might then be expelled into interstellar space. Fifth, the molecules might be formed in the dense environs of a "presolar nebula," that is, in the final phases of the collapse of a protostar into a self-luminous star.

The molecules must be produced as fast as they are removed or the lifetime of a molecular cloud would be limited, and it might be very short indeed. The two principal processes that destroy interstellar molecules are photodissociation by ultraviolet radiation and freezing out on the surface of dust particles. Recently L. J. Stief and his associates at the Goddard Space Flight Center in Greenbelt, Md., have studied the rates at which molecules are photodissociated, utilizing the results of elegant laboratory work and measurements of the intensity of interstellar ultraviolet radiation made during rocket flights. The important result is that molecules such as water, ammonia, formaldehyde and carbonyl sulfide dissociate in somewhat less than 100 years if they are unshielded by dust and are exposed to the average flux of interstellar ultraviolet radiation. This is a very short time by astronomical standards.

Dust clouds, however, are an extremely effective shield against ultraviolet radiation. A typical dust cloud with a diameter of one light-year and a density of 1,000 molecules of hydrogen per cubic centimeter will attenuate light by about four magnitudes and ultraviolet radiation by a factor perhaps as large as $10^{24}$. As a result the lifetime of the molecules can be increased to as much as 10 million years. We must therefore conclude that interstellar molecules have been created within the dust clouds where they are now observed or that they have been transported into these clouds with some kind of protection, perhaps on the surface of dust particles.

Within the large dust clouds the density of the dust particles is such that a molecule will condense out of the gas onto the surface of a dust particle in about 100,000 years. The temperatures in these clouds are so low that it appears highly unlikely that the molecules would evaporate back into the gaseous state. A period of 100,000 years is considerably shorter than the lifetime of such clouds. Therefore, regardless of how the molecules originated, they must somehow be regenerated from the dust particles many times during the lifetime of the cloud. It is possible that this regeneration is accomplished by the few cosmic rays or photons of ultraviolet radiation that manage to penetrate the cloud or by invisible sources of infrared radiation such as protostars within the cloud.

The situation is much more critical in the dense molecular clouds associated with H II regions. Because of the much higher density of the cloud, molecules will stick to dust particles in as short a time as 100 years. The higher temperatures within these clouds, however, probably allow significant numbers of the molecules to evaporate back into the gaseous state. While the molecules are on the dust particle, surface reactions may well occur that give rise to new and more complicated molecules. Such effects might possibly explain why a wider variety of complex molecules is found near H II regions than is found in the relatively rarefied dust clouds.

It is only fair to state that there is no consensus among molecular astronomers in favor of any of the five mechanisms proposed for the formation of molecules. Perhaps no one of them fully applies. Nonetheless, there are a number of ways that these processes could act to form molecules. Some choices between the various possibilities now seem to be emerging from the observations. It does seem that the atmosphere of stars cannot be the direct source of the bulk of the interstellar molecules. Complex molecules could not survive passage through the hot corona of the stellar atmosphere on their way to the interstellar medium, even if a suitable mechanism for their ejection could be found. Furthermore, the very types of stars whose atmospheres are cool enough to form molecules are known from theory (and in a few cases from observation) to have a ratio of carbon 12 to carbon 13 of between 4 : 1 and 10 : 1, as against the interstellar ratio of 89 : 1.

Some evidence, perhaps not overly conclusive, also suggests that molecules do not come solely from the breakup of interstellar dust particles. If they did, the dust particles themselves would not have been originally formed in interstellar space but would need to have been ejected from stars. Although such ejection is almost certainly possible in principle, again it is only those stars with a low ratio of carbon 12 to carbon 13 that can produce dust particles efficiently. Interstellar molecules that resulted from the breakup of these dust particles would share the low ratio of carbon 12 to carbon 13, which is contrary to the observations. In addition, molecules are not observed primarily in regions showing evidence of shock waves, energetic subatomic particles or other agencies that might decompose the dust particles by heating.

Therefore it seems that one of the first, second and fifth mechanisms, or some combination of them, is responsible for the interstellar molecules. Different processes may well occur in different types of molecular cloud. In principle, reactions on the surface of dust particles can proceed in any type of cloud, at a rate that depends on the density, temperature and chemical constitution of the particles. Still other primary mechanisms may well dominate the formation of molecules on dust particles. For example, George H. Herbig of the Lick Observatory has suggested that molecules might be created in presolar nebulas within the dense clouds surrounding H II regions. H II regions are well known to be regions of star formation, so that many presolar nebulas would be present. Such nebulas can expel both molecules and dust quite efficiently into the surrounding molecular clouds. The molecules and dust are expelled by the pressure of infrared radiation from the central protostar or by an excess of rotational momentum developed during the latter stages of the collapse of the presolar nebula. In protostars nuclear reactions have not yet begun; in such objects the isotope ratios would be the same as those observed in interstellar space.

In the less dense dust clouds, well away from H II regions, only the first and second mechanisms seem applicable to the formation of molecules. These mechanisms are not, however, without their problems. If surface reactions alone are considered, the question remains: How did the dust particles originate? Conceivably they could have been manufactured by stars elsewhere and have come to rest in clouds that slowed their passage because they were denser than the surrounding interstellar medium. Or the dust particles could have been built up over long periods of time by accretion, starting with pairs of atoms forming diatomic molecules. The theoretical details of this process are quite complicated. It appears that diatomic molecules can form at an adequate rate in this way, but subsequent collisions with the atoms in the surrounding gas would be likely to destroy the molecules already formed rather than creating new ones with more atoms. Only when the molecules already consist of at least a dozen atoms do collisions in the gas appear able to create larger molecules.

Whatever the mechanism responsible for molecules in the dust clouds may be, it seems certain that, in view of the low densities and temperatures in such clouds, the rate at which the molecules interact with the dust particles must be much lower than it is in the denser clouds near H II regions. Moreover, in dust clouds there are probably no efficient primary sources of molecules such as presolar nebulas. In these low-density clouds the more complex molecules (such as cyanoacetylene and methyl alcohol) appear to be less abundant with respect to the simpler molecules (such as the hydroxyl radical, carbon monosulfide and formaldehyde) than they are in the denser clouds near H II regions. If this suspicion is confirmed by additional observation, such a difference in chemical constitution may indicate that the mechanism of molecular production in one of the two types of source differs from the mechanism in the other.

## Further Questions

The mechanisms I have discussed so far appear capable of forming interstellar molecules in the observed quantities in regions adequately shielded from destructive ultraviolet radiation. They do not, however, explain either the relative abundances of the observed molecules or the absence of certain others. What is known about the rate at which these molecules are dissociated by ultraviolet radiation indicates that there is no relation between the rates and the observed amounts. The inescapable conclusion is that very specific and very poorly understood formation mechanisms are at work. They almost certainly involve reactions on the surface of dust particles. Such reactions are likely to be important even in the clouds near H II regions where presolar nebulas may be the primary source of molecule formation; after molecules are ejected from the protostar into the surrounding cloud they must continue to interact rapidly with the dust particles. Unfortunately, little is known about surface reactions even when the surfaces are well defined. The composition of interstellar dust particles is unknown, and that composition may vary from region to region depending on local conditions. Over the years proposals for the composition of the dust particles have included graphite, graphite covered with frozen water and ammonia and methane, and a mixture of graphite, silicate and iron. It goes without saying that the particular molecules whose formation would be catalyzed on these surfaces would depend strongly on the composition of the dust particles.

If I have ended this discussion on a somewhat negative note, it is intentional. It is now clear that many scientific disciplines are needed to solve the problems of interstellar chemistry. It can be said that a new discipline has been born: astrochemistry. Our knowledge of interstellar molecules, full of gaps as it is, has already been advanced immeasur-

ably by cooperative work in traditionally nonastronomical fields. Microwave spectroscopists have contributed information about the precise frequencies needed to search for new molecules. Photochemists have measured the probability that ultraviolet photons will destroy particular molecules. Experiments on the formation of simple organic molecules have been conducted for many years by workers interested in the problem of how life itself arises. With the discovery of at least four interstellar molecules that are considered to be necessary for the subsequent formation of biological molecules (water, formaldehyde, hydrogen cyanide and cyanoacetylene) these experiments obviously apply to the fascinating question of whether primitive life, or at least the amino acids that are the essential building blocks of it, can be detected in interstellar space.

In at least two areas further laboratory work is critically needed. One area involves identifying unknown spectral

lines in interstellar space that are apparently molecular in origin. Two lines in the radio region of the spectrum are in this category; one is "X-ogen" and the other may possibly be a very unstable isomer of hydrogen cyanide that is unknown on the earth. A broad, diffuse spectral line of unknown origin has also been observed at the visible wavelength of 4,430 angstroms. These lines originate in many regions of interstellar space and may accordingly be quite basic to our understanding of cosmic chemistry. If they arise from molecules that are very unstable on the earth, as is often supposed, then an understanding of how these molecular species form and survive so well in the interstellar medium but not on the earth may bear on some fundamental aspects of molecular structure. The second area requiring much laboratory work is the chemistry of surfaces. Although the subject has been well investigated on an empirical basis in some kinds of catalytic processes that are im-

| ATOM | TERRESTRIAL ISOTOPE RATIO | INTERSTELLAR ISOTOPE RATIO | INTERSTELLAR MOLECULE AND REGION |
|---|---|---|---|
| CARBON 12/CARBON 13 | 89 : 1 | ABOUT 50 : 1 IF OXYGEN 16/OXYGEN 18 = 488 : 1 | $H_2CO$, GALACTIC CENTER |
| | | ABOUT 89 : 1 IF OXYGEN 16/OXYGEN 18 = 870 : 1 | $H_2CO$, GALACTIC CENTER |
| | | ABOUT (89 ± 15) : 1 | $H_2CO$, OUTSIDE GALACTIC CENTER |
| | | (82 ± 15) : 1 | $CH^+$, ZETA OPHIUCHI |
| OXYGEN 16/OXYGEN 18 | 488 : 1 | 488 : 1 IF CARBON 12/CARBON 13 = 50 : 1 | $H_2CO$, GALACTIC CENTER |
| | | 488 : 1 IF CARBON 12/CARBON 13 = 50 : 1 | CO, GALACTIC CENTER |
| | | 870 : 1 IF CARBON 12/CARBON 13 = 89 : 1 | $H_2CO$, GALACTIC CENTER |
| | | (390 ± 100) : 1 | OH, GALACTIC CENTER |
| OXYGEN 16/OXYGEN 17 | 2,700 : 1 | AT LEAST 2,700 : 1 | OH, GALACTIC CENTER |
| NITROGEN 14/NITROGEN 15 | 270 : 1 | GREATER THAN 70 : 1 | $NH_3$, GALACTIC CENTER |
| | | (230 ± 70) : 1 | HCN, ORION |
| SULFUR 32/SULFUR 34 | 22.5 : 1 | (24 ± 5) : 1 | CS, GALACTIC CENTER AND ELSEWHERE |
| SULFUR 32/SULFUR 33 | 125 : 1 | GREATER THAN 100 : 1 | CS, ORION |

RATIOS OF ISOTOPES of various atoms in interstellar space compare well to those same ratios on the earth. When molecules other than carbon monoxide and formaldehyde are used, the comparison can be made relatively simply for oxygen, nitrogen and two isotopes of sulfur. In the galactic center, however, the very large abundances of carbon monoxide and formaldehyde confuse the interpretation. The spectral lines become "saturated," that is, the abundance is no longer proportional to the intensity of the spectral line. In this case it is possible only to obtain the ratio of carbon and oxygen together, that is, the ratio of carbon 13 and oxygen 18 to carbon 12 and oxygen 16. This dual ratio is found to be 1.8 times the terrestrial ratio. No known nuclear processes in stars affect the ratio of oxygen 16 to oxygen 18; moreover, this ratio seems to have a terrestrial value, as measured by the hydroxyl radical. It is therefore believed that the nonterrestrial value of the dual ratio of carbon and oxygen is due to a nonterrestrial value of 50 for the carbon ratio carbon 12 to carbon 13 in the region of the galactic center. Such a ratio could be produced by nuclear burning of carbon in giant stars, which may be more numerous in the galactic center. Elsewhere in the galaxy even the carbon ratio appears to be the same as that on the earth. Since all the interstellar isotope ratios seem to be near the terrestrial value, the chemistry of the interstellar medium must have been remarkably constant for a time at least equal to the age of the earth: some five billion years.

portant to industry, fundamental understanding of the elementary processes occurring at surfaces is largely lacking and thus predictions about them are generally not possible.

Whether astronomers are working at visible, infrared or radio wavelengths, they still must contribute vital information. The composition of the interstellar dust particles will most likely be determined from their properties of attenuation and polarization at visible and ultraviolet wavelengths and from the way they emit infrared radiation. Many molecules that are currently undetected and that must play central roles in interstellar chemistry can be observed only in the visible or infrared regions of the spectrum. Three of these molecules are molecular nitrogen ($N_2$), molecular carbon ($C_2$) and methane ($CH_4$).

Higher resolution is possible with infrared wavelengths than is possible with radio waves. Thus observations in the infrared will be vital in studying the fine structure of the interstellar sources where much of the activity may be going on. In the near future most new molecules will probably be detected at radio wavelengths, where the technology is more highly developed. The need for increasingly detailed study of the spatial distribution of the molecules within the sources, however, is at least as important as the discovery of new molecules. Recent observations have indicated that there is a wealth of unexpected detail in the structure of the molecular clouds. If such structure is not taken into account, serious errors may result in the estimation of the physical conditions and the relative abundances

of different molecules within the clouds.

Notwithstanding these problems, which promise to maintain the field of astrochemistry in a highly excited state for some time, molecular spectroscopy has emerged as a major astronomical tool. The energy levels of molecules are typically much more closely spaced than the energy levels of individual atoms, so that molecules are ideally suited for the study of the cool regions between the stars. This fact, in addition to the ability of the longer wavelengths of the spectral lines of molecules to penetrate regions of dust and gas, has enabled us to explore the dust clouds and the early stages of star formation within them. Soon we should also be able to study the cool regions of galaxies outside our own.

# Updating Galactic Spiral Structure

Bart J. Bok

Five years ago, the Editors of *American Scientist* published my article "The Spiral Structure of Our Galaxy" (*1*). Now, I am happy to have the opportunity to bring my story up to date. It is not necessary for the reader to consult the earlier article before reading the present one, but I shall note at various places what sorts of new observational evidence have become available in the past five years and how, as a result of this and of new theoretical work, our views on galactic spiral structure are changing.

By 1967 a picture had emerged according to which our galaxy appeared to possess a rather tightly wound spiral pattern. The "Companions in Zealous Research" have not been idle these past few years, and many new research workers have been attracted to the field. "Progress" and "controversy" are the bywords for the total effort over the past decade.

Our galaxy consists basically of three parts: the nuclear region, extending to a distance of about 5,000 parsecs from our sun (1 parsec = 3.26 light-years); the thin disk, not more than 500 parsecs thick at the sun, which contains the most spectacular Population I objects and in which spiral structure prevails; and the outer halo, mostly inhabited by Population II stars. We shall be exclusively concerned with the parts showing evidence for spiral structure. The spiral region extends in the galactic plane from about 5,000 to 15,000 parsecs from the sun. We note that our sun is placed rather centrally in this spiral belt, at a distance of about 10,000 parsecs from the center.

The interplay between observational research for our own and neighboring spiral galaxies has increased markedly during the past five years, and the anomalies found among our neighbors find critical application in studies of our own galaxy. New spiral tracers, optical as well as radio, have been proposed. Optical astronomers are gradually pushing out the limits to which spiral features can be studied in and near the absorption-polluted central plane of our Milky Way system, so much so that we now talk about features at 8 kiloparsecs (kpc) from the

sun with the same confidence that we had in discussing features at 3 or 4 kpc from the sun five years ago (one kpc = 1,000 parsecs = 3,260 light-years). Radio astronomers are not only contributing increasingly to the total volume of data available for spiral analysis, but there has been some deep soul-searching in this field, and the methods of analysis have become more flexible and less dogmatic than they were earlier.

In a way, the most remarkable activity has occurred in the theoretical field. Five years ago, we suggested that the Lin-Shu density wave theory of spiral structure seemed to provide useful guidance in the theoretical interpretation of observed regularities; this theory has now developed into a respectable theoretical edifice. The value of the Lin-Shu theory is that it gives us specific predictions that can be subjected to observational tests. In addition, it leads to certain firm predictions about processes of star birth in relation to spiral features. It is a very healthy development that during the past year or so quite a few papers have been written questioning the validity of the Lin-Shu approach, and there have been many suggestions for completely different theories, most of them not yet fully developed.

The interplay between theory on the one side and optical and radio observations on the other is providing much activity. During the past few years theoreticians working on the physics of the interstellar medium have gained increased insight into the physical conditions of the gas and dust in spiral arms and in interarm regions. The whole field is in turmoil and development, and this seems like a good time to examine just where we stand and what the prospects are for the future.

## Neighboring Galaxies

It has become increasingly clear that the Milky Way astronomer interested in the problems of the spiral structure of our own galaxy must at all times have in mind the properties of spiral arms and related features in galaxies outside our own. From our local position immersed in the galactic disk, 10,000 parsecs distant from the galactic center, we simply cannot

From *American Scientist*, vol. 60, pp. 709–722, 1972. Reprinted with permission.

238

**Figure 1.** Messier 31, the Andromeda Galaxy, photographed
with the 48″ Schmidt telescope, Mount Palomar. [Photograph
courtesy of Hale Observatories. © by the California Institute of
Technology and the Carnegie Institution of Washington.]

obtain an overall view. By studying neighboring galaxies, for example Messier 31 (Fig. 1) and others, we can view the whole of the spiral pattern from a single photograph or from a comprehensive analysis of neutral atomic hydrogen (HI) 21-cm profiles. Photographs of neighboring galaxies such as Messier 31, 33, 51, 81, and 101, some of which are illustrated in these pages, show beautifully the sweep of the structure in the appearance of the spirals closest to us. But there are several properties known about these galaxies, which are recognized as the finest of spirals, that do not fit nicely into a simple, all-inclusive spiral picture.

Let us take Messier 31, the Great Spiral in Andromeda, as an example. In a study published in 1964, H. C. Arp re-analyzed some of Walter Baade's material on the spiral structure of this galaxy. He rectified the photographs and drawings for a tilt of the central plane of the galaxy to the line of sight of 11°. In other words, he attempted to obtain a face-on view of Messier 31. His results are reproduced in Figure 2. The Arp diagrams show that the emission nebulae found by Baade do give patterns that can be adjusted to logarithmic spirals, but we note that the basic pattern found by Arp has almost a closer resemblance to a ring structure than to true spirality. If we want to see spiral arms, we can certainly find them, and most of us would have no doubt that the basic structure in Messier 31

is of a spiral nature. Yet the ring structure seems to have quite a bit to recommend itself.

Several astronomers have attempted to trace the radio spiral structure from 21-cm profiles for the Andromeda spiral. The first study was made by Morton S. Roberts, and, more recently, Vera C. Rubin and W. Kent Ford have again studied it, combining the results of 21-cm and optical studies. There are marked HI-21-cm peaks at the distances from the center where the Baade-Arp distribution charts of ionized hydrogen regions, observed optically, also show peaks. However, the highest concentration of neutral atomic hydrogen appears to be in a ring structure that falls mostly beyond the distances from the center of the Andromeda galaxy where the HII (ionized hydrogen) regions are found. This result bears an important similarity to one found from the analysis of hydrogen distribution in our galaxy. It was shown some years ago by Gart Westerhout that in our galaxy the ionized hydrogen distribution peaks at 5,000 parsecs from the sun, whereas the neutral hydrogen distribution peaks at distances of the order of 13,000 parsecs from the center. We find a similar situation in Messier 51.

A close examination of photographs of normal galaxies with spiral structure shows many worrisome features. In most of the beautiful spirals shown in the photographs of the

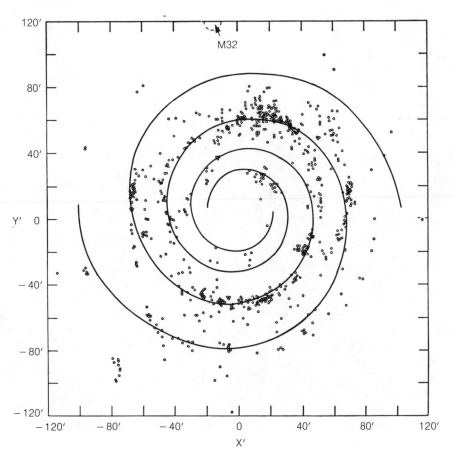

**Figure 2.** The Andromeda Galaxy. The positions of 688 emission nebulae (HII regions) in this galaxy are marked as they would appear corrected for a tilt i = 11° to the line of sight for the plane of M31. A logarithmic spiral has been fitted to the points. Note how difficult it is to distinguish between possible ring and spiral structures. [Diagram by H. C. Arp, Hale Observatories.]

**Figure 3.** Messier 51, the great spiral in Canis Venatici.
[Photograph courtesy of Hale Observatories.]

*Hubble Atlas* (*2*), the finest collection of galaxy photographs, the arms are by no means smooth and continuous. There are often holes in them, and bifurcations, loops, and fringe structures occur in abundance. Even with the best of optical and radio telescopes, the task of analysis for astronomers living and working in our galaxy—or in certain sections of Messier 31, 33, 51 or 81—is certainly quite difficult! Paul W. Hodge and others have traced the distribution of HII regions in several neighboring galaxies. While, on the whole, their spiral patterns stand out with reasonable clarity, Hodge has found several in which supreme confusion seems to reign.

Messier 51 (Fig. 3) is one of the most beautiful nearby spiral galaxies with two developed arms. Seen almost face-on, it can be examined in great detail. Optically it shows very well how the most prominent spiral tracers—blue-white supergiant stars and their associated nebulosities—fix the overall spiral pattern. Figure 4 shows two negative photographs of Messier 51 and its companion. The photograph on the left shows beautifully the distribution of the hydrogen H-alpha concentrations, which clearly outline the basic spiral pattern and help to define the manner in which one of the spiral arms merges into the companion. The negative picture on the right repre-

sents a normal blue photograph of the galaxy. The two photographs show that the spiral arms contain mostly gas and young stars and that the luminous bridge connecting the main spiral and its companion has the same basic composition.

Beverly T. Lynds has studied the distribution of obscuring matter in Messier 51 and has concluded that HII regions most frequently occur at the outer edges of the continuous dust lanes. Her diagrams for the distribution of the dust are shown in Figure 5. The spiral galaxy Messier 51 has been one of the first to be studied by the Dutch radio interferometer, the Westerbork Array, with its mile-long series of 12 interconnected radio telescopes. Figure 6 shows the result of the Dutch studies at a wavelength in the radio continuum slightly shifted with reference to the 21-cm line. D. S. Mathewson, P. C. Van der Kruit, and W. N. Brouw find that the two strong continuous radio spiral arms coincide with the inner edges of the luminous optical arms. The radio spiral arms apparently follow very well the dust arms delineated by Lynds. To complete the picture for Messier 51, we should mention that R. B. Tully has made an optical study of the motions of the gas in its inner spiral regions and has found some clearly defined streaming characteristics, which are in accordance with

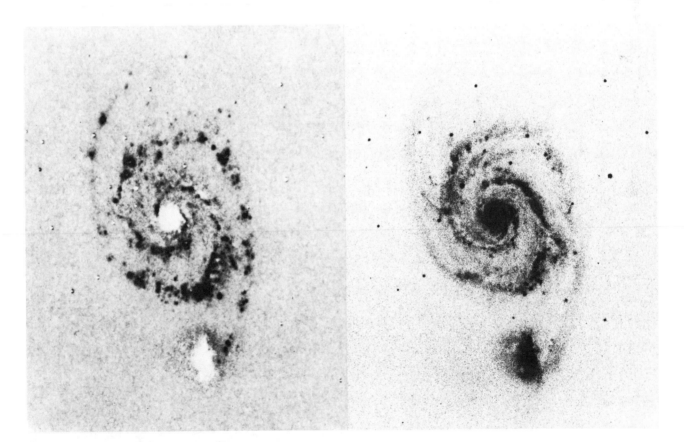

**Figure 4.** Two negative prints of Messier 51 (see Fig. 3). *Left,* a superposition of two photographs: the first is a negative print of M51 taken with an H-alpha color filter with a pass band of 100 Å; the second was taken with a comparable filter centered at a wavelength to the blue of H-alpha. The positive print of the second plate has been superposed on the negative print of the first. The result is a picture of M51 in the light of H-alpha only. The H-alpha spiral features appear very distinctly. *Right,* M51 is shown in normal blue light, for comparison. [Photographs by H. C. Arp, Hale Observatories.]

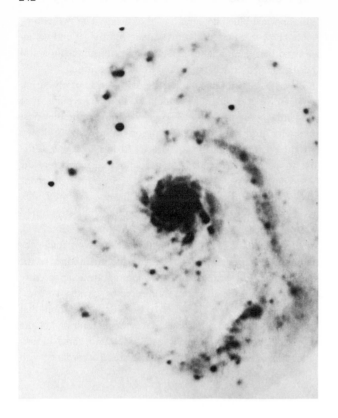

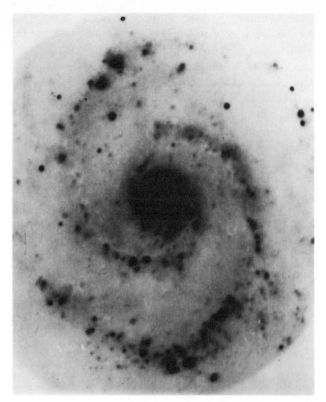

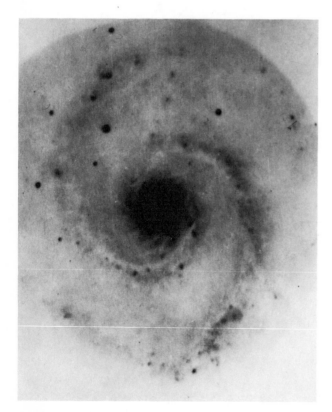

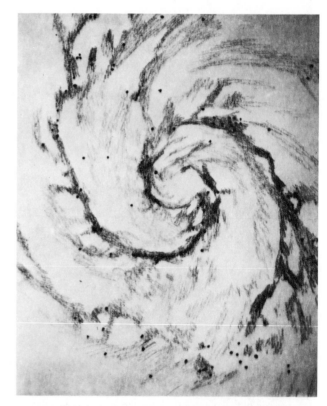

**Figure 5.** Obscuring matter in Messier 51. Beverly T. Lynds has prepared a series of photographs and drawings illustrating the relation between the lanes of obscuring matter and the spiral arms. The photographs at the top represent negatives of exposures through (*left*) a broad-band filter in the blue and (*right*) an H-alpha filter. The spiral arms are clearly shown. *Lower left,* a negative of a plate taken with a narrow-band red filter (eliminating H-alpha) centered at 6,650 Å. *Lower right,* a drawing based upon one of Milton Humason's long-exposure photographs with the 200-inch Hale reflector. Lynds has drawn the dark lanes by proper shading, showing the HII (ionized hydrogen emission) regions as black dots. The photographs at the top were made with the Steward Observatory-University of Arizona 90-inch reflector.

**Figure 6.** Radio emission from Messier 51. The ridges of 1415 MHz radio continuum emission radiation (solid lines) are shown superposed on Humason's photograph of M51 taken with the 200-inch Hale reflector. Note that the ridge lines of radio emission coincide nicely with the absorption lanes of Lynds shown in Figure 5. Dashed lines A–E show interarm radio links and F shows the link with the companion galaxy NGC 5195. [After a paper by D. S. Mathewson, P. C. van der Kruit, and W. N. Brouw.]

**Figure 7.** Messier 81, the spiral nebula in Ursa Major. This galaxy has an intricate pattern of young stars, HII emission regions, and dark matter. [Photograph courtesy of Hale Observatories.]

the predictions of the density-wave theory of Lin, Shu, and Yuan, to be described below.

While Messier 51 is one of the prize spiral galaxies on view, it possesses one structural feature that is very disturbing—the companion galaxy that marks the end of one of its two major spiral arms. Arp has provided good evidence to show that this galaxy is truly associated with Messier 51, and he has expressed the opinion that it was probably ejected from the nucleus of Messier 51 as recently as 10 to 100 million years ago. He considers this to be observational evidence for the theory of V. Ambartsumian, which will be discussed later in this article, according to which mass ejection from the nucleus may be the source of all spiral structure.

The student of galactic structure should always bear in mind the great variety in the spiral galaxies that have been photographed. While it is true that the overriding property of spiral galaxies is that they show two trailing arms (Fig. 7), many are quite different from this simple pattern. The nearby spiral Messier 101 (Fig. 8) is a multiple-arm spiral in its outer parts, and some of the more distant spirals show fireworks patterns with as many as six sections of arms or arm-like features. We should also bear in mind that there is a class of

barred spirals (Fig. 9), objects showing near-circular spiral arms that seem to emanate from a central bar. Some astronomers have suggested that there may be evidence for a bar structure in our own galaxy, but others—including the author— have not accepted this suggestion. The presence of an outer and inner ring structure appears to be quite common in spiral galaxies, with outer rings being shown beautifully in many barred spirals.

One of the most important lessons to be learned from an inspection of spiral galaxies is that the spiral phenomenon seems to possess a good deal of permanence. If it were a fleeting property, then we would not expect to find so many of our neighbors showing spiral structure of a very similar type. If the arms were in the habit of winding up, we would expect to see many more galaxies with tightly wound inner spirals than we seem to observe.

During the past year, J. H. Piddington of Australia has drawn attention to peculiar and interacting galaxies that do not fit into the basic two-arm spiral pattern shown in most of our illustrations. Almost 15 years ago, B. A. Vorontsov-Velyaminov, of Moscow, published a remarkable *Atlas and Catalogue of Interacting Galaxies* (*3*), and, 7 years ago, Arp

**Figure 8.**   Messier 101. A famous spiral galaxy in Ursa Major. [Photograph courtesy of Hale Observatories.]

**Figure 9.** NGC 1300, a barred spiral galaxy in Eridanus. [Photograph courtesy of Hale Observatories.]

brought out his very useful *Atlas of Peculiar Galaxies* (*4*). Piddington has expressed the opinion that the exceptions may be the rule and that there is no basic two-arm design among spiral galaxies. One might indeed be tempted to throw up one's hands in despair! To me, it seems more effective to describe and understand first the most commonly observed features and to interpret these as fully as can be done.

Piddington has attacked vigorously the assumption made by many, especially in the researches by the author of this article, that spiral features are long, connected streamers of hydrogen gas and young stars. He doubts that there is any marked concentration of interstellar gas associated with spiral features. I disagree, and I have good evidence to support my statements on the subject, not only from data of our own galaxy but also on the basis of data for several of the spirals illustrated. The Westerbork Array of radio telescopes is ideally suited to throw light on this question, and within the next few years the Dutch results should give us full information on the distribution of neutral atomic hydrogen in all nearby galaxies observable from Holland. These observations should go a long way toward settling the basic problem of the gaseous content of spiral features.

### Optical Spiral Tracers

Baade, and W. W. Morgan and his associates, considered the O to B2 stars to be good spiral tracers. These stars continue to hold this place of honor. They are often found associated with HII regions, which are gaseous emission nebulae shining brightly in hydrogen, H-alpha radiation. The stars responsible for exciting HII regions are single stars of spectral class O, or of classes B0, B1, or B2, or they are clusters or associations of O to B2 stars. A star of spectral type O to B2 is a blue-white giant or supergiant star which, according to present evolutionary theories, has been in existence for only a few million years, generally less than ten million years. Photography with image conversion tubes makes it possible to reach the faintest accessible stars in reasonably short exposure times. Intermediate and narrow band filters, now available for photoelectric and photographic work, prove to be a great help in obtaining magnitude data and precise colors, which make possible fair distance estimates.

The *cepheid variables* with periods greater than ten days are fair-to-good spiral tracers. They are supergiant stars that presumably have gone through the OB star stage, and they have probably also been red supergiants for cosmically brief intervals before entering the phase of long-period cepheid variability. Robert P. Kraft has called attention to their important potential as spiral tracers, stressing that they can be detected and studied to far greater distances than are within reach for O and B stars, even with the largest reflecting telescopes. As of now, it seems very likely that the most luminous of these cepheid supergiants, those with the longest periods, are among the youngest stars, and they seem to be the best potential spiral tracers. However, their numbers are not as large as those of the O to B2 stars.

*Dark nebulae* are as yet one of the least-studied groups of objects, but the time has come to use them more extensively. Good distance estimates must be made for many dark nebulae seen along the band of the Milky Way. Most distant nebulae are small objects found in very crowded regions where the scale of Schmidt-telescope photographs is not sufficient for careful and detailed study. Hence we must look to the large reflectors for precision work on them. Large telescopes now under construction almost always have provisions for wide-field, open-scale astronomical photography, and these instruments offer excellent opportunities for further research on galactic dark nebulae.

There are many special groups of stars, other than O and early B types, that are useful as spiral tracers. The *B-emission stars* are, according to Th. Schmidt-Kaler, among the best for this purpose. Lindsey F. Smith has shown that the hot emission-line *Wolf-Rayet stars* are also fairly good.

In all work relating to O and B stars, longest-period cepheids, and dark nebulae, we are beginning to pay attention to their relative distribution within specific spiral features. We noted already that the dark nebulae are often found concentrated along the insides of the spiral arms, and there are indications that the cepheids and O or B stars trace arms or spiral features that do not coincide precisely, for that matter, with the 21-cm features. At the inside of a typical arm we find the strongest concentration of gas and dust, which are the building blocks for stars not yet born. The youngest stars, and the nebulae that are made to shine by the ultraviolet radiation emitted by them, are concentrated along the central backbones of the spiral arms. In a way we are looking for the equivalent of "tree rings" inside spiral arms.

*Supergiant stars* of all varieties, from the blue-white OB supergiants to the reddest M supergiants, are exceedingly useful spiral tracers. The discovery of such stars from objective prism spectral plates for both the northern and southern Milky Way is not a difficult task, even for faint and distant stars; any star of interest can now have its spectrum photographed in intimate detail. Most of our larger spectrographs are fitted with image conversion equipment that makes it possible to obtain high-quality spectra of fairly high dispersion in relatively short exposure times, even for quite faint stars. From these spectra one can make estimates of absolute magnitudes for the stars under investigation, and radial velocities of most can be determined with precision. In recent years Roberta M. Humphreys has been especially active in this area. Her results have demonstrated that the supergiant stars may be the key by which we can penetrate the great distances of many sections of the Milky Way. Prominent among the supergiants are the longest-period cepheid variable stars. Radial velocities for southern cepheids determined by M. W. Feast are contributing basic data for the study of spiral kinematics.

It seems established that *galactic magnetic fields* are not the major controlling fields of force producing the observed spiral patterns. Such fields must, however, have considerable organizing influence on details of spiral structure. The observed *polarization effects* in the light of distant stars prove that mag-

netic fields are associated with spiral features. The observed percentage polarizations are generally largest, and the alignment of the polarization vectors appears to be most regular, when we view a star across a spiral arm; small percentage polarizations, combined with haphazard orientations of the vectors, are observed when we look longitudinally along a spiral arm. Hence, present indications are that polarization effects are helpful in tracing spiral features, even though we are far from understanding just what roles large-scale magnetic fields play in theories of spiral structure.

We now possess extensive and precise material on the strengths and orientations of the polarization vectors for stars along the entire Milky Way, north as well as south. The original data, which had been contributed by the discoverers of interstellar polarization, W. A. Hiltner and J. S. Hall, have been supplemented by new catalogues by D. S. Mathewson, of Mount Stromlo Observatory, and by G. Klare and Th. Neckel, of Hamburg. Figure 10 shows the Mathewson polarization map. We saw earlier that continuum studies on the distribution of thermal hydrogen radiation at short radio wavelengths yield much useful information for studies of radio spiral structure. Synchrotron radiation, which dominates at longer wavelengths in the radio range, has been helpful in defining certain edges in the longitudinal distribution of radiation emitted by free electrons of high energy moving under the influence of galactic magnetic fields. These directions are identified as ones oriented more or less lengthwise along a spiral feature. Pulsars and X-ray sources have been suggested as potential spiral tracers; however, their precise status is still very much in doubt, and nothing definite can be said about them at this time.

Clouds of interstellar gas produce *interstellar absorption lines* in the spectra of distant stars, the best known being the interstellar K-line, produced by ionized calcium, and the D-lines, from neutral sodium. The most extensive work in this field has been done by Guido Münch for stars within reach from northern latitudes; the southern Milky Way still awaits full exploration. High-dispersion spectra often show multiple interstellar absorption lines, and for each observed component we can readily measure the radial velocity.

The interstellar gas clouds are presumably most concentrated in and along the spiral arms, and the radial velocities found for the interstellar absorption lines make it possible to pinpoint their places of origin. To derive the distances to individual clouds, we require a model of rotational velocity for the galaxy that transforms these radial velocities for each particular direction in the galactic plane into distances from the sun and, hence, into distances from the galactic center. However, it does help to know that the interstellar cloud producing the absorption lines must lie between the sun and the star in whose spectrum it is observed.

The interplay between kinematical and structural observations promises to become increasingly important in future years. Very often we have extensive data from the spectra, magnitudes, and colors of the stars, star cluster, or association responsible for a given HII region. We can then derive the distance of the cluster or association on the basis of known absolute magnitudes and colors of embedded stars and from

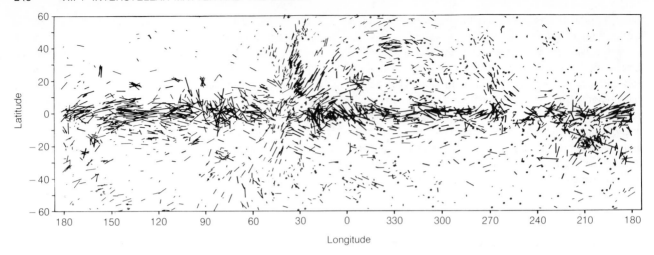

Figure 10. Optical polarization. This diagram summarizes all available data on interstellar galactic polarization. The plot shows the E-vectors of polarization and represents data for 7,000 stars. Measurement for 1,800 stars was made by Mathewson at Siding Spring Observatory; the remainder are from the catalogs by Hall, Hiltner, Behr, Lodén, Appenzeller, Visvanathan, and E. van P. Smith. *Small circles*, positions of stars with percentage polarization P < 0.08; *thin lines*, vectors for stars with 0.08 < P < 0.60; *thick lines*, vectors for stars with P < 0.60. [From a paper published by Don S. Mathewson in the *Memoirs of the Royal Astronomical Society*.]

available knowledge of the amount of intervening absorption.

If we can measure as well the radial velocities for the stars and the associated gas, we may then proceed to check whether or not the observed radial velocities of the gas clouds agree with the values predicted from an accepted model of rotational velocities in our galaxy. A difference between the observed and predicted velocities will suggest that either there is something wrong with our basic velocity model or, alternatively, it will give an indication of local streaming. We approach the time when we can say that our basic velocity model is reasonably well established, and we can then use any observed differences as indicators of large-scale streamings. Stellar and nebular optical radial velocities have shown in many places the presence of large-scale regional departures from circular motion, which suggest large-scale streaming of gas.

We need optical radial velocities for HII regions and their associated stars to complement those obtained by radioastronomical techniques. In recent years the results of many studies of radial velocities for HII regions have been published, most notably by G. Courtès, Y. P. and Y. M. Georgelin, G. Monnett, and P. Cruvellier, of Marseille Observatory, and by J. S. Miller and M. J. Smith, associated with Kitt Peak National Observatory and Cerro Tololo Inter-American Observatory. Without establishing any distance scales, we can find for many directions in the galactic plane whether or not identical radial velocities are obtained from radioastronomical and optical studies of the same gas clouds. In addition, we may compare radial velocities from gas clouds with those from associated stars or star clusters.

## Radio Structure

Radio astronomy is contributing mightily to the study of spiral structure in our galaxy, the most important information being still obtained through the study of the 21-cm radio spectral line of neutral atomic hydrogen (HI). The transition takes place near the lowest level of the neutral hydrogen atom, which emits 21-cm radiation in detectable amounts from regions of low-temperature gas. Early in 1951, Harold Ewen and Edward Purcell, following up Hendrik van de Hulst's suggestion, discovered the 21-cm line. Thus they found a way to pinpoint hydrogen gas located not only in our local spiral arms but also in very distant spiral arms, most of which are hidden from the optical astronomer's view by thick cosmic dust clouds.

The basic gathering and analysis of observational material from a 21-cm line survey of the distribution of neutral atomic hydrogen in the galaxy follows a rather simple pattern. First we select the section of the Milky Way that we wish to study. For example, it may stretch from galactic longitude 275° to 305° and cover the zone in galactic latitude from −10° to +10°. We establish within this area a fine network of points. Along the galactic equator we may place these positions $\frac{1}{2}$° or 1° apart in galactic longitude, and we may have rows of additional positions, perhaps not quite so tightly spaced, at every degree of galactic latitude from −10° to +10°.

When the network of points is as close as indicated, we should have access to a radio telescope with an angular resolution of about $\frac{1}{2}$° or better; a 100-ft aperture radio telescope with a surface precise to 1 cm or better serves very well for the purpose. We should use this instrument with receiver equipment of high frequency resolution, so that features in the line profile separated by frequency differences corresponding to a radial velocity of 1 or 2 km/sec will stand out clearly as separate features.

The observation at each of these positions consists of pointing the radio telescope in the right direction, something that must be done with high precision, and then recording the profile for the 21-cm line. This is generally done with multichannel receiving equipment, in which each channel successively records the intensity of the radiation in a narrow band of frequencies. As we noted, the resolution in frequency must be sufficiently great so that features in the line profile with radial velocity differences of 1 or 2 km/sec stand out clearly as separate features. The observation for each position consists basically of a trace in which the intensity of the hydrogen 21-cm radiation is plotted as a function of frequency. Since the profile for a cloud of hydrogen gas at rest would be very narrow indeed, each frequency in the observed profile corresponds to a certain radial velocity of approach or recession for the cloud. The rest frequency of 21-cm hydrogen radiation is very precisely known, and it is, therefore, not difficult to translate each observed frequency into a radial velocity of approach or recession for the cloud to which the observation refers. Hence the final data for each center can be shown as a plot of intensity of 21-cm radiation vs radial velocity. A typical profile is shown in Figure 11.

Once the network of profiles has been obtained, the radio astronomer faces the difficult problem of assigning approximate distances to the various features in the profile that occur at certain radial velocities of approach or recession. In the original interpretation of such profiles by radio astronomers in Leiden and in Sydney, the assumption was made that there exists a single, well-defined rotation curve of circular velocities applicable to all of our galaxy. The basic curve, derived from a dynamical model of the galaxy by Maarten Schmidt, is the most dependable one available; it relates circular velocity of rotation and distance from the galactic center, making it possible to fix for any direction of galactic longitude precisely what the distance would be of a cloud that showed a certain radial velocity of recession or approach. This follows, assuming of course that all hydrogen clouds move around the center of our galaxy with the precise circular speeds assigned to them by the basic rotation curve.

However, it soon became clear that the situation is not so simple. First, not unexpectedly, the clouds show motions of their own, mostly of the order of 5 to 10 km/sec. This may cause us to assign an erroneous distance to a gas cloud on the basis of its observed radial velocity. Wholly apart from problems of streaming and random motions, we encounter grave difficulties in our attempts to fix a good basic rotation curve. At one time it was even suggested that there are two basic rotational curves, one applying to the parts of the Milky Way visible from northern latitutdes, another for the sections observable from southern latitudes. By now, there is pretty general agreement that the differences in the two curves are caused mostly by large-scale streaming effects. Such effects are predicted by the Lin-Shu density wave theory, which is described in the section that follows.

The best available approach to the study in depth of the neutral atomic hydrogen (HI) distribution for a section of the

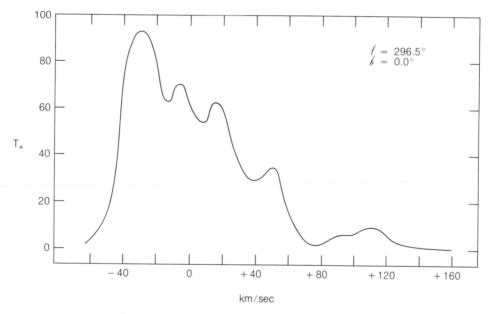

**Figure 11.** A typical 21-cm (HI) velocity profile, this one for a center in the southern Milky Way at galactic longitude 296.5°. The abscissa gives the radial velocities of the HI gas. The velocities have been corrected for the effects of the sun's local motion. The ordinate gives the antenna temperature, $T_A$, in degrees Kelvin. If we assume that all the peaks are due to concentrations of HI (not to velocity crowding effects), then we find for the HI clouds at l = 296.5° the following radial velocities: −30 km/sec, −7 km/sec, +15 km/sec, +55 km/sec, and +114 km/sec. The Schmidt velocity model would then predict the following kinematical distances for the five clouds: (1) the −30 km/sec gas is probably a streaming effect since this negative velocity does not correspond to any model velocity for l = 296.5°; (2) the −7 km/sec gas is 1.5 kpc from the sun; (3) the +15 km/sec gas is 9 kpc from the sun; (4) the +55 km/sec gas is 12 kpc from the sun; (5) the +114 km/sec gas is 17 kpc from the sun.

Milky Way is to assume various reasonable models for the clouds' density and random velocity distribution. If we make use of the best available gravitational rotation curve for the galaxy, we can then predict the profiles for each model at each galactic longitude. The model that yields the closest representation of the observed profiles in the section presumably has the best chance of being closest to the truth.

Probably the most important lesson to be learned from our current difficulties with the analysis of 21-cm profiles is that, to proceed with some degree of confidence, we shall need to analyze together the optical and radio data for each section of the Milky Way. It is only from optical studies that we can obtain independently distances to features that are associated with spiral structure. In other words, the optical data give us the anchor points that will determine the distance scale. The 21-cm profiles can then give us exceedingly useful supplementary information about the distribution of the hydrogen gas along the line of sight.

To show the present state of results obtained from 21-cm analysis, we can probably do no better than to refer to three of the more recent diagrams of the spiral structure of our galaxy. Figure 12, a diagram presented by Kerr and published in my 1967 *American Scientist* article, is based on his joint analysis with Westerhout of the radio data available for the northern and southern Milky Way. Figure 13, by W. Becker and Th. Schmidt-Kaler (also reproduced in my earlier article) presents optical data to support spiral patterns. Figure 14, a working diagram recently prepared by Harold F. Weaver, is based on new observational data of the spiral structure exhibited by the northern Milky Way and by a small part of the southern Milky Way. Since the section of the southern Milky Way between galactic longitudes 250° and 0° is inaccessible from California, Weaver left the fourth quadrant blank. We see from the two diagrams that there are good areas of agreement and some marked areas of disagreement. The overall picture presents a rather tightly wound pattern of spiral features leading to trailing spiral arms with average pitch angles in the range of 5° to 12°. The pitch angle is defined as the angle between the tangent to the spiral feature and the direction of circular motion.

The radio equivalents of optical interstellar absorption lines have been used effectively in attempts to study neutral atomic, HI, features and other radio absorption features. The work of W. Miller Goss and of Frank J. Kerr and Gillian R.

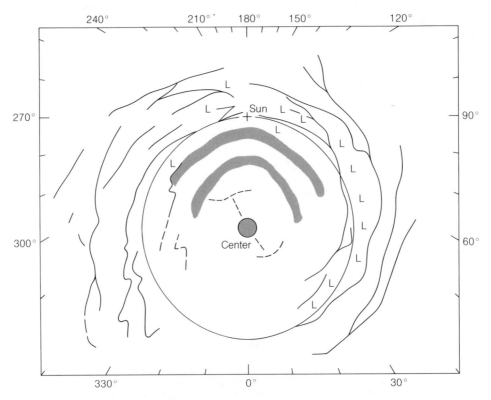

**Figure 12.** Radio spiral structure of our galaxy, 1967. This diagram was prepared by F. J. Kerr from 21-cm profiles observed at southern latitudes with the 210-ft radio telescope at Parkes, Australia (observations by Kerr and J. V. Hindman), and by A. P. Henderson from profiles observed from northern latitudes by G. Westerhout with the 300-ft radio telescope at Green Bank, West Virginia. The distribution in the inner parts is only roughly sketched in the diagram. The most noteworthy feature is the near-circular patttern of spiral structure that emerges, with pitch angles of the order of 5°. The trough of low hydrogen density that can be traced from galactic longitude 20° to 80° is another remarkable feature. Regions of low hydrogen density are indicated by *L*. [Diagram from the University of Maryland.]

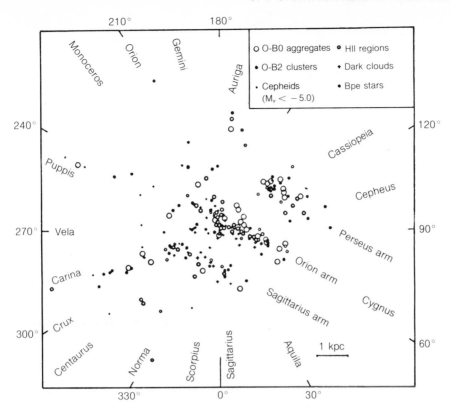

**Figure 13.** Optical spiral structure of our galaxy. The sun is at the center of the diagram and the galactic center is in the direction toward 0° galactic longitude at a distance of 10 kpc from the sun. The principal observed sections of the Sagittarius, Orion, and Perseus arms are shown. The directions from the sun toward some of the key constellations along the band of the Milky Way are marked along the periphery of the diagram. [Original diagram by W. Becker and Th. Schmidt-Kaler, 1964. Figures 12 and 13 were also published in (*1*).]

Knapp is providing much useful information. The HI 21-cm absorption features that they have observed assist us in assigning improved distances for objects in the radio spectra, since any identified absorbing cloud must lie between the object and the sun and earth.

Radio astronomy has developed several techniques for the study of HII regions. First came the techniques that make use of the effects produced by free-free transitions. These occur in regions of ionized hydrogen, where the passage of a free electron close to a positively-charged nucleus of the hydrogen atom, the proton, produces continuum radiation in the centimeter and decimeter range. The observed strength of the radiation depends on the density of protons and on the speeds at which the free electrons move; these speeds depend, of course, on the temperature of the free-electron gas. Continuum radiation was first observed by radioastronomers in the early 1950s, and it has been mapped for the entire Milky Way.

The second way in which ionized hydrogen can be studied by radio techniques is through the so-called alpha transitions, a method first suggested by N. S. Kardashev, of Moscow. In regions with much ionized hydrogen, the capture of a free electron by a proton may well take place in one of the very high energy levels of the neutral hydrogen atom. After capture, the electron will not stay in this high level for more than a

minute fraction of a second; it will then cascade down toward levels of lower energy, ultimately ending up in the level of lowest energy, the Lyman level. The first transitions in the cascading process may well be from, say, level 158 to 157, or 110 to 109, the frequency of each transition depending upon the probability of capture in the higher level multiplied by the probability of a subsequent transition to the next level of energy.

Many high-level transitions fall within the wavelength and frequency ranges of modern radioastronomical receiving equipment. Since the emission lines should be quite sharp, any shift in frequency between the observed and rest frequencies may be attributed to the radial velocity of approach or recession of the gas cloud in question. Thus we not only have a tool for detecting HII regions but we are also in a position to measure the radial velocities of the gas clouds. As with the 21-cm line, these radial velocities may be translated into distances from the sun for any particular direction, and we can thus plot the positions of the gas clouds in the galactic plane. Strictly speaking, we observe in this case an effect produced in neutral hydrogen atoms. However, it is the direct aftereffect of the capture by a proton of a free electron; these captures at high levels, and the subsequent alpha transitions, are observed only in regions of high ionization, hence

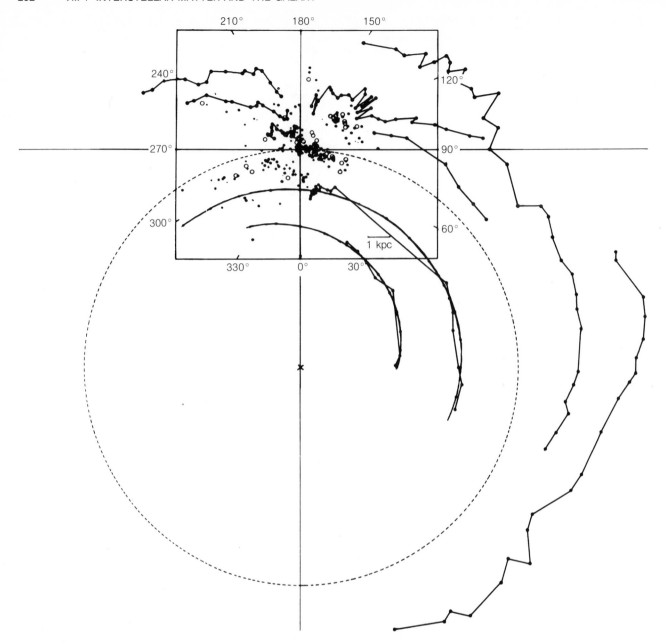

**Figure 14.** A preliminary map of radio spiral structure. Harold F. Weaver made this radio map on the basis of the Hat Creek 21-cm survey. It represents only the preliminary analysis (1972) of an extensive body of basic data. For comparison, Weaver has entered the optical data (see Fig. 13) of W. Becker and Th. Schmidt-Kaler. The probable extensions of the observed features to the southern Milky Way (unobservable from Hat Creek Observatory, in California) have been sketched by Weaver.

in HII regions. Alpha transitions* thus serve to pinpoint distant HII regions.

Entirely new radio possibilities are now becoming available through the study of molecular radiations. One of the most remarkable developments of the past decade has been the detection and study of radio emission and absorption lines produced largely by interstellar molecules associated with cosmic dust. The discoveries have been so plentiful (close to 30 molecules have been detected already) that it is becoming possible to undertake surveys of the distribution and kinematics of some of the most widely distributed molecules. For example, surveys for formaldehyde and carbon monoxide are now quite feasible, and they should make possible the further tracing of distant spiral features. Data gathering for our own galaxy is very much in progress.

### Gravitational Theories

Ten to fifteen years ago, it was widely hypothesized that galactic spiral structure is caused and maintained by large-

---

*These lines belong to the class commonly known as *radio recombination lines.*—Ed.

scale magnetic fields, which were supposed to align the gas in spiral patterns. However, it was then discovered that because the longitudinal magnetic fields average in all probability only two or three microgauss in strength, they appeared too weak to produce the major controlling effects that are required. It is more likely that gravitational forces are responsible for the observed overall spiral patterns and that the magnetic fields are frozen into these patterns as a by-product of the distribution of the ionized gas, mostly hydrogen in proton form. Gravitational concentration of this ionized gas may locally produce magnetic fields of unusual strength. The behavior patterns of polarization vectors like those in Figure 10 can thus be considered a result of the concentration of ionized gas in spiral features.

Bertil Lindblad had for many years been the lone champion of gravitational theories to explain spiral structure. In the middle 1960s, C. C. Lin and his associates Frank H. Shu and Chi Yuan began a series of studies according to which the observed spiral features would be produced by a density wave passing through the interstellar gas. This was the beginning of the famous Lin-Shu density wave theory.

Lin starts out by assuming that there is a possibility that the field of gravitational potential near and in the central plane of a highly flattened, rotating galaxy can have a spiral component. (We must make it clear that, in a way, this represents an ad-hoc hypothesis.) He found that gas and stars moving with near-circular velocities can, because of the effects of this spiral component, be subject to density waves, with the gas and the associated stars piling up for shorter or longer times in the gravitational spiral-shaped troughs of low potential. In other words, gas and young stars linger longer in the spiral-potential grooves than in between, and thus material in the arm is constantly being replenished.

Lin investigated for various types of spiral patterns in the gravitational potential whether or not this piling up is a transitory phenomenon, and he found certain strict conditions that must be satisfied if the spiral potential field is to be a semipermanent one. If it is to persist, then it must be sustained by the gravitational effects produced by the stars and gas that participate in the patterns of spiral behavior.

Lin's theory predicts that in most highly flattened galaxies there will be a region in which spiral structure can prevail. For our own galaxy this region lies roughly in the range of distances between 5,000 and 15,000 parsecs from the galactic center, which means that our sun is, as it should be, right in the part of the galaxy where spiral structure may be expected. The density wave theory predicts that the only stable and semipermanent spiral configurations will be those showing two trailing spiral arms emanating from opposite sides of the nucleus. Further, it predicts quite precisely the spacing between the arms and their pitch angles. The latter are determined by the nature of the underlying force field of the galaxy. The predicted pitch angles are about 6°, very much in line with the radio-astronomical results of Kerr, but out of line with the Becker-Morgan results. A typical calculated spacing between arms separates them by 3,000 parsecs, which is in fair average agreement with what we find in our galaxy. Lin's theory predicts some large-scale streamings of gas and young stars, with average stream velocities possibly as high as 8 to 10 km/sec relative to pure circular motion. As we shall see later, such deviations have actually been observed and are of the right order. The observed deviations from pure circular motion fit quite well into the pattern of the streamings predicted for our galaxy by Lin's theory.

In the earlier picture that most of us had (with the notable exception of Bertil Lindblad), the spiral arms were considered fixed loci of objects such as clusters, associations, and emission nebulae. It seemed inevitable that, after one or two galactic revolutions, they should wind up, since the outer objects would move at slower speeds than those closer to the nucleus. This problem ceases to exist if one accepts Lin's view that spiral arms do not contain the same stars and gas indefinitely, that they are the product of density waves, and that the only semipermanent feature is the underlying spiral component of gravitational potential. We should note that the spiral pattern of gravitational potential in Lin's theory does not rotate as fast as the galaxy for regions near the sun. Hence we find that the instantaneous spiral arms do not persist, and the locus of maximum spiral features shifts with time and with respect to existing spiral arms.

There is, as yet, no good theory to explain how the spiral potential fields of force come about in the first place. They rotate in the central plane of the galaxy, resembling in some ways a two-armed boomerang! In the outer parts of the spiral structure, the galaxy and the field move at about the same speed, but in the inner parts, the boomerang does not move as fast as the stars, the cosmic dust, and the interstellar gas that go around the galactic center at the speed of the general galactic rotation.

The stars are not bothered very much by the spiral potential field. However, as noted, in troughs of low potential the gas will have a tendency to pile up, slow down temporarily, and then move on at the regular rotation speed, leaving the new gas behind it to go through the same performance. A spiral feature will be observed stretched along the curve of greatest density, and, in the course of repeated galactic rotations, the gas will pass several times through these potential troughs, thus producing a steady spiral pattern of constantly varying gas clouds.

In Lin's picture, at the sun's position, the spiral potential field, or boomerang, moves at a rate a little more than half that of the normal galactic rotation. Agreement between rotational and pattern speed is found at a distance of 14,000 to 15,000 parsecs from the center of the galaxy. It will be noted that the shape of our boomerang fixes the shape of the spiral arms, and since the boomerang form does not change, the spiral arms should have the same appearance for as long as the density-wave pattern persists. There is no winding-up problem in Lin's theory. It is worth noting that in our galaxy the outer ring structures of neutral atomic hydrogen are found approximately at the positions where the pattern moves at the normal rate of galactic rotation. Studies of the stability of the density-wave patterns have shown that a two-armed trailing spiral boomerang of potential has the best chance of surviving for several galactic revolutions.

The predictions of Lin's theory can be tested by mathematical calculations relating to the behavior of the wave pattern, and they can also be checked through modern computer calculations for stellar systems with large numbers of gas clouds or with hundreds of thousands of stars. Phenomena similar to density waves have been produced through computer calculations by Per Olaf Lindblad in Sweden (Bertil's son) and by Richard H. Miller, Kevin H. Prendergast, William J. Quirk, and others in the USA.

An associate and former student of Lin, William W. Roberts, has studied in some detail the processes that take place when the gas passes through the regions of lowest potential—that is, through the strongest part of the boomerang. He has found that sudden compression of the gas will take place and that galactic shock waves will be produced along a wide front. Narrow lanes of very high density gas and interstellar dust can be expected, and it seems as though star formation might well take place under these conditions of high compression. The dust lanes would be at the inside of the spiral feature, and the stars would migrate outward from it, with the youngest stars located closest to the shock front and the slightly older ones a bit farther removed. Star formation from interstellar gas and dust is a very complex phenomenon, and the passing of a density wave through the gas and dust alone cannot guarantee that it will take place on an extensive scale. However, if other conditions are right, then the density wave may well provide the required trigger action.

While we should not leave the false impression that the Lin-Shu theory is the accepted gospel in spiral structure, it does represent a major breakthrough in a field where theory had made little headway. Its greatest asset to observing astronomers is that it presents firm predictions for the behavior, distribution, and kinematics of interstellar gas, cosmic dust, and young stars in and near spiral features. The most severe of Lin's critics has been the Australian physicist J. H. Piddington, who favors complete rejection of the density wave theory and especially of the shock-front hypothesis contributed by Roberts. Lin and Shu have firmly rejected Piddington's criticisms. In place of the Lin-Shu theory, Piddington suggests a hydromagnetic theory of his own making, but its details are not now available.

All workers in the field realize that there are potentially many theoretical and observational approaches to the study of spiral structure. The density wave theory does not exclude the presence of important effects produced by large-scale galactic magnetic fields, and it is quite reasonable that the importance of such fields be fully explored. Nuclear ejection of the type favored by Ambartsumian and Arp may provide a basic motive force for initiating and maintaining spiral structure. The researches of A. and J. Toomre have shown that encounters between galaxies may produce calculable spiral structure-like phenomena. And, in the end, the barred spirals may well provide the major key to an understanding of spiral structure. There is obviously no such thing as a simple theory of spiral structure, nor is it feasible to describe in a general way all the complex phenomena shown by spiral features in our own and other galaxies.

Since the late 1950s, Ambartsumian has urged astronomers to look to the central regions of our own and other galaxies for the impetus for the spiral structure we observe in the outer parts. High-energy jets of ionized gas may enter the outer parts of the galaxy after having been generated by giant explosive phenomena inside and very near the galactic nucleus. The principal regions of impact may be at 3,000 to 5,000 parsecs from the center, at which interface we may expect to find clouds of relativistic electrons entering into the more peaceful sections of the galactic disk. It is at 5,000 parsecs from the center of our galaxy that we find the great ring of giant HII regions, noted by Peter Mezger, Gart Westerhout, and others. Large-scale magnetic fields may play more of a role than we have assigned them in our treatment.

## Spiral Structure of Our Galaxy

It would be pleasant if we could offer for the end of this article a good diagram for the spiral structure of our galaxy. Alas, we cannot do so at the present time. We do not yet possess for the galaxy the Grand Design which C. C. Lin began to seek a decade ago.

Why are there differences between the Kerr-Westerhout and the Weaver diagrams of radio spiral structure (Figs. 12 and 14)? One reason is the different interpretations used by the analysts for those sections of the Milky Way for which sufficient 21-cm data are not yet available. Other differences arise through variations in the method of analysis of the basic 21-cm profiles and because of the use of different velocity models for the galaxy. One of the most important sources of discrepancies appears to lie in the manner in which different authors connect the major neutral atomic hydrogen concentrations, on which there often is agreement, into a single all-inclusive spiral pattern.

The optical astronomer is faced with equally difficult problems in the analysis of his material. For example, most of us working in the field are agreed that there is a major concentration of young stars, gas, and dust in the direction of the southern constellation of Carina (Fig. 15), and other spiral-like features have been studied in Cygnus, in Orion, and in Norma. The Perseus arm is well established over a considerable range of distance beyond the sun, and the Sagittarius arm is equally well delineated at distances from the galactic center 2,000 parsecs less than that of the sun. Becker and associates tend to interconnect these features into an overall spiral pattern with a pitch angle of 25°, whereas to Bok and others it has seemed more reasonable to connect these same features into a pattern with a pitch angle of 6°, more or less.

The game of connecting recognized radio and optical spiral features into an overall spiral pattern for our galaxy will undoubtedly continue in the years to come. In the end, the basic pattern will probably be revealed by analyses of radio astronomical data, especially 21-cm line data. However, the analyses will by no means be straightforward. It is relatively easy to obtain definitive results relating to neutral atomic hydrogen concentrations in various directions of galactic longitude and latitude, but, for each direction, these concentrations will be observed as coming from clouds that have specific observed radial velocities of approach or recession. We shall require an intricate model of galactic rotation and of streaming motions applicable to the gas clouds in the galaxy before we can trans-

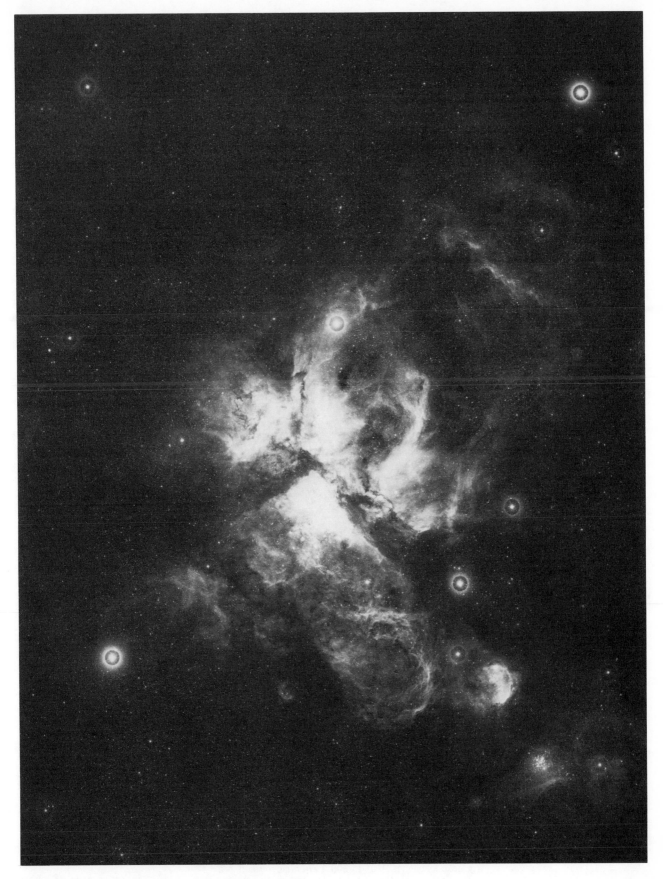

**Figure 15.**   The Eta Carinae Nebula. [Photographed from the
Cerro Tololo Inter-American Observatory, La Serena, Chile.]

late, for each direction, the observed velocities into distances from the sun.

To make life even more difficult, we encounter certain disturbing, purely geometrical effects in galactic 21-cm analysis that complicate the situation. These have been pointed out by W. Butler Burton and Whitney W. Shane. At some galactic longitudes, the radial velocity of approach or recession changes very slowly with distance. This has a very disturbing consequence for the 21-cm profile at that galactic longitude. Even if the distribution of the neutral atomic hydrogen were perfectly uniform along the line of sight, there still would be an unduly large amount of it in the direction that possesses a radial velocity close to the critical value. At the critical radial velocity, the signal builds up in intensity, thus becoming abnormally strong for that particular direction. The consequence of this simple geometrical effect is that we get a strong signal that, at first sight, may be wrongly attributed to radiation from a cloud of neutral atomic hydrogen. Thus, velocity crowding is one of the most serious concerns of the radio astronomers studying the distribution of neutral atomic hydrogen in our galaxy.

We noted that it is quite possible to isolate and study certain definite spiral features. At Steward Observatory in recent years, B. J. Bok, Roberta M. Humphreys, Ellis W. Miller, and others have made comprehensive studies of the distribution of gas, cosmic dust, and young stars for the Carina section of the Milky Way. The optical data alone, especially studies of OB stars by John A. Graham, show that in Carina we are probably looking edgewise along a major spiral feature.

Graham showed that the sharp outer edge of this feature is at galactic longitude 283°, and there is evidence for some sort of an inner edge near galactic longitude 300°. The feature has been traced optically over a distance of 7,000 parsecs, starting at about 1,000 parsecs from the sun and stretching to at least 8,000 parsecs from it. Figure 16 shows the diagram produced by Miller. The same sort of studies have been made for a feature in Cygnus, where Hélène Dickel, H. J. Wendker, and J. H. Bierritz have studied the distribution and distances of 90 HII regions, all associated with a Cygnus spiral feature stretching to at least 4,000 parsecs from the sun in the direction of galactic longitude 75°. Sections of the Perseus arm are equally well delineated, and the inner Sagittarius arm is also clearly marked.

These detailed studies are exceedingly useful for an understanding of the properties of the spiral arms of the galaxy. For a feature like that observed in Carina, one may obtain the distribution of dust and of gas and young stars of various ages across the spiral feature, and thus hope to arrive at useful conclusions about the processes of star birth. The Lin theory predicts certain systematic motions at the inner and outer edges of spiral arms. Observational evidence for deviations from pure circular motion associated with the Carina spiral feature has been obtained by Humphreys, who finds velocities of the order of 6 to 8 km/sec against galactic rotation on the inner side of the spiral feature and velocities of comparable amounts with galactic rotation on the outer side. This is of the amount and in the sense predicted by the Lin-Shu theory.

For the Perseus feature some very interesting phenomena

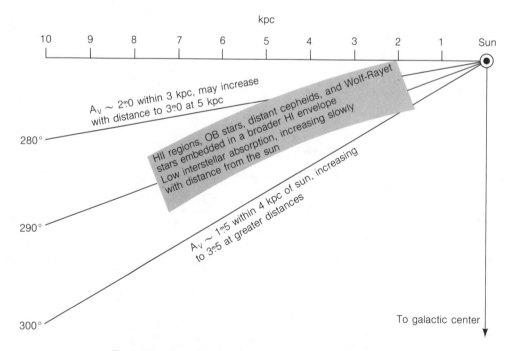

**Figure 16.** A working diagram of the Carina spiral feature. Ellis W. Miller prepared this diagram to show how objects in the southern constellations of Vela, Carina, Crux, and Centaurus (galactic longitudes 270° to 300°) are concentrated in a curved spiral feature that can be traced between 2 and 8 kpc from the sun. The interstellar obscuration reaches its greatest value on the inside of the feature, but there is marked obscuration on the outside as well. The center of the galaxy is below the sun outside the diagram at a distance of 10 kpc from the sun.

**Figure 17.** Messier 104, the spiral galaxy in Virgo, seen edge-on. [Photograph courtesy of Hale Observatories.]

have been found. Some of the gas clouds associated with the Perseus arm appear to be rushing toward us at a rate of 20 to 30 km/sec, while the stars appear to move more or less in the circular orbits expected from basic galactic rotation.

By continuing our attempts to present diagrams showing the overall spiral structure of the galaxy, we shall try to obtain the Grand Design. However, in the meantime, much valuable research can be done on the detailed properties of isolated spiral features.

The happiest aspect of present-day studies of galactic spiral structure is that the problems are being studied simultaneously from many angles. Theorists and users of large computers have their pencils sharpened and are hard at work; observing optical and radio astronomers consult and argue with each other uninterruptedly; and astrophysicists are probing as best they can into the physical conditions of interstellar matter in spiral arms and the regions between them. The full study of the spiral structure of the galaxy is providing a stimulus for astronomical research on all fronts.

**References**

1.  A. R. Sandage. 1961. *The Hubble Atlas of Galaxies.* Washington, D.C.: Carnegie Institution of Washington.
2.  B. A. Vorontsov-Velyaminov. 1959. *Atlas and Catalogue of Interacting Galaxies,* Vol. 1. Moscow: Sternberg Astronomical Institute.
3.  H. C. Arp. 1966. Atlas of Peculiar Galaxies. *Astrophysical Journal* (suppl.), Vol. 14, No. 123.
4.  W. Becker and G. Contopoulos, Eds. 1970. *The Spiral Structure of Our Galaxy.* International Astronomical Union Symposium No. 38. Dordrecht: D. Reidel.
5.  Cornelis de Jager, Ed. 1971. *Highlights in Astronomy,* Vol. 2. (14th General Assembly of the International Astronomical Union.) Dordrecht: D. Reidel.

# 32

# Interstellar Grains in Museums?

A. G. W. Cameron

Abstract. *It is argued that carbonaceous chondrites, particularly of type I, are probably collections of interstellar grains which have been mildly transformed through exposure to higher than normal temperatures, resulting in a loss of volatile materials.*

During this conference there have been many discussions of methods for determining the properties of interstellar grains by means of electromagnetic observations and by investigation of the properties of possible laboratory prototypes. It has not been realized that it may be possible to study interstellar grains directly within the laboratory.

Theoretical investigations of the star formation process indicate that the collapse of an interstellar cloud is likely to take place under very low temperature conditions. The calculated temperature in the interior of a collapsing gas cloud is about 10 K (Larson, 1969). The temperature in the gas starts to rise only when the gas density becomes high enough so that the gas becomes opaque to the transmission of its own radiation, and the compression goes over from an isothermal stage to an adiabatic stage. M. R. Pine and I have recently constructed models of the primitive solar nebula which may have resulted from such an interstellar collapse process (Cameron and Pine, 1973). I believe that these models are sufficiently general in indicating that the temperature will not rise high enough to evaporate completely the interstellar grains, contained within the gas, beyond about one or two astronomical units from the central spin axis of the primitive solar nebula (Cameron, 1973). The important aspect of this conclusion is that small bodies formed at very much larger distances from the central spin axis are likely to be composed of collections of interstellar grains whose properties have been transformed by varying degrees. Naturally, if we are interested in the laboratory examination of interstellar grains, we must seek material which has suffered the least rise in temperature during the adiabatic compression of the interstellar gases.

Ideally, it would be nice to be able to examine a piece of a comet, which has probably formed at very great distances from the central spin axis of the primitive solar nebula, and in which the temperature has always stayed very low. Unfortunately, it is unlikely that we will obtain any sample of cometary material for some time to come, and therefore we must seek the next most primitive material.

This material is presumably carbonaceous chondritic material. Studies of elemental abundances in such meteorites, as well as of the ratios of the oxygen isotopes, tend to show that carbonaceous chondrites have been accumulated in the primitive solar nebula at a temperature in the vicinity of 350 K. Large amounts of water and carbon compounds have been retained, although it is obvious that most of the condensed ice has disappeared, along with ammonia and methane, with only the more complex carbon compounds and water of crystallization being retained at this temperature. The more refractory materials, silicates and metallic oxides, have probably been retained with very little change of properties.

One of the most common substances contained in carbonaceous chondrites is magnetite. Huffman and Stapp (1973) have already pointed out to us that magnetite has some interesting polarization properties which suggest interstellar grain material. It is also interesting that the small magnetite grains tend to have remanent magnetism. Even if this were not the case, the magnetite grains would be interesting as a possible subset of the interstellar grains because the larger ones would be ferromagnetic, and the smaller ones would be superparamagnetic. Thus those interstellar grains which have magnetite cores may play a large role in the polarization of interstellar starlight.

It is likely that interstellar grains are nucleated in cool stellar atmospheres, being condensed and expelled by radiation pressure, or possibly precipitated in rapidly expanding stellar atmospheres such as in nova explosions. Studies of the remanent magnetism of magnetite grains in carbonaceous chondrites tend to indicate that the magnetism was acquired in the presence of a field having a strength of 0.1 to 1 G* (Brecher, 1972). Such fields naturally exist in the vicinities of stellar surfaces, but the precipitating substance is likely to be metallic

From *Interstellar Dust and Related Topics*, IAU Symposium, vol. 52, pp. 545-547. International Astronomical Union. Reprinted with permission.

*Gauss.—Ed.

iron rather than magnetite. However, I have been informed by Aviva Brecher, who has carried out a number of these studies, that in her opinion the remanent magnetism would be retained if a magnetized iron grain is oxidized to magnetite.

Thus it appears to me to be natural that the nucleation of some of the interstellar grains may provide magnetic particles, upon which other materials can condense in interstellar space. These particles should be alignable by relatively weak interstellar magnetic fields, and they will be part of any collapsing interstellar cloud which forms stars and planetary systems. Those grains which remain rather far out in any primitive gaseous nebula which is formed, will assemble into material resembling carbonaceous chondrites, with only a loss of the more volatile materials, and without a loss of magnetic properties. Of course, I refer here only to the matrix material in types II or III carbonaceous chondrites, and not to chondrules or inclusions. All of the material in type I carbonaceous chondrites should qualify.

Thus it is possible that meteorite collections in our museums may be very valuable for research on interstellar grains. Thin sections of the matrix material in carbonaceous chondrites cannot be expected to reproduce precisely the optical properties of interstellar grains, because there has been some loss of material by heating, and possibly some mineral changes in the remaining material, but some of the properties should undoubtedly resemble those of interstellar grains if this hypothesis is correct, and if so, then the materials responsible for these properties could be identified with somewhat greater certainty than is done at the present time. I therefore recommend that investigations of the properties of carbonaceous chondrites be carried out with this hypothesis in mind.

## Acknowledgements

This research has been supported in part by grants from the National Science Foundation and the National Aeronautics and Space Administration.

# References

Brecher, A.: 1972, On the Primordial Condensation and Accretion Environment and the Remanent Magnetism of Meteorites, in *Proc. IAU Symposium on the Evolutionary and Physical Properties of Meteoroids.*

Cameron, A. G. W.: 1973, *Icarus* 18, 407–450.

Cameron, A. G. W. and Pine, M. R.: 1973, *Icarus* 18, 377–406.

Huffman, D. R. and Stapp, J. L.: 1973, in *Interstellar Dust and Related Topics*, ed. J. M. Greenberg and H. C. van de Hulst, p. 297.

Larson, R. B.: 1969, *Monthly Notices Roy. Astron. Soc.* 145, 271–295.

The Coma cluster of galaxies, photographed with the Mayall telescope. [Kitt
Peak National Observatory.]

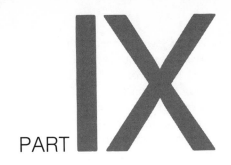

PART IX

# GALAXIES AND THE UNIVERSE

". . . in locations where the spectrograph slit crosses dust patches in M31, the observed velocities suggest that clouds of gas and dust are falling into the central plane of the galaxy." Vera C. Rubin reports how the use of electronic image intensifier tubes, in combination with large telescopes and spectographs, has enabled her to explore "The Dynamics of the Andromeda Nebula." As the author explains, M31 is the designation of the great spiral galaxy in Andromeda in an eighteenth-century catalogue of celestial objects compiled by the French astronomer Charles Messier. On a clear autumn night at a dark location, you can see the Andromeda galaxy with the naked eye if you know where to look (consult any good star map, such as those that appear monthly in *Sky and Telescope* magazine, Sky Publishing Corporation, 49 Bay State Road, Cambridge, Mass. 02138). Its diffuse appearance explains why, historically, it was known as a "nebula" although recognized as a galaxy in modern times.

Although the motions of individual parts of a galaxy such as M31 can be measured at a given point in time by techniques such as those of Rubin, systematic changes in the shape of a galaxy or the pattern of a close pair or group of galaxies must be explored by computer calculations, for "life is too short to watch galaxies move . . ." Alar and Juri Toomre explain how they have used computers to study "Violent Tides between Galaxies."

The fundamental parameter that astronomers investigate to determine the size and age of the universe is the famous *Hubble constant*, which tells the rate at which galaxies are receding from each other with increasing distance; the more distant the galaxies, the faster they are receding. It is not an easy quantity to measure, and it happens that improved observations have gradually tended to give lower values for it through the years, as described by William D. Metz in "The Decline of the Hubble Constant: A New Age for the Universe."

In "Cosmology To-day," William H. McCrea reports on the current state of investigation of the nature and origin of the universe. This article is the text of a lecture given in honor of the cosmologist Georges Lemaître, who made important contributions to the expanding-universe theory. (Professor McCrea states that Einstein "predicted no interesting observable behaviour of the astronomical universe." He is referring to the universe as a whole; Einstein did predict astronomical phenomena involving the motion of Mercury, the bending of starlight near the sun, and gravitational redshifts in dense stars.) McCrea also mentions large quasar redshifts; since he wrote this article, even larger redshifts have been found for some quasars.

# The Dynamics of the Andromeda Nebula

Vera C. Rubin

All the stars that can be seen with the unaided eye from the earth belong to our galaxy. They are members of a flat spiral system that rotates around a massive center some 10,000 parsecs from the sun. (One parsec is 3.26 light-years.) It is difficult to study the internal motions of the galaxy directly because we are located in its central plane, which is clogged with interstellar dust and gas that at visible wavelengths obscure the galactic center and most of the more distant stars. Thus although we can study some of the motions of the galaxy from the observation of relatively nearby stars and from the radio waves emitted by distant clouds of hydrogen, in order to learn more about the dynamical behavior of galaxies we must turn to other systems. The nearest galaxy closely resembling our own is the

**GREAT NEBULA IN ANDROMEDA, also known as M31, is the nearest spiral galaxy that resembles our own. It was photographed with the 48-inch Schmidt telescope on Palomar Mountain. Rectangle outlines region shown in greater detail on page 264.**

Great Nebula in Andromeda, and its internal motions have recently been studied in considerable detail. Some of the results of these studies are quite unexpected.

On a clear night away from city lights in the Northern Hemisphere the Andromeda nebula is just barely visible to the unaided eye as a faint elongated patch of light. It was described by the Persian astronomer Umar al-Sufi Abd-al-Rahman in the 10th century; it appeared on Dutch star charts in 1500. It was first observed with a telescope in 1612 by Simon Marius of Germany, who described it as resembling the light of a candle flame seen through translucent horn. In 1781 Charles Messier of France listed it as No. 31 in his catalogue of nebulous objects, and to this day the Andromeda nebula is also commonly known as Messier 31, or M31.

William Parsons of England, better known as the Earl of Rosse, began observing M31 in 1848 with his 72-inch speculum-metal reflecting telescope. His journal of observations was published some 40 years later, in 1885. That same year a brilliant new star—a supernova—appeared near the center of M31. Ultimately this star served as a link in a chain of reasoning that established that spiral nebulas were not nearby clusters of stars or clouds of gas but were stellar systems outside our own. Another link was provided by Sir William Huggins, who obtained the first spectrogram of M31 in 1890, and by Julius Scheiner, who first discussed the spectrum of M31 in 1899. Scheiner recognized that the spectrum arose from the light of many stars rather than from a glowing cloud of gas.

Modern observations of M31 date from 1914, when V. M. Slipher, using the 24-inch refracting telescope at the Lowell Observatory in Flagstaff, Ariz.,

determined that the solar system and the center of M31 are approaching each other at a speed of 300 kilometers per second. It is now known that most of this observed velocity reflects the motion of the sun around the center of our galaxy. The sun is rotating around the galactic center at a velocity of about 250 kilometers per second in the direction of M31. If we could make observations from the galactic center, they would show that our galaxy and M31 are actually approaching each other at a rather modest speed: some 50 kilometers per second.

Soon after Slipher had made his observations F. G. Pease obtained two spectra of M31 with the 60-inch reflector on Mount Wilson. For the first spectrum he lined up the slit of the spectrograph along the long axis of the tilted galaxy; for the second one he lined up the slit along the short axis. Each spectrum was the product of some 80 hours of exposure time spread out over a period of three months! The absorption lines of the stars in M31 were inclined at an angle in the spectrum taken along the long axis, but they were not inclined in the spectrum taken along the short axis [see illustrations on page 266]. Pease correctly inferred that the shift of the lines was due to the fact that the galaxy was rotating.

Over the next three decades further observations of the rotation of M31 were conducted by Horace W. Babcock and Nicholas U. Mayall at the Lick Observatory, but the extreme faintness of the individual stars made the exposure times almost prohibitively long and made it impossible to determine velocities very far from the center of the galaxy, where the density of stars is low. In the 1950's and 1960's, however, the situation changed.

From *Scientific American*, vol. 228, pp. 30–36, June 1973. Copyright © 1973 by Scientific American, Inc. All rights reserved. Reprinted with permission.

First, photography had advanced to the point where photographic plates sensitive in the red region of the spectrum had been developed. Walter Baade, working with the 100-inch reflector on Mount Wilson, detected faint nebulous patches on a red-sensitive plate during a long exposure he was making of M31 for another purpose. He inferred that these nebulous patches, now known as H II regions, were clouds of gas ionized by the ultraviolet radiation of hot stars. The red color of the patches arises from the fact that the gas and dust in M31 absorb the blue light of the galaxy more than red light. Moreover, a significant portion of the light from the H II regions is emitted at two lines in the red region of the spectrum: the line designated hydrogen alpha and a "forbidden" line of ionized nitrogen. (Forbidden lines arise when an electron within an atom drops

from a state of higher energy that is metastable, or long-lived, to the ground state. In the diffuse gas of interstellar space an oxygen atom spends approximately $10^4$ seconds, or some three hours, in such a metastable state before radiating a strong forbidden line in the visual region of the spectrum. Under normal conditions on the earth the atom would be deexcited by many collisions during that length of time and would never lose its energy by radiating a forbidden line.) During Baade's lifetime he identified 688 H II regions in M31.

We now know that young, hot Type O and Type B stars and their surrounding ionized H II regions are a major constituent of the spiral arms of galaxies. In the middle and late 1930's, however, Edwin P. Hubble of the Mount Wilson Observatory had been unable to detect any nebulous regions in M31 with photo-

graphic plates sensitive in the blue region. He had concluded that the absence of H II regions meant the absence of Type O and Type B stars in the galaxy. It was the availability of photographic plates sensitive in the red that made Baade's later observations successful.

A second change in the situation was that in the early 1960's electronic image-intensifiers, or image tubes, had come into their own. When photons of starlight enter an image-intensifier, they strike a photoemissive surface that ejects electrons. The electrons are accelerated and multiply, and finally they produce an amplified image of the star field or spectrum. Image-intensifiers in the early 1960's were comparable to photographic plates in their resolution and their capacity to distinguish a signal from noise, and they could produce an image 10 times faster. Moreover, their relative gain was

SOUTHERN REGION OF M31 is resolved into individual stars in this photograph taken with the 100-inch reflecting telescope on Mount Wilson. The interstellar dust and gas and the stars in this portion of the galaxy are rotating toward the earth as they orbit around the nucleus; in the northern region of M31 (*not shown*) they are rotating away from the earth. Their velocities can be determined by amount by which lines in their spectra are Doppler-shifted, owing to their motion, relative to their normal positions.

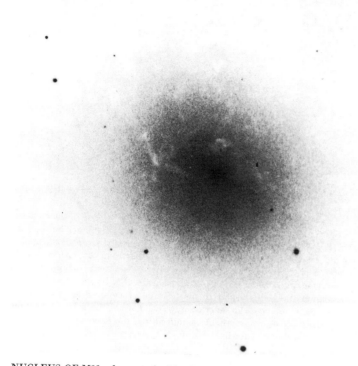

NUCLEUS OF M31, photographed by Walter Baade with the 200-inch Hale reflecting telescope on Palomar Mountain, contains clouds of dust and gas. In this negative print the clouds of dust show up as wispy white tendrils. They show that there are lanes of dust in the center of the Andromeda galaxy as well as in the spiral arms, a discovery that is unexpected and surprising. The observed velocities and directions of the dust clouds (*see illustration below*) fall into a pattern that suggests that they may be falling into the galactic plane.

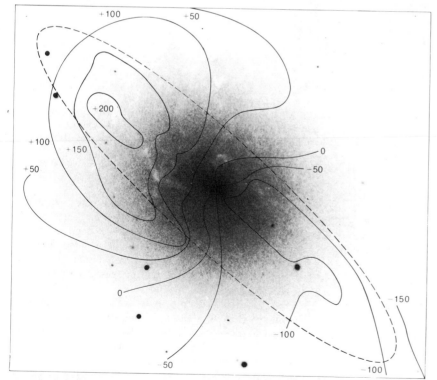

PATTERN OF OBSERVED VELOCITIES of the excited gas in the nucleus of M31 is shown superimposed on the photograph of the nucleus. The contours represent lines of equal velocity in kilometers per second. Positive values indicate that the gas is receding in the line of sight; negative values indicate that it is approaching. The gas is rotating and expanding asymmetrically. In northern quadrant of the galaxy (*top*) positions of excess positive velocities are correlated with positions of the dust clouds.

even greater in the red, where photographic plates are relatively insensitive and where the H II regions emit their principal radiation. Hence the study of H II regions in M31 with image-intensifiers was full of promise.

The spectrum of an H II region consists primarily of bright emission lines. Some are recombination lines of hydrogen and helium, which arise as an electron combines with an ionized atom and drops to lower energy levels. Also present in the spectrum are forbidden lines of un-ionized oxygen and lines of singly and doubly ionized oxygen (oxygen stripped of one or two electrons) and singly ionized nitrogen and sulfur. With a telescope of moderate size and an image-tube spectrograph (an image-intensifier attached to a spectrograph) it is possible to record the bright lines in the spectrum of an H II region in M31 in about an hour.

The velocity of each H II region along the line of sight results in a Doppler shift of the emission lines from their normal unshifted position. By measuring the location of the hydrogen-alpha line on the photographic plate to the nearest micron (thousandth of a millimeter) with respect to comparison lines put on the photographic plate at the telescope, one can obtain a velocity for each H II region. From the observed velocity one can compute the H II region's circular velocity around the center of M31. In principle one could observe the spectra of individual stars and obtain their velocities instead of the velocities of the H II regions. However, because the spectrum of a star is continuous, that is, spread out over all wavelengths, a star whose magnitude over all these wavelengths is equal to the magnitude of an H II region will require an exposure that is many times longer. H II regions emit almost all their light in just a few spectral lines. My colleague W. Kent Ford, Jr., and I exploited this fact when we chose to obtain spectra of these regions.

For the past six years Ford and I have been obtaining spectra from selected areas within M31: from 70 individual H II regions that define the spiral arms, from the integrated starlight of the bulge of the galactic nucleus and from a disk of diffuse excited gas within the nuclear region. From these spectra we have mapped regions of differing velocity within M31 and have learned about the variation of the abundance of the chemical elements as a function of distance from the center of the galaxy. We have been working on the spectra of the H II regions since 1966, using a spectrograph that was designed and built in the Department of Terrestrial Magnetism of the Carnegie Institution of Washington.

The spectrograph has been attached to the 72-inch Perkins reflector of Ohio Wesleyan University and Ohio State University at the Lowell Observatory and to the 84-inch reflecting telescope at the Kitt Peak National Observatory near Tucson, Ariz. All the H II regions are too faint to be visible in telescopes of this size. In order to obtain a spectrum of each H II region we had to offset the telescope from brighter visible stars on the basis of positions that had been determined beforehand from long-exposure photographs.

We have not identified any emission regions closer to the galactic nucleus than 3,000 parsecs. On spectra taken of regions near the nucleus, however, a weak forbidden line of singly ionized nitrogen emitted by the diffuse gas does appear superimposed on the spectrum of the integrated starlight. For the gas we have been able to measure velocities to within a few parsecs of the galactic center. We plotted the velocities of the gas in the nucleus and of the H II regions against their distance from the center of M31. The most striking feature of the resulting rotation curve is a deep minimum in the gas velocities at about 2,000 parsecs from the center [see illustration on next page]. If we assume that the motions we observe arise from particles of dust and gas moving in circular orbits in the gravitational field of M31, then it is possible to map the distribution of mass within the galaxy from the velocities. In this way we determined the distribution of mass within M31 with respect to distance from the nucleus. That curve too displayed a deep minimum at a radius of 2,000 parsecs from the galactic center. The minimum implies that there is a region of very low mass at that distance.

The total mass of M31, out to the last observed H II region some 24,000 parsecs from the galactic center, is $1.8 \times 10^{11}$ times the mass of the sun. The form of the rotation curve near the nucleus does not affect the determination of the total mass or the distribution of the mass outside the nuclear region of M31. Even if the observed velocity minimum is due to some local disturbance, and does not indicate the overall normal circular gravitational velocity at that point, there is definitely a peculiarity in the velocities of the gas at a radius of 2,000 parsecs from the center.

If the only velocities observed are from stars or gas moving in normal cir-

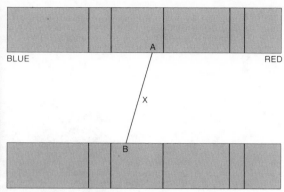

SCHEMATIC SPECTRUM ALONG LONG AXIS of M31 (left) yields spectral lines that are inclined at an angle because of the rotation of stars and gas within the galaxy. White rectangle represents the slit of the spectrograph. With respect to the observer, the northern portion (A) of M31 is receding and the southern portion (B) is approaching. The resulting spectrum (right) shows an emission line from the gas in the northern portion Doppler-shifted toward longer wavelengths, or the red region of the spectrum (A) with respect to the undisplaced center of the galaxy (X); the same line from the southern portion is shifted toward the shorter-wavelength, or blue, region of the spectrum (B). By measuring the displacement of the emission line with respect to the position of comparison lines placed on the spectrum at the telescope, one can determine the velocity of the galaxy's rotation. In actual practice spectrograph slit covers about a hundredth of the long axis of M31. Only for more distant (and hence smaller) galaxies can a single spectrum cover the entire galaxy.

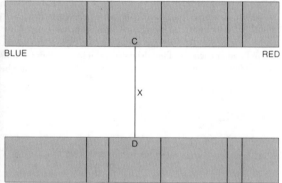

SCHEMATIC SPECTRUM ALONG SHORT AXIS of M31 (left) yields undisplaced spectral lines. The western portion (C) and the eastern portion (D) of the galaxy move across the field of view and have no motion along the line of sight toward or away from the observer. Thus the emission line (from C to D) in the spectrum (right) shows no Doppler shift with respect to the undisplaced center of the galaxy (X) and is not inclined at an angle.

cular orbits around the nucleus of the galaxy, then all the velocities we see along the short axis should be equal to the velocity of the center of M31; there would be no component of the rotational velocity along the line of sight. This is not, however, what the observations show. We have measured the velocities of the gas in the nucleus along the short axis. These measurements indicate that at some points very close to the nucleus, gas is flowing out of it at velocities that range as high as 135 kilometers per second. Only a very small amount of mass, however, is leaving the central region: less than 1 percent of the mass of the sun per year.

Ford and I were intrigued by the presence of this diffuse ionized gas near the

nucleus of the galaxy, and so we extended our observations to nearby regions. The emission line of singly ionized nitrogen was difficult to detect above the background radiation from the stars. We attempted to increase our system's sensitivity by increasing the dispersion of the spectra from 135 angstroms per millimeter to 28 angstroms per millimeter. The result of this procedure is that the light from the background stars is spread out by a factor of five and is therefore dimmed, while the sharp forbidden emission line of singly ionized nitrogen is left unaltered. The technique was so successful that we were able to detect five emission lines from the gas where formerly we had seen only one. The only penalty for working at a higher dispersion is one of exposure time: even with our image-tube spectrograph, exposures as long as six hours were necessary.

From spectra taken at 16 different angles across the nucleus of the galaxy we have deduced how the diffuse nuclear gas is distributed. Within some 400 parsecs of the nucleus the gas is concentrated into a very flat disk, perhaps 25 parsecs thick. The disk is rotating with velocities that reach 200 kilometers per second about 200 parsecs from the nucleus. In addition to the rotation, gas is streaming outward from the nucleus, principally in two directions about 180 degrees apart near the short axis. The velocities of the streaming gas reach 135 kilometers per second near the far side of the short axis. Moreover, in locations where the spectrograph slit crosses dust patches in M31, the observed velocities suggest that clouds of gas and dust are falling into the central plane of the galaxy. This gas could have been shed by evolving stars, and it could contribute to the mass of the disk. Although this model simplifies some of the complex motions we observe, it does account for their major features.

In 1963 Bernard F. Burke, Kenneth C. Turner and Merle A. Tuve of the Department of Terrestrial Magnetism began studying the motions of the un-ionized hydrogen gas in M31 at the radio wavelength of 21 centimeters. More recent radio investigations have been conducted by Morton S. Roberts of the National Radio Astronomy Observatory and S. T. Gottesman, V. C. Reddish and R. D. Davies of the Nuffield Radio Astronomy Laboratories at Jodrell Bank in England. Since the angular resolution of a radio telescope is lower than the resolution of an optical telescope, the actual detail that can be resolved in M31 is much less. In studies covering all of M31 to date, the diameter of the radio telescope beam on the sky has been 10 minutes of arc or greater. That corresponds to an ellipse some 2,000 parsecs wide by 9,000 parsecs long on M31 because of the foreshortening produced by the fact that we see the galaxy tipped only 13 degrees from edge-on. Our spectrograph slit covers about five parsecs by 400 parsecs on the galaxy. In spite of the difference in resolution, however, the agreement between the rotation curves resulting from the radio studies and those resulting from the visual studies is impressive for distances greater than 3,000 parsecs from the center of M31. (Closer than

3,000 parsecs from the nucleus there is too little un-ionized hydrogen to be detected easily by radio telescopes.)

The radio-wavelength observations have extended our knowledge to the outer limits of M31. I have mentioned that at visible wavelengths we have been able to determine velocities within the galaxy only out to the outermost known H II region 24,000 parsecs from the nucleus. Just last year Roberts and Robert Whitehurst of the University of Alabama extended the rotation curve out to 34,000 parsecs from the center with 21-centimeter observations. They found that the rotational velocity of un-ionized hydrogen remains constant at 200 kilometers per second between 24,000 and 34,000 parsecs from the nucleus. The mass contained between those limits is equal to $10^{11}$ times the mass of the sun, yielding a total mass for M31 out to a radius of 34,000 parsecs of $3 \times 10^{11}$ times the mass of the sun. Since we still know very little about the boundaries of galaxies in general, this extension of the diameter of M31 is a matter of some importance.

So far I have discussed only the motions of the gas within M31 and by inference the motions of the hot young stars that ionize the H II regions. The study of the motions of stars themselves is more difficult. Part of the reason is, as I have mentioned, that the individual stars are faint and that their spectra are continuous. We do see broad, diffuse stellar absorption lines in the continuous spectra from regions near the nucleus. Indeed, it is possible to measure the velocities of stars from these lines, although the measurements are less precise than those that can be made with the sharp emission lines from the interstellar gas. Moreover, the analysis is complicated by the fact that stars near the nucleus of the galaxy are distributed not in a flat disk as the gas is but in a nearly spherical bulge. Hence we must obtain the spectra across a long projected path. At large distances from the galactic nucleus the velocities of individual stars are still too difficult to obtain.

We have measured the velocities of the stars on both the long and the short axes near the nucleus of M31 with a single absorption line. We anticipated that the velocities would indicate that the stars are following simple circular orbits around the center of the galaxy, and that they would exhibit none of the complexities of the gas velocities. This, we have found, is not the case. The general features of the stellar motions resemble those of the gas motions. There is a steep

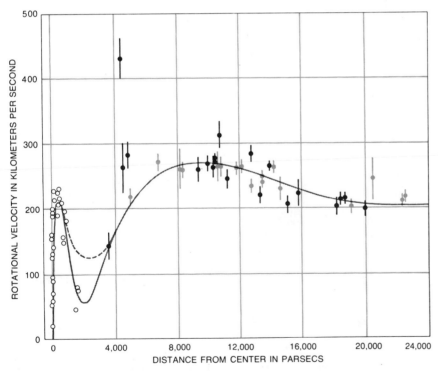

ROTATION CURVE of M31 shows the velocities of the gas within the galaxy as a function of its distance from the galactic center. The open circles represent the velocities determined from the diffuse gas within the nuclear disk. The black dots represent the velocities of the H II regions receding in the line of sight in the northern portion of M31; the gray dots represent the velocities of the H II regions approaching in the line of sight in the southern portion. Vertical lines indicate the amount of observational uncertainty for each determination. Although there are two possible versions that have been calculated for the rotation curve near the nucleus (*broken line and solid line*), both alternatives show a peculiar deep minimum in the velocities of the gas at a radius of about 2,000 parsecs from the galactic center.

gradient in the velocities of the stars across the galactic nucleus reaching a minimum at 2,000 parsecs, and the stars seem to be moving outward along the short axis. In 1939 Horace Babcock had observed a minimum in the stellar velocities near 2,000 parsecs, but his observations had never been confirmed or widely accepted.

For several reasons these results are both unexpected and not understood. On the basis of our present understanding of stellar evolution it is believed the stars whose spectra we are observing near the nucleus of M31 are some four billion years old. In contrast, the irregularities in the motions of the gas should smooth out in less than 10 million years. Thus it is not clear how the old stars, which during their long life should have moved several times around the galaxy, can have the same motions as the young gas. Moreover, although it is relatively easy to work out mechanisms that impart irregularities to the gas motions, it is difficult to do the same for stars. Our present suggestion is that we are observing stars traveling in noncircular orbits that are dynamically stable. One possibility is that some of the mass of the nucleus of M31 is distributed asymmetrically. The inner spiral arms of the galaxy are decidedly asymmetrical, but it is unlikely that their mass is great enough to distort the orbits of the stars. Perhaps there are resonances, or other cooperative effects, that stabilize the orbits (as gravitational theories of galactic structure now predict). Stars are known to shed mass as they evolve; it is possible that the gas we observe has come from old stars. Detailed understanding must await more observations.

Beyond the region in M31 where the velocities of stars and gas fall to a deep minimum at 2,000 parsecs from the galactic nucleus, the H II regions have circular velocities that rise to 250 kilometers per second at 10,000 parsecs from the nucleus and remain at that velocity out to 24,000 parsecs. We can deduce a remarkably similar pattern of velocities for our own galaxy. A flattened disk of un-ionized hydrogen extends several hundred parsecs from the center, rotating with velocities of up to 200 kilometers per second and expanding at velocities as high as 135 kilometers per second. At a distance of 800 parsecs from the galactic nucleus both the velocity and the density of the gas are very low. Near the position of the sun, 10,000 parsecs from the nucleus, the stars and the gas are rotating around the nucleus at a velocity of some 250 kilometers per

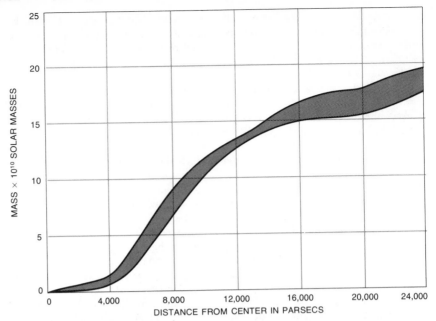

**DISTRIBUTION OF MASS** within M31 can be determined from the rotation curve of the velocities (*see illustration on preceding page*). The region in gray shows the range of values that the actual curve of the distribution of mass might have, and reflects the observational uncertainty in the rotation curve. The mass distribution displays the same deep minimum at a radius of 2,000 parsecs from the center as the rotation curve does, implying that there is a region of very low mass at that distance. The total mass of M31, out to the last observed H II region some 24,000 parsecs from the center, is about $1.8 \times 10^{11}$ times the mass of the sun.

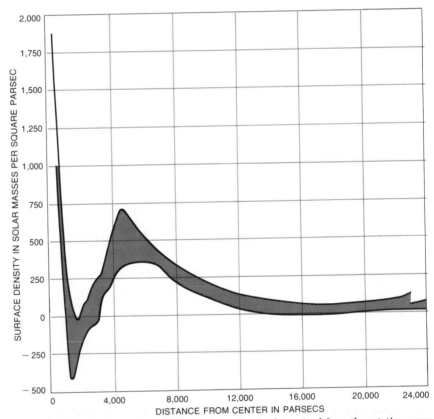

**DISTRIBUTION OF DENSITY** within M31 can also be determined from the rotation curve of the gas velocities. It too displays the same deep minimum, indicating that there is a region of low density some 2,000 parsecs from the center. Negative densities, which arise from the very steep fall of the rotation curve near 2,000 parsecs, are unrealistic and raise the question of whether the observed velocities of the gas near 2,000 parsecs arise from explosive or other nonequilibrium phenomena. For purposes of calculation the flat galaxy was approximated by a disk with no thickness; therefore the density (*vertical axis*) is given in terms of the number of stars per square parsec (area) instead of per cubic parsec (volume).

second. That velocity decreases to approximately 175 kilometers per second at a distance of 24,000 parsecs. The overall resemblance between these figures and those for M31 implies that the distribution of mass in our galaxy is similar to that in M31.

Spectrographic studies of M31 that we have carried out with C. Krishna Kumar of the Department of Terrestrial Magnetism also provide information about the relative abundances of the chemical elements in the galaxy. The abundances can be determined from the strength of the spectral lines emitted by atoms of the elements with respect to the strength of the hydrogen-alpha line. The abundances, like the velocities, vary with distance from the nucleus. In the diffuse gas of the nucleus the spectral lines of oxygen, nitrogen and sulfur are all stronger than the hydrogen-alpha line. Some of the H II regions 4,000 parsecs away from the nucleus also show these anomalously strong lines, although in the normal H II regions at that distance the lines of all three elements are weaker than the hydrogen-alpha line.

Between 5,000 and 15,000 parsecs from the nucleus the nitrogen lines decrease in strength with increasing radius and the oxygen lines increase in strength. We infer from this observation that the abundance of nitrogen is decreasing with respect to the abundance of hydrogen by a factor of about one-half with increasing distance from the nucleus. The abundance of oxygen is also decreasing, but by a small factor: perhaps 1/1.2. The decrease in oxygen has a curious result, because under the conditions of temperature and density in an H II region, radiation by oxygen atoms acts to cool the gas in which they are embedded. Therefore a decrease in oxygen leads to an increase in temperature and to a strengthening of the line of doubly ionized oxygen with respect to the line of hydrogen alpha.

In our galaxy we can observe H II regions only near the sun, that is, at a distance of 10,000 parsecs from the galactic nucleus. In these regions the strongest spectral line is always hydrogen alpha. Leonard Searle of the Hale Observatories has studied giant H II regions in external galaxies other than the Andromeda nebula, and he finds an even larger variation in the strength of the lines of oxygen and nitrogen than Ford and I see in M31. A decrease in the abundance of oxygen, nitrogen and sulfur with increasing distance from the galactic nucleus is probably a general feature of normal spiral galaxies. Perhaps the interstellar gas near the nucleus has been enriched over several generations in the formation of stars. During its lifetime a star transforms by thermonuclear reactions some of the hydrogen in its interior into heavier elements. These elements are returned to the interstellar gas if the star explodes at the end of its life cycle, resulting in an interstellar gas that is richer in heavy elements than it was initially.

Our study of M31 has been a satisfying one. There have been surprises, such as the existence of the disk of gas in the galactic nucleus and the complexity of its motions. There have been puzzles, such as the motions of the stars close to the nucleus. There has been controversy, such as the interpretation of the minimum in the velocities observed near the radius of 2,000 parsecs. Overall, however, we have been able to determine the details of motions in a galaxy much like our own. These results suggest new problems to examine in order that we may better understand the dynamics of spiral galaxies in general and of our own galaxy in particular.

SPECTRUM VERY NEAR THE NUCLEUS OF M31 taken at the original low dispersion of 135 angstroms per millimeter (*top band*) is contrasted with three spectra taken at a higher dispersion of 28 angstroms per millimeter (*bottom three bands*). Middle two spectra are laboratory comparison lines of neon; the numbers are wavelengths of certain emission lines in angstroms. In the low-dispersion spectrum the forbidden emission line of singly ionized nitrogen is indicated by an arrow; all other emission lines are due to radiation from the earth's atmosphere. Gray background continuum is from the light of the stars in the nucleus. The exposure time was four hours. Increasing the dispersion of a spectrum spreads out the continuous background radiation and reduces its brightness, while leaving the sharp emission lines from the diffuse gas unaltered. Thus it increases the signal-to-noise ratio for a sharp feature, although longer exposures are necessary. In the three bands of the high-dispersion spectra the emission line of hydrogen alpha ($H\alpha$), the two weak lines of singly ionized sulfur (S II) and one additional forbidden line of singly ionized nitrogen (N II) is visible as well as the original one (*arrow*) in the low-dispersion spectrum. The lines are curved because of the rotation of M31. The exposure time was six hours for each spectrum.

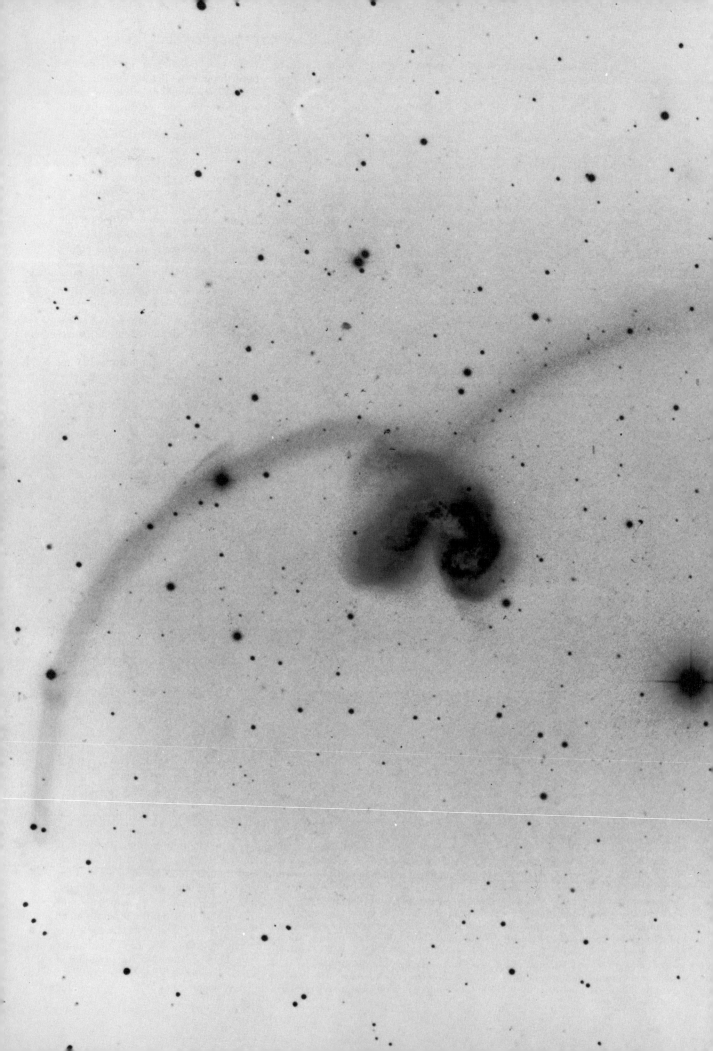

# Violent Tides
# Between Galaxies

Alar Toomre and Juri Toomre

Almost every crowd includes a few charming eccentrics or confounded exceptions. This is true of the "crowd" of galaxies. Most are objects of majestic regularity and symmetry and can be readily classified. One or 2 percent, however, do not conform. Because of their bizarre appearance or unusual spectra they are known to astronomers as "peculiar" galaxies.

Many of the galaxies that are peculiar in shape are members of multiple-galaxy systems, and it is only natural to suppose that their unusual and sometimes even grotesque forms may have resulted from the interaction of two or more galaxies. The nature of this interaction has been a matter of controversy, however. In the 1950's, when large numbers of galaxies with strange appendages were first discovered, it was immediately proposed that these morphological anomalies were the aftereffects of gravitational forces exerted during near collisions between galaxies. In the 1960's this idea fell into disrepute, although no alternative theory won general acceptance. In the 1970's computer experiments such as those described here have begun to reaffirm that gravitation may in fact be responsible for the appearance of some of the most peculiar galaxies.

One of the most strikingly peculiar galaxy pairs was discovered before it had been demonstrated that galaxies other than our own exist and can be seen from the earth. In 1917 C. O. Lampland

noted that long, faint filaments were visible on improved photographic plates of a double nebulosity listed in the 1888 New General Catalogue (NGC) as entries 4038 and 4039. The objects were photographed again in 1921 with the newly completed 100-inch Hooker telescope on Mount Wilson. J. C. Duncan, who made the Mount Wilson photographs, was particularly impressed by "faint extensions of extraordinary appearance," rather "like antennae" [see illustration on opposite page].

The curving filaments do resemble the antennae of an insect, and the system has become known as the Antennae. It is about 50 million light-years from our galaxy, which among galaxies is not a very great distance. Only about 1,000 easily recognized galaxies lie closer to ours.

The only other notably peculiar object found in this era was discovered by Heber D. Curtis in 1918; he observed a small, luminous "jet" protruding from Messier 87, later identified as an elliptical galaxy at about the same distance as the Antennae. (The Messier catalogue was compiled by Charles-Joseph Messier of France about 1800.)

By 1924 Edwin P. Hubble had deduced that many of the objects then called nebulae were in fact galaxies. The following two decades were an important period of discovery for extragalactic astronomy, yet almost none of the many galaxies identified seemed as bizarre as the Antennae or M87. The only exception, perhaps, was the "faint but definite

band of nebulosity" that Philip C. Keenan found to connect two rather widely separated galaxies, NGC 5216 and 5218 [see top illustration on page 280].

One reason that scant attention was paid to peculiar galaxies was that an intense effort was being made to understand "ordinary" galaxies. The diversity of forms was great even among those galaxies that clearly fitted Hubble's broad categories of elliptical, spiral and irregular galaxies. In addition, basic questions of the size, distance and velocity of the galaxies remained to be answered. Hubble himself knew of the Antennae and urged Duncan to take an interest in them, but he never systematically searched for similar galaxies.

There is another reason, however, that few peculiar galaxies were discovered in this period: the instruments in use were ill-suited to finding them. As we now know, the "links," "wisps," "plumes," "streamers," "filaments" or "extensions" exhibited by one or two galaxies in a hundred are usually quite faint; they are often little brighter than the night glow of the earth's atmosphere. Such dim, extended features are not easily detected with the type of telescope that was then available. Telescopes such as the 100-inch reflector on Mount Wilson, which has a focal ratio of $f/5$, are quite "slow," that is, they require long exposure times. To photograph the filaments of the Antennae, for example, would take an entire night. In addition, the field of view of such instruments is narrow: about a quarter of a degree, or half the apparent diameter of the moon. For the observation of the faint, extended features of galaxies a "faster" instrument with a wide field of view was needed. A suitable telescope was invented by Bernhard Schmidt in 1931; it provides a focal ratio of about $f/2$ for faster exposures.

A 48-inch Schmidt telescope went into

LONG, FAINT FILAMENTS curve away from the pair of galaxies NGC 4038 and NGC 4039, also known as the Antennae. These two galaxies, whose "tails" extend across more than a third of a degree of arc in the sky, are almost certainly genuine companions and not merely objects superimposed in our line of sight. The photograph was taken in 1956 by Fritz Zwicky with the 200-inch Hale reflecting telescope on Palomar Mountain. It is printed as a negative rather than in the conventional white on black to accentuate the faint features.

From *Scientific American*, vol. 229, pp. 39–48, December 1973. Copyright © 1973 by Scientific American, Inc. All rights reserved. Reprinted with permission.

operation on Palomar Mountain in 1949 as a major partner to the just installed 200-inch Hale telescope. For the next seven years this large Schmidt telescope was used to photograph all the northernmost three-quarters of the sky visible from California in a pattern of almost 900 overlapping square fields, each seven degrees on a side. This was the Palomar Sky Survey; when it was complete, the number of known peculiar galaxies had grown from a few to an untidy and baffling multitude.

By 1956 Fritz Zwicky of the California Institute of Technology wrote that "a surprisingly large number of rather widely separated galaxies appear connected by luminous intergalactic formations." In addition, particularly among close pairs and other multiple galaxies, "many were found to possess long extensions not previously known." Zwicky had somewhat anticipated these discoveries through his work with an 18-inch Schmidt, and he went on to make detailed photographs of many of the galaxies with the 200-inch telescope.

By 1959 the Russian astronomer B. A. Vorontsov-Velyaminov compiled an illustrated catalogue of 355 "interacting" galaxies based on the Sky Survey. Others discovered still more, rephotographed the known ones to larger scale and studied them spectroscopically [see "Peculiar Galaxies," by Margaret and Geoffrey Burbidge; SCIENTIFIC AMERICAN, February, 1961]. Today a particularly fine collection of more than 300 photographs

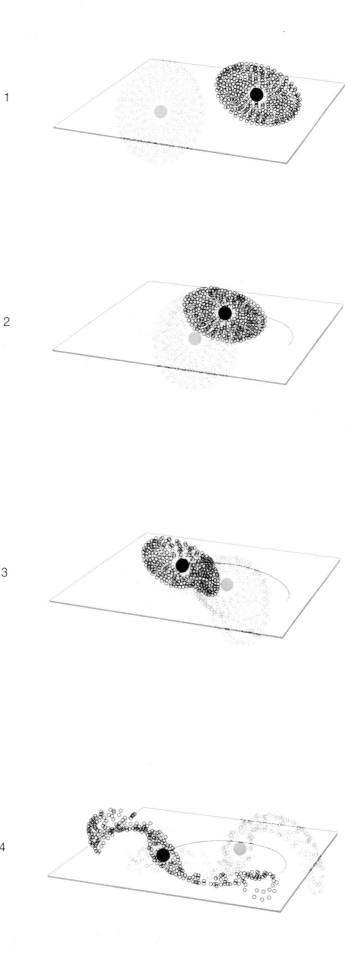

**SEQUENCE OF DRAWINGS** by computer shows the very close passage of two identical model galaxies and demonstrates how the encounter between them can produce two tidal tails similar to the tails in the photograph of the Antennae (*see illustration on page 270*). In the computer sequence the mass and gravitational force of each galaxy are concentrated at its center (*large central dot*). Around this central mass rotates a disk of some 350 massless test particles. The two central masses approach each other in elliptical orbits in one plane (*1 and 2*), bringing their disks of particles with them. Even before they reach the instant of closest approach (*3*) they feel a severe pull from the other central mass. Soon the effects due to tidal forces are dominant (*4*), and as more time passes, the material from the far side of each former disk stretches into an ever lengthening tail of debris (*5–7*). The two narrow, arching tails would appear to be crossed, and the bodies of the galaxies would almost overlap if the end product of the encounter (*8*) were viewed from the direction of the arrow. Successive frames are separated by an interval of 100 million years.

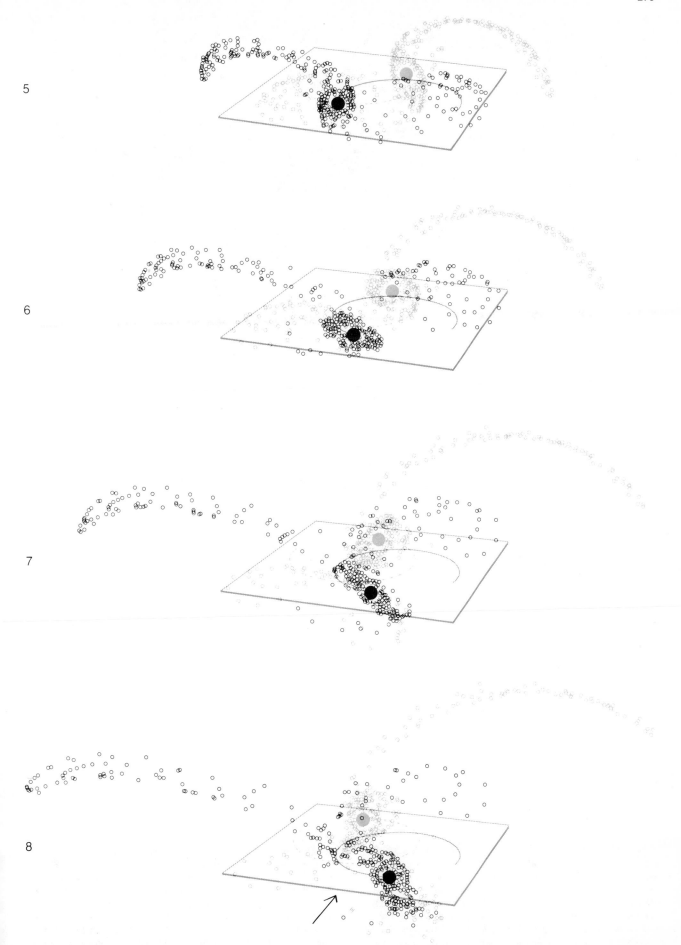

5

6

7

8

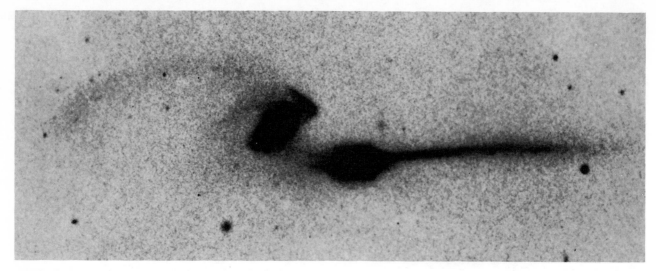

ANOTHER PAIR OF TAILS dominates the pair of galaxies NGC 4676A (*right*) and NGC 4676B (*left*), nicknamed "the Mice." The long tail from NGC 4676A is intense, narrow and almost straight; the one from NGC 4676B appears fainter, more diffuse and much more curved in this photograph made by Halton C. Arp with the 200-inch telescope. The Mice seem roughly one-quarter the angular size of the Antennae yet they are also four times more distant. Hence the true dimensions of the two systems are comparable.

COMPUTER-SIMULATED CLOSE PASSAGE of two identical galaxies yielded the long tails of tidal debris viewed from two separate directions. The top diagram models the actual view of the Mice in the photograph at the top of the page; the bottom diagram repeats the scene as if it were viewed from a direction nearly at right angles. The pericenter is point of closest approach. The central cross marks the center of mass of the system for reference. Long before the encounter each galaxy was again taken to be a circular disk of test particles revolving around a central mass. The two disks grazed each other in elliptical orbits. The tails and surviving bodies of the galaxies appear dissimilar only because each disk was tilted at a different angle with respect to the plane of its orbit around the other. The less tilted disk on the right spilled the material from its far side into a nearly flat tail that lies almost in its orbital plane; the more tilted disk on the left produced the tail that arches high above the plane in the top view.

of these objects exists in the *Atlas of Peculiar Galaxies* published in 1966 by Halton C. Arp of the Hale Observatories.

None of these later discoveries closely resemble the Antennae, but a number do exhibit luminous arcs extending far into space from at least one galaxy in a group. All such curving filaments have come to be known as "tails"; two are evident in NGC 4676, a pair of galaxies named the Mice [*see illustrations on opposite page*].

The connecting filaments, or "bridges," also assume many forms. One type is represented by the narrow link discovered by Keenan in 1935; Arp's *Atlas* shows many others. One class of bridges seems particularly distinctive. It consists of galactic systems in which a much elongated spiral arm seems almost to grope for a neighboring galaxy. In many cases another limb projects from the opposite side of the deformed galaxy.

Whatever gave rise to these intergalactic bridges and such formations as the Mice and the Antennae? The answer is not yet certain. It appears, however, that the intuitions of the observers of the 1950's may have been for the most part correct.

The obvious clue available to workers such as Zwicky was that most of the distorted galaxies come in pairs. It was also evident from the shapes and rotational motions of many normally formed galaxies that gravitation remains significant even on a galactic scale, over distances of tens of thousands of light-years. Hence it was natural to wonder if the bizarre forms might represent damage that adjacent galaxies had inflicted on one another by their mere presence and gravity. In other words, could we be viewing colossal tides?

Zwicky believed that tides could in fact explain the tails and bridges. He considered as being particularly good evidence the far-side features, or "counterarms," that usually accompany intergalactic bridges. These offered a crude two-sided symmetry analogous to that of more familiar tides, such as those induced in the terrestrial oceans chiefly by the moon. On the earth high tide comes every 12 hours rather than every 24 because the water level is raised not only at the moving point closest to the moon but also at the point diametrically opposite [see "Tides and the Earth-Moon System," by Peter Goldreich; SCIENTIFIC AMERICAN, April, 1972].

Unlike the orbit of the earth and moon, however, the orbits of most galactic partners probably are not almost circular. For a few years in the 1950's it

was widely supposed that many such pairs were not even true satellites or companions. They were believed instead to be mere passersby that had almost collided in their separate courses through space.

One reason this close-approach hypothesis became popular was that only grazing passages seemed capable of producing tides of sufficient magnitude: the tide-raising force varies approximately as the inverse of the cube of the separation of the two bodies, and so it decreases with distance even more rapidly than gravity itself. A second and more enticing reason, however, was that actual collisions between galaxies gave promise of explaining the powerful sources of radio-frequency signals, such as Cygnus A, that had recently been discovered [see "Colliding Galaxies," by Rudolph Minkowski; SCIENTIFIC AMERICAN, September, 1956].

Cygnus A was the first discrete source of radio waves detected outside the solar system, and it remains the most powerful extragalactic radio source perceived here. It was identified with a visible object by Walter Baade and Rudolph Minkowski in 1951. They found that it is at an enormous distance, by modern estimates about a billion light-years, and that it must therefore radiate prodigious quantities of radio energy. Its optical spectrum was also found to be unusual; it suggests gases in a high state of excitation.

Photographs made with the 200-inch telescope showed two large galaxies that appeared to be almost overlapping. Baade and Minkowski reasoned that the observed radio and optical emissions could be produced if the galaxies were interpenetrating at high speed. In such a collision the probability that the stars of the galaxies will collide is nil, since even within galaxies the distance between stars is vast. Interstellar gases would interact, however, causing mechanical commotion and, it was presumed, copious radiation of the wavelengths detected.

For a few years this explanation of extragalactic radio sources seemed nearly confirmed by observations of two nearer radio sources approximately coincident with the galaxies NGC 1275 and 5128. Both of these sources are very peculiar elliptical galaxies rich in dispersed gases and dust. It was thought once again that their curious features represented other galaxies in transit through them. One can thus understand why Zwicky keenly noted in 1956 that even the Antennae are "a weak radio source and therefore probably a system

of two galaxies in the process of a close collision." Minkowski agreed.

Nevertheless, by 1960 the colliding-galaxies theory of radio sources had all but expired. For one thing, the theory proved unable to account for numerous newly discovered sources whose appearance and internal motions gave no hint of any collision. It was also found that the radio emissions of several of the distant bodies come not from the observed galaxies but from regions on each side of them, many thousands of light-years beyond the areas where the impact of gas clouds would be most vigorous. Finally, the radio static was recognized as a type of emission known as synchrotron radiation (because it resembles the radiation produced by the particle accelerators called synchrotrons). This radiation is emitted by charged particles moving in a magnetic field at speeds close to the speed of light; there is no known mechanism by which the collision of galaxies could yield such particle speeds.

When the collision theory of radio galaxies thus became discredited, tidal effects between galaxies were obviously deprived of an important ally. Left to fend for themselves, they met a barrage of postponed criticism.

It was pointed out, for example, that encounters between unrelated galaxies might be common enough to account for the rare radio sources, but that the space between galaxies is too vast for 1 or 2 percent to have grazed another in a chance meeting. It was also noted that tails emanate from some galaxies that appear to be isolated from their neighbors. Moreover, even if the occurrence of many near collisions could be explained, it was remembered that no one had demonstrated that such encounters could produce the narrow filaments seen in the Antennae, the Mice and certain other distorted galaxies.

In fact, tides seemed plain wrong on two counts. First, Vorontsov-Velyaminov noted that in double galaxies tails are more common than bridges. How can this be reconciled, he asked, with the two-sided symmetry of all known tides? Or even if that symmetry were imperfect, why should the distortion of the far side be greater than that of the side most exposed to the external gravitational field during an encounter? The other and even more worrisome point was that some bridges and tails are strikingly narrow. This seemed very odd, since known tides raise masses over wide areas. Zwicky himself was troubled by this objection; it seemed to demand that the galaxies behave almost like taffy.

It was not only such intrinsic weaknesses of the tidal theory, however, that caused its widespread abandonment. Also involved was a change of scientific mood or fashion. Even before the first quasars were discovered in the early 1960's, astronomers had developed respect for some highly energetic and even explosive phenomena in galaxies. Supernovae and radio galaxies are but two of many examples. Another is the jet of M87, which, as we have noted, had been known about as long as the Antennae; the discovery in the mid-1950's that even its visible light is synchrotron radiation startled many astronomers. Although poorly understood, all these processes seem to have little to do with gravity. Hence there was much open-mindedness toward invoking other forces, such as electromagnetism, in efforts to interpret various puzzling phenomena.

In this intellectual climate Geoffrey and Margaret Burbidge speculated that the "tubular forms" of such galaxies as the Mice might belong to systems still in the process of formation, probably in the presence of magnetic fields. Vorontsov-Velyaminov wondered if the strange shapes might have resulted from some novel "force of repulsion." More recently it has been suggested that the jet of M87 and some unusual features of other galaxies may have exploded from galactic nuclei. Arp has gone on to ask whether such ejecta might not include some of the small galaxies that seem to be connected to larger ones by bridges.

None of these alternatives of the 1960's were presented as full-fledged theories. They were only hopes or suspicions, no doubt nourished in large part by despair of accounting for objects such as the Mice and the Antennae primarily by gravitation. In this despair were echoes of older remarks, such as "a tidal perturbation can alter the shape of a galaxy but cannot draw out a long narrow filament." Such sentiments had one flaw: it had never been established that gravity could not do it.

Theoretical work since about 1970 has shown that tides can in fact account for some very peculiar structures. As our computer models illustrate, it seems possible after all for a slow near collision to rip the outer parts of a disk into thin and taillike ribbons by gravity alone. It also appears to be possible for other such tides to evolve into remarkably slender bridges and counterarms.

These conclusions are mostly our own, based on extensive experiments we have conducted during the past four years with mathematical models of some double-galaxy systems using the computer and graphic displays of the Goddard Institute for Space Studies in New York. To be sure, like many other overdue ideas, the need for such calculations dawned on several workers more or less simultaneously. Seven workers have recently reported results at least vaguely like ours. One, the Russian physicist N. Tashpulatov, definitely preceded us. Yet the roots of this work go all the way back to an inspired but long forgotten study made by the German astronomer Jorg Pfleiderer about 1960, at a time when chance flybys of galaxies still retained some promise.

Pfleiderer's models, like ours, were intended only as efficient caricatures of the loose confederations of orbiting stars, dust and gas that are the real galaxies. Pfleiderer was no less aware than we are that the mass of an actual galaxy is dis-

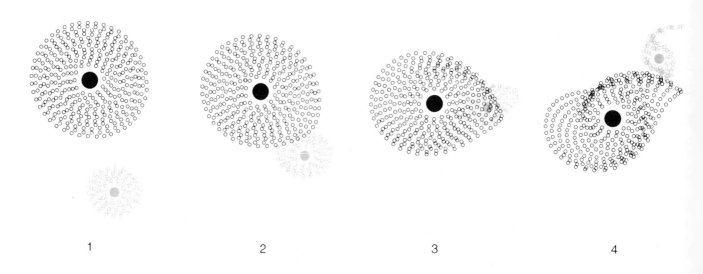

1    2    3    4

**FLYBY OF A SMALL GALAXY** produces the striking, if transient, spiral structure seen in this sequence of eight time frames of equal intervals from another computer simulation. Here the body at the center of the larger disk (*black*) is four times more massive than the one at the center of the smaller disk (*gray*). Both disks are viewed face on and spin counterclockwise; they encounter each other in parabolic orbits. The plane of their orbits is inclined 45 degrees from the vertical and the disks do not actually interpenetrate. In

persed over its disk; yet all our models pretend boldly that the matter in their disks is of such small mass that the entire inverse-square gravitational field derives from a single point at each center.

The rationale for using such highly idealized models is twofold. Pfleiderer reasoned that tidal effects should be much the strongest in the exposed and relatively slowly rotating outer parts of the galactic disks. Out there, at least, the mass is small and its self-gravity must be weak. The models should thus remain basically valid even if the peripheral mass is neglected entirely.

Second, as Pfleiderer was particularly conscious in those early days of electronic computers, the idealization of mass as a single point greatly simplifies the computational task of predicting the successive shapes of a disk composed of many particles. For a model galaxy consisting of $n$ particles these numerical economies are roughly $n$-fold. In the most realistic models possible every particle in a collision of two equal galaxies would be influenced directly by $2n - 1$ other particles; in the simple models the motion of each particle is influenced by only the postulated central masses of the two disks. As it is, the motions of the $2n$ test particles constitute a set of $2n$ distinct "restricted three-body problems"; such three-body equations have no known analytic solutions of practical value, but they are easily solved in tandem by computer. One can gauge the efficiency of this process by noting, for instance, that our entire simulation of the Antennae (with $n = 350$) could be rerun in less than five minutes on any fast modern computer.

Incidentally, because the test particles are without mass it makes no difference to the motions of one galaxy whether the other arrives with or without its own retinue of massless points. Each of our simulations is thus an anthology of two stories, calculated separately but in the diagrams superposed.

One might well ask why, if these calculations were so inexpensive, we did not adopt more elaborate mathematical models taking some account of self-gravity. The reason is mainly that the construction of even these few examples required hundreds of trial encounters for the purpose of gaining an understanding of the effects of various mass ratios, orbital parameters and times and directions of viewing. These fairly systematic searches revealed that the results are insensitive to changes in certain of the parameters. Yet three conditions seem consistently vital. To produce narrow bridges and tails (1) the galaxies must approach in parabolic orbits or in even slower, elliptical orbits; (2) they must penetrate each other, but not too deeply; (3) the approach of the "attacking" galaxy must be in roughly the same direction as that in which the "victim" disk rotates. Bridges result if the passing galaxy is of fairly small mass, whereas tails require that the two galaxies be nearly equal.

The above may be the crucial ingredients for making bridges and tails, but why do tides assume such forms at all? One reason is that the galaxies themselves are already spinning; the other and less obvious reason is that they experience the intense tidal force only over a relatively short interval. If the sequences portrayed in these computer models actually occurred, they would

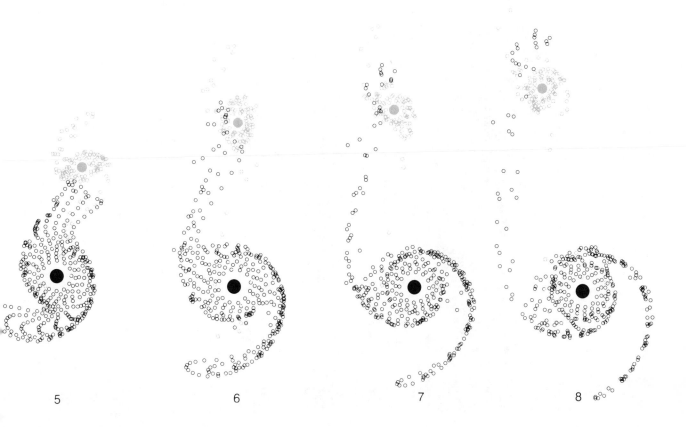

5                    6                    7                    8

time frames *1* and *2* the barely distorted small galaxy is still rising toward the viewer. At its closest approach to the larger galaxy (3) it passes as much in front of it as to the right of it. The tidal effects in both disks (4) are distinctly two-sided. As the smaller galaxy re-
cedes (5-7) the tide it raised on the side of the larger disk closer to it evolves into a narrow bridge connecting the two galaxies. The similar bulge that it caused on the far side wraps into a fine counter-arm that will become sparse and eventually disappear.

NORMAL PHOTOGRAPH OF THE WHIRLPOOL NEBULA in the constellation Canes Venatici, also known as Messier 51, exemplifies the interior spiral pattern of the stars, dust and gas. The smaller, irregular galaxy appears to be a genuine companion to the larger one.

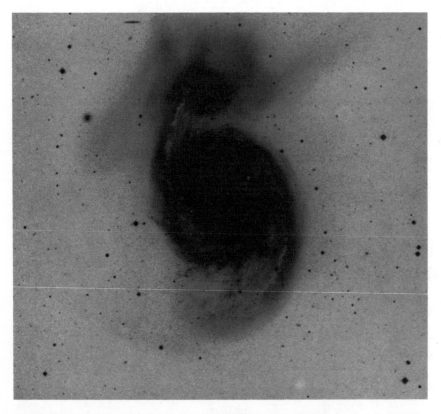

OVEREXPOSED PHOTOGRAPH OF THE WHIRLPOOL NEBULA, taken in 1969 by Sidney van den Bergh with the 48-inch Schmidt telescope on Palomar Mountain, reveals the confusion of faint material that surrounds the smaller galaxy and the two long streamers that extend from it in the directions of two o'clock and eight o'clock. The broad lower arm of the spiral galaxy, prominent in this photograph, is almost invisible in most others.

have required hundreds of millions of years; to the galaxies involved, however, these encounters would have seemed fairly sudden.

The distorting force is not the gravitational field itself but the difference between the fields perceived in the near and far parts of the galaxy. This is in fact the tidal force; as we have noted, it varies inversely not as the square of the distance but as the cube, and therefore it does not become significant until the galaxies are really close. In these computer models it is strong only during the three time frames that bracket the instant of pericenter.

Because of this "hit and run" nature of the tidal force, by the time the tidal damage looks impressive the model galaxies have almost ceased to interact. Their further evolution is merely kinematic; they drift on independently, like two armadas of spacecraft coasting after a brief, simultaneous firing of engines. Hence the spiral forms (and even the tails) develop not because the imposed gravitational field had a spiral structure but because particles assigned "low" orbits always tend to overtake those in "high" orbits. Those nearer the galactic nucleus simply shear more and more out of alignment with those farther away.

This qualitative reasoning goes a long way toward explaining the rather two-sided bridge and counterarm that become evident in our computer simulation of the passage of the small companion. The question remains, however, of why the tidal damage in the simulation of the Antennae should have become so much more pronounced on the far side of each galaxy. Part of the answer is that such equal partners obviously damage each other more than the ones in the bridge-building sequence. In fact the near-side tidal forces in that encounter are so great that the material pulled out of one galaxy, rather than forming a bridge, falls in an avalanche into an amorphous mass in the general vicinity of the other. At the same time the far-side material is "launched" vigorously, if indirectly, by having its parent mass practically yanked out from under it. Much of this debris eventually escapes from the influence of both galaxies, resulting in counterarms that grow ever more grotesque and soon dominate the appearance of the galaxy pair. Evidently they are the tails that puzzled Vorontsov-Velyaminov and others.

The reason the orbits must be parabolic or slower is that otherwise the bridges would not connect, nor would there be any avalanching. These failings, if they can be called that, were illustrat-

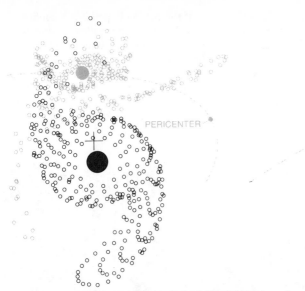

PERICENTER

PERICENTER

**COMPUTER MODEL OF THE WHIRLPOOL NEBULA** explores in two views the probable geometry of the encounter that seems to have deformed at least the outer parts of both galaxies. The first view (*left*) shows that the deflected particles from two distinct disks whose central masses are in the ratio of three to one can indeed mimic several of the faint outer features visible in the overexposed photograph at the bottom of the opposite page. The computer model shows both an open-spiral structure in the nearly face-on larger disk and long tidal streamers drawn out from the smaller companion. Second view (*right*) is at right angles to the first and can be thought of as being edge on to the sky; it shows that the arm of the larger galaxy is not a true bridge to the companion.

ed by Pfleiderer's calculations with fast but massive passersby. He obtained some fine transient spirals but no bridges or tails.

Since life is too short to watch galaxies move, one cannot be sure that real galaxy pairs orbit each other in the parabolas or elongated ellipses demanded by our models. It should be noted, however, that the statistical objections previously voiced against chance hyperbolic encounters do not apply here. Galaxy pairs in highly eccentric elliptical orbits would necessarily spend a large fraction of their orbital period near maximum separation and would descend only occasionally, like comets, for brief but spectacular displays. At any given moment we would expect to see most such partners well separated from each other, and, unless we knew otherwise, we might never suspect that they were destined to come close. There are in fact many such loose double systems in the sky.

In this discussion devoted to the bizarre it may seem odd to mention M51, a pair of galaxies dominated by the well-known Whirlpool. On most photographs the form and regularity of the Whirlpool show it to be an almost idealized specimen of the spiral galaxy [*see top illustration on opposite page*]. It has probably appeared in more textbooks, articles and even advertisements than any other galaxy. Indeed, it was the first galaxy in which a spiral structure was detected (by the Earl of Rosse in 1845).

In spite of the magnificence of its inner spiral structure, the Whirlpool is a peculiar galaxy. The evidence is abundant in a recent photograph made by Sidney van den Bergh; it shows the faint outer features of both members of the M51 pair about as well as they can be recorded with the telescopes and photographic emulsions available today [*see bottom illustration on opposite page*].

Two anomalies of the outer structure revealed by the photograph have been known for some time. They are the "horns" above the companion galaxy and the arm of the Whirlpool that seem to link the two objects. These clues have long suggested tidal damage, but they are inconclusive. Much more significant are the two "plumes" that seem to emanate from the companion. Their importance was first recognized by van den Bergh, although they had been noted two decades ago by Zwicky. Also very interesting is the broad lower extremity of the Whirlpool. This feature, almost invisible in the standard photograph, is most likely a counterarm still in the process of development.

We have calculated the geometry of an encounter between the Whirlpool and its companion that seems qualitatively to explain all those observed outer shapes and yet is consistent with the substantial known speed of recession of the companion [*see illustration above*]. If this hypothetical encounter is realistic, one fact is clear: the connection between the galaxies is an illusion; the apparent bridge and the more distant companion merely lie in the same line of sight.

Like the vast majority of spiral galaxies, the Whirlpool probably developed most of its fine spiral structure through processes that are intrinsic and have little to do with tides. Yet one wonders: Were the presumed tidal effects such as the outermost arms only superposed on that preexisting structure, or was even the interior of the galaxy somehow rendered more photogenic by the violent tides? These questions remain largely unanswered.

As we have seen, tidal models promise to explain some strange features of galaxies, but they by no means account for all. Of the original threesome of the Antennae, the jet of M87 and Keenan's system that were known to Hubble, only the first now seems to have been plausibly explained by tides. The second appears to be far more esoteric, and as for Keenan's system, no one as yet seems to have any inkling of whether its origin was mundane or exotic.

Many other geometrically peculiar galaxies also remain to be explained. One particularly baffling object is NGC 3921, a galaxy with multiple streamers somewhat like the tails formed in our models [*see bottom illustration on next page*]. Its deformities are large, yet no second galaxy has been detected anywhere in the vicinity. The hypothesis that we are seeing two galaxies with streamers in the same line of sight might explain this particular object if it were unique, but there are at least five other NGC galaxies in Arp's *Atlas* known to have multiple tails.

Even the existence of a "black hole" nearby would not suffice; it might well cause tides, but there is no known way it could cause multiple streamers.

Yet an explanation even of NGC 3921 may not be outside the realm of gravitational dynamics. In our models collisions were assumed to be perfectly elastic; in reality, as the Swedish astronomer Erik B. Holmberg pointed out in the 1930's, collisions between galaxies would involve some frictional forces. It simply costs orbital energy to raise all those violent tides. In the same spirit we suspect that in NGC 3921 and similar objects one is witnessing the vigorous tumbling together or merger of what until recently were two quite separate galaxies.

If this merger hypothesis is confirmed, it will raise the possibility of similar recent goings-on in those peculiar elliptical galaxies that now exhibit either double nuclei or strange interspersed material. It may also provide further impetus to the study of sudden "refuelings" of the very centers of galaxies with fresh interstellar matter.

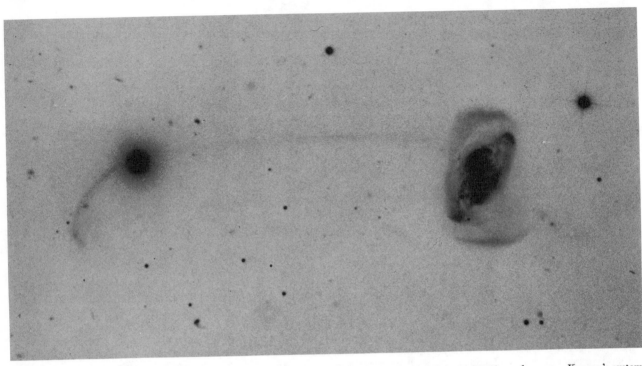

INTERGALACTIC BRIDGE not only connects these two galaxies but also seems to extend through and beyond the one at left. The galaxies, NGC 5216 (*left*) and 5218, are known as Keenan's system after Philip C. Keenan, who discovered the filament in 1935.

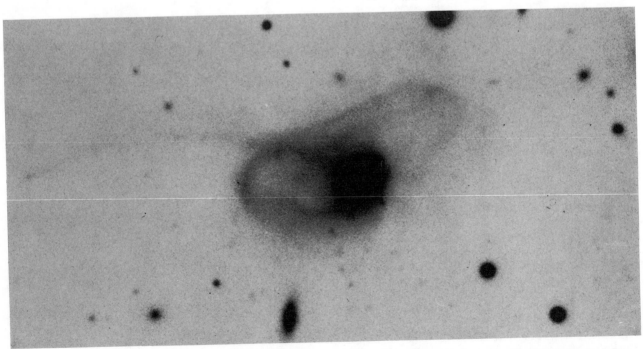

MULTIPLE STREAMERS extend from the object designated NGC 3921. One extends well to the left; another starts in about the same direction but soon turns sharply and seems to end in the faintly luminous region on the right. Authors speculate that two galaxies may here be permanently merging. Both this photograph and the one above were made by Halton C. Arp of the Hale Observatories.

# The Decline of the Hubble Constant: A New Age for the Universe

William D. Metz

*It is my enormous pleasure to ask Allan Sandage to take us on a trip through that enormous dimension of time and space in which he feels at home*—Martin Schwarzschild, introducing the Henry Norris Russell lecturer, 138th meeting of the American Astronomical Society, Michigan State University, 15–18 August 1972.

The universe may have started with a big bang, or it may have always been in a steady state. There are few measurements of the nature of the universe, and the most important has been found greatly in error. Allan Sandage presented evidence for the big bang, and announced that new data on the time lapse since the initial explosion give an age for the universe that is consistent with the ages of life, the earth, and the stars. The well-known astronomer from the Mt. Wilson and Palomar Observatories further predicted that within the next 10 years it may be possible to tell whether the universe will keep expanding forever or eventually slow down and contract.

Following the tradition of eminent astronomers such as Russell himself, who laid the groundwork for the understanding of stellar evolution, Sandage spoke eloquently and authoritatively. His presentation touched almost every point in modern cosmology; indeed, it seemed to signal that the study of the evolution of the universe had progressed a step closer toward becoming a full-fledged empirical science. However, some of the arguments made at the end of the talk were clearly speculative, arguments thought by some of the younger astronomers in the hall to be reminiscent of a grand but perhaps less rigorous age of astronomy.

Since Edwin Hubble established, in 1921, that the universe is expanding, it has been known that more distant galaxies recede faster. The constant of proportionality in the relation between the velocity and the distance of a galaxy (the Hubble constant) indicates an age for the universe under certain assumptions about the expansion. With the best techniques of his day, Hubble determined a constant which indicated an age

of only 1.8 billion years. But even in the late 1930's this was known to be less than the age of the earth's crust. Either the simple "big bang" model was incorrect, or the Hubble constant was wrong. This famous discrepancy was a prime motivation for the "steady state" model developed by Hermann Bondi, Thomas Gold, and Fred Hoyle, which describes a universe that has no beginning or end, but continuously remakes itself according to a fixed and immutable pattern.

The original measurement of the Hubble constant was in error. In fact, the Hubble constant has changed so often that it is a notable example of a mutable constant. According to Sandage, "It has gone down linearly with time," and has now reached a value that makes the age of the universe consistent with the age of its constituents. The most important announcement at the Russell lecture was that the new age of the universe, estimated from the remeasured Hubble constant, is 17.7 billion years, an age remarkably close to the best estimated age of the galaxies (12 to 15 billion years).

The Hubble constant is difficult to measure because there are random velocities of galaxies in addition to the velocities of expansion. Galaxies receding at such great speeds that these perturbations are insignificant are so far away that their distances are extremely difficult to measure. According to Sandage, "You have to look so far in order to see cosmological velocities that individual stars cannot be seen. So you have to devise a technique to bridge the gap between the place where precision indicators [of distance] exist and where the universe is really expanding without any perturbing effects."

Measurement of distance must be done in many successive steps, beginning with the calibration of Cepheid stars in our galaxy (a peculiar class of variable stars whose brightness can be determined by the cycle of variation in their intensities), next measuring the angular sizes of certain hydrogen regions in galaxies near ours, then using distances of further hydrogen regions to calibrate the absolute luminosities of galaxies having a cosmological velocity. Distance can be determined from absolute luminosities by the inverse square law.

In 1932, Hubble established a value of 530 kilometers per second per megaparsec as his constant (a megaparsec is about 3.3 million light years), but the scales of optical magnitude

From *Science*, vol. 178, pp. 600–601, 10 November 1972. Copyright 1972 by the American Association for the Advancement of Science. Reprinted with permission.

were not accurate for faint objects because certain nonlinearities in photographic plates were not understood. Furthermore, the absolute scale of Cepheid brightness was in error, as discovered by W. Baade in 1952. Correcting these two errors reduced the Hubble constant to about 265. In 1956, it was stated to be 180, then after corrections of Hubble's data for other errors, Sandage estimated in 1958 that the best value was 75. The value Sandage announced at the Russell lecture, based on the first complete remeasurement, was 55 ± 7. Sandage commented quite candidly on the contrast between his estimated error and the enormity of mismeasurement over the years.

> Now that's an incredibly small error, 15 percent of the value. Hubble said his value was good to 15 percent also. HMS [Humason, Mayall, Sandage] said their value was good to 15 percent, and the value of 75 is good to 15 percent. Almost everybody, when quoting distances . . . quotes 15 percent. So that's kind of unrealistic, but Martin Schwarzschild said today that one should always underquote the errors so as to give himself some enthusiasm to continue on with the problem.

The problem of the next 10 years, as Sandage sees it, is to look out to greater distances to see whether the linear relationship between the distance and velocity changes. The largest red shift used by Hubble or Sandage was 0.46, but Sandage thinks it will be possible to find the distances of certain galaxies with red shifts of 0.8. (The red shifts of some quasi-stellar objects (QSO's) are almost as great as 3.) The point of measuring objects with larger red shifts is that they may be far enough away so that the time for light to travel to us is a significant fraction of the age of the universe. If the universe is slowing down because of the braking action of its own gravitational forces, then the speeds of very distant galaxies will be observed as larger than one would expect because they would be observed at the expansion rate of an earlier age; in other words, the Hubble relationship would not be exactly linear. Thus, better data at large red shifts will allow astronomers to determine a second constant, called the deceleration parameter. In Friedmann's equation that describes many cosmological models, a deceleration parameter of $-1$ indicates a steady state universe, a value of $+\frac{1}{2}$ indicates a flat Euclidean universe, and a value greater than $+\frac{1}{2}$ indicates a universe that is decelerating and will eventually contract. The best value

available from the present data $(1 \pm 1)$ is not definitive, but slightly favors a "big bang" history for the universe.

After stating so succinctly the outstanding problem that must be solved to ascertain what the future of the universe will be, Sandage ventured the suggestion that there is already enough evidence to determine the past. Though many scientists have questioned whether the very large red shifts of QSO's are really indicative of velocities near the speed of light, Sandage presented some arguments in favor of the traditional interpretation. He then estimated that the light from QSO's with the largest red shifts was emitted before 89 percent of the history of the universe had elapsed. Furthermore, data that Sandage presented in his talk suggested that the 200-inch Mount Palomar telescope should be able to detect QSO's with red shifts larger than 3, but searches for objects listed in the 4th and 5th Cambridge catalogs of radio sources have not revealed any.* Looking further out in space is equivalent to looking further back in time, and Sandage suggests that suddenly the objects run out.

> If one could substantiate that a red shift limit of 3 is real, have we actually observed the edge of the universe or the horizon of the universe in time? If so, this would be a fairly decent proof that the universe has not always been the way it is now, that it has evolved. This plus the agreement of time scales [the age of galaxies and the age of the universe] would surely be an indication of an evolution: the world did begin.

While astronomers reared in the oriental cultures express very little interest in cosmology, scientists educated within the western Judeo-Christian tradition continue to be fascinated with questions about the origin of the world. The Russell lecture ended with a powerful allusion to the religious overtones of that fascination.

> The best text that could be indicated here would be that in the beginning there was darkness upon the deep. There was light, and out of that light came everything that we now observe.

Astronomers, of course, will continue making observations.

---

*A few quasar redshifts larger than 3.5 have since been measured.—Ed.

# Cosmology Today

W. H. McCrea

The University of Louvain has done me a great honour; I value it most profoundly because of the fame of your University. But I value it for more personal reasons as well. In my extreme youth, my father used to tell me of the sufferings of Louvain in the war of 1914–1918 and of the heroic exertions of your people in their post-war recovery. He showed me pictures of the building in which we meet to-day, and so it is with singular emotion that I stand here half a century later. Then, as a student, I first learned to appreciate the logical elegance of mathematical analysis from the writing of your distinguished professor, Ch.-J. de la Vallée Poussin. Finally, for many years I had the privilege of knowing that great and devoted son of your University, and pupil of de la Vallée Poussin—Georges Lemaître.

Lemaître did some of his early work in England in association with another famous man of our science, Sir Arthur Eddington, who declared that Lemaître's discoveries marked the turning point in his own researches. Indeed, it was from Eddington that I first learned of Lemaître's work. Most appropriately, in 1953 Lemaître was the first recipient of the Eddington Medal of the Royal Astronomical Society.

All of us who knew Lemaître had deep and affectionate admiration for him as a man. He breathed refreshment and encouragement and enjoyment of life. Whenever I think of him, I hear in recollection that wonderful hearty laugh he always had.

## The Beginnings of Modern Cosmology

Modern cosmology began with Einstein's work in 1917. His great contribution was to show that general relativity could self-consistently treat a model of the universe as a whole. Unlike classical theory, it required no specially contrived boundary conditions or special additional assumptions in order to do this. But Einstein took it for granted that the uni-

verse in the large must be unchanging with time, and so his theory predicted no interesting observable behaviour of the astronomical universe. Also in 1917, de Sitter produced his model and this did have interesting—but disconcerting—features.

Ten years later, Lemaître transformed the scene with his bold and illuminating ideas. They were bold, because he admitted that the universe in the large *may* be changing with time; they were illuminating, because Lemaître showed that the theory of relativity *demanded* such behaviour. He demonstrated that the Einstein model is unstable so that, when disturbed in the direction of expansion, it will go on expanding for ever, and tending towards a de Sitter model in the limit. Thus Lemaître made the best of both worlds, Einstein's and de Sitter's, by showing that they are initial and final states of a single more general model. But even this model was only one amongst infinitely many other models admitted by Lemaître's formulation. The most interesting of these models started at a finite time in the past in a state of infinite congestion.

Thus Lemaître inferred that according to the theory of general relativity the universe must be expanding and must have started either as an Einstein universe or, more likely, in a state of extreme congestion.

Here I recall that A. Friedman (Leningrad) had in 1922 published similar results in an equally remarkable paper. Lemaître did not know this until after he had done his own work. Unfortunately, Friedman did not live to pursue the subject. But it does seem fair to say that Lemaître showed himself more aware than Friedman, or anyone else at the time, of the astronomical bearing of the results.

## Man's Comprehension of the Universe

In 1929 E. P. Hubble (Mount Wilson) announced his discovery of the general law of redshifts for remote galaxies. The only feasible interpretation, then and now, is that these galaxies are receding from us (and from each other) with speeds proportional to their distances, that is to say, the whole astro-

From *Revue des Questions Scientifiques*, vol. 141, no. 2, pp. 223–241, April 1970. Inaugural lecture of the Chaire Georges Lemaître, Catholic University of Louvain, October 28, 1969. Société Scientifique de Bruxelles. Reprinted with permission.

nomical universe is expanding. The observation and its interpretation were of a sort never before encountered by man.

Let us therefore contemplate man's main advances in his comprehension of the universe around him. They have been his acceptance of the universality of

Order
Law
Composition
Coherent behaviour.

The acceptance of order—that the universe is not merely capricious—is indeed the acceptance of the universe as something to be comprehended; we associate this with the Greeks. The acceptance of law—that the order in the cosmos may be inferred from a few universal physical laws—we associate with Galileo and Newton and their times. The realization that the matter in the cosmos is everywhere the same matter composed of the same chemical elements we associate with the nineteenth century. The acceptance that the cosmos as a whole is behaving in a coherent manner we associate above all with Hubble's discovery.

The last advance was the most surprising. As regards each of the others, once the concept had been proposed, it would have been surprising had it *not* been valid. As regards Hubble's discovery, it was profoundly surprising to most scientists of the time that it should be valid.

Einstein in 1917 had taken it for granted that the universe is everywhere and always statistically the same. So he had postulated one form of universal behaviour, and nobody seemed to be surprised at this form. It was the appearance of a concerted behaviour—an evolution—of the entire cosmos that was so startling.

Now the most surprising feature of this most surprising advance was that it had been *predicted* on the basis of an existing theory. In this prediction, Lemaître had played a major part, in his work of 1927, and he played the major part in elucidating the position.

The happenings of about 1930, when all this became generally known, changed the course of physical science: *a)* Einstein's theory of general relativity came into its own as a physical theory. Hitherto it had predicted minute departures from classical physics. Now it had successfully made the greatest prediction in the history of science. *b)* This turned the attention of physicists generally to the cosmic laboratory. Besides arousing unprecedented interest in astrophysics as a whole, it caused physicists to appreciate that astrophysics placed significant constraints and requirements upon their theories. *c)* It marked the beginning of modern cosmology, the study of the universe in the large, as a subject of the most tremendous and profound *astronomical* interest.

It has been only in meditating upon this lecture that I have come to recognize that, while Einstein was maybe the grandfather of modern astronomical cosmology, Lemaître was the father. Before Lemaître, the astronomer's universe was a collection of galaxies; with Lemaître it became the *system* of the galaxies. The study of this system as a whole, as we have just said, was now of paramount interest in astronomy.

We salute also the vital contributions of Hubble, Eddington and other great pioneers. But it was Lemaître who, some 40 years ago, caught the spirit of the most recent phase of cosmology in our own day. He believed that cosmology ought to account for the expansion of the universe, the origin of the elements, the origin of cosmic rays, the formation of the galaxies, and so on. The way has proved longer and more devious than he expected. Some of the objectives have become more clearly defined, but it was Lemaître who first discerned them and we are still striving towards them.

### Signposts

I now recall a few of the signposts we have found along the way we have travelled since Lemaître directed our first tentative steps.

*Age.* Hubble's observations, and most of the theoretical models constructed to fit them as well as could be, originally indicated that the universe was in a singular state of some sort not more than 2 billion ($2 \times 10^9$) years ago. At first, this appeared to be an attractive feature since astronomers thought it might explain the origin of various properties of stellar systems and so on. However, the shortness of the time soon became an embarrassment—for example, the age of the Earth appeared to be more than twice the age of the universe!

Lemaître himself propounded one escape from the dilemma. In order to describe it, we return briefly to Einstein and his model universe. The general form of Einstein's theory of general relativity involves a certain constant known as the cosmical constant $\Lambda$, which is unrelated to other physical constants. Einstein's original model universe exists if and only if $\Lambda$ has a positive value. Friedman-Lemaître models exist for all values of $\Lambda$. When Einstein realized this, he wanted to discard $\Lambda$ and to adopt one of the Friedman-Lemaître models having $\Lambda = 0$. All the $\Lambda = 0$ models start with a big bang, and they are the models that are most embarrassing as regards shortness of age. Lemaître remarked, however, that a model having $\Lambda$ just greater than Einstein's original value would start with a big bang, expand to nearly the Einstein state, pass slowly through the Einstein state, then expand more rapidly to the "present" state, and would tend finally to the de Sitter state. We now call such a model the "Lemaître model." By taking $\Lambda$ sufficiently close to the critical Einstein value, the model will linger as long as may be required in the vicinity of the Einstein state. Thus by this simple device Lemaître could produce a model universe having apparently as long a life as we need.

Note the perversity of the situation! Einstein had needed $\Lambda$ in order to get a model, but had then rejected it on account of the work of Friedman and Lemaître. Lemaître did not need $\Lambda$ in order to get a model, but Lemaître insisted upon retaining $\Lambda$ in order to get what he regarded as a satisfactory model. We shall have to mention this model again.

The other escape from the age dilemma was proposed in 1948 by Bondi and Gold and by Hoyle. It was *steady-state cosmology* based on the hypothesis of continual creation. It avoided the age difficulty because it prescribed only an average

age for the contents of the universe, not an age of the universe itself.

This theory played a crucial rôle for nearly 20 years. *a*) It provided clear-cut predictions to be tested by observation. Observers had a better-defined task in seeking some violation of the predictions, rather than in simply amassing information in the hope of being able to fit a model to it. *b*) It stimulated much work of lasting value on the origin of the chemical elements—but that is another story.

I shall shortly mention why most cosmologists have now rejected steady-state cosmology in its original simple form. I shall not have time to discuss further possible developments of the basic notions involved. But I must remark that big-bang cosmology and steady-state cosmology may be regarded as extreme cases and, maybe, a more satisfactory treatment will ultimately be found between these extremes.

Before leaving the question of age, we must recall how the work of Baade, announced in 1952 and followed by that of Sandage and others, has greatly changed the numerical estimates. The question of age is not now so embarrassing. There are indications that the remaining difficulties will be solved by very recent revised estimates of the ages of stars and galaxies, rather than by cosmology—that again is another story.

### Counts of Radio Sources.
After Hubble, the next greatest advance in observational cosmology was undoubtedly the discovery, from about 1950 onwards, by M. Ryle (Cambridge) and his colleagues of *radio galaxies* and their development of methods of counting such sources completely and reliably down to various limiting flux-densities (or radio apparent magnitudes). Unless there is some gross misinterpretation of these counts, they show conclusively that the universe is *not in a steady-state* on the scale on which it is explored by this means.

### Quasi-stellar Objects.
The radio-galaxies just mentioned appear to be objects similar to optical galaxies but undergoing some kind of upheaval. In recent years, radio astronomy has led also to the discovery of quasi-stellar radio sources and thus to the so-called *quasi-stellar objects* generally. These are also observable optically, and they show redshifts up to more than 10 times the greatest shown by optical galaxies (such as those studied by Hubble). If they are what they appear to be, they enable astronomers to explore a volume of the universe about 1000 times greater than they could do before.

Nature thus shows wonderful consideration for cosmologists! When they reach the limit of what they can explore by observations of ordinary galaxies, nature provides another type of object that can be observed ten times as far away. Unfortunately, cosmologists have not been able to keep pace with nature. Owing apparently to the great spread of intrinsic properties of quasi-stellar sources, astronomers have not yet learned much more about the universe in the large by studying these objects.

There is, however, one feature of special interest to us here. A particular value of the redshift ($z = 1.95$ in the conventional way of measuring the redshift) is found to occur in unexpectedly many cases in both the emission and the absorption spectra of quasi-stellar objects. Several authors have remarked how this suggests that the universe does indeed behave like the Lemaître model, supposing that the long pause ("stagnation") in the model occurred at expansion-factor corresponding to $z = 1.95$. Thus a quasi-stellar object seen at any epoch during the pause would show $z \approx 1.95$ in emission; any such object seen as it was before the pause would show redshift $z \approx 1.95$ in absorption, if the object happens to be viewed through clouds of material at distances corresponding to the pause. The fact that it is a pause means, of course, that we have a bigger chance of seeing things as they were during the pause rather than at other epochs in the past.

If the occurrence of an apparently favoured redshift is not a matter of extraordinary chance, then the explanation in terms of Lemaître's model is the simplest that has been proposed. Much further investigation is needed before a decision is reached. In any case, however, it is surely a token of the fruitfulness and genius of a man's ideas that they should play a key rôle in the discussion of new discoveries made several decades later.

### Microwave Background Radiation.
Radio astronomy has led also to yet another remarkable discovery. Again supposing the observations to have been correctly interpreted, it is the most significant of all for cosmology since the discovery of Hubble's law. For it supplies us with a new *quantitative parameter* of the cosmos, like Hubble's constant or like the mean density of matter in the universe. It is the *microwave background radiation;* apparently, the universe is filled with radiation that in our cosmic vicinity at our cosmic epoch is thermal (black-body) radiation of temperature about 3°K.

The existence of such background radiation was predicted, about 20 years before its discovery, by Gamow and his school on the basis of Friedman-Lemaître cosmology. But 20 years earlier still, in his first paper on cosmology, Lemaître had explicitly allowed for its existence by the manner in which he chose to formulate his treatment. This was still another example of the sureness of Lemaître's insight.

### Helium Problem.
The discovery of the background radiation has far-reaching implications. It implies that the universe began with a hot big bang ("universal fireball"). From the present local values of the temperature of the radiation and of the mean density of matter, relativistic cosmology enables us to infer the sequence of events in the first minutes of the history of the cosmos! In particular, it follows that matter would have emerged from this stage with about a 25 percent helium content by mass. Such matter would then provide the raw material for the later formation of galaxies. This is about the value of the helium content actually inferred from observation. It is much more than the amount of helium that could have been produced in the interiors of known stars since they were first formed. In fact, the existence of so much helium in the universe had by about 1964 presented astrophysicists with what they began to call the *helium problem.*

The accidental discovery (of the background radiation) that turned out to have been predicted by the theory and that promptly solved another major problem (the helium problem) was what finally convinced most cosmologists of the basic validity of big-bang cosmology.

I understand from Professor O. Godart that this was about the last scientific topic he discussed with Monseigneur Lemaître a few days before his death, and that Lemaître was deeply interested. He had good cause to be. For Lemaître himself had long ago contemplated the big bang as the spontaneous decay of a "primeval atom" that comprised all the material of the universe. In its day it was an astonishing anticipation of the up-to-date picture of the "universal fireball."

## Present Position in Cosmology

In the years since Lemaître and his early contemporaries started their work, hundreds of other investigators have written thousands of contributions on cosmology. What has all this achieved? In broad outline, we may claim: a) It has established the ability to discuss the universe in the large in a meaningful way. b) It has demonstrated that the behaviour of the universe in the large is governed by gravitational processes alone (in the sense of general relativity). That is to say, in proceeding from terrestrial physics to cosmic physics there appears to be no demand for new physical effects—provided we regard terms involving the cosmical constant $\Lambda$ as part of the treatment of gravitation in general relativity. In particular, the work shows there must be a simple relation between the mean density of matter in the universe, the Hubble time, and the familiar gravitational constant. c) It has demonstrated, further, that the universe started with a hot big bang that determined the present radiation density in the universe and also the value of the helium abundance (apart from effects of nucleosynthesis in stars). d) It has activated important work in other fields, relativity, theory of gravitation, nuclear physics, stellar and galactic evolution and so on. Also it has provoked fresh thought about the nature and scope of physical theory.

Outstanding weaknesses remain, however, and that is why I say we may claim all this—not that we are bound to accept it. For instance, the discovery of quasars and their properties has revived the question as to whether the redshift is truly "cosmological," or is due to some undetected or unknown physical process in space or in the sources. As another example, the black-body character of the background radiation is still in some doubt on the observational side. Or we may be troubled simply by the fact that cosmology has not accounted for more known features of the astronomical universe.

## Astronomical Problems

Rather than explore the criticisms themselves, it seems better here to mention some definite problems in astronomy, observational and theoretical, whose solution would in any case be important but which might also serve to allay the misgivings to which I have just made reference. I shall mention only three problems which happen to be closely related to each other and to a particular interest of Lemaître.

*Intergalactic Material.* Observational astronomy studies directly only the "luminous" matter in the universe, that is to say, matter that transmits some detectable radiation. We naturally ask, Is there a significant amount of non-luminous matter? This might be diffuse material throughout the whole universe, or it might be associated with clusters of galaxies. Or there might be gravitationally collapsed bodies from which radiation can no longer escape. And, of course, there are other possibilities.

Now most attempts to fit a cosmological model to observations have in fact implied that the total mean density of matter in the universe is much greater (maybe 100 times) than the mean density of luminous matter. Also, studies of the dynamics of clusters of galaxies imply that in most cases the total mass of a cluster considerably exceeds the sum of the masses of its luminous member-galaxies. Thus the question as to whether the theories are mistaken or whether the universe does indeed contain a vast amount of "missing mass" is perhaps the most important of all for present-day astronomy. It can be settled, presumably, only by new observations.

*Matter and Anti-matter.* Still more fundamental is the question as to whether the universe is made of the same sort of matter as that of which the Earth, and evidently the entire Milky Way Galaxy, is made, or whether the universe as a whole is made of comparable amounts of "matter" and of "anti-matter."

*Formation of Galaxies.* Whether or not there now remains much material besides the galaxies, we must infer that the galaxies were formed out of material in some other, supposedly more diffuse, state. There is no accepted theory as to how this took place.

Lemaître was, I think, the first to seek to relate this problem to the general evolution of the universe as studied by relativistic cosmology. He considered that the time spent near the Einstein state in the Lemaître model was the time when galaxies and clusters of galaxies were formed out of gas-clouds. For in this phase there is a near-balance between gravitational attraction and cosmical repulsion (represented by the $\Lambda$-terms) that produces the sort of instability that may lead to condensations.

The two essential features of Lemaître's ideas were a) that the expansion of the universe plays an essential rôle in galaxy-formation, and b) that the formation and evolution of clusters must be deemed an essential ingredient in the problem. These are widely accepted. But no one has yet had much success in deriving quantitative predictions. For one thing, we probably need much more extensive observational studies of the composition of clusters in order to define the problem in more precise terms.

Thus we have inherited from Lemaître these and other great problems that are of fundamental importance for astronomy itself and any progress in solving them would be of the utmost value in the further advance of cosmology.

## Why Is Cosmology Possible?

I wish now to review the situation in a way that will, I hope, give some fresh insight.

Nearly half a century ago, Friedman and Lemaître produced cosmological models that were the simplest possible models admitted by the theory. They were homogeneous and isotropic with no irregularities and each model admitted at most one singularity in the whole universe (in space and time). Such models could have been cast aside because the actual universe is infinitely more complex—the models could be no more than toys. This is what did almost occur. However, thanks to some seeming accidents of history, the work of Friedman and of Lemaître was rescued from oblivion. During more than 40 years, astronomers have been striving to interpret the large-scale behaviour of our incredibly complicated universe in terms of these simple models—these toys. There has been little serious attempt to construct more elaborate models.

Evidently Friedman and Lemaître had wrought better than they knew or claimed. We must surely ask why this is the case.

We begin by recalling that in recent years S. W. Hawking (Cambridge) and his associates have proved certain remarkable theorems of great generality in the theory of general relativity. They show that any cosmological model that is in accordance with the theory must possess one or more singularities. Any such singularity would be interpreted physically as some sort of big bang. Of course, we cannot assert that the actual universe *must* obey general relativity. As a result of Hawking's work, however, we can say that if our ideas about space, time and matter are generally valid, then the existence of the universe is inconceivable without the occurrence of one or more big bangs. After all, this is the intuitive view of most people.

On very general grounds, therefore, we "expect" a cataclysmic start (or starts) to the observed universe. This implies that the universe on the very large scale is evolving. But that brings us to our next problem—is it reasonable to suppose that we can know how it is evolving? For we know that, if we have any unique ordinary physical system, it is in general impossible from observations of its present behaviour to infer much about its past behaviour.

Nevertheless, the case of the universe in the large is different. If the universe as a whole is evolving, then by observing parts distant in space, we observe them early in cosmic time. Thus the evolution of the universe is [in] principle observable and, consequently, we presume that it is subject to scientific treatment.

But there are difficulties still. The further away the parts of the universe we observe, the less information we can get about them. In particular, we certainly cannot observe the big bang itself. After all, therefore, we might conclude that we should be unable to construct a useful cosmological model because we can never know enough about initial conditions.

We must not lose hope, however! It is now beginning to appear that the properties of the universe as we know it are to a considerable degree independent of initial conditions. Nature seems to operate a principle of compensation in our favour. Let us consider the available examples of this feature.

*a*) E. A. Milne long ago remarked that, if we have a localized cloud of particles in ordinary space, whatever be their initial motions (so long as these motions are not precisely identical), sooner or later any one of the particles will see every other receding directly away from himself with speed proportional to distance. Thus, after a sufficient time, the particles must be moving in accordance with "Hubble's law," independently of any special initial conditions.

*b*) In 1967, C. W. Misner made the important discovery that, even were the universe highly anisotropic just following the big bang, the phenomenon of *neutrino viscosity* would rapidly produce a high degree of isotropy. Thus, after a sufficient time, the universe must be isotropic, independently of the initial conditions.

*c*) A universe that is everywhere isotropic must also be homogeneous. So I think we must conclude that the universe must be homogeneous, independently of the initial conditions.

*d*) R. V. Wagoner, W. A. Fowler and F. Hoyle have recently shown that the hot big bang has two most important implications regarding the relative abundances of the chemical elements in the hot big-bang universe (apart from the effects of nuclear reactions in the stars). First, there is no sense in an initial chemical composition. The present composition depends only upon what we may call the amount of matter available, and not upon what that matter, in any sense, was originally made of. Secondly, the present composition is not at all sensitive even to that one parameter. In particular, the hydrogen to helium ratio is about 3 to 1 by mass for a very wide range of initial conditions. Thus we can say that the universe must have a chemical composition with major features of the sort actually found, almost independently of special initial conditions.

To summarize, we may claim that the universe must have had a cataclysmic start and that, almost independently of initial conditions, it must be homogeneous and isotropic, it must exhibit a Hubble behaviour, and it must have a particular chemical composition. All those properties appear to be in general agreement with experience.

## Principle of Compensation

We may now tentatively propound a principle of compensation in cosmology:

> The less information we can get, the less we need in order to make predictions that are testable by observation.

In other words, if in regard to some feature of the universe we require information of a certain sort, then if in principle only a little information can be obtained, then that which can be obtained should nevertheless suffice to predict the present state of the universe in regard to this feature.

The principle is not at present demonstrable in any precise form. As I present it, it is obviously speculative and inexact. But the underlying idea is borne out by the examples I have just described. If it is correct, it has an important converse:

> From our observations of the universe we can infer little about its very early state.

Clearly, if we are right in asserting that the universe must exhibit a Hubble behaviour almost independently of its "initial" state, then when we do in fact observe the universe to exhibit this behaviour we can infer almost nothing about the "initial" state of motion. And so on as regards the other properties we have been discussing.

The principle here presented appears to make sense of so much in existing work on cosmology that I think it has to be regarded as significant. The following are some relevant considerations:

*a*) The principle indicates why the simple Friedman-Lemaître models have proved to be so significant. For it now appears that even if we had started with more general models we should subsequently have inferred that they would have to satisfy conditions (of isotropy etc.) that would then restrict them to the Friedman-Lemaître cases.

*b*) The principle justifies much of the effort of current cosmological research. Tremendous labour is being lavished upon the problem of selecting from all the Friedman-Lemaître cases available the model that best fits all the available empirical information. Although no particular selection has earned general agreement, there is no indication that the work is leading to a nonsense. Our principle makes us optimistic partly by suggesting that if we do our best with whatever information we can get, we may hope to reach a significant outcome, and partly because, if we are right in what we have just said in *a*) then the Friedman-Lemaître cases are more significant that they originally appeared to be.

*c*) The principle begins to indicate how we may be led to a *unique* cosmological model. Since there is by definition only one universe, a satisfactory theory ought to admit only one acceptable theoretical model universe. On the other hand, we might say that according to relativistic cosmology we could start with infinitely many possibilities as regards anisotropy, inhomogeneity, initial chemical composition, and so on. *A priori*, all these possibilities seem to have the same status. However, it is surely most unsatisfying to say that the actual universe might be any of these, and that it just happens to be some particular one! But in our discussion we have seen reason to assert that the universe must not be anisotropic or inhomogeneous (in the large) and that an initial chemical composition is meaningless, and so on. Thus on general grounds we are already able enormously to reduce the range of possibilities. It is reasonable to believe that we shall find further general grounds for reducing the range still further.

*d*) The present point of view gives added confidence in the general character of the laws of modern physics. We do not believe that the universe started with a big bang because general relativity says it must. Rather we gain confidence in general relativity because it yields this plausible inference! Again, it should be noted that the inferences I mentioned about isotropy, etc., come from various different parts of physics—the theory of nuclear reactions, the theory of neutrinos, and so on. It seems to be a highly gratifying feature of these theories that they do enable us thus to make acceptable inferences about the universe in the large.

While I do not wish to labour this point, I must say that it seems to me to provide a new test for physical theory. Besides asking whether a theory makes accurate predictions regarding particular experiments, we can ask if it tells us that the universe in the large must be expected to have certain general features. If the universe does have such features, we do not say that the theory has explained why it has them, but we infer that in using the theory we are thinking about the universe in a satisfactory way.

*e*) Some years ago I formulated a principle of uncertainty in cosmology. This depended upon what appears to be an inherent difficulty in obtaining sufficient information to make predictions with certainty. It now seems that the principle of compensation is complementary to this in showing how the uncertainty is distributed. Apparently we can hope to infer the present state of the universe with considerable confidence but we can never have precise knowledge of the earliest stages.

*f*) The principle illuminates the potentialities of cosmology. We have noticed certain features of the cosmos that must hold good, according to various theories, almost independently of special initial conditions. Perhaps it is obvious that these are precisely the properties that we must *expect* the universe to possess. Now natural philosophers through the ages (from Aristotle to Eddington and Milne) have striven in various ways to show by pure reason that the universe of logical necessity must be what we see it to be. This possibility seems to be exceedingly remote. What we do seem to be finding is that from certain physical theories we can infer that the universe almost necessarily has certain properties. These theories are highly sophisticated and are not themselves derived by pure reason. Nevertheless, the possibility of making such inferences is a big intellectual advance and may point the way to further advances.

## Prospects

Can we say what any of these advances might be?

We should look for other features of the cosmos that might be almost independent of initial conditions. The masses (mean masses or maximum masses) of galaxies might possibly be an example.

We hope that in due course cosmology will tell us why all electrons are alike, why all protons are alike, and so on, and why matter is composed of such elementary particles.

These are just indications of future possibilities. In this lecture I have struck a note of optimism partly because Lemaître was an optimistic man. This optimism is based upon our finding deep reasons for accepting some of the apparently simple ideas of Lemaître (and of some of his contemporaries). In spite of this, the universe must surely be more complicated than anyone has yet been able to apprehend—and I think Lemaître would agree.

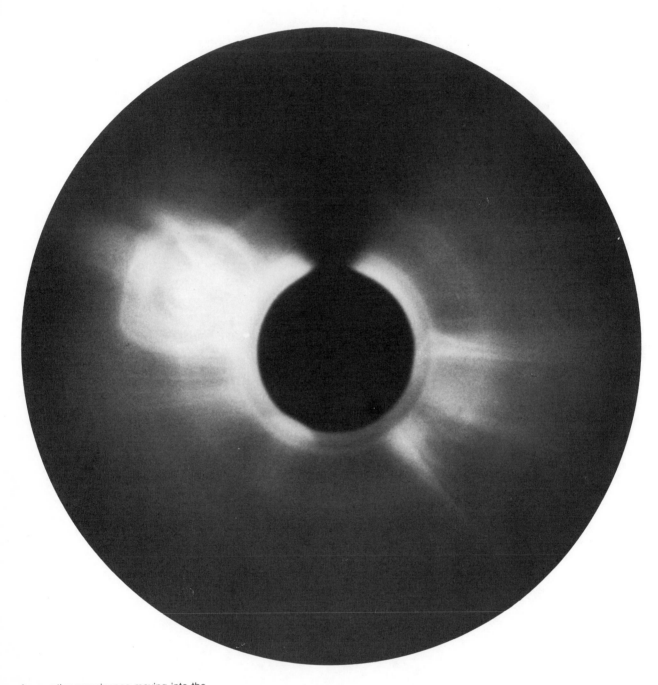

An eruptive prominence moving into the solar corona with an apparent speed of 450 km/sec, as photographed from Skylab's Apollo Telescope Mount (10 June 1973, 09ʰ43ᵐ GMT, 9 second exposure, unpolarized). The occulting disk in the center of the photograph blocks light from the photosphere, and it has a projected diameter of three solar radii. [High Altitude Observatory, Boulder, Colorado.]

PART

# SPACE-AGE ASTRONOMY

"The largest clouds, each roughly 30 times the earth's diameter, sped outward at velocities reaching 1,100 kilometers per second." Editor Maran and solar physicist Roger J. Thomas describe the major findings of a sun-watching satellite in "The OSO-7 Year of Discovery." The next major spacecraft after *OSO-7* to conduct solar research was the manned *Skylab*, which also investigated Comet Kohoutek, as mentioned in Section 3, and other scientific subjects. Space science reporter Everly Driscoll relates some preliminary results from *Skylab's* battery of solar telescopes in "Peeking Beneath the Sun's Skirts."

According to radio astronomer Robert M. Hjellming, studies of the celestial x-ray and radio source Cygnus X-3 may hold the record for "more investigation of a single object, by more astronomers, in a shorter period of time, than ever before in the history of astronomy." He tells why in "An Astronomical Puzzle Called Cygnus X-3." Hjellming mentions *thermal bremsstrahlung*, which is the normal process by which continuum radiation (whether in the x-ray, visible light, infrared, or radio region of the spectrum) is emitted by a hot ionized gas.

Cygnus X-3 is only one of a wide variety of objects, many of them unusual, that have been detected with x-ray satellites, as is made clear by a pioneer in this field, Herbert Friedman of the Naval Research Laboratory, in "Cosmic X-Ray Sources: A Progress Report." In addition to thermal bremsstrahlung as a method by which x rays are produced in space, Friedman discusses the *inverse Compton* process. In this process, x rays are generated when lower-energy photons such as those of the radio portion of the spectrum collide with high-energy cosmic ray particles; this produces high-energy (i.e., x-ray) photons and lower-energy cosmic rays.

Of even higher energy than x rays are gamma rays. The most important discovery in the new field of gamma-ray astronomy thus far is that presented in the original research paper "Observations of Gamma-Ray Bursts of Cosmic

Origin," by physicists Ray W. Klebesadel, Ian B. Strong, and Roy A. Olson. They tell how the *Vela* satellites, which were designed to detect gamma rays from nuclear weapons tests in space, found naturally occurring, strong, short-lived bursts of gamma rays that come from locations in our galaxy (or beyond) that are still unknown. Among the technical terms that they use are MeV (million electron volts), a unit of energy of a gamma ray that corresponds to a wavelength of 0.0124 angstrom; CsI (cesium iodide), a type of crystal that scintillates (i.e., emits visible light) when struck by an energetic particle or a gamma ray; high-Z shield, meaning a protective layer of a substance having a large atomic number; and anticoincidence shielding, which is an instrumental provision that aids in distinguishing scintillation light produced by gamma rays from that produced by cosmic-ray particles.

# The OSO-7 Year
# of Discovery

Stephen P. Maran and Roger J. Thomas

On September 29, 1971, a launch-vehicle malfunction sent the seventh Orbiting Solar Observatory into an unplanned eccentric orbit. Spinning rapidly and apparently unable to lock onto the sun, OSO 7 seemed to be doomed as its battery voltage dropped steadily. But the spacecraft was saved, and on September 5-8, 1972, some 125 astronomers gathered at Stanford University in California to learn of the remarkable observations obtained during OSO 7's first year in space.[1]

At this third OSO workshop, John Donley of Goddard Space Flight Center told how ground controllers struggled to save OSO 7 on the day of launch. Larger and heavier (1,400 pounds) than any previous satellite in the OSO series, this was the first one whose trajectory required a coasting period between two burns of the Delta rocket's second stage. The second firing miscarried, sending OSO 7 into an orbit with perigee and apogee of 329 and 575 kilometers, respectively. A circular orbit at 550 kilometers above the earth's surface had been intended.

On separation from the second stage, the spacecraft was in a flat spin at roughly 60 turns per minute. Although it became stabilized, the orientation was such that the sun shone on the bottom portion of the OSO "wheel," instead of on the sail section that bears the solar cells. Without electrical-power generation, the batteries began to drain, and on the satellite's first pass over the Johannesburg tracking station in Africa, urgent instructions were sent to "power down" nonessential systems. In the next eight hours, 2,352 commands were sent.

Success was achieved when the sun was acquired by a sensor with a 90-degree field of view. A gyro system provided stability as bursts of control gas pitched the OSO over until the sail faced sunward and electricity began to flow from the solar cells. The spacecraft bus voltage had dropped to only 17.0 volts, just 0.2 volt above the failure threshold. The two systems that saved the day—the wide-angle sun sensor and the gyro—were flown on this OSO spacecraft for the first time.

From *Sky and Telescope*, vol. 45, no. 4, 1973. Reprinted with permission.

For directing the emergency efforts that saved OSO 7, Goddard's John M. Thole has been awarded the NASA exceptional service medal.

Equally as exciting as the launch drama are the OSO 7 scientific findings, which reveal hitherto unknown processes in solar flares and some remarkable properties of the corona.

## Nuclear Reactions on the Sun's Surface

Gamma-ray emission lines provide evidence of nuclear reactions in solar flares, according to David J. Forrest of the University of New Hampshire. The measurements were made during the large flares of August 4 and 7, 1972, with an instrument designed by Edward L. Chupp and carried in the wheel of OSO 7. The data clearly show line emission at energies of about 0.5 MeV (million electron volts) and 2.2 MeV, with weaker enhancements at 4.4 MeV and 6.1 MeV.[2]

What processes cause this radiation? At the lowest energy, what has been observed is undoubtedly the 0.511-MeV line produced by the mutual annihilation of electrons and positrons. Positrons are antimatter particles and they cannot long survive in the ordinary-matter environment of the sun. Thus, the positrons involved in the events of August 4th and 7th must have been produced in the flares themselves. Theory suggests that short-lived radioisotopes of carbon, nitrogen, and oxygen were formed by nuclear reactions in the flares, then decayed to stable isotopes by releasing positrons.

The other strong emission can be identified with the 2.22-MeV line of deuterium (heavy hydrogen). This radiation occurs when an excited deuterium nucleus, produced by the collision of a neutron and a proton in the flare, decays to a lower state, emitting a gamma ray.

We are used to thinking of nuclear reactions deep within the sun, where they are responsible for its basic energy generation. At the enormous temperatures and pressures that characterize the solar core, thermonuclear reactions are induced. However, Dr. Chupp emphasizes that the nuclear processes in flares as revealed by OSO 7 are probably of another type—

On September 29, 1972, by coincidence one year after OSO 7's launch, NASA officials visited the OSO control room at Goddard Space Flight Center. Author Stephen P. Maran points to a hydrogen-alpha television picture, transmitted from a nearby ground-based telescope to help astronomers select interesting regions for study by the satellite. At right is an extreme-ultraviolet map of the sun relayed from OSO 7. From left to right, the visitors are John F. Clark, director of Goddard; John E. Naugle, NASA associate administrator for space science, and James C. Fletcher, administrator of NASA.

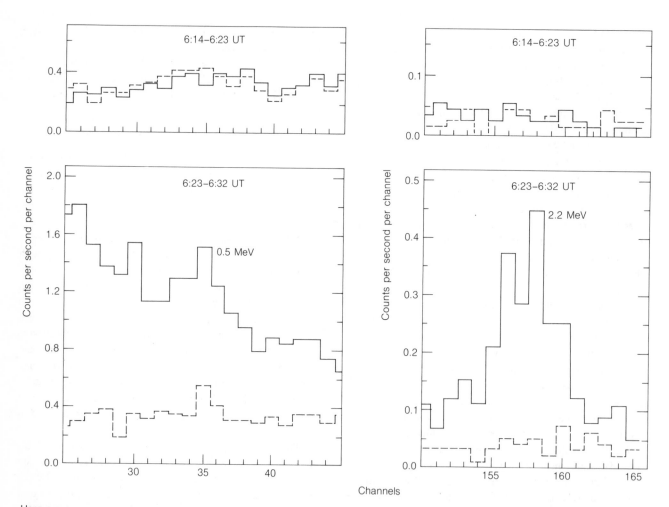

Here are gamma-ray records obtained by OSO 7 when the great solar flare occurred on August 4, 1972. The upper pair shows preflare conditions for the sun (solid lines) and the sky background (dashed lines). In the lower pair, the peak centered in the left-hand spectrum has been identified as the 0.511 MeV line caused by positron-electron annihilation, while the right-hand peak is the 2.22-MeV line from deuterium formation when hydrogen nuclei absorb thermal neutrons. [University of New Hampshire chart.]

*nonthermal reactions.* They resemble the effects observed in an accelerator laboratory when a beam of fast protons strikes a target. Thus, there appear to be beams of fast protons in solar flares, possibly complementing the fast electron streams that have long been indicated by observations of radio bursts from flares.

Some flare electron streams also produce bursts of hard X-rays, on which another OSO 7 instrument has concentrated. Although we do not know very much about the stream properties, several models have been proposed. Two ideas in particular are the "thick target" and "thin target" bremsstrahlung theories. In the thick-target case, the fast electrons lose essentially all of their energy in the region where the X-rays are produced; in the thin-target case, the electrons escape from the emission region. David L. McKenzie of Aerospace Corp. reported relevant observations with the hard X-ray telescope designed by Laurence E. Peterson of the University of California at San Diego.

A detailed analysis was made of a burst of hard X-rays that occurred on November 16, 1971, at 5:19 Universal time. It was probably associated with a subflare in the southeast quadrant of the sun's disk. The burst lasted one minute, as observed at energies of 30 keV (thousand electron volts) to 44 keV, that is, at wavelengths of 0.41 to 0.28 angstrom. Associated with it was a longer burst at low energies, with the characteristic spectrum of thermal radiation from a plasma at 20,000,000°

Kelvin. Calculations show that if the thick-target theory is correct, the energy carried by the fast electron stream was sufficient to heat the ambient plasma to the temperature inferred from the low-energy burst. It is perhaps too soon to be sure that this theory is right, but we are encouraged by the emergence of a consistent picture of these flare phenomena.

## Holes and Poles in the Corona

An early OSO 7 finding was the observation by Goddard's Werner M. Neupert of high-latitude regions in the corona where there is reduced emission at X-ray and extreme-ultraviolet wavelengths. Such an effect might be due either to a lower electron density or a lower temperature in these regions. To test these possibilities, Dr. Neupert obtained maps of the corona's radiation at the wavelengths of several emission lines, each arising mainly from matter at different characteristic temperatures, in the range from 800,000° to 3,000,000° K. He noted that the emission was fairly evenly distributed on the lower-temperature maps, but increasingly uneven at the higher temperatures, where it was concentrated toward moderate and lower latitudes.

Typical coronal temperatures out to about 1.5 solar radii above the surface are 3,000,000° to 4,000,000° in active regions, about 1,700,000° in the quiet corona, and roughly 1,000,000° near the "poles." Occasionally, however, tempera-

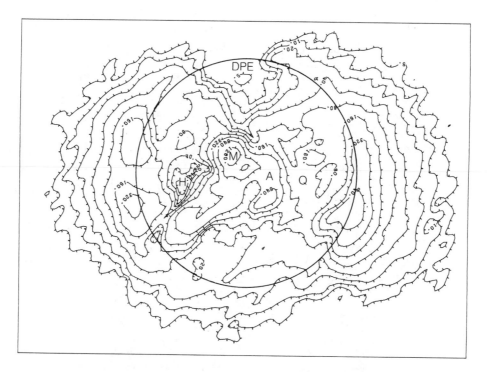

This map of the sun's corona, observed by OSO 7 at 14:04 Universal time on August 4th, has been plotted by NASA computer in the light of Fe XV (iron ionized 14 times) at 284.15 angstroms in the extreme ultraviolet. The contours represent photoelectron counts per unit time, 5 being outermost; they double each time and reach 1,280 at *M*, which labels active region 11976, where several large flares originated. Next to it *A* marks a coronal arch extending to another active region, while *Q* is a region of quiet corona. At the north pole of the sun, *DPE* indicates strongly depressed polar emission. [NASA chart from Goddard Space Flight Center.]

tures close to 1,000,000° are found in small regions at low latitudes. Called *coronal holes,* they are becoming recognized as a basic phenomenon in the corona, with a distinct set of properties that can be determined from satellite observations.

The first extensive study of a coronal hole was performed with Harvard Observatory's ultraviolet spectrometer on OSO 4. Further data came from OSO 6 and OSO 7, and from a recent sounding-rocket experiment. This was one of the liveliest topics discussed at the Stanford meeting.

George L. Withbroe of Harvard summarized physical conditions in a typical coronal hole as follows. The temperature is roughly 600,000° lower than in the quiet corona, and the gas pressure is roughly one-third of the normal value. The temperature gradient (rate of temperature increase with altitude) near the bottom of the hole, at the chromosphere-corona transition zone, is 10 times smaller than in the transition zone of the quiet corona. This last result implies that the transfer of heat by electron conduction downward from the corona to the chromosphere is also 10 times smaller in a hole.

At Harvard, G. Noci has deduced that the solar wind escaping from the corona should be enhanced in the region of a hole. This agrees with findings by Allen S. Krieger and Adrienne F. Timothy of American Science and Engineering and Edmond C. Roelof of the University of New Hampshire. These three scientists studied X-ray coronal photographs obtained by a rocket on November 24, 1970. A coronal hole that appears in these pictures was compared with data on the solar magnetic field and the solar wind. Uncertainties were necessarily introduced by inferring solar-wind trajectories near the sun from measurements made by earth-orbiting satellites. Nevertheless, it appears likely that a recurrent high-velocity stream in the solar wind arose from the vicinity of the coronal hole, and that the magnetic field lines were arranged in a diverging radial pattern in the hole.

Stanford scientist Peter A. Sturrock noted that there seem to be no unusual structures or phenomena in the photosphere beneath coronal holes, and the magnetic field strengths there are not abnormal. Hence, the diverging pattern of field lines in a hole probably explains the apparent solar-wind enhancement. The basic mechanism for the escape of the solar wind from the corona has been compared to the acceleration of exhaust gases in a supersonic rocket, resulting from the constriction associated with the rocket nozzle. A rapidly diverging magnetic field in a coronal hole would move the "throat" of the "nozzle" closer to the surface of the sun.

This would produce an enhanced solar wind, and if the energy lost in this way were sufficiently great the power budget of the corona would be significantly affected, leading to reduced coronal temperature and density. Thus, the determination of coronal-hole properties by Dr. Withbroe and his associates is consistent with the relationship to the solar wind and magnetic field reported by Drs. Krieger, Timothy, and Roelof.

However, is it certain that no underlying photospheric

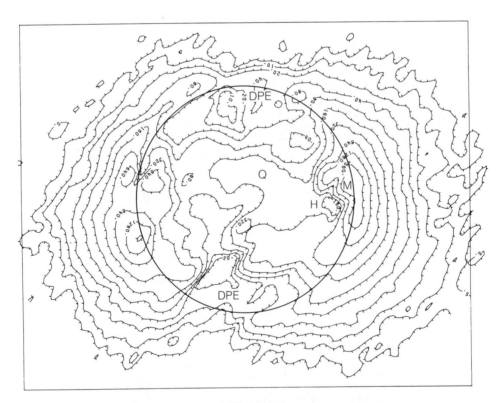

On August 10th at 15:15 UT, this 284-angstrom map was obtained. In six days McMath region 11976 had moved to the sun's west limb, but the associated coronal enhancement was still intense (contour maximum 2,560) and east of it was the same coronal hole, marked *H,* seen on the August 4th map. The quiet corona *Q* was quite extensive, and there was depressed polar emission at both poles. [NASA chart from Goddard Space Flight Center.]

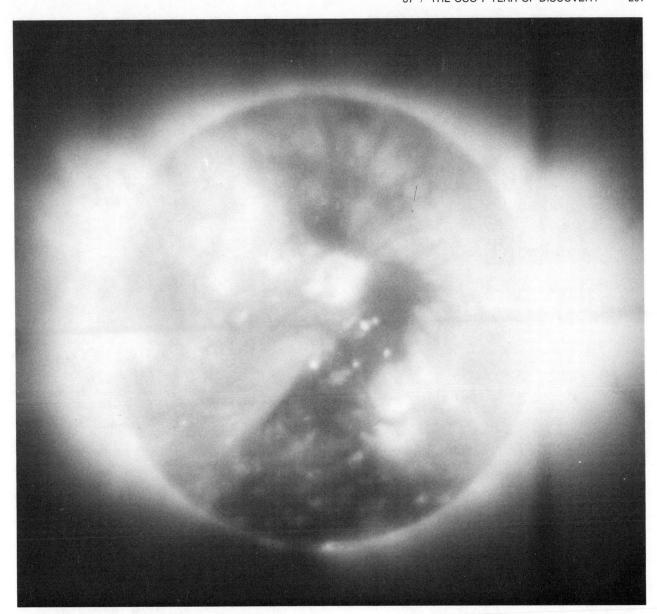

This picture of the sun at X-ray wavelengths of 3-35 and 44-51 angstroms was obtained by rocket on November 24, 1970. A. S. Krieger, A. Timothy and E. C. Roelof classify the entire dark region as a coronal hole; it extends from latitude 10° north, longitude 20° west of the central meridian, almost to the sun's south pole. [American Science and Engineering photograph.]

structure is associated with a coronal hole? Relatively little has been done to check this conclusion in detail. We need careful study of the photosphere in these areas, coordinated with the satellite observations that reveal the holes. This became clear in the discussions at Stanford and, as a result, arrangements have been made by author Thomas to notify solar astronomers at Kitt Peak National Observatory in Arizona whenever a new hole is spotted by OSO 7.

### Coronal Streams and the Interplanetary Blast

The sun's most spectacular attribute, the white-light outer corona that delights total-eclipse viewers, is also among its least understood regions. This results from lack of opportunity, not lack of interest, since significant observations have been made only during the precious moments of eclipse and from a few balloons and rockets. (Ground-based coronagraphs ordinarily observe only the innermost part of the corona.)

A white-light coronagraph, with an external occulter and an SEC vidicon imaging detector, was flown on OSO 7 to remedy this situation. Developed by Naval Research Laboratory scientists under Richard Tousey, it is systematically monitoring the outer corona. For the first time we can study coronal streamers as they form, evolve, and dissipate. In addition, several kinds of violent dynamic events have been seen.

Besides these morphological studies, physical parameters of coronal phenomena can be determined from the calibrated

Above. A candid photograph shows E. Tandberg-Hanssen (left) of the High Altitude Observatory in Colorado conversing with R. Tousey, leader of the group of OSO 7 experimenters at the Naval Research Laboratory in Washington, D.C.

Right: Visiting scientist Z. Svestka of the Fraunhofer Institute in West Germany presents the Stanford audience with thoughts on the nature of solar flares. [Maran photographs.]

measurements of surface brightness and polarization, after instrumental effects such as vignetting are removed. Mercury, Venus, the moon, and Saturn have passed through the coronagraph's sun-centered field of view, providing discrete geometrical targets that facilitate in-orbit tests of the equipment's performance. The 1st-magnitude stars Spica and Regulus have also been observed in the same manner.

From the first year's observations, Dr. Tousey reported that almost all coronal streamers are straight, being directed outward from the sun in the region surveyed by the coronagraph, three to 10 solar radii. Arched and loop structures, sometimes seen in the inner corona, are very rare in the outer. One of the few exceptions, a streamer that possibly was curved, and which did not point backward to the solar disk, was recorded on July 21, 1972.

Events in the lower atmosphere of the sun may affect the coronal streamers. Thus, Tousey's group found that a streamer broadened and (for several hours) brightened on June 16th, during a period of enhanced flare activity. Earlier, on February 8th, an eruptive prominence occurred on the sun's northeast limb. Several hours after this event was noted at Sacramento Peak Observatory in New Mexico, the ejected material had reached the outer corona, attaining velocities of 250 to 400 kilometers per second, as deduced from the OSO 7 coronagrams.

A completely different phenomenon was seen on August 2, 1972, when a general expansion of the corona in the northeast quadrant was observed in association with a large flare. This phenomenon resembled one aspect—the "interplanetary blast"—of the great coronal events of December 13–14, 1971.

As reported in SKY AND TELESCOPE (March, 1972, page 158), a bright streamer was disrupted and bright plasma clouds were ejected from the sun. The largest clouds, each roughly 30 times the earth's diameter, sped outward at velocities reaching 1,100 kilometers per second. The clouds originated in the low corona or the high chromosphere, which are hidden by the coronagraph occulter. But by projecting the motions of the four largest clouds backward to the sun, Guenter Brueckner of NRL was able to identify them as the stimuli of the four Type-II solar bursts that were recorded by the radioheliograph at Culgoora, Australia.

It has long been known that such radio bursts signify the outward motion of solar electron clouds; now at last we have recorded their visual counterparts. But in addition to the clouds, there was an overall outward motion of the gas in a large region of the corona—the interplanetary blast. Although the plasma clouds carried more than $1.6 \times 10^{31}$ ergs of kinetic energy, Dr. Brueckner concluded that the dominant process was the interplanetary blast. While much less obvious on the video pictures than the bright clouds, the blast involved at least 12 times as many particles as were concentrated in the clouds, and released more than five times as much energy. Later, on December 15th, the streamer had reappeared, but apparently it contained much less matter than on the 13th.

Shortly after the conference, on September 23, 1972, the NRL instrument recorded a giant transient loop in the corona. With its top five solar radii above the solar surface, the loop extended downward at least to the three-solar-radii level, where its "feet" were separated by a full solar diameter. The

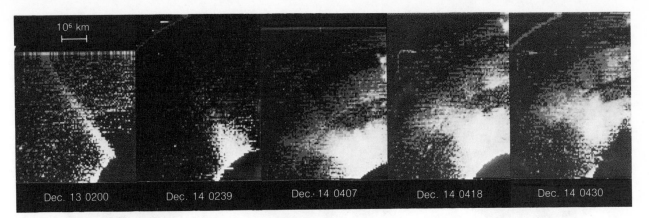

This series of white-light coronagrams covers the region of the great coronal events on December 13-14, 1971, beginning with the long straight streamer at extreme left. Universal times are indicated. The remaining frames may be compared with those on page 159 of *Sky and Telescope* for March, 1972. Here the detail has been considerably improved by electronic processing to remove the background of the quiet corona. [Naval Research Laboratory photograph.]

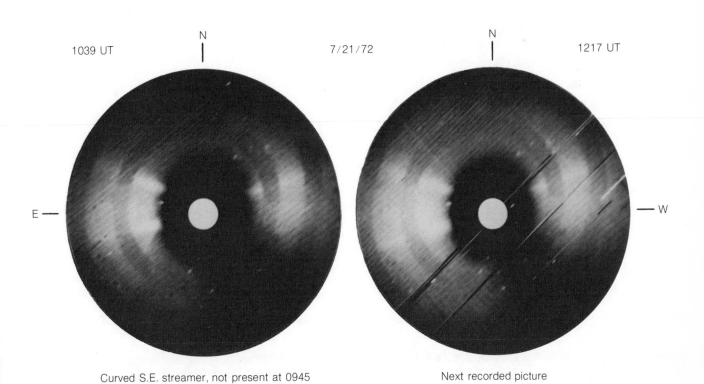

The transient curved streamer at 8 o'clock in the left picture extends outward from about one quarter inch outside the white disk representing the sun itself. This streamer may have origi-nated behind the solar limb, and it disappeared in about 1.6 hours. [Naval Research Laboratory coronagrams.]

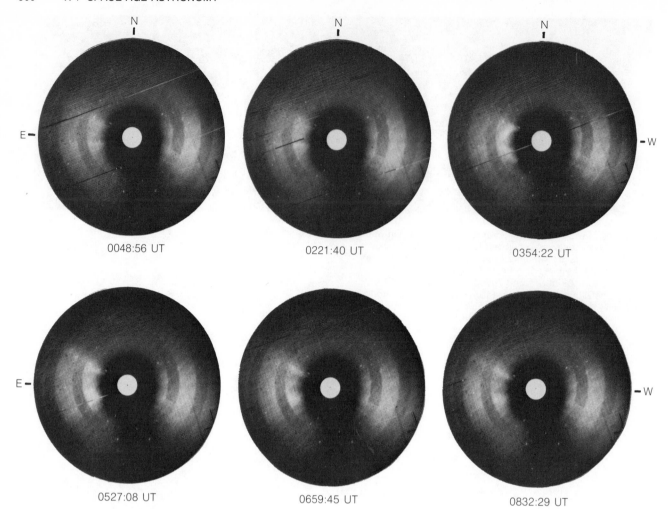

0048:56 UT

0221:40 UT

0354:22 UT

0527:08 UT

0659:45 UT

0832:29 UT

Naval Research Laboratory coronagrams of September 23, 1972, 1$^h$ 33$^m$ apart, showing the giant coronal loop at 3:54:22 UT.

The OSO 7 spacecraft. On top are the solar cells, which are directed at the sun. Within this sail section are the Goddard spectroheliograph (left box) and the NRL white-light corona-graph (right box), with its occulting disk extended on a rod. The lower wheel section constantly rotates to provide gyro-scopic stability. Four of its compartments are seen. The left one contains the celestial X-ray experiment, while the right one shows an aperture for the solar X-ray instrument (both from the University of California at San Diego). The two central com-partments are for tape recorders and support systems. Not seen are Massachusetts Institute of Technology's celestial X-ray telescope and the University of New Hampshire's gamma-ray spectrometer. [NRL photo.]

only such structure observed since OSO 7's launch, it suggests that more surprises are in store for us during this satellite's second and last full year in orbit.

In addition to the highlights reviewed above, many other interesting OSO 7 observations have been made. For example, spectroheliograms in the extreme ultraviolet have been re-corded by Donald Michels of NRL, and ultraviolet solar oscillations have been discovered by Robert D. Chapman of Goddard. Extensive observations of X-ray sources in the galaxy and beyond have been made with OSO 7 wheel instru-ments provided by George Clark of Massachusetts Institute of Technology and Laurence Peterson of the University of California at San Diego.

In its elliptical path, the OSO 7 spacecraft experiences significant atmospheric drag at perigee, and as a result it will reenter the lower atmosphere and be destroyed in late 1973 or early 1974.[3] OSO 7 was constructed by Ball Brothers Research Corp., Boulder, Colorado.

## Notes

1. For earlier SKY AND TELESCOPE information about this satellite, see November, 1971, page 271; December, 1971, page 346; and March, 1972, page 158.
2. 1 MeV corresponds to a wavelength of 0.0124 angstrom, 2 MeV to 0.0062 angstrom, and so on.
3. In fact, it did not re-enter until July, 1974. —Ed.

# 38

# Peeking Beneath the Sun's Skirts

Everly Driscoll

Solar scientists are getting an eyeful this summer—more than they had ever hoped for.

Their sensitive solar instruments on board Skylab are clicking away in unison like finely tuned machines. (Such flawless performance is not guaranteed on any space mission let alone on a patched-up space station.) The resulting photographs of the sun have a clarity and resolution never before achieved. And that monstrous, churning, ever-changing ball of gas spewing out enough energy to feed a solar system (*Science News:* 1/27/73, p. 61) is turning out to be much more complex than they ever imagined.

After only two months of looking through the telescopes' eyes at some 40,000 pictures of their favorite star, the scientists sit knee-deep in unanalyzed data, speak of mind-boggling phenomena and structure, and confess they don't know what it all means—yet. But the answers about how the sun works are all there just awaiting discovery, and they know it.

The sun is full of mystery, the true teaser of science.

The six telescopes in orbit photograph the sun simultaneously over a broad range of the electromagnetic spectrum (the first time this has been done). The photographs reveal features on the sun only hinted at before from rocket flights and ground based studies: features in the corona (the hot, thin outer atmosphere), the transition region, the chromosphere (the lower atmosphere) and the photosphere (the surface).

The corona has yielded some of the biggest surprises. Before Skylab, most scientists studying the corona called it the quiet, homogeneous outer layer of the atmosphere. No more. "The quiet homogeneous corona doesn't exist anymore," says Giuseppe A. Vaiana of American Science and Engineering.

Robert MacQueen of the High Altitude Observatory in Boulder agrees. "We see changes—dramatic, large-scale changes day by day and even orbit by orbit. We are impressed with the bewildering array of structure. The corona is a dynamic beast."

The photographs reveal clearly for the first time the whole range of coronal features, from the intensely active regions to the weakest bright points, filament boundaries and the limb brightening in coronal holes.

But the most impressive scenes in the corona are the ribbon-like structures that look like piles of spaghetti. These are the magnetic fields believed to trigger much of the sun's spectacular events. The fields themselves cannot be seen. But the plasma that follows the field lines can. These looped features, says Vaiana, emphasize that the major force controlling and shaping the corona is the magnetic fields. The ribbons change with time, both in shape and in spectrum (what frequencies of light they emit). This was a surprise. "For decades we've been looking at the magnetic fields during flares and the fields never seemed to change," says Goetz Oertel, chief of solar science at NASA headquarters. Not only do the fields change, they sometimes change drastically. During a flare they may get unstable, blow up or combine. The larger ribbons become prominences that arch for [thousands] of kilometers out in the corona.

On the other hand, during the June 15 flare (*Science News:* 6/23/73, p. 402), the changes were not large-scale at all. "The increase of surface brightness registered during the event [flare], from its moment of triggering to its peak, was in excess of a thousand," says Vaiana. But "the resulting structural changes in the region which flared are minimal in comparison with those shown previously. This indicates that large-scale restructuring [of the fields] is not a *necessary* consequence of large flares." Why the fields change drastically at one time and not at another is still not understood. It could be that restructuring is related to changes in some way on the surface.

Another dramatic mystery is the huge areas of the sun called coronal holes. These are visible in the pictures as dark areas that spread up to half a million kilometers across the sun. The holes—[two of them] sometimes called polar caps because

From *Science News*, vol. 104, nos. 7 and 8 (August 18 and 25), 1973. Copyright 1973 by Science Service, Inc. Reprinted with permission.

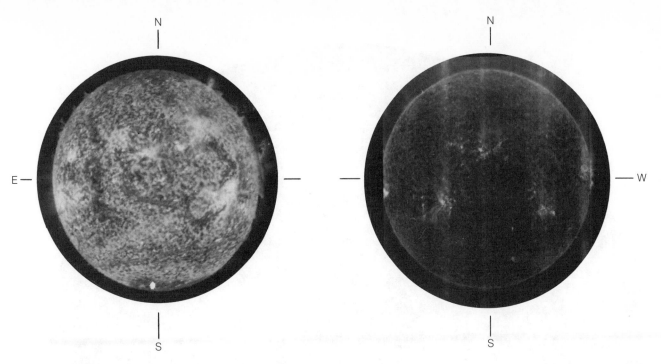

Structure at transition zone (left) disappears suddenly in corona
(right). [Naval Research Laboratory.]

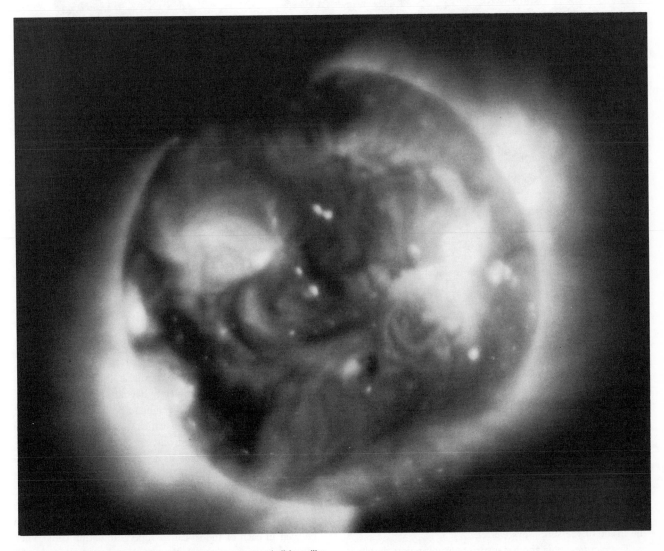

Corona in X-rays: Holes, bright dots, active zones, and ribbon-like
structure. "A dynamic beast." [NASA.]

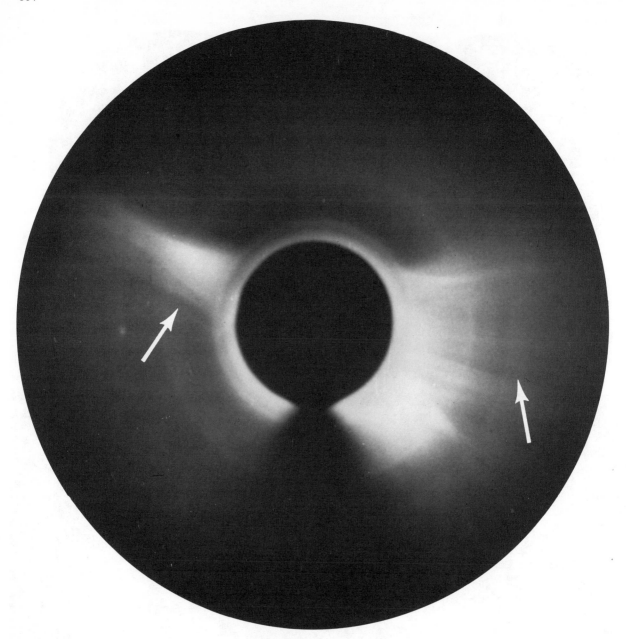

Instrument disk blocks out sun to show coronal streamers and voids (arrows). [High Altitude Observatory.]

one is at each pole—are usually about a million degrees cooler than the rest of the corona. Previously they were thought to be regions of little activity. They are most likely connected with unipolar magnetic fields rather than bipolar fields. It is thought the solar wind escapes through these holes.

"It turns out these are not simply coronal holes," says Richard Tousey of the Naval Research Laboratory. The holes extend down well into the chromosphere where the temperatures are only 50,000 degrees C. Why is the chromosphere different in these holes? "We may well be getting a clue about the inside of the sun," says Oertel.

The pictures show another surprise—bright points that look like forest fires scattered over a planet as viewed from space.

The bright points also dot the coronal holes, which may indicate the holes are not completely devoid of activity after all. One would expect the bright points to be little active regions (some of them could be flaring, which would be an even *greater* mystery) and thus be found only in the equatorial zone between 40 degrees north and 40 degrees south where much of the sun's violent activity occurs. But they aren't. They are distributed all over the disk, which is a puzzle.

Oertel thinks the bright points may be a new kind of solar activity. In the lower chromosphere, such bright points can be seen in each of the so-called supergranulation cells. Then as one moves up through the chromosphere and transition zone via the telescopes, most of the bright points disappear.

In the corona there is about one bright point for every 10 or 20 cells.

The X-ray photographs show the corona head on. The white-light coronograph, however, uses an occulting disk, and reveals the outer corona as its material surges into space. Although the material is seen above the limb, or edge of the sun, it is actually coming from all over the sun—front and back. These finely structured, thin lines are called streamers. They are free electrons that probably follow magnetic field lines out into space. "We really don't know what streamers are," says Oertel. He thinks they could mark the boundaries between the magnetic field layers.

What causes streamers, where they originate and how they are shaped are questions MacQueen hopes to answer by observing the streamers daily as they rotate in the corona. They could be shaped like a cylinder, the Eiffel tower, or a thin sheet.

One picture of the streamers reveals black strips running almost parallel to the streamers. MacQueen calls them "coronal voids" because the regions are void of free electrons. They could be areas where the streamers have moved from, and in the process, swept out the electrons. On the other hand, the presence of coronal voids may be an indication that there is actually no background corona. Most scientists have thought a background corona—a general distribution of free electrons—surrounds the sun. If there is no such background, the outer corona would be composed merely of a lot of streamers.

Tousey has been studying the sun now for 27 years—first with captured German V-2 rockets and then with other rockets. He regards the Skylab work as the culmination of his career. His instrument photographs the sun in the extreme ultraviolet. Both it and the ultraviolet scanning spectroheliometer of E. M. Reeves of Harvard College Observatory look down into the sun—through the hot corona and the atmosphere to the surface. The two instruments can actually distinguish and photograph six layers of the sun: the visible surface where the temperatures are 6,500 degrees C; just above [this], the surface where the temperature *falls* to its minimum of 4,200 degrees; then the lower and upper chromosphere where the temperatures rise in a 1,000-kilometer distance from 4,200 degrees up to 50,000 degrees; then the transition zone, a thin layer [of] only 100 kilometers or so where the temperatures soar from 50,000 degrees up to 700,000 degrees; and last, the corona, where the temperatures jump to a million degrees in the holes and 3 million to 4 million in active regions.

"The theories about how atmospheric temperatures should change with height just do not apply anywhere in the solar atmosphere," says Reeves of his findings. "Our assumptions about the sun just aren't good enough. They will have to be improved."

One surprise concerned the supergranulation structure: Tousey and Reeves found it was not restricted to the surface or the lower chromosphere at all. "It preserves itself all the way up through the chromosphere and the transition zone," says Reeves. "Then poof. At the corona the structure becomes smeared and almost disappears. That tells us something."

What changes occur so abruptly at the corona to make this structure disappear? Reeves thinks it might have something to do with the energy transfer from the surface to the corona. That same energy transfer is what is believed to cause the high corona temperatures.

This heating of the corona could be due to shock waves that propagate up from the surface and dump energy in the corona. Reeves has a chance to see the effects of these shock waves with his instrument. They would show up as fluctuations in the intensity of the radiation and density of the plasma.

Another enigma is that while most of this chromospheric network smears out, not all structure does. Some structures, such as the bright points, "are like soda straws that stick up all the way to the corona," says Reeves.

As in the early days of lunar exploration, the solar explorers may well have found more questions than answers so far. But they admit they are only on the ground floor looking up with their Skylab 1 data. The astronauts of Skylab 2 are now doubling the time and photography at the solar telescopes. The Skylab 3 crew will get even more.* (They will also have a chance to train these sensitive instruments on the planet Mercury and on the comet Kohoutek.)†

"We have just had a quick peek-a-boo," says Reeves. "Now we have to go back and do a lot of hard work."

---

*The third Skylab crew completed its mission in early February, 1974. —Ed.
†And did so very successfully!—Ed.

# An Astronomical Puzzle Called Cygnus X–3

R. M. Hjellming

Before the summer of 1972 the name Cygnus X-3 "merely" identified one of the compact x-ray sources in the constellation of Cygnus. Now this name, usually abbreviated Cyg X-3, represents a presently unique x-ray, infrared, radio, and cosmic ray source in what is probably an unusual double star system. It was the search for and subsequent study of radio emission from Cyg X-3, aided by a spectacular application of serendipity, that revealed that it could temporarily become one of the strongest compact radio sources in the sky. This discovery has led to what may have been more investigation of a single object, by more astronomers, in a shorter period of time, than ever before in astronomy. Astronomers now are asking, what natural phenomena, occurring in a double star system containing what may be a neutron star or black hole, can turn this system on the far edge of our galaxy into a cosmic accelerator for just a day or two, causing some $10^{49}$ electrons and presumably protons to be ejected with relativistic energies? In addition, why is such a large-scale phenomenon inoperative most of the time in Cyg X-3, and does it occur often in other objects without our noticing it? These are some of the many intriguing questions posed by the x-ray, infrared, and radio behavior of Cyg X-3.

## Cyg X-3 as an X-Ray Source

Cyg X-3 was first discovered as an x-ray source by Giacconi et al. (1) during a survey of the Cygnus region made with an x-ray instrument on an Aerobee rocket launched in 1966. The most notable feature of Cyg X-3 at that time was the prominent low energy cutoff of the x-ray spectrum (2, 3). The spectrum (3) is shown in Fig. 1 together with the theoretical curves for two models: a blackbody source with a temperature of $2 \times 10^7$ °K, but without attenuation at low energies, and a thermal bremsstrahlung source with a temperature of $7.4 \times 10^7$ °K, [and with] substantial low energy attenuation.

From *Science*, vol. 182, pp. 1089–1095, 14 December 1973. Copyright 1973 by the American Association for the Advancement of Science. Reprinted with permission.

The latter interpretation was believed (2) to be most likely, with the low energy absorption being blamed on absorption by the interstellar gas between Cyg X-3 and the earth. It was first estimated (2) that the amount of absorption corresponded to a line of sight column density of hydrogen of $3 \times 10^{22}$ atoms per square centimeter. Other x-ray observations have indicated (4, 5) that more than $10^{23}$ atoms per square centimeter are needed. Such large column densities indicated that Cyg X-3 is at least several kiloparsecs (1 parsec = $3.08 \times 10^{18}$ cm) from the earth.

Following the radio outbursts of Cyg X-3 in September 1972, the x-ray data for this source were subjected to considerably more scrutiny than before. The x-ray satellite Uhuru had been periodically observing Cyg X-3 during the time of the radio outbursts. Early analysis (5) of part of the x-ray data showed no signs of x-ray flaring in Cyg X-3 at the time of the radio outbursts. However, when all available data were examined it was found (6) that Cyg X-3 was emitting at roughly twice its normal x-ray flux the day before the radio outbursts were detected, at a time when no radio data were being taken.

It was also discovered (5,7) that the x-ray flux from Cyg X-3 normally varied sinusoidally with a period of 4.8 hours. The 4.8-hour periodicity suggests that Cyg X-3 is associated with a double star or binary system with this period; the observed variations would then be produced by an eclipsing or occultation phenomenon.

In addition, variations observed in the low energy x-ray cutoff of Cyg X-3 (6) correspond to changes in the hydrogen column density from $3 \times 10^{22}$ up to $2 \times 10^{23}$ atom cm$^{-2}$. While there may be $3 \times 10^{22}$ atom cm$^{-2}$ in the interstellar gas between Cyg X-3 and the earth, the changes are most likely due to variations in the intrinsic x-ray emission of the source or in absorbing gas near the source.

## Cyg X-3 as an "Ordinary" Radio Star

Since 1970 considerable effort has been made to find radio counterparts of compact x-ray sources. These searches began after it was discovered that Scorpius X-1, the strongest x-ray

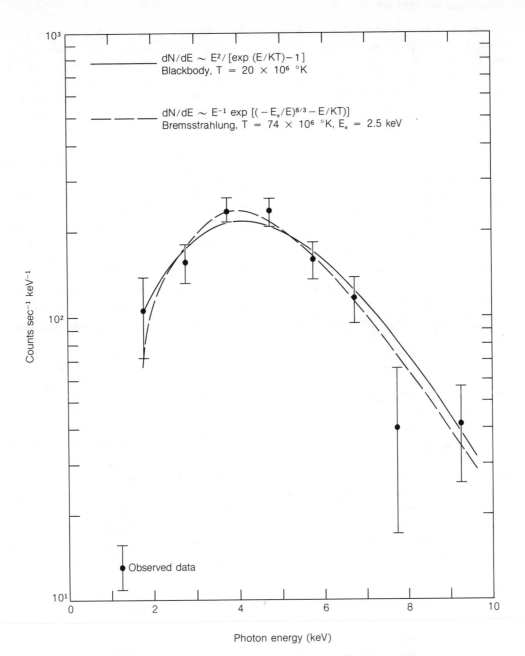

**Figure 1.** X-ray spectrum (3) of Cyg X-3 in terms of observed counts per second per kilovolt (dN/dE) as a function of photon energy (E). Also shown are theoretical curves for a blackbody source and a bremsstrahlung source (cutoff energy $E_a$) that fit the data. The Boltzmann constant is K.

source in the sky (besides the sun), was associated with a variable radio source (8) with unusual properties. A radio source variable by as much as a factor of 2 in hours was coincident with the star identified with the x-ray source, and this radio and x-ray star was surrounded by a double radio source such that all three sources lay on a straight line, a phenomenon common in radio galaxies and quasars.

In addition to being intrinsically interesting as radio sources, radio counterparts of x-ray sources provide important positional information. Since x-ray source positions usually have rather large uncertainties, the detection of an unusual radio variable inside an x-ray "error box" could be taken as prima facie evidence that the x-ray source had been located, but with positional accuracy down to a fraction of an arc

second. With this technique the interferometer of the National Radio Astronomy Observatory (NRAO) in West Virginia and the Westerbork array in the Netherlands have detected radio counterparts of the x-ray sources Cyg X-1 (9, 10), GX17 + 2 (9), Cyg X-2 (11), and, of greatest interest to us at the moment, Cyg X-3 (12).

The Cyg X-3 radio counterpart was first detected with the Westerbork array at a frequency of $1.415 \times 10^9$ hertz (1.415 Ghz). These and subsequent observations with the NRAO interferometer at 2.695 and 8.085 Ghz (13, 14) established that Cyg X-3 was usually a radio source at the level of a few tenths of a flux unit [1 flux unit (f.u.) = $10^{-26}$ watt per square meter per hertz] with a variability of up to a factor of 2 or so on a time scale of hours. This made Cyg X-3 the strongest of the known radio counterparts of x-ray sources, but not so strong that it was anything but a very faint object in the radio sky.

The early radio data on Cyg X-3 were such that the nature of the radio emission mechanism could not be definitely established. However, it was noted (15) that the only other well-studied radio source in the sky that behaved like Cyg X-3 was the radio star Algol, a triple star system 27 parsecs away and easily visible as a second magnitude star.

## Cyg X-3 the Weak Becomes
## Cyg X-3 the Very Strong

Serendipity is one of the most helpful allies of the astronomer, and the events of August and September 1972 provided another example of this. The first episode in what has become one of the most intensively studied series of events in a radio source began on 30–31 August 1972, when coordinated radio (14) and x-ray (16) observations of Cyg X-3 found that in the x-ray region it was so weak as to be below detection limits, while at radio frequencies it was present but weaker than ever observed before (or since): 0.01 f.u. Figure 2 shows the time-dependent behavior of the Cyg X-3 radio source at 2.695 and 8.085 Ghz at that time.

With normal human lack of prescience, astronomers were not observing Cyg X-3 at radio wavelengths between 31 August 0800 and 2 September 2000 universal time (U.T., which is the same as Greenwich mean time), but serendipity fortunately intervened.

On 2 September 1972 radio observations of the star Algol were begun simultaneously by a group of Canadian astronomers (17) using the 150-foot antenna at Algonquin Park in Canada, operating at 10.5 Ghz, and a group of American astronomers (14) using the NRAO interferometer, operating at 2.695 and 8.085 Ghz. On the evening of 2 September

P. C. Gregory of the Canadian group found himself waiting for Algol to rise above the horizon and decided to take a peek at Cyg X-3 to "see what it was doing." The result was unprecedented—Cyg X-3 appeared to be such a strong radio source that the instrumental gain had to be reduced to allow a proper measurement. When this was done Cyg X-3 seemed to measure 21 f.u. at 10.5 Ghz. This made Cyg X-3 appear to be one of the strongest point sources in the radio sky. Such circumstances quickly caused the suspicion that the equipment was not working properly. Knowing that the NRAO group was preparing to observe Algol, Gregory quickly telephoned them asking in words to this effect, "Would you believe Cyg X-3 at 20 f.u.?" Within moments the NRAO interferometer was finding Cyg X-3 to be about 10 f.u. at 2.695 Ghz and about 18 f.u. at 8.085 Ghz. Thus began an unprecedented international campaign to study Cyg X-3.

Figure 3 shows the time variation of Cyg X-3 at 8.085 (14, 18) and 10.5 (17) Ghz on the night of 2–3 September 1972. At 2345 U.T. on 2 September Cyg X-3 reached its peak at 22 f.u. at 10.5 Ghz and 20 f.u. at 8.085 Ghz. At this time it was a factor of 2000 stronger than it had been on 30–31 August. During the next 10 hours the radio emissions at the two frequencies varied up and down together, peaking again at 0600 U.T. on 3 September at 21 f.u.

## Cyg X-3, the Most Intensively Studied Source

As Cyg X-3 was rising to its first and highest peak on 2 September 1972 there was considerable discussion over the phone between Gregory and myself. Our first idea was that this might be the radio equivalent of a supernova—the rare but well-known event where a collapsing star explodes with such energy and brilliance that it can temporarily shine as bright as an entire galaxy of stars. We therefore immediately decided

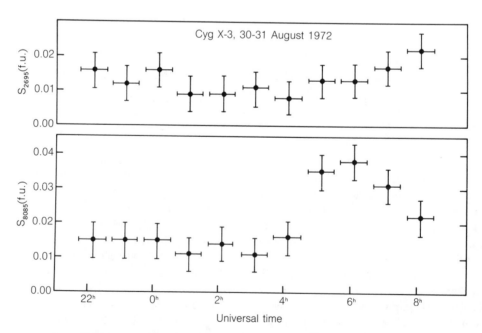

**Figure 2.** Cyg X-3 radio flux densities (S) in flux units (f.u.) at 2.695 and 8.085 Ghz on 30–31 August 1972 plotted as a function of time (14).

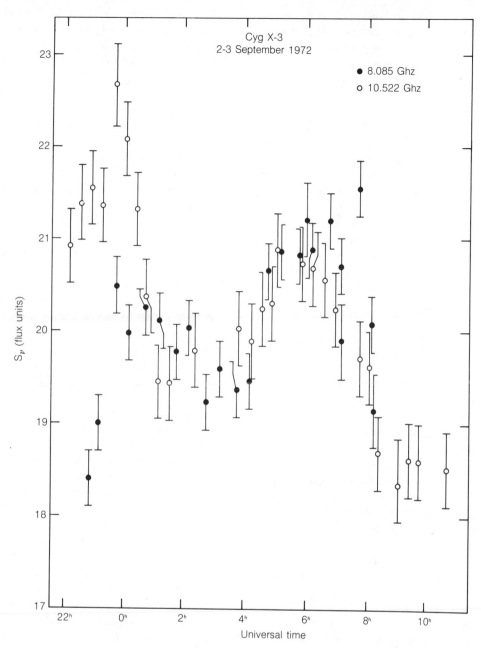

**Figure 3.** The beginning of the unprecedented radio outburst of Cyg X-3 as observed on 2–3 September 1972 at frequencies of 10.5 (*17*) and 8.085 (*14*) Ghz.

that everyone possible should be notified by telephone, and that we would ask everyone to pass on to others the name Cyg X-3, the information about the event, and the exact co-ordinates of the radio source (*13*).

Very soon Cyg X-3 was being observed by the interferometer at the California Institute of Technology at 1.42 Ghz (*19*), the Haystack antenna in Massachusetts at 15.5 Ghz (*20*), the 85-foot antenna at the University of Michigan at 8.085 Ghz (*18*), the 300-foot antenna at NRAO at 2.695 Mhz (*21, 22*), the millimeter-wavelength antenna of Aerospace Corporation in California at 80 Ghz (*23*), and the 48-inch Schmidt camera and 200-inch telescope at Mount Palomar, California, at optical wavelengths (*24*). In the midst of all the excitement, with telephone calls crisscrossing the North American continent,

many places that should have been called were inadvertently neglected. One radio astronomer in the Netherlands was reached and asked to notify all the European observatories so that they could observe Cyg X-3 as it rose over their horizon. Eventually many astronomers observed Cyg X-3 on the basis of fourth- or even fifth-hand information. More than a hundred individuals and dozens of observatories, including a few x-ray satellites (*5, 7, 25*), became involved in observing Cyg X-3 during the week following 2 September.

One of the topics discussed over the phone between Gregory and myself the night of 2 September was, now that we had tried to get everyone involved in observing Cyg X-3, how should the problem of putting the information together and getting it published be handled? It was obvious that no two people would agree on what should be done, but it was felt that the results should appear as soon as possible because it was hard to tell where these studies might lead. It was therefore decided that as other people were informed about Cyg X-3 they should also be asked to write independent papers describing their own results and to submit them to *Nature,* and that we would ask the editors of *Nature* if it would be possible to publish them all together. In the following days the editors agreed with this plan. The majority of those involved in observing Cyg X-3 during the first weeks of September 1972 agreed to the plan, with the result that the editors were deluged with papers of assorted lengths and content. Eventually the discovery papers of the Canadian (*17*) and NRAO groups (*14*) appeared in the 20 October issue of *Nature,* and 21 papers appeared in a special Cyg X-3 issue of *Nature Physical Science* on 23 October. In this first burst of publication there were 23 papers involving 91 authors. A large number of other papers eventually appeared, mostly in *Nature.*

## Cyg X-3 as a Radio Source During September and October 1972

It is impossible to fairly describe all the radio data gathered on Cyg X-3 during September and October 1972 without writing a book. However, there were a number of radio frequencies for which the coverage was most extensive during that time period, and for which the data have been published. In Fig. 4 the extensive radio measurements at 8.085 (*14, 18, 26*), 2.695 (*14, 22, 26, 27*), 1.4 (*19, 26, 28-30*), 0.408 (*30*), and 0.365 (*31*) Ghz are plotted as a function of time, together with smooth curves connecting data for the same frequencies. As can be seen from Fig. 4, the radio peaks of 2-3 September at the highest frequencies were preceded by rises where higher frequencies were strongest and were followed by steady decay of a radio source strongest at the lower frequencies (nonthermal). The lower the frequency the lower the flux levels reached, and the more delayed was the peak in time. This behavior is the signature of a synchrotron radiation event in which relativistic particles are ejected in an expanding region of relativistic particles and coupled magnetic fields; the radio emission is produced by interaction between the cosmic ray electron component and the magnetic fields, with the particles rapidly losing energy. Such phenomena are familiar in quasars, although in quasars such events have time scales of years, whereas the first Cyg X-3 event had a half-life of a little more than a day. The most significant result of the identification of the Cyg X-3 event of 2 to 13 September as a synchrotron radiation event is the conclusion that Cyg X-3 is capable of ejecting bursts of cosmic rays in copious amounts for short periods of time.

An obvious question being posed in the middle of September 1972 was whether such radio outbursts were common. A partial answer came when a second and rather different series of Cyg X-3 events began on 18 September. A number of observatories were routinely monitoring the radio behavior of Cyg X-3, so these events were caught early and their complexity was fairly well recorded.

Up until 21 September the second Cyg X-3 event appeared to be a similar, but less intense, version of the event of 2 to 13 September. The increase in radio flux looked like a typical rise from a self-absorbed source which reached a peak on 20 September, then began a decay as a relatively transparent nonthermal synchrotron event. However, the decay was interrupted on 21 September by a reenergization of the source, which appeared to produce two broad, major peaks at the high frequencies during the next week. As can be seen in Fig. 4, only the 365-Mhz curve (*31*) shows a single rise and decay. Because the third and fourth peaks show no significant signs of self-absorption, and show only small variations on short time scales, the general interpretation is relatively simple. The expanding "bubble" of coupled relativistic particles and magnetic fields that was observed from 18 to 21 September largely swept away the magnetic fields in the vicinity of the cosmic ray source. Thus, particles ejected in later events expanded in relatively field-free space until they caught up with the more slowly expanding bubble containing magnetic fields associated with the first ejection (18-19 September). The increased supply of cosmic ray electrons in a relatively large volume produced a transparent nonthermal source that increased and then decreased its strength under the influence of both a variable resupply of cosmic rays and the usual energy loss mechanisms. After roughly 27 September the major resupply ceased, and the normal decay due to energy losses more or less proceeded. The 365-Mhz variations are then of particular interest because the source remained self-absorbed all during the major ejection events and merged together on about 29 September to produce one peak at 365 Mhz, which then decayed away.

Thus, the four major peaks observed at the higher radio frequencies represent four successive periods during which Cyg X-3 succeeded in ejecting relativistic particles or cosmic rays. The short time scale variations seen in Figs. 3 and 4 are probably due primarily to fluctuations in the cosmic ray ejection, but in some cases also to inhomogeneities in the strength and structure of the magnetic fields that must be present to produce the radio source. The most striking short time-scale variations appear as quasi-sinusoidal modulations of the radio flux, particularly at 8.085 Ghz for the peak on 2-3 September, 25 September, 26-27 September, and 27-28 September, with lesser examples also occurring. Strikingly, the minima and maxima for this modulation seem to appear with a periodicity of 0.5 or 1.0 day.

Among the radio observations that could be made during the Cyg X-3 outbursts a few are of particular interest. At least three major efforts (*27-29*) were made to measure the absorption in the radio spectrum due to neutral hydrogen (HI) in the line of sight to the source. The most detailed results were obtained with the Caltech interferometer by Chu and Bieging (*29*), who made a very sensitive absorption measurement exactly at the 1.42-Ghz peak on 21 September 1972. Their absorption spectrum and the associated HI emission profile for this region are shown in Fig. 5. It is well established that the three major bumps in the HI emission profile are due to emission from gas in three different regions: (i) the broad bump near 0 velocity is due to gas in the local region of the spiral arm of our galaxy where the sun is located; (ii) the bump at

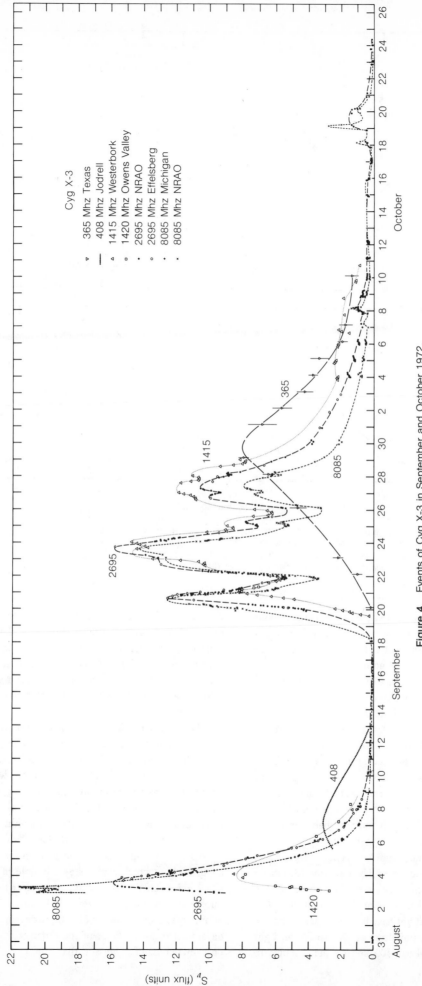

**Figure 4.** Events of Cyg X-3 in September and October 1972. Radio flux densities measured at five frequencies by eight observatories are plotted as a function of time. The lines are meant only to sketch the continuity of the variation at each frequency.

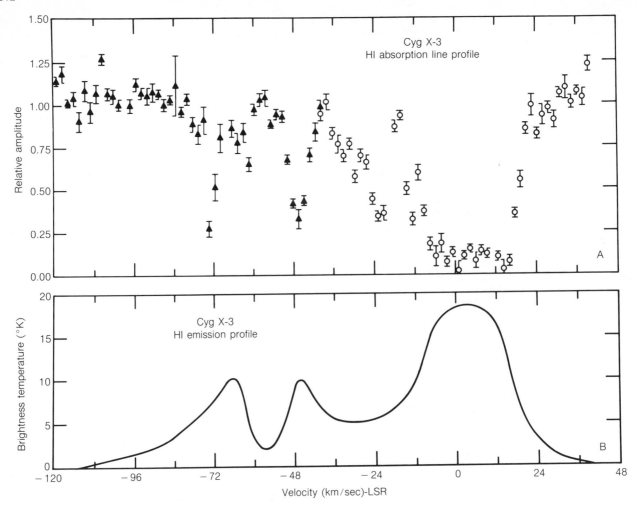

**Figure 5.** (A) HI absorption line profile measured by Chu and Bieging (*29*) on 21 September 1972 and (B) emission line profile for this region of the sky. Both are plotted against velocity with respect to the local standard of rest (LSR).

−47 kilometers per second is due to a spiral arm feature about 8.3 kpc away; and (iii) the bump at −73 km/sec is due to another feature about 10.4 kpc away. The fact that all three emission line peaks have corresponding absorption peaks in Fig. 5 means that Cyg X-3 is at least 10.4 kpc away. Three less solid arguments support the conclusion that Cyg X-3 is probably in the spiral arm feature about 10 kpc away. The first is that the absorption peak at −73 km/sec is not as broad as might be expected, so perhaps not all the hydrogen in the feature is in the line of sight. The second is that in this direction a distance of 10 kpc places Cyg X-3 on the far edge of our galaxy, or else outside it. The third, complementary to the second, is based on the evidence that Cyg X-3 is not significantly outside our galaxy, hence it most probably is in the farthest feature in the spiral arm in that direction.

The hardest evidence that Cyg X-3 is not very distant comes from a measurement at very high resolution (*32*) made on 22–23 September, which showed that at that time the radio source was greater than 0.01 arc second in size. Because the radio source would not have begun expansion before about 18 September, the limitation of the speed of light places an upper limit of about 100 kpc on the distance to Cyg X-3. Perhaps equally powerful is the simple observation that, with a decay half-life of roughly 1 day largely due to adiabatic expansion losses, the radio source could not have been larger than 1 light day ($10^{-3}$ pc) on, say, 4 or 21 September; with the speed of light as the limit for the expansion velocity, an upper limit of roughly 50 kpc can be placed on the distance.

The HI absorption measurements (*29*) also determine the hydrogen column density along the line of sight to be $1.7 \times 10^{20} \, T_s$ atom $cm^{-2}$, where $T_s$ is the excitation temperature for the hydrogen. This is compatible with the smallest densities inferred from the low energy x-ray cutoff if one can take $T_s$ as approximately 150°K, a value which is only slightly larger than usual for the interstellar medium.

The general description of the Cyg X-3 events to be derived from the radio data involves a source at a distance of about 10 kpc, which expands as a bubble of relativistic particles and magnetic fields. A few days after the beginning the bubble is expanding at a few tenths the speed of light, attaining a size of the order of a light day in a few to several days, with magnetic fields of the order of several hundredths or a few tenths of a gauss at that time.

## Cyg X-3 as an Optical and Infrared Source

The first optical observations of Cyg X-3 were at visual wavelengths and consisted of a few radio astronomers strolling outside their well-lit telescope control rooms to see if there was a new bright star in the sky at the right position, as might be possible if it had been a supernova event. The results were negative. Properly scientific optical studies were carried out at Mount Palomar by a Caltech group (24) using the 48-inch Schmidt camera and the 200-inch telescope. The group found no optical object on the position of the radio source down to a limit of $2.5 \times 10^{-5}$ f.u. on 3 September and $2 \times 10^{-6}$ f.u. on subsequent days ($10^{-6}$ f.u. corresponds to a visual magnitude limit of 23.9). Considering the possible amounts of optical absorption, these results do not rule out the presence of even a relatively bright star at a distance of 10 kpc.

The first search for Cyg X-3 in the infrared was carried out on 3 October 1972 by a Caltech group (33) using the Hale 200-inch telescope and a very sensitive infrared photometer. An infrared source was discovered at wavelengths of 1.6 and 2.2 micrometers that was exactly coincident with the radio source. At this time the infrared source was not examined for variability or other unusual characteristics, and it was simply noted that the results were consistent with observations of a blue supergiant at a distance of 10 kpc if there were essentially complete absorption at visual wavelengths by the interstellar dust.

The most exciting contribution (34) of the infrared studies occurred during July 1973, when Cyg X-3 was found to be very variable in the infrared on many different time scales. First of all, when relatively weak in the infrared Cyg X-3 was found to vary sinusoidally with a period of 4.8 hours—exactly the same as for the x-ray source. This absolutely clinches the identification of the radio and infrared source with the x-ray source Cyg X-3. An example of the simultaneous x-ray and infrared sinusoidal variations of Cyg X-3 is shown in Fig. 6 (on page 314).

In addition to clinching the identification, the infrared result indicates that Cyg X-3 is the first object to show presumably binary or rotation phenomena in the infrared. Furthermore, the Cyg X-3 infrared source shows very active flaring on time scales as short as minutes. Such correlations promise to reveal a great deal about Cyg X-3. The possibilities are good that similar studies will reveal interesting relationships between the variations in x-rays and the infrared, and the usual variation of Cyg X-3 in the radio spectrum at the level of a few tenths of a flux unit, or less.

## Cyg X-3 as Interpreted by Synchrotron Radiation Theory

Theoretical studies of synchrotron radiation sources have been extensively developed because of their application to well-observed objects like supernova remnants, quasars, and radio galaxies. The event of 2 to 13 September and the beginning of the event of 18 to 21 September in Cyg X-3 appear qualitatively as expected (35) for synchrotron radiation from an expanding bubble of relativistic particles and magnetic fields.

TABLE 1
*Parameters for the radio events of Cyg X-3.*

| Parameter | Value |
|---|---|
| Distance | $\sim 10$ kpc |
| Radio luminosity | $\sim 2 \times 10^{24}$ erg sec$^{-1}$ hz$^{-1}$ |
| Radio energy per event | $\sim 5 \times 10^{39}$ erg |
| Number of cosmic ray electrons per event | $\sim 10^{49}$ |
| Energy in cosmic ray electrons per event | $\sim 5 \times 10^{44}$ erg |
| Velocity of expansion | $\sim 0.1 \times$ speed of light |
| Size ($t \sim 4$ days) | |
| Angular size ($t \sim 4$ days) | |
| Magnetic field ($t \sim 4$ days) | |

When cosmic ray particles are ejected from a very small region these relativistic particles begin an expansion which carries along magnetic fields that originally surrounded the system. The evolution of this expanding bubble of relativistic particles is then determined by: (i) the strength of the magnetic fields, (ii) the time variation of the particle ejection, (iii) the energy spectrum of the newly ejected particles, and (iv) the energy losses of the particles. The observable radiation is produced by synchrotron emission when cosmic ray electrons interact with the magnetic fields, but because of the compact nature of the initial Cyg X-3 source the major energy losses are due to adiabatic processes as the bubble increases its volume.

The Cyg X-3 radio events behave as if the initially ejected particles obey a power law in energy described by $E^{-\gamma}$, where $\gamma$ is a constant. The initial increases observed on 2 September and 18–19 September show a source growing in size with considerable reabsorption of the emitted radio photons; because absorption processes are strongest at lower frequencies, a self-absorbed source is initially strongest at higher frequencies, and then becomes transparent in order of decreasing frequency (see Fig. 4).

Detailed analyses of the first Cyg X-3 radio outbursts have allowed the major parameters of the expanding radio sources to be determined. Many can be derived from the simplest of models (35), and more sophisticated models such as those discussed by Peterson (36) take into account more complex possibilities for the particle ejection and energy losses. One of the major conclusions that can be drawn from the radio data is that major particle ejection with $\gamma$ approximately 2 lasted for 1 to 2 days for the event of 2 to 13 September, but three more successive episodes of particle ejection produced the three successive major peaks later in September. The major parameters for the radio events of Cyg X-3 are summarized in Table 1 where determinations are good to a factor of 3 or better.

## Cyg X-3 as a Star System

The 4.8-day periodicity appearing in both the x-ray (5, 7) and infrared (34) emission of Cyg X-3 strongly suggests that the object is basically a double star system. From Kepler's law the knowledge of the period provides a relation between

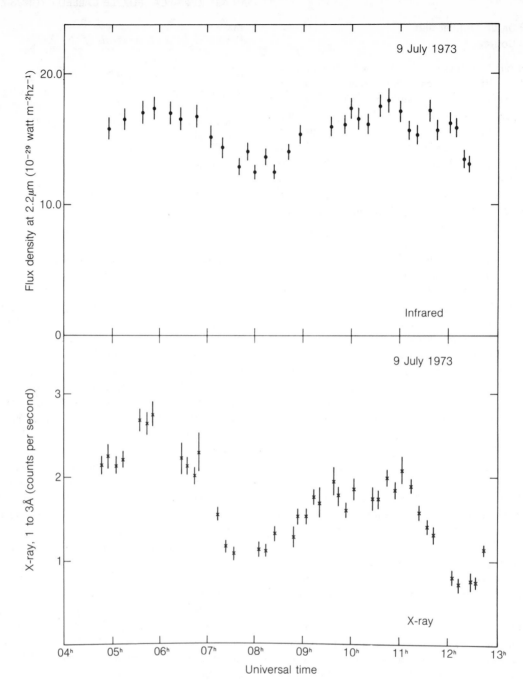

**Figure 6.** The sinusoidal variations of Cyg X-3 in the infrared (wavelength, 2.2 μm = 2.2 × 10⁻⁶ cm) and x-ray (1 to 3 Å, 1 Å = 10⁻⁸ cm) spectra on 9 July 1973, as measured by Becklin *et al.* (*34*).

the separation of the two stars ($a$) and the sum of the masses ($M_1$ and $M_2$) such that (*35*)

$$a = 1.4[(M_1 + M_2)/M_\odot]^{1/3}R_\odot \quad .$$

where $M_\odot$ and $R_\odot$ indicate units of one solar mass and one solar radius, respectively. For any reasonable masses for the stars the separation cannot be more than one to a few solar radii, which is a very close binary system. In such systems the stars are usually in contact or nearly in contact with each other.

The question of how stars can evolve to this state has been discussed by van den Heuvel and De Loore (*37*). The process of evolution is as follows. In the beginning there are two

normal stars in a binary system; the more massive of the two evolves with mass loss, attaining a stable final state as a neutron star. At this point a neutron star of 1 $M_\odot$ is in, say, a 5.0-day orbit with a 15-$M_\odot$ star, which then evolves until it fills the critical gravitational equipotential surface that is usually called a Roche lobe. This is shown in Fig. 7A. There are then two extreme possibilities, depending on whether the subsequent mass loss from the 15-$M_\odot$ star is transferred to the neutron star or escapes from both stars. In the first case (Fig. 7B) the star losing mass attains a stable state at 3.8 $M_\odot$, while the addition of more mass·forces the neutron star to continue collapse into a black hole. With the other option the lost mass escapes from the system, leaving a 3.85-$M_\odot$ stable

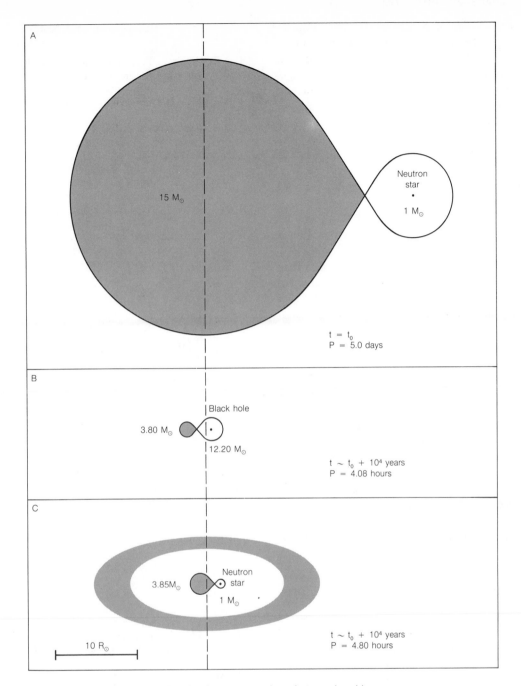

**Figure 7.** Schematic representation of stages in a binary system that can evolve to produce very short period systems, as calculated by van den Heuvel and De Loore (*37*). Abbreviations are t, time; P, period. (A) Onset of the second stage of mass exchange. (B) Final system in the case of conservation of total mass and orbital angular momentum. (C) Final system in the case where mass is ejected from the binary—this state may represent Cyg X-3.

star and an unchanged neutron star (Fig. 7C). In the first case the final orbit has the stars separated by 3.5 $R_\odot$, while the second case leaves the stars separated by 2.4 $R_\odot$. The option resulting in a neutron star is probably closest to being applicable to Cyg X-3 because of the probable importance of mass transferring to the compact object, though the truth may be somewhere in between.

All the evidence suggests that Cyg X-3 as an x-ray and infrared source is produced by an extensive hot plasma surrounding a much more compact object. Presumably the x-ray

and infrared source is eclipsed by the companion star once every 4.8 hours. In this context the radio emission probably results from interesting transient events in the unstable outer parts of the x-ray and infrared emission regions.

## Cyg X-3 as a Cosmic Ray Accelerator

The clear interpretation of the Cyg X-3 radio events of September 1972 in terms of synchrotron radiation means that the penultimate cause of the radio events is processes of acceleration and ejection of cosmic ray electrons from the central system. In the context of what is known about the x-ray, infrared, and radio source this suggests that the basic acceleration mechanisms operate in the unstable outer portions of the very hot plasma filling the Roche lobe around the compact star. It is not clear whether the basic processes are similar to the flare events known to accelerate particles on the sun or to a relativistic and variable stellar wind analogous to the well-known solar wind (which has a lower velocity). The essentially continuously variable nature of the weak radio emission from Cyg X-3 shows that such processes are operating continuously.

The outbursts of Cyg X-3 in September 1972 are unusual, even for Cyg X-3; therefore one must ask for the ultimate cause of the modification of "normal" behavior which produced such unusual events. Since the ejection of such large numbers of relativistic particles must be transient, one must seek some sort of transition from one moderately stable state to another, with the large outburst being a result of the transition itself. The most likely possibility is a transient change in structure in one or the other of the two stars—a "starquake." Such starquakes presumably are transient alterations on the natural dynamical time scale of the star: $t_{\rm dyn} \sim (R^3/GM)^{1/2}$, where $R$ and $M$ are the radius and mass of the star, and $G$ is the gravitational constant. For $M \sim 1 M_\odot$ and $R \sim 1 R_\odot$, $t_{\rm dyn} \sim 27$ minutes. Successive events could be associated with "aftershocks" of such alterations of structure. In the case of a change in the mass ejection of the normal star, the result could be a temporary increase in mass infall for the other star with its sensitive hot plasma "atmosphere." On the other hand, a change in structure (continued gravitational collapse?) in the compact object could also have drastic effects on its atmosphere and on the processes that continually produce the radio emission.

## Cyg X-3, a New Phenomenon in the Sky

The object known as Cyg X-3 is in the process of altering many of our preconceptions. For the radio astronomer there is now the realization that the radio sky could be filled with interesting, transient radio sources blinking on and off—and without a fantastic stroke of luck or serendipity we would never know about most of them. For the infrared astronomer there is the surprising correlation with x-ray and radio astronomy found in the first infrared object showing probable binary characteristics. For the x-ray astronomer there is the puzzle of the shortest period x-ray binary known (by a factor of 10), which is only the second known x-ray source showing coupling between x-ray and radio events. Finally, for those interested in cosmic rays and their acceleration mechanisms, Cyg X-3 requires that such phenomena be produced in a stellar system with interesting x-ray and infrared emission. All of this certainly makes Cyg X-3 one of the most interesting of astronomical puzzles.

## References and Notes

1. R. Giacconi, P. Gorenstein, H. Gursky, J. R. Waters, *Astrophys. J. Lett.* 148, 119 (1967).
2. H. Gursky, P. Gorenstein, R. Giacconi, *ibid.* 150, 75 (1967); P. Gorenstein, R. Giacconi, H. Gursky, *ibid.*, p. 85.
3. R. Giacconi, *Non-Solar X- and Gamma-Ray Astronomy*, L. Gratton, Ed. (Reidel, Dordrecht, Netherlands, 1970), p. 107.
4. R. D. Bleach, E. A. Boldt, S. S. Holt, D. A. Schwartz, P. J. Serlemitsos, *Astrophys. J.* 171, 51 (1972).
5. D. R. Parsignault, H. Gursky, E. M. Kellogg, T. Matilsky, S. Murray, E. Shreier, T. Tananbaum, R. Giacconi, *Nature Phys. Sci.* 239, 123 (1972).
6. S. Murray, H. Gursky, C. Jones, E. Schreier, H. Tananbaum, R. Leach, D. R. Parsignault, *Bull. Amer. Astron. Soc.*, 5, 394 (1973).
7. P. W. Sandford and F. H. Hawkins, *Nature Phys. Sci.* 239, 135 (1972).
8. J. G. Ables, *Astrophys. J. Lett.* 155, 27 (1969); R. M. Hjellming and C. M. Wade, *ibid.* 164, 1 (1971); C. M. Wade and R. M. Hjellming, *Astrophys. J.* 170, 523 (1971).
9. R. M. Hjellming and C. M. Wade, *Astrophys. J. Lett.* 168, 21 (1971).
10. L. L. E. Braes and G. K. Miley, *Nature* 232, 246 (1971); C. M. Wade and R. M. Hjellming, *ibid.* 235, 271 (1972).
11. R. M. Hjellming and L. Blankenship, *Nature Phys. Sci.* 243, 81 (1973).
12. L. L. E. Braes and G. K. Miley, *Nature* 237, 506 (1973).
13. R. M. Hjellming, M. Hermann, E. Webster, *ibid.*, p. 507.
14. R. M. Hjellming and B. Balick, *ibid.* 239, 443 (1972).
15. R. M. Hjellming, *X- and Gamma-Ray Astronomy*, H. Bradt and R. Giacconi, Eds. (Reidel, Dordrecht, Netherlands, 1973), p. 98.
16. A. Brisken, thesis, University of Maryland (1973).
17. P. C. Gregory, P. P. Kronberg, E. R. Seaquist, V. A. Hughes, A. Woodsworth, M. R. Viner, D. Retallack, *Nature* 239, 440 (1972).
18. H. D. Aller and W. A. Dent, *Nature Phys. Sci.* 239, 121 (1972).
19. D. B. Shaffer, G. A. Shields, B. Schupler, *ibid.*, p. 131.
20. W. A. Dent, J. E. Kapitzky, B. G. Leslie, G. Kojoian, M. L. Meeks, H. H. Danforth, J. J. Kollasch, E. J. Chaisson, D. F. Dickinson, L. E. Goad, C. J. Lada, *ibid.*, p. 126.
21. W. A. Dent, *ibid.*, p. 127.
22. A. H. Bridle, M. J. L. Kesteven, A. E. Niell, *ibid.*, p. 121.
23. R. B. Pomphrey and E. E. Epstein, *ibid.*, p. 125.
24. J. A. Westphal, J. Kristian, J. P. Huchra, S. A. Shectman, R. J. Brucato, *ibid.*, p. 134.
25. J. P. Conner, W. D. Evans, D. E. Mook, *ibid.*, p. 125.
26. R. M. Hjellming and B. Balick, *ibid.*, p. 135.
27. L. L. E. Braes, G. K. Miley, W. W. Shane, J. W. M. Baars, W. M. Goss, *ibid.* 242, 66 (1973).
28. R. Lauqué, J. Lequeux, Nguyen-Quang-Rieu, *ibid.* 239, 119 (1972).
29. K. W. Chu and J. H. Bieging, *Astrophys. J. Lett.* 179, 22 (1973).
30. B. Anderson, R. G. Conway, R. J. Davis, R. J. Peckman, P. J. Richards, R. E. Spencer, P. N. Wilkinson, *Nature Phys. Sci.* 239, 117 (1972).

31. F. N. Bash and F. D. Ghigo, *ibid.* 241, 93 (1973).

32. H. F. Hinteregger, G. W. Catuna, C. C. Counselman, III, R. A. Ergas, R. W. King, C. A. Knight, D. S. Robertson, A. E. E. Rogers, I. I. Shapiro, A. R. Whitney, T. A. Clark, L. K. Hutton, G. E. Marandino, R. A. Perley, G. Resch, N. R. Vandenberg, *ibid.* 240, 159 (1972).

33. E. E. Becklin, J. Kristian, G. Neugebauer, C. G. Wynn-Williams, *ibid.* 239, 130 (1972).

34. E. E. Becklin, G. Neugebauer, F. J. Hawkins, K. D. Mason, P. W. Sanford, K. Matthews, C. G. Wynn-Williams, *Nature* 245, 302 (1973).

35. P. C. Gregory, P. P. Kronberg, E. R. Seaquist, V. A. Hughes, A. Woodsworth, M. R. Viner, D. Retallack, R. M. Hjellming, B. Balick, *Nature Phys. Sci.* 239, 114 (1972).

36. F. W. Peterson, *Nature* 242, 173 (1973).

37. E. P. J. van den Heuvel and C. De Loore, *Astron. Astrophys.* 25, 387 (1973).

38. The National Radio Astronomy Observatory is operated by Associated Universities, Inc., under contract with the National Science Foundation.

# Cosmic X-Ray Sources: A Progress Report

Herbert Friedman

The recent discoveries of x-ray astronomy are stimulating new interest in the physics of highly condensed matter and in cosmology. Models of the evolution and structure of neutron stars can now be compared with detailed observations of their behavior as pulsars and as accretion-type x-ray sources in binary star systems. The possibility that the x-ray source Cygnus X-1 is a black hole is strongly indicated; observations of the emission properties of such a black hole can test general relativity in novel ways. In the field of cosmology the existence of intergalactic matter has been a crucial observational unknown. X-ray astronomy introduces unique possibilities for revealing the invisible mass of the universe and establishing its physical parameters.

A decade of observations with instruments carried on rockets and balloons, capped by the launching of the first small x-ray astronomy satellite (SAS-1), UHURU, has provided evidence of some 125 discrete sources* (1). The UHURU instrumentation covers the spectral range 2 to 20 thousand electron volts. At lower and higher energies the obervational evidence has been derived primarily from rockets and balloons. About 80 sources lie within 20 degrees of the galactic plane and most likely exist within the galactic disk; the remainder are possibly extragalactic. Underlying the strong discrete sources is a diffuse background that has been spectrally mapped from 100 to 100 million electron volts. This background is characterized by high isotropy at all energies, but its spectral character is complex and probably reflects a variety of different sources, both intrinsically diffuse and so abundantly discrete as to appear diffuse when observed with the modest spatial resolution of present instrumentation. Truly diffuse x-ray emission could originate in hot intergalactic gas as thermal x-ray bremsstrahlung, or as nonthermal x-rays if a universal distribution of cosmic rays is scattered by inverse Compton collisions with photons of the microwave background at 3 K.

In this article I review the present knowledge of discrete sources, treating a few examples in some detail, and estimate the importance of discrete and diffuse contributions to the background.

## X-Ray Stars and Nebulas

Only a small fraction of the galactic x-ray sources have been identified with optical or radio counterparts. At least two major classes exist—those associated with supernova remnants and those associated with binary pairs. Among the x-ray emitting supernova remnants are the Crab Nebula, SN Tycho Brahe, Cassiopeia A, Vela X, and the Cygnus Loop. Although some 70 radio pulsars have been discovered,[†] only the Crab pulsar (NP 0532) and possibly the Vela pulsar (PSR 0833-45) emit x-ray pulses. Several eclipsing binaries have been detected in x-rays and must be members of a large class of x-ray sources that derive their power by mass transfer from a visible primary to an invisible compact companion. It is somewhat surprising that the number of galactic x-ray sources thus far discovered does not exceed 100 since it is estimated that the galaxy contains about 70,000 close binary pairs.

The discovery of pulsars and binary x-ray stars has led to renewed theoretical effort in modeling the supernova process and, more generally, the evolution of collapsed cores and the shedding of stellar envelopes. According to present concepts, stellar evolution depends on initial mass in roughly the following way. Stars with masses originally less than 3.5 solar masses ($M_\odot$) evolve to white dwarfs and, in the process, shed enough mass to leave less than 1.4 $M_\odot$ (the Chandrasekhar limit) in the compacted star. If the initial mass is 3.5 to 10 $M_\odot$ the cores collapse to neutron stars with residual mass less than 3 $M_\odot$, which is the maximum that can be supported by neutron degeneracy pressure. The excess mass is dispersed by the accompanying supernova explosion. When the original mass lies in the range 10 to 60 $M_\odot$, collapse is not accompanied by explosion of the outer layers. Material continues to rain down

From *Science*, vol. 181, pp. 395–407, 3 August 1973. Copyright 1973 by the American Association for the Advancement of Science. Reprinted with permission.

*By 1974, the number of UHURU sources had increased to about 160. —Ed.

[†]By early 1975, about 150 radio pulsars had been discovered.—Ed.

on the collapsing core, overwhelms neutron degeneracy pressure, and thus forces collapse to a black hole.

Two processes are of paramount importance in gravitational collapse to a neutron star. First, conservation of angular momentum requires that the gravitational energy derived from contraction be converted largely into kinetic energy of rotation. If the sun were compressed to a radius of 10 kilometers, its 1-month period of rotation would decrease to a millisecond. Second, the original stellar magnetic field strength at the surface would be amplified by compression. A field of about 1 gauss at the solar surface would transform to $10^{10}$ gauss at a radius of 10 km. The energy derived from gravitational collapse far exceeds the nuclear binding energy, so that the compact spinning neutron star begins its life with more kinetic energy of rotation ($>10^{50}$ ergs) than its parent star could derive from nuclear fusion.

Slowdown of a spinning neutron star and the accompanying dissipation of energy may be associated with magnetic dipole radiation, provided the magnetic axis is inclined to the spin axis. Electrons and protons become locked in phase with the expanding spherical electromagnetic waves and are swept along to relativistic energies. An alternative acceleration mechanism is homopolar magnetic induction. Plasma surrounding the neutron star conducts current between the poles and the equator of the spinning magnetized sphere. If the spin axis and the magnetic dipole axis are coaligned, an electric field as great as $10^{16}$ volts per centimeter may be generated close to the star and can accelerate charged particles to relativistic energies.

The details of the supernova implosion-explosion process are still theoretically unclear. One currently held theory requires that implosion of the core be followed by an outgoing shock wave, which reaches relativistic velocities in the thin outer layers of the star. This shock powers the explosive dispersion of debris and may accelerate particles to energies of $10^8$ to $10^{21}$ ev, accompanied by intense neutron fluxes. Pulsars could play a major role in generating the highest-energy cosmic rays ($>10^{12}$ ev), but would be inadequate to supply the much larger energy flux observed in the range $10^9$ to $10^{12}$ ev. The pulsar mechanism may also work for white dwarfs. If they evolve by collapse of normal stars to about 1 percent of their original radii, their magnetic fields must reach strengths of about $10^4$ gauss and their rotation periods must reduce to 10 to 100 seconds. An electric field of about $10^6$ volt/cm can be generated near the surface, sufficient to accelerate particles to energies greater than $10^8$ ev, possibly as high as $10^{14}$ ev. The large number of white dwarfs—of the order of $10^{10}$—which are believed to populate the galactic disk with a scale height of about 1 kiloparsec (kpc) would satisfy the requirements of isotropy observed in the cosmic rays. Compton scattering in the high-density photon field near the surface of a white dwarf would produce gamma rays with energies in the million electron volt range, perhaps in sufficient intensity to account for the diffuse cosmic background flux above 1 Mev (2).

Accretion of gas onto a compact star is a very efficient means of developing hot plasma and x-ray emission (3). For example, a proton falling onto a neutron star can acquire 10 Mev of kinetic energy. At the rate of $10^{-9}$ $M_\odot$/year raining onto a neutron star or black hole of about 1 $M_\odot$, conversion of thermal energy to x-rays would generate about $10^{36}$ ergs per second. A white dwarf could produce a comparable luminosity by collecting $10^{-7}$ $M_\odot$/year. It is well established that some variable binary stars such as Beta Lyrae eject about $10^{-5}$ $M_\odot$/year and that roughly half of the ejected gas may impact on the companion star. The rate of accretion onto a neutron star or black hole in interstellar space would be many orders of magnitude less than in a close binary system. In the spiral arm regions an isolated black hole of 1 $M_\odot$ might accrete about $10^{-15}$ $M_\odot$/year and the resulting luminosity would reach about $10^{31}$ erg/sec. If a supermassive black hole ($\sim 10^8$ $M_\odot$) existed near the galactic center, accretion could reach $10^{-3}$ $M_\odot$/year and the power generated would exceed $10^{43}$ erg/sec.

Several galactic x-ray sources have been identified with binary pairs in which accretion takes place by mass transfer from a visible primary to an invisible compact secondary, such as a neutron star or black hole. Figure 1 illustrates a hypothetical binary situation. The visible primary may be a blue supergiant which fills its Roche critical equipotential surface. Gas leaks through the inner Lagrangian point and is gravitationally drawn toward the compact companion star. Because the gas has high angular momentum relative to the compact star, accretion is not spherically symmetrical. The gas enters into Keplerian orbits and accumulates in a disk somewhat resembling Saturn's rings, but with much greater density. The disk circulates about the star and gas slowly spirals inward. Viscosity removes angular momentum and heats the gas so that it radiates bremsstrahlung x-rays. The formation of an x-ray emitting region—and its spectrum, power, stability, and geometrical emission pattern—are determined by various parameters of the compact object.

Accretion onto a neutron star (4) will depend on the magnetic field strength and the spin rate. If the magnetic field is only moderately strong and the spin is relatively slow, accretion can progress steadily onto the gas disk and down to the surface of the star at the magnetic poles. The spectrum would be composed of contributions from both the gas disk and

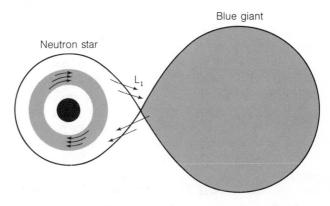

**Figure 1.** Schematic representation of a typical close binary pair: an early-type supergiant and a neutron star. The figure eight defines the Roche critical equipotential surface. Gas overflows from the giant star through the region of the inner Lagrangian point (L₁) and collects in a disk that revolves about the neutron star.

the stellar surface. If the field is very strong and the spin rate is high, accretion becomes more difficult and the observed emission must then come from a pulsar-like process as in the case of NP 0532. If accretion is toward a black hole, a modest flow to the surrounding gaseous disk could produce thermal emission with a range of temperatures corresponding to different depths of accretion. The resulting composite spectrum would resemble a power law even though the basic process is thermal. Accretion at a very high rate could disrupt the simple symmetry and channel the mass flow into localized hot spots. The resulting x-ray emission would then be sporadic and flare-like. Symmetrical accretion is ultimately limited by radiation pressure. The photon scattering cross section is some $10^6$ times greater for electrons than for protons, whereas gravity works equally on the masses of electrons and protons. Charge separation takes place and an electric field is created, which drags the protons after the electrons. Gravitational accretion is balanced by this radiation pressure when the luminosity reaches about $10^{39}\, M/M_\odot$ erg/sec, where $M$ is the mass of the accreting star. This critical luminosity is known as the Eddington limit.

***Crab Nebula and NP 0532.*** The Crab Nebula and its pulsar have been studied intensively over the wide spectral range from radio waves to about 1-Gev gamma rays (*5–7*). The supernova was seen in A.D. 1054 and the present extent of the nebula (~6 light-years) fits the observed rate of expansion of the explosion debris into interstellar space and the elapsed time. Its nebular spectrum is highly polarized and is most likely produced by the synchrotron radiation of relativistic particles spiraling in a field of about $10^{-4}$ gauss. Before the discovery of the pulsar it was difficult to understand the energy source of the visible and x-ray synchrotron emission, because particles with energies high enough to produce the radiation ($10^{12}$ to $10^{14}$ ev) must decay with lifetimes of less than a few hundred years for the optical radiation and less than a few years for the x-rays. It now appears that the pulsar can supply relativistic particles to the nebula for tens of thousands of years. The stellar collapse that triggers a supernova may endow the spinning neutron star with $10^{49}$ to $10^{52}$ ergs of rotational kinetic energy. Pulsar NP 0532 is dissipating this energy at the rate of $10^{38}$ erg/sec ($10^5$ times the energy radiated by the sun), thus lengthening its period about 15 microseconds per year. Conversion of this energy to particles and radiation with 10 percent efficiency can account for the total pulsed and nebular x-ray emission, which is about $10^{37}$ erg/sec.

Pulsar NP 0532 pulses 30 times per second. Figure 2 shows the pulse profile in various spectral regions from optical to gamma ray. If the magnetic dipole axis is inclined to the spin axis of the neutron star, the north and south poles will alternately be visible from the earth and radiation will beam past the earth twice per cycle. Although the profile does show two pulses, the secondary pulse follows the principal pulse by 13.4 msec, which is 3 msec sooner than half the full period of 33 msec. The asymmetry may be explainable in terms of time delay if the magnetic dipole is inclined at 45 degrees to the spin axis and the direction to the earth is at a latitude of about 20 degrees, so that the emission region linked to one magnetic

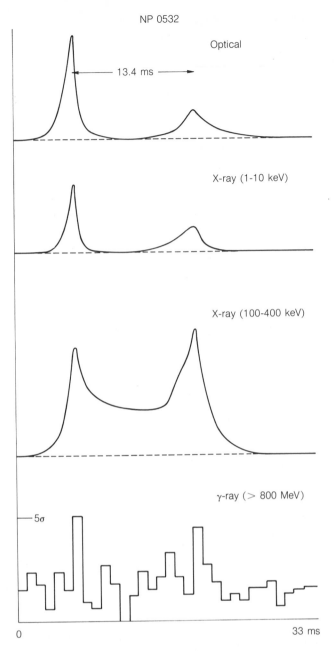

**Figure 2.** Pulse profiles of NP 0532 at different energies. The soft x-ray (1 to 10 kev) and optical patterns are very similar. With increasing energy, the secondary pulse gains in intensity and width relative to the principal pulse, and the interpulse region fills in markedly. At the highest energy (> 800 Mev) the double pulse pattern appears to persist (*6*) although the statistics are somewhat marginal. (5σ = 5 standard deviations.)

pole always comes about 3 light-milliseconds (~1000 km) closer to the earth than the region of opposite polarity when its beam sweeps by (*8*).

Throughout the spectrum to the billion electron volt range, the principal pulse remains sharp (full width at half-maximum about 1 msec) but the secondary pulse broadens in the range 10 to 400 kev and gains in intensity relative to the principal pulse. The 13.4-msec region between the two pulses fills with a substantial flux of radiation at the higher energies. At the very highest energies (~1 Gev) the observations are still marginal but the basic pulse pattern seems to persist. Pulsar models have not been sufficiently developed yet to explain these detailed features and their spectral dependence.

The Crab pulsar is the fastest and, presumably, the youngest of the roughly 70 pulsars thus far detected by the radio astronomers. With increasing age the pulsed power must decline rapidly. Although the pulsations may continue for several tens of thousands of years, the spectral distribution should shift toward longer wavelengths. Why have not x-ray pulsations been identified with supernova remnants still younger than the Crab? Continuous x-rays have been detected from Tycho SN 1572 and from Cassiopeia (Cas) A (estimated occurrence about 1700) but not from Kepler SN 1604, which is perhaps too distant (7 to 10 kpc) to have been found with the sensitivity of UHURU. The x-ray emission from Cassiopeia A may result from thermal bremsstrahlung of the hot plasma created behind the shock front of the expanding nebula, because the interstellar density there is as high as 30 $cm^{-3}$. Weak pulsations could be masked by the high thermal flux. In the vicinity of Tycho SN 1572 the interstellar gas is very thin, less than 0.1 $cm^{-3}$, and the nebula is quite extended so that thermal emission is inefficient. The observed x-ray emission is therefore most likely associated with the neutron star itself. Failure to detect pulsing may simply be a matter of orientation of the beam away from the earth. Shklovsky (9) proposes that the observed x-rays may still be attributed to a pulsar, but may be blurred by emission from a broad interpulse region rather than a sharply peaked pulse. He notes that the region between the principal pulse and the secondary pulse in NP 0532 fills in with radiation as the energy increases until it accounts for as much flux as the sharp pulses in the range 100 to 400 kev. Thus, a broad interval of "quasi-isotropic" x-ray emission may exist in some pulsars, perhaps associated with energetic particles trapped in the neutron star's magnetosphere. For Tycho 1572, the emission associated with a hypothetical pulsar 2.5 times as powerful as NP 0532—consistent with its younger age—could explain the observed continuous emission. A search for a small component of rapid modulation with more sensitive instruments would be desirable.

Neutron stars older than the Crab pulsar must be very abundant. Their pulsations having died out, all that remains to identify them is the extended nebula, thermal radiation generated by accretion, or possibly the blackbody emission of the stellar surface. The Veil Nebula in Cygnus (about 50,000 years old) is an example of a supernova remnant in which all the remaining emission is confined to the shock front of the nebula, where it expands against the interstellar gas. Its x-ray spectrum (10) is consistent with free-free emission from a thin plasma at about $5 \times 10^6$ K. In theory, the surface of a neutron star cools rapidly, to about $10^6$ K within about a year after its formation. At that relatively low temperature it is difficult to observe the flux of very soft x-rays against the diffuse background with detectors that have the large fields of view typical of the mechanical baffles presently employed. The spectrum of Circinus X-1 is consistent with a blackbody source at $10^7$ K, but that temperature is much higher than theoretical models of neutron stars allow (11). Table 1 lists the various supernova remnants that have been identified with x-ray emission.

TABLE 1
*Supernova remnant x-ray sources.*

| Supernova remnant | Age (years) | X-ray power (erg/sec) | Energy (kev) |
|---|---|---|---|
| Crab Nebula | 900 | $10^{37}$ | 1–200 |
| NP 0532 [pulsar] | 900 | $10^{36}$ | $1-10^6$ |
| Cassiopeia A | 300 | $5 \times 10^{36}$ | 1–10 |
| Tycho 1572 | 400 | $5 \times 10^{36}$ | 1–10 |
| Puppis A | $10^4$–$10^5$ | $10^{36}$ | 0.2–3 |
| Vela X, Y, Z | $10^4$–$10^5$ | $10^{36}$ | 0.2–3 |
| Cygnus Loop | $10^4$–$10^5$ | $2 \times 10^{36}$ | 0.2–1 |
| IC 443 | | $2 \times 10^{34}$ | 2–10 |
| MSH 15-52A | | $5 \times 10^{35}$ | 2–10 |

**Hercules X-1.** The class of x-ray stars associated with mass transfer to a spinning neutron star in a binary pair is best illustrated by Hercules X-1. The optical counterpart is the binary HZ Herculis (12), an irregular optical variable with a light difference from maximum to minimum of 1.5 magnitudes. It exhibits x-ray pulsations with a period of 1.24 seconds, attributable to a spinning neutron star which is eclipsed by its giant companion for about 6 hours every 1.7 days. Over the course of half a year of observation, the pulsation period decreased by 4.5 $\mu$sec. This spin speedup may result from transfer of angular momentum of the infalling mass to the spinning star. The fast pulsation undergoes a Doppler shift in synchrony with the eclipse period. From the parameters of the system it is estimated that the mass of the x-ray source is 0.3 to 1.0 $M_\odot$.

The x-ray emission appears to be further modulated with a 35-day period during which the x-ray flux is on for 10 or 11 days and off during the remaining 25. While the orbital period remains very constant, the 35-day period is somewhat irregular. When first observed, the turn-on time was precise to about 1 hour in 35.7 days. Fourteen months later, the period was 34.9 days with a scatter of plus or minus 1 or 2 days. Precession of the spinning neutron star may cause the beam to oscillate into and out of the direction of view from the earth with the 35-day period. The asymmetry that induces the precession might result from accretion of matter onto the star in the regions of the magnetic poles (13).

The 1.24-second pulse profile determined from UHURU data shows a principal pulse and an intermediate pulse somewhat resembling those of NP 0532. However, the widths are much greater, indicating a broad beaming, and the x-ray emission is almost fully modulated by the pulsation. A recent rocket observation (14) resolved a double-peaked structure in the principal pulse (Fig. 3) and revealed variations in amplitude and shape in only a matter of seconds. Broad beaming might be expected from a thermal source, generated at the surface of the star if geometrical shadowing and optical opacity were the limiting factors. However, it has also been proposed that the emission mechanism is cyclotron radiation (15) in higher harmonics of the cyclotron resonance frequency, and that the emission is beamed in a wedge pattern at right angles to the magnetic field direction with an opening angle of about

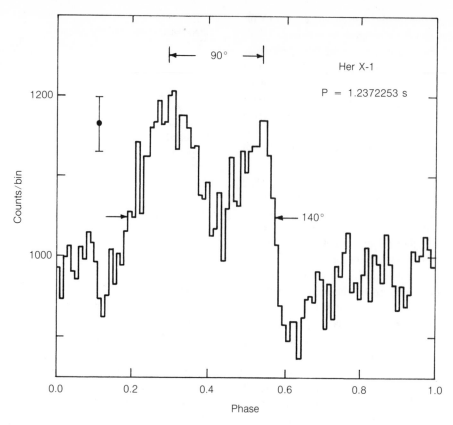

**Figure 3.** Pulse profile of Hercules X-1 obtained from rocket flight (*14*). The principal pulse shows a double structure which varied markedly in pattern and amplitude during the few minutes of flight. The period (*P*) is in seconds; a phase of 90° corresponds to one-quarter of a period.

30 degrees. The observed spectrum up to 40 kev is fit well by a model of thermal bremsstrahlung at approximately $6 \times 10^7$ K.

With the beginning of each 10- to 11-day on period, the x-ray onset is abrupt; the intensity increases for 3 to 4 days and then decreases to the end of the on period. The x-ray flux appears sharper at emergence from eclipse, and disappears more gradually at entry into eclipse. Such behavior may be related to the geometry of the gas flow in the accretion region and the opacity of intervening gas.

The 35-day period can be traced in optical data, which show a sinusoidal variation of 1.5 magnitudes synchronized with the 1.7-day eclipse, with minimum light at the middle of the x-ray eclipse. The stellar brightening, when the x-ray source is on the visible side of the star, is produced by absorption and conversion of the x-ray energy in the stellar photosphere. Examination of Harvard Observatory plates dating back to 1900 shows that the variations disappear for years at a time and then return.

***Centaurus X-3.*** Centaurus X-3 (*16*) resembles Hercules X-1 in many respects. Its pulsation period is 4.8 seconds and the eclipse period is 2.1 days. No longer periodicity is present, but the x-ray signals disappear from time to time for periods of days. No optical counterpart has been found, perhaps because of high obscuration in the galactic plane. Other eclipsing binary sources detected by the UHURU satellite and

their eclipse periods are: 2U(0900-40), 8.95 days; 2U(1700-37), 3.4 days; and 2U(0115-73), 4 days, in the Small Magellanic Cloud. The power of the last source is about $10^{38}$ erg/sec, close to the Eddington limit.

***Cygnus X-1 (A Black Hole?).*** Special interest attaches to Cygnus X-1 because of the possibility that it may be a black hole. The x-ray emission shows extreme variability (*17*) down to the shortest time scale detectable, approximately 0.1 second, but there is no evidence of steady, rapid pulsation, or of a binary eclipse. It does, however, very nearly coincide with a radio and optical variable. The optical source is a 9th-magnitude B0I spectroscopic binary, HDE 226868, with a 5.6-day period. The mass of the blue supergiant, inferred from its apparent spectral class, should be about 20 $M_\odot$. From the orbital parameters, the mass of the companion star is then estimated to be approximately 13 $M_\odot$. In theory, a neutron degeneracy pressure cannot support more than about 3.2 $M_\odot$; accordingly, Cygnus X-1 should be a black hole. However, some reservation should be attached to the estimate of the mass of the primary.

Theorists have been developing detailed models of disk-type accretion onto a black hole. Rapid fluctuations in x-ray emission are predicted on a variety of time scales down to tens of microseconds. The shortest fluctuations may result from focusing and defocusing of the x-rays by the curvature of space close to the black hole. Doppler shifts for an orbiting hot spot near the inner edge of the gas disk would be very pronounced. According to general relativity the minimum period should be approximately 70 $\mu$sec. The study of such short-lived bursts of x-rays will require highly sensitive large-area detectors with effective apertures of the order of square meters, such as were planned for the High Energy Astronomy Observatory (HEAO) program of the National Aeronautics and Space Administration.

**Cygnus X-3.** Cygnus X-3 is a unique x-ray variable (18) with a period of 4.8 hours—much longer than can be attributed to rotation of a pulsar and very short for an eclipsing binary. It is difficult to believe that a compact star could be orbiting a giant companion at such high speed. Over an extended period of observations with UHURU, its spectrum has shown strong variability in long-wavelength absorption that may be associated with a thick and variable hydrogen cloud either surrounding it or nearby. On 2 September 1972, its radio flux increased a thousandfold* and then slowly returned to normal in about a week. Four repeats of such a radio outburst were registered by 6 October, but no x-ray variations were recorded (19).

It cannot be assumed for certain that the x-ray and radio sources are the same object. The radio behavior somewhat resembles that expected from the expanding cloud model of active radio quasars (quasi-stellar objects) and Seyfert galaxies. Each outburst of Cygnus X-3 appears to expel clouds of relativistic electrons in tangled magnetic fields (about $10^{40}$ ergs of particle energy per cloud), but there is no evidence relating these outbursts to the x-ray source and its 4.8-hour period.

**Scorpius X-1.** Scorpius X-1 is the brightest x-ray star and was the first to be discovered. It has been identified with a 12th-magnitude blue star and with a radio star. In spite of a wealth of observational detail in all parts of the spectrum accumulated over a decade of observations and many efforts to develop theoretical models, its fundamental nature is still mysterious (20).†

**Transient Sources.** Three transient sources‡ with intensity variations similar to those of optical novas have been observed in the past few years. The source 2U 1543-47 was seen in Lupus in late 1971 when its brightness increased a thousandfold to match the flux of the Crab Nebula, the second brightest x-ray source in the sky (21). It slowly died away over the course of a year. The other two transients were Centaurus X-2 and Centaurus X-4 observed for about 80 days each. Each reached a brightness greater than that of Scorpius X-1 at maximum.

**Summary of Galactic Sources.** The sources described above are the best characterized of the approximately 80 sources near the galactic plane. Additional conclusions about the nature of x-ray stars can be sought from statistical considerations of the number distribution with respect to brightness and from spectral evidence of interstellar attenuation. Several of the strongest sources show large, nonvarying absorption at low energies, which implies that they are at about the distance of the galactic center. Accordingly, their x-ray luminosities must be of the order of $10^{38}$ erg/sec, close to the Eddington limit for accretion onto 1 $M_\odot$. The 17 brightest sources in the UHURU survey lie within 40 degrees longitude of the galactic center and group about 3 degrees median latitude. Analysis of the distribution of the remaining sources indicates that they occupy the spiral arms at an average distance of 2 kpc, corresponding to x-ray powers of about $10^{36}$ erg/sec. The conclusion drawn from these distributions is that nearly all of the sources detected thus far radiate at approximately $10^{36}$ erg/sec or more (1 to 10 kev).

## X-Ray Galaxies and the Diffuse Background (1 to 10 kev)

The roughly 45 discrete x-ray sources which lie at high galactic latitudes (>20 degrees) may include many distant galaxies and clusters of galaxies. Underlying these sources is the diffuse background, which is highly isotropic. The relative contributions of discrete sources and diffuse radiation processes to the composite background are not clear. Observations of x-rays from Andromeda and the Magellanic clouds, as well as our own galaxy, indicate that most normal galaxies radiate at a few times $10^{39}$ erg/sec. Integrating the contribution of all such galaxies over a Hubble radius (22), about 10 billion light-years, would lead to a background flux (1 to 10 kev) about 50 times less than observed.

A new pulsar may outshine the total of all other x-ray sources in the galaxy. If it is endowed with $10^{52}$ ergs of kinetic energy of rotation, conversion to x-rays at 1 percent efficiency would provide $10^{50}$ ergs over the lifetime of the pulsations. If we assume that one supernova per 100 years leads to a pulsar, the time-averaged contribution to galactic x-ray luminosity would be $10^{41}$ erg/sec. Thus, the combined x-ray emission from all galaxies could equal the observed diffuse background. Estimates of the initial energy of rotation differ widely, however, from as low as $5 \times 10^{49}$ to as high as $10^{53}$ ergs. So far, only upper limits of $10^{50}$ to $10^{51}$ ergs are available from a few x-ray observations of distant galaxies at times of visible supernova events (23).

X-rays have been detected from powerful radio galaxies, Seyfert galaxies, one quasar, and several rich clusters of galaxies. Although the statistics of the small sampling of known sources hardly justifies any great confidence in estimating the integrated cosmic contribution to the background, the evidence does support the possibility that discrete extragalactic sources can supply much of the background radiation (1 to 10 kev). The brightest x-ray galaxies are also powerful nonthermal sources of radio and infrared emission. Among the best-studied objects are the radio galaxies M 87 (NGC 4486,

---

*See Article 39 by R. M. Hjellming.—Ed.

†In 1975, improved observations finally showed that this x-ray source is a binary system.—Ed.

‡Many other transient sources have been found since 1975, when new satellite instruments suited to large-area sky scanning and monitoring went into operation. They are now recognized as a common phenomenon among x-ray sources.—Ed.

Virgo A), NGC 1275 (Perseus A), and NGC 5128 (Centaurus A); the Seyfert galaxy NGC 4151; and the quasar 3C273. The first three are among the most spectacularly active galaxies and have many morphological similarities.

**M 87.** The first extragalactic x-ray source to be discovered by rocket astronomy was M 87 (*24*). Optically, it is a large elliptical galaxy characterized by an extended jet, which emerges from the nucleus to a distance of about 1500 pc (Fig. 4). A counter jet is also faintly discernible, formed of two approximately parallel bands of irregular intensity stretching about 10 kpc from the core of the galaxy (*25*). A similar structure can be seen in NGC 5128, but the filaments reach a length of about 30 kpc. The orientation is perpendicular to the dark band (dust lane) across the equatorial plane of the galaxy. The source NGC 1275, when photographed in the H$\alpha$ line of hydrogen, is highly filamentary, out to a distance of about 60 kpc. The filaments form a radial pattern emerging from the nucleus to about 3 kpc, and then turn abruptly, presumably along the direction of the outer magnetic field (Fig. 5).

The bluish light of the jet of M 87 is highly polarized, which is indicative of synchrotron radiation. Several knots, or condensations, dot the length of the jet, and various authors have suggested the possibility of synchrotron x-ray emission from these knots (*26*). The relativistic electrons could be generated by (i) proton-proton collisions, (ii) clouds of millions of pulsars, or (iii) the spin-off from large rotating magnetoids (*27*) (magnetized plasma bodies of about $10^4\ M_\odot$). Alternatively, the x-ray source may be concentrated toward the nucleus. Radio astronomers have found an intense compact core only 4 light-months in diameter, which could be the origin of x-ray emission generated by inverse Compton interactions between relativistic electrons and the high density of radio photons in the nuclear region.

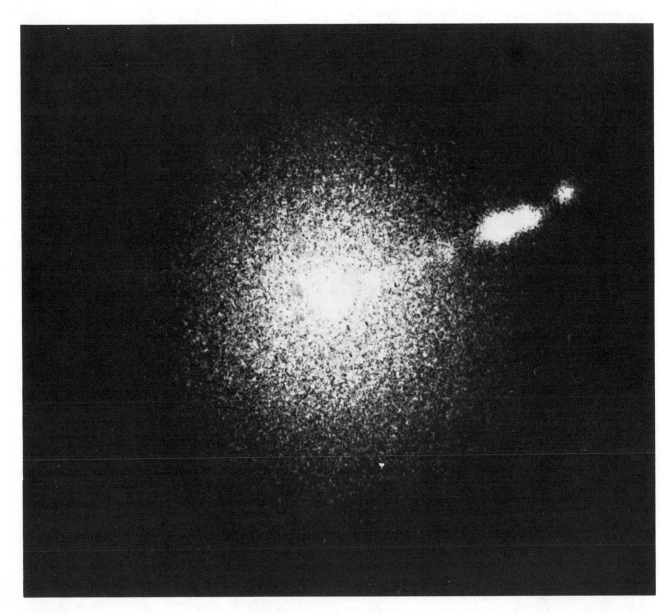

**Figure 4.**  M 87, showing jet emerging from the core of the elliptical galaxy. [U.S. Naval Observatory, Flagstaff, Arizona.]

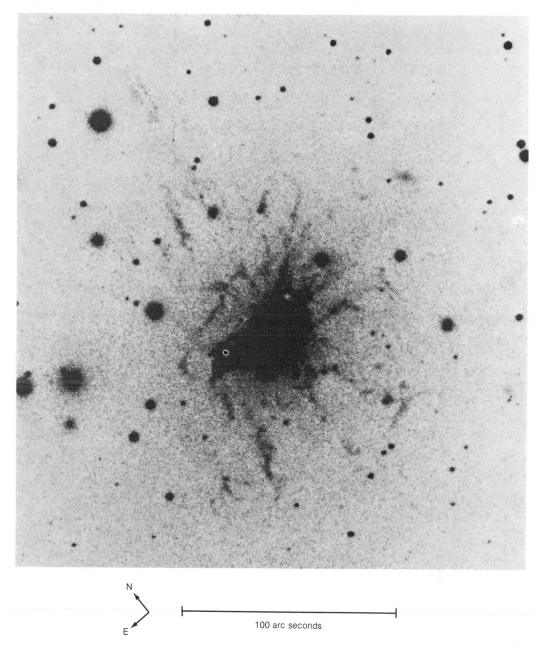

N

E

100 arc seconds

**Figure 5.** NGC 1275 photographed in Hα to show extended filamentary structure. [Kitt Peak National Observatory.]

Additional features of the radio image are an extended halo and a fan jet. The optical jet and counter jet probably describe the major magnetic field direction about which the galaxy is rotating, with the fan jet spreading in the equatorial plane over approximately one-third of a galactic revolution. The fan jet suggests that gas clouds are expelled from a compact rotating body. Shklovsky (*25*) estimates that repeated releases from the central body supply a few million solar masses to the various jet forms. To sustain a reservoir of material and energy, gas may be constantly accreting onto a large rotating mass in the nuclear region; the origin of this gas could be planetary nebulas separated from old-population red giants. By analogy with our galaxy, the rate of evolution of planetary nebulas in M 87

should be about 30 per year, releasing about 0.1 $M_{\odot}$ of gas each in clouds that fall toward the core of the galaxy. In an elliptical galaxy such as M 87, these infalling clouds will have little angular momentum relative to the nucleus and gas cloud collisions will inhibit star formation, thus permitting the gas to fall steadily onto the central rotating magnetoid. To account for the observed x-ray emission of $10^{43}$ erg/sec there must be a compact nuclear mass of about $10^{9}$ $M_{\odot}$ onto which 3 $M_{\odot}$ per year of gas released in planetary nebulas is steadily falling.

The UHURU observations of M 87 suggest that the x-ray source is somewhat extended (*28*) or that a localized emission from the galaxy is superposed on a more diffuse emission region. It is not possible to characterize the x-ray spectrum as uniquely thermal or nonthermal. There is some evidence for short-term variability (of the order of months) in both the x-ray flux and the spectrum, but the observations need to be extended. Radio observations over a span of 18 months have shown no significant flux variations or changes in size of the core (*29*).

The galaxy NGC 5128, which is located at the center of the extended radio galaxy Centaurus A, radiates about $6 \times 10^{41}$ erg/sec in the range 1 to 10 kev (*30, 31*). The radio galaxy is of the common double source type produced by the ejection of matter from the central region of the galaxy in two roughly comparable streams oppositely directed. Two distinct explosions are defined by two pairs of radio plasmons, symmetrically placed (Fig. 6) about $2 \times 10^4$ and $10^6$ light-years apart on a line running through the nucleus of the galaxy. The distance of Centaurus A is only 4.5 Mpc and it covers about 10 degrees of arc, so that even the 0.5-degree resolution of the x-ray detector of UHURU is adequate to set rather narrow limits on the extent of the source of x-ray emission, which is confined to the central region of the optical galaxy and is limited to a diameter smaller than the separation of the two innermost radio lobes. Most likely, the x-ray source coincides with a very compact radio (*32*) and infrared source (*33*) in the nucleus. The upper limit on x-ray emission from the extended radio lobes does not exceed the flux of inverse Compton x-rays that should be expected from collisions of relativistic electrons of the synchrotron radio source with the universal background photons at 2.7°K. The total x-ray power from NGC 5128 is only about $10^{42}$ erg/sec; if this is typical of most radio double galaxies it suggests that they are not adequate to make up the background radiation. However, x-rays have been detected from the direction of a rich cluster of galaxies that include the most powerful radio double galaxy, Cygnus A (*1*). If the x-rays originate in Cygnus A, its power is approximately $8 \times 10^{44}$ erg/sec.

***Seyfert Galaxies.*** One or two percent of all galaxies appear to be Seyferts. They are characterized by compact, brilliant nuclei which contain $10^9$ to $10^{10}$ stars within a diameter of about 1000 light-years. Strong, widened optical emission lines and a polarized continuum indicate high-velocity gas clouds, hot gas, and nonthermal processes. The source NGC 1275 exhibits many of the properties of a Seyfert and is also one of the most violently active radio galaxies. Its radio structure is of the core halo type, with a core less than 1 light-year in diameter. Repeated explosions in the nucleus on a time scale of about 10 years release clouds of relativistic particles into the halo, where they accumulate over subsequent millions of years. The nuclear region is orders of magnitude brighter in the infrared than in visible light, and its x-ray power (about $4 \times 10^{45}$ erg/sec) approaches its infrared power (about $3 \times 10^{46}$ erg/sec) (*34*).

The more typical Seyferts—NGC 4151, NGC 1068, and NGC 4051—have been scanned by UHURU. Only NGC 4151 is a detectable x-ray source. If, on the average, one-third of all Seyferts were as powerful x-ray sources as NGC 4151, the integral contribution would come close to satisfying the background flux and isotropy requirements.

***Quasars.*** The source 3C273 is the nearest and brightest of the quasars thus far discovered (*30*). Quasars behave like hyperactive Seyferts; the continuum radiation almost swamps the line emission. Typically, a quasar is about a hundred times as luminous in visible light as a normal galaxy, even though its diameter is nearly an order of magnitude smaller. But the infrared luminosities are even more spectacular. For 3C273 the infrared luminosity is about $6 \times 10^{48}$ erg/sec, approximately three orders of magnitude greater than its visible luminosity. Its x-ray power is about $10^{46}$ erg/sec. The x-ray observations are still too crude to indicate spectral character or short-term variability comparable to that observed in the infrared. If the space density of quasars is about $10^{-8}$ Mpc$^{-3}$ and 3C273 is assumed to be typical, the contributions to the diffuse background would be only about 2 percent of the observed flux.

***Clusters of Galaxies.*** Galaxies may cluster in groups of as many as 10,000, and hundreds of such clusters may agglomerate into superclusters. The density of galaxies toward the center of a rich cluster may reach $10^3$ to $10^6$ times the average galactic density of the sky. The three rich clusters closest to us are Virgo, Coma, and Perseus. Their x-ray luminosities are $10^{43}$, $2 \times 10^{44}$, and $4 \times 10^{44}$ erg/sec, respectively. Several of the weaker x-ray sources found at high galactic latitudes appear to be centered on dense clusters, so it is reasonable to assume comparable x-ray powers for all rich clusters. The mean x-ray fluxes expected from just the integrated fluxes of normal galaxies in rich clusters is at least an order of magnitude less than observed. Hot, intergalactic gas may be the source of the observed x-ray emission; alternatively, nonthermal interactions between relativistic particles released by a single active galaxy and the ambient photon field may produce x-rays.

***Coma Cluster.*** The Coma cluster is a rich, centrally condensed, spherically symmetrical system for which the virial theorem suggests a binding mass as much as seven times that present in normal galaxies (*35*). Since the cluster appears to be bound, there has been much attention to the search for the missing binding mass in the form of intracluster gas. If a binding mass of gas were present, the upper limit on the temperature would be about $10^8$ K, which corresponds to the escape velocity. The absence of 21-cm radio emission or absorption (*36*) implies that the gas is hotter than $3 \times 10^4$ K. An upper limit on the flux of soft x-rays below 1 kev sets the temperature somewhat higher, above $5 \times 10^5$ K (*37*). The observed x-ray flux in the range 2 to 10 kev can be interpreted as bremsstrahlung from a gas of density $6 \times 10^{-4}$ cm$^{-3}$ in a volume with a radius about 0.5 Mpc at a temperature of $7 \times 10^7$ K, but the total mass is only a few percent of the binding mass (*38*). If a binding mass of ionized gas exists in the temperature range $3 \times 10^4$ to $5 \times 10^5$ K, it should be detectable as hydrogen Lyman-alpha recombination emission. At the red shift of Coma, Lyman-alpha appears at about 1250 Å, which places it well outside the intense geocoronal Lyman-alpha background at 1216 Å. Rocket observations (*39*) of the Lyman-alpha emission from Coma almost suffice to rule out any appreciable fraction of a binding mass of gas in the tem-

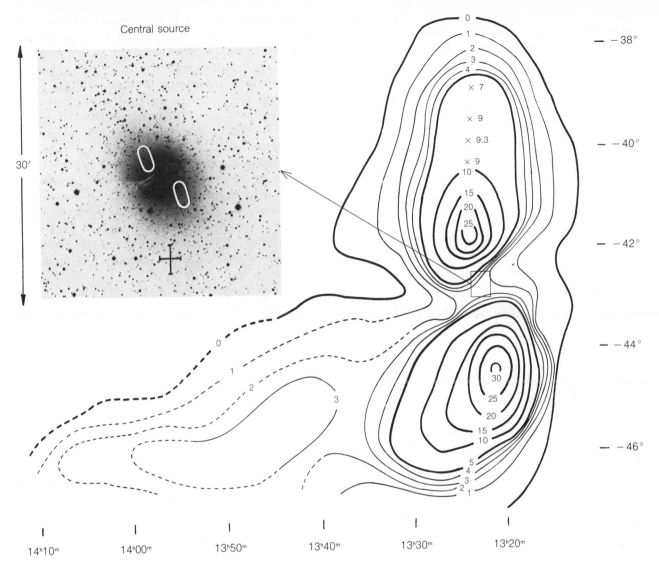

**Figure 6.** NGC 5128 (Centaurus A); the optical galaxy is shown in the inset. The radio contours (labeled in relative intensities) show the typical double source structure of a radio galaxy. Evidence of a more recent explosion is given by the double radio source still located within the optical galaxy, with a component separation of 25,000 light-years. The abscissa gives the right ascension; the vertical measure on the inset is 30 arc minutes (negative print).

perature regime below $5 \times 10^5$ K. The combined evidence of all spectral observations, therefore, rules against the existence of a binding mass of gas and raises questions about previous estimates of the mass-to-light ratio of the galaxies, unless one resorts to more esoteric explanations such as an abundance of black holes.

The size of the x-ray emitting region gives additional clues to the nature of the source. It extends over a diameter of about 45 arc minutes (about 1 Mpc). Internal cluster gas may be accreted from intergalactic gas outside the cluster, or it may be ejected from centrally located active galaxies. Whether the gas is accreted or ejected, its distribution could be expected to follow the spatial distribution of the galaxies, which appear to be concentrated within 10 arc minutes at half-maximum intensity, a region considerably smaller than the x-ray source (40). A better size match is found with Coma C, an extended radio source about 40 arc minutes in diameter at the center of the cluster. It is structureless on a scale smaller than 30 arc minutes and is presumably diffuse rather than a superposition of discrete sources. Compton scattering of the microwave background photons by the relativistic electrons in Coma C may account for the extended x-ray emission. X-ray synchrotron radiation may be ruled out because the observed flux is well above the extrapolated radio spectrum power law.

***Perseus Cluster.*** Perseus, in contrast to Coma, is highly irregular. The velocity dispersion of its galaxies is very large, and either most of the mass required by the virial theorem is invisible, or the system has positive energy. X-ray emission from the Perseus cluster is centered on NGC 1275, but the source is extended to a diameter of about 0.5 degree, about the same size as the radio halo 3C84B. The resolution of x-ray measurements is not sufficient to distinguish between emission from NGC 1275 and from the extended region. The spectrum is not clearly thermal or nonthermal. It may be a composite of various temperature regimes from $10^7$ to $10^8$ K, but could just as well be fitted to inverse Compton or synchrotron power laws. The relativistic electrons needed to generate nonthermal x-rays may be abundantly produced by the violent activity in the compact nucleus of NGC 1275. An inverse Compton model requires similar spectral indexes for both the radio and x-ray portions of the spectrum. The spectral index of a power law fit to the x-ray data for Perseus A is 1.1, but the radio synchrotron emission fits a power law with a spectral index of only about 0.7. However, there is enough uncertainty in both indexes that they cannot be assumed to disagree.

In summary, several possible processes might contribute to the x-ray emission from rich clusters:

1) Thermal bremsstrahlung from gas at $10^7$ to $10^8$ K, with a density of about $10^{-4}$ to $10^{-3}$ cm$^{-3}$;

2) Inverse Compton scattering of the microwave background radiation in the halos of active galaxies by relativistic electrons escaping from the nuclear regions;

3) The combined emission of a few very powerful x-ray galaxies, each radiating about $10^{43}$ erg/sec.

In addition to Coma, Virgo, and Perseus, other cluster candidates for identification with x-ray sources at high galactic latitudes are Abell 401, 3C129, Abell 1367, Abell 2256, and the cluster centered on Cygnus A. Their x-ray powers range from $5 \times 10^{43}$ to $8 \times 10^{44}$ erg/sec. The space density of rich clusters is about $10^{-6}$ Mpc$^{-3}$. At an average x-ray power of about $10^{44}$ erg/sec, they would come close to supplying the diffuse background.

## Diffuse Radiation

The full spectrum of the diffuse background radiation is shown in Fig. 7. Inflections in the curve indicate that different processes must be effective in different spectral ranges. In the range of lowest energies ($E = 100$ ev to 1 kev) the spectral energy dependence follows $E^{-1}$ to $E^{-2}$. The spectrum flattens to $E^{-0.4}$ from 1 to 10 kev and then bends progressively to a steeper dependence, approximately $E^{-0.75}$ from 20 to 40 kev and $E^{-1.3}$ from 40 kev to about 1 Mev. Considerable uncertainty still exists about the shape of the spectrum beyond 1 Mev. Several observers have claimed a flattening in the range 1 to 6 Mev, but instrumental effects may be responsible. At higher energies the spectrum follows $E^{-2}$.

From 1 to 10 kev, all the evidence from discrete sources reviewed above shows that the integral flux from galaxies and clusters may add up to the required background. However, the uncertainty is still an order of magnitude, and a truly

diffuse cosmological source may also be operating throughout intergalactic space. Both thermal and nonthermal processes have been proposed to explain the spectrum above 10 kev. The earliest proposal to receive general support was based on inverse Compton scattering of the microwave background photons by cosmic-ray electrons (*41*). It would require $10^{-4}$ ev/cm$^3$ of ultrarelativistic electrons throughout intergalactic space to provide the observed background. Could powerful radio galaxies provide such copious sources of cosmic-ray electrons? The estimated mean life of radio galaxies is about $10^9$ years (*42*). If each produced $10^{60}$ ergs of relativistic electrons, as deduced from synchrotron radio fluxes of the most active galaxies, the electron energy density still would average only $10^{-5}$ ev/cm$^3$. To compensate for the deficiency evolutionary effects can be invoked, for example by assuming a higher density of radio sources or greater luminosities in early epochs.

The diffuse x-ray background may have a thermal origin that involves the nature of the intergalactic medium and the universal distribution of cosmic rays. A critical average density of all matter and radiation in the universe is expressed by

$$\rho_c = \frac{3H_0^2}{8\pi G}$$

where $H_0$ is the Hubble constant and $G$ is the gravitational constant. At this mean density, expansion of the universe approaches a halt as time apppproaches infinity. For densities less than $\rho_c$ the universe is open and expands forever; if the density exceeds $\rho_c$ the universe is closed and the expansion will stop and reverse after some finite time. A recent assessment (*43*) indicates a value for $H_0$ of approximately 50 km sec$^{-1}$ Mpc$^{-1}$, which corresponds to $\rho_c \approx 4.7 \times 10^{-30}$ g/cm$^3$ (or a particle density of about $3 \times 10^{-6}$ cm$^{-3}$). Estimates of the density of galactic matter averaged over all space out to a Hubble radius give about $7.5 \times 10^{-32}$ g/cm$^3$, or about 2 percent of the critical density. Certainly there must be additional mass in the form of intergalactic gas, and a substantial number of subluminous objects such as white dwarfs, neutron stars, and black holes may contribute significantly to the mass budget. Attempts to detect cool intergalactic gas by virtue of its absorption or emission of electromagnetic radiation have all been negative; the indicated upper limit is about $10^{-11}$ cm$^{-3}$ for the density of neutral gas. A hot, fully ionized medium could exist at much higher densities and would be detectable by its thermal x-ray emission. The identification of such diffuse emission is one of the great challenges to x-ray astronomy.

Recently, Cowsik and Kobetich (*44*) have performed a more precise calculation of the inverse Compton scattering process and shown that it cannot match the observed shape from 10 kev to 1 Mev. Instead, they find a close fit to bremsstrahlung at $3 \times 10^8$ K (Fig. 8), which could be produced by universal, hot, intergalactic gas at a density of $3 \times 10^{-6}$ cm$^{-3}$. For $H_0 = 50$ km sec$^{-1}$ Mpc$^{-1}$, this density is close to that required for closure of the universe. At the very high temperature, the gas escapes freely from clusters and would be so uniformly dispersed throughout the universe as to satisfy the high degree of isotropy of the observed x-ray background. Underlying this thermal background at higher energies there

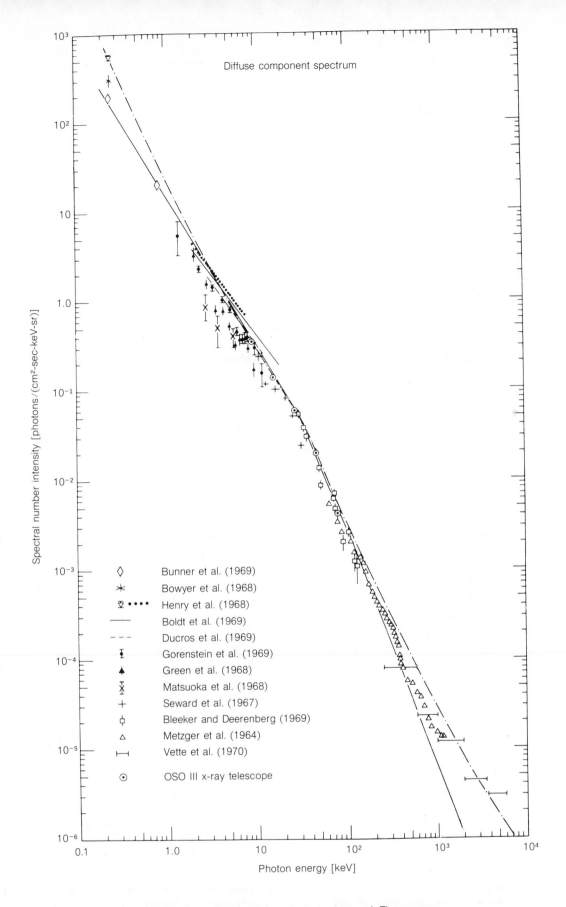

Figure 7. Spectrum of diffuse x-ray background. The sources of the data are: Bunner *et al.* (*60*), Bowyer *et al.* (*70*), Henry *et al.* (*59*), Boldt *et al.* (*71*), Ducros *et al.* (*5, 72*), Gorenstein *et al.* (*73*), Green *et al.* (*74*), Matsuoka *et al.* (*75*), Seward *et al.* (*76*), Bleeker and Deerenberg (*77*), Metzger *et al.* (*78*), Vette *et al.* (*47*), and (OSO III) Schwartz *et al.* (*79*). [Courtesy of L. E. Peterson, University of California at San Diego.]

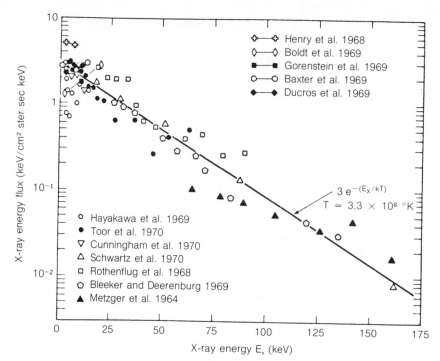

**Figure 8.** Fit of observations of the diffuse background (10 to 175 kev) to a thermal spectrum ($3.3 \times 10^8$ K). The sources of the data are: Henry *et al.* (59), Boldt *et al.* (80), Gorenstein *et al.* (73), Baxter *et al.* (81), Ducros *et al.* (5), Hayakawa *et al.* (82), Toor *et al.* (83), Cunningham *et al.* (84), Schwartz *et al.* (79) Rothenflug *et al.* (85), Bleeker and Deerenberg (77), Metzger *et al.* (78). [Courtesy of Cowsik and Kobetich (44).]

would be an order of magnitude weaker inverse Compton flux. It would surpass the thermal flux at about 1 kev and account for the very soft x-ray background (100 ev to 1 kev). If the thermal model for $3 \times 10^8$ K fits the observations, it leaves unknown the source of energy for the hot gas. It is interesting to recall Gold and Hoyle's steady-state model of the hot universe (45), in which matter was created in the form of neutrons which quickly underwent beta decay (46), delivering about 300 kev of kinetic energy to the electron. When fully converted to thermal energy, this would raise the temperature of intergalactic gas to $10^9$ K.

A semblance of a flattening in the diffuse background spectrum at 1 Mev was first reported on the basis of data obtained with the Ranger III and ERS-18 spacecraft. This feature implies an additional component of background flux in the range 1 to 6 Mev (47). Since the measurements were made from large distances above the earth, atmospheric background introduced no contamination. However, several authors have attributed the observed effect to locally produced radioactivity in the NaI(T1) crystal detector and its surroundings. More recent measurements, including those made on the Apollo 15 and Apollo 16 missions and from Cosmos satellites and balloons, are contradictory, and the validity of the results is still uncertain (48). Several theoretical interpretations have been offered for the feature in the 1- to 6-Mev range, if it is real. Neutral pions ($\pi^0$) decay with the emission of a spectrum of gamma rays that peak at 70 Mev. If they are produced at red shifts of the order of 100, the observed gamma rays would concentrate in the range 1 to 6 Mev (49). Other proposed

mechanisms are electron bremsstrahlung [either galactic (50) or extragalactic (51)]; proton bremsstrahlung (52); cosmological matter-antimatter annihilation (53); and nuclear line emission (54). But recent theoretical work discounts all but the hypotheses of $\pi^0$ decay with a high red shift and matter-antimatter annihilation.

Apollo 15 and Apollo 16 experimenters have extended the spectrum to 30 Mev, where the intensity obtained is $5 \times 10^{-5}$ photon per square centimeter per second per steradian per million electron volts, in rough agreement with balloon results (55). Above 50 Mev one data point from the satellite OSO-3 is available (56), and it indicates a steepening of the spectrum, although not significantly in conflict with a power law extrapolation from lower energies. At these high energies, the contribution from the thermal model for $3 \times 10^8$ K is insignificant. Theorists suggest Compton scattering of a universal distribution of cosmic-ray electrons on the 2.7 K background (57), or interaction of relativistic electrons generated by rapidly spinning white dwarfs with the high photon densities near the stellar surfaces (2).

Consider finally the background of very soft x-rays below 1 kev. Observations are difficult in this range and there is considerable spread in the published results. However, there is general agreement that the flux near 44 Å exceeds any simple extrapolation of the spectrum from 1 to 10 kev. Most observers find an excess toward the galactic pole (58–60); and it is also well established that a substantial part of the soft flux originates in the galactic disk (61, 62). Figure 9 illustrates the results of a rocket survey of a large area of the sky at 44 Å and

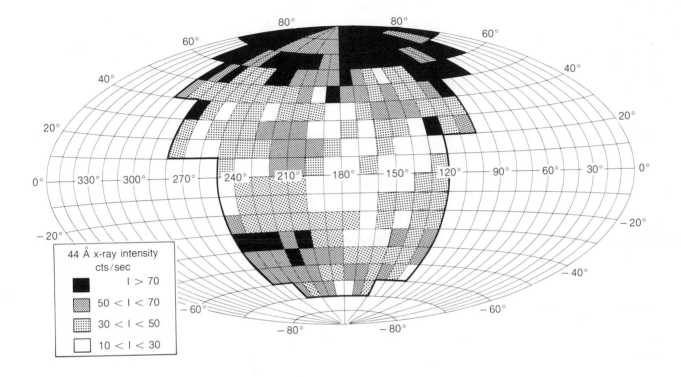

44 Å x-ray intensity
cts/sec

$I > 70$

$50 < I < 70$

$30 < I < 50$

$10 < I < 30$

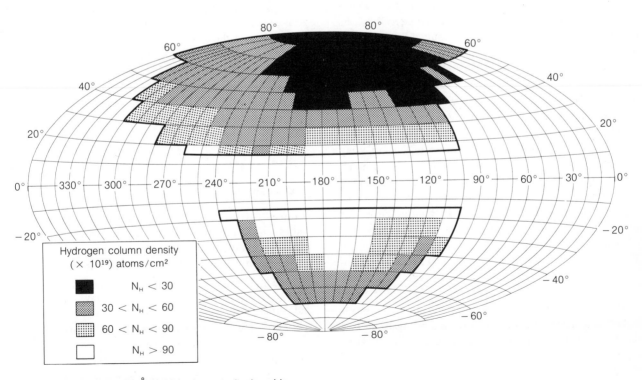

Hydrogen column density
$(\times 10^{19})$ atoms/cm²

$N_H < 30$

$30 < N_H < 60$

$60 < N_H < 90$

$N_H > 90$

**Figure 9.** Sky maps in 44-Å (0.28-kev) x-rays (top) and in hydrogen column density (bottom) (62). The maps are in new galactic coordinates. The portions of the sky surveyed are outlined by heavy lines.

includes a map of the hydrogen column density for comparison (62). In general, the x-ray flux is highest where the hydrogen column density is smallest. About two-thirds of the x-ray flux at the poles can be considered extragalactic and the remainder appears to originate from discrete or diffuse sources within the disk.

Some discrete features seem to be present in these data and in the data of other observers, who have attempted to identify them with radio remnants of supernovas, such as the North Polar Spur and the Vela supernova remnant. However, other spurs, such as Loop III, were scanned in the rocket survey with negative results at x-ray limits well below that of the North Polar Spur. It does not appear, therefore, that all extended radio spurs are sources of soft x-rays or that they contribute substantially to the soft x-ray background.

A good fit to the observed soft x-ray flux is obtained if the sources in the disk are distributed with the same scale height as interstellar gas. The variation of flux from pole to plane then follows the expected attenuation of an extragalactic flux by a smooth distribution of hydrogen in the disk. Sources that scale with the neutral hydrogen in the disk include supernova remnants (63) and defunct pulsars (64). For a supernova remnant model, the density of sources is small and the average distance must therefore be great. Very large soft x-ray luminosities would be required to overcome interstellar absorption because the optical thickness of neutral hydrogen is only about 100 pc in the disk. The isotropy of the observed x-rays argues against the relatively small number of sources in a supernova remnant model. Defunct pulsars, however, should occur in relatively great densities. They would produce thermal x-rays by accretion of interstellar gas.

The soft x-ray background has been attributed by some authors (58, 59) to thermal bremsstrahlung from gas at a temperature of about $3 \times 10^6$ K. If the gas is intergalactic, its density would be of the order of that required for a "closed" universe; if it is condensed in clusters, a much smaller universal mass is indicated since the bremsstrahlung flux varies with the square of the density. However, it has not yet been established whether the gas is associated with the local cluster, is condensed in clusters of galaxies, or is smoothly distributed in extragalactic space.

Evidence against a cosmological source of the soft x-ray background (65) comes from a rocket scan across the Small Magellanic Cloud. If the origin of the background x-rays were more distant, the Cloud would be expected to mask the background. No such effect was found, and a limit of 25 percent was set on the fraction of the radiation in the band from 44 to 60 Å that could originate beyond the distance of the Cloud. The data of Fig. 9 led to the conclusion that 67 percent of the observed x-rays in the direction of the pole was extragalactic. The Small Magellanic Cloud is at a galactic latitude of about −40 degrees, where the hydrogen column density is much higher than at the poles and the expected extragalactic flux should be attenuated to 40 percent of the observed background—within a factor of 2 of the limit set by the observation of the Cloud. A further possibility is that soft x-ray emission from the Cloud itself fills in the shadow. The observation is certainly crucial and should be improved on.

If the source of the soft x-ray background is closer than the Small Magellanic Cloud it may be associated with a hot halo around the galaxy as had been proposed by Spitzer (66). His highest-temperature model ($10^6$ K) required a density of $5 \times 10^{-4}$ cm$^{-3}$ and a radius of 8 kpc. This amount of gas would not provide the x-ray background. The x-ray observations require a somewhat higher temperature, about $3 \times 10^6$ K, and an emission measure, $\int n_e^2 dl$ (67), of about 0.1 pc cm$^{-6}$. In order to avoid excess pressure from the halo at the interface with the neutral disk, the gas density there must not be greater than about $10^{-3}$ cm$^{-3}$ and the radius of the halo must approach a megaparsec. The halo would therefore more correctly be described as an intracluster gas.

The x-ray emission from a regular cluster of galaxies such as Coma also provides evidence about the density of intergalactic gas if the x-rays are thermal bremsstrahlung from gas within the cluster (68). Where did this gas originate? It may be residual from the original cloud out of which the cluster formed, some of it may have been swept out of the galaxies, and some may have accreted from intergalactic space. If all the gas in Coma has fallen in from intergalactic space in the course of $10^9$ years, the external gas density would not exceed $10^{-7}$ cm$^{-3}$. The intergalactic gas therefore contributes less than (or at most an amount comparable to) the mass in galaxies to the total mass budget of the universe. According to this argument the universe is open, unless a great deal of hidden mass is present in other forms such as black holes or dead galaxies.

Another interesting bit of evidence for the ratio of intracluster to extragalactic gas comes from studies of double radio sources (69). The sources within clusters have only about one-half the separation of those outside clusters. If the motions of the expanding clouds are opposed by the pressure of the ambient gas, the estimated intracluster densities are 15 times as great as those in intergalactic space.

In summary, the x-ray background data are not inconsistent with thermal models of an intergalactic gas at near-critical density. A very hot gas, about $3 \times 10^8$ K, fits the harder x-rays very well and accounts for the spectral bump at about 30 kev. Gas at about $4 \times 10^6$ K fits the soft x-ray spectrum below 1 kev. However, this soft x-ray component may originate within the local group.

## Conclusion

X-ray astronomy has been carried on with balloons, rockets, and small satellites for the past decade. Great progress was expected from the HEAO program, which fell victim to budget limitation in 1973. With its 9000-kilogram payload capability in near-earth orbit, HEAO was designed to carry a new generation of instruments for detecting x-rays, gamma rays, and particles with very large gains in detection capability and speed of response. Large detectors with high sensitivities are required to characterize more precisely the physical properties of the wide variety of galactic and extragalactic sources. Higher sensitivity coupled with very fast response is essential for detecting the microsecond variations predicted by models of black holes and their associated gas disks. At the time of this writing, NASA is attempting to reconstruct* the HEAO program with a lower budget, to begin flights in 1977. Although this observatory will be substantially reduced in capability from the original one, it will go a long way toward satisfying the urgent observational demands in high-energy astronomy.

## References and Notes

1. R. Giacconi, S. Murray, H. Gursky, E. Kellogg, E. Schreier, H. Tananbaum, *Astrophys. J.* 178, 281 (1972).

2. R. Cowsik, *Proc. 12th Int. Conf. Cosmic Rays* 1, 334 (1971).

3. I. D. Novikov and K. S. Thorne, preprint.

4. J. E. Pringle and M. J. Rees, *Astron. Astrophys.* 21, 1 (1972).

5. G. Ducros, R. Ducros, R. Rocchia, A. Tarius, *Nature* 227, 152 (1970).

6. B. McBreen, S. E. Ball, Jr., M. Campbell, K. Greisen, D. G. Koch, *Astrophys. J.* 184, 571 (1973).

7. G. J. Fishman, F. R. Harnden, Jr., W. N. Johnson, R. C. Haymes, *Astrophys. J. Lett.* 158, 61 (1969); R. R. Hillier, W. R. Jackson, A. Murray, R. M. Redfern, R. J. Sale, *ibid.* 162, 177 (1970); G. Fritz, J. F. Meekins, T. A. Chubb, H. Friedman, R. C. Henry, *ibid.* 164, 55 (1971); J. D. Kurfess, *ibid.* 168, 39 (1971); C. E. Fichtel, R. C. Hartman, D. A. Kniffen, M. Sommer, *Astrophys. J.* 171, 31 (1972); S. Rappaport, H. Bradt, W. Mayer, *Nature Phys. Sci.* 229, 40 (1971); A. J. M. Deerenberg and J. A. M. Bleeker, *ibid.*, p. 113; L. E. Orwig, E. L. Chupp. D. J. Forrest, *ibid.* 231, 171 (1971); D. Brini, C. Cavani, F. Frontera, F. Fuligni, *ibid.* 232, 79 (1971); R. Browning, D. Ramsden, P. J. Wright, *ibid.*, p. 99; H. W. Smathers, T. A. Chubb, D. Sadeh, *ibid.*, p. 120; J. Vasseur *et al.*, *ibid.* 233, 46 (1971); C. Cavani, F. Frontera, F. Fuligni, D. Brini, *ibid.*, p. 155; D. Boclet, G. Brucy, J. Claisse, P. Durouchoux, R. Rocchia, *ibid.* 235, 69 (1972); E. A. Boldt, U. D. Desai, S. S. Holt, P. J. Serlemitsos, R. F. Silverberg, *Nature* 223, 280 (1969); F. W. Floyd, I. S. Glass, H. W. Schnopper, *ibid.* 224, 50 (1969); R. L. Kinzer, R. C. Noggle, N. Seeman, G. H. Share, *ibid.* 229, 187 (1971); P. Albats, G. M. Frye, Jr., A. D. Zych, preprint; R. L. Kinzer, G. H. Share, N. Seeman, preprint.

8. F. G. Smith, *Nature* 231, 191 (1971).

9. I. S. Shklovsky, *ibid.* 238, 144 (1972).

10. R. J. Grader, R. W. Hill, J. P. Stoering, *Astrophys. J. Lett.* 161, 45 (1970); P. Gorenstein, B. Harris, H. Gursky, R. Giacconi, R. Novick, P. Vanden Bout, *Science* 172, 369 (1971).

11. B. Margon, M. Lampton. S. Bowyer, R. Cruddace, *Astrophys. J. Lett.* 169, 23 (1971).

12. H. Tananbaum, H. Gursky, E. M. Kellogg, R. Levinson, E. Schrier, R. Giacconi, *ibid.* 174, 143 (1972); W. Forman, C. A. Jones, W. Liller, *ibid.* 177, 103 (1972); A. Davidsen, J. P. Henry, J. Middleditch, H. E. Smith, *ibid.*, p. 97; J. N. Bahcall and N. A. Bahcall, *Astrophys J.* 178, 61 (1972).

13. K. Brecher, *Nature* 239, 325 (1972); I. Novikov, preprint.

14. R. Doxsey, H. V. Bradt, A. Levine, G. T. Murthy, S. Rappaport, G. Spada, *Astrophys. J. Lett.* 182, 25 (1973).

15. Yu. N. Gnedin and R. A. Sunyaev, *Astron. Astrophys.*, in press.

16. E. Schreier, R. Levinson, H. Gursky, E. Kellogg, H. Tananbaum, R. Giacconi, *Astrophys. J. Lett.* 172, 79 (1972).

17. S. Shulman, G. Fritz, J. Meekins, H. Friedman, *ibid.* 168, 49 (1971); E. Schreier, H. Gursky, E. Kellogg, H. Tananbaum, R. Giacconi, *ibid.* 170, 21 (1971).

18. D. R. Parsignault *et al.*, *Nature Phys. Sci.* 239, 123 (1972); P. W. Sanford and F. H. Hawkins, *ibid.*, p. 135; C. R. Canizares, J. E. McClintock, G. W. Clark, W. H. G. Lewin, H. W. Schnopper, G. F. Sprott, *Nature*, in press.

19. L. L. E. Braes and G. K. Miley, *Nature* 237, 506 (1972); P. C. Gregory, *ibid.* 239, 439 (1972); R. M. Hjellming and B. Balick, *Nature Phys. Sci.* 239, 135 (1972).

20. W. A. Hiltner and D. E. Mook, *Annu. Rev. Astron. Astrophys.* 8, 139 (1970); K. Davidson, F. Pacini, E. E. Salpeter, *Astrophys. J.* 168, 45 (1971).

21. T. A. Matilsky, R. Giacconi, H. Gursky, E. M. Kellogg, H. Tananbaum, *Astrophys. J. Lett.* 174, 53 (1972).

22. The Hubble radius is the distance at which a galaxy would be receding from us with a velocity equal to the speed of light, according to the distance-velocity relation for galaxies.

23. M. Ulmer, V. Grace, H. S. Hudson, D. A. Schwartz, *Astrophys. J.* 173, 205 (1972).

24. E. T. Byram, T. A. Chubb, H. Friedman, *Science* 152, 66 (1966); H. Bradt, W. Mayer, S. Naranan, S. Rappaport, G. Spada, *Astrophys. J. Lett.* 150, 199 (1967).

25. I. S. Shklovsky, *Astrophys. Lett.* 10, 5 (1971).

26. J. E. Felten, *J. Roy. Astron. Soc. Can.* 64, 33 (1970).

27. I. Shklovsky, *Nature* 228, 1174 (1970).

28. H. Gursky, A. Solinger, E. Kellogg, S. Murray, H. Tananbaum, R. Giacconi, A. Cavaliere, *Astrophys. J. Lett.* 173, 99 (1972).

29. K. I. Kellermann, B. G. Clark, M. H. Cohen, D. B. Schaffer, J. J. Broderick, D. L. Jauncey, *ibid.* 179, 141 (1973).

30. C. Bowyer, M. Lampton, J. Mack, F. de Mendonca, *ibid.* 161, 1 (1970).

31. E. Kellogg, H. Gursky, C. Leong, E. Schreier, H. Tananbaum, R. Giacconi, *ibid.* 165, 49 (1971).

32. C. Wade, R. Hjellming, K. Kellermann, J. Wardle, *ibid.* 170, 11 (1971).

33. E. Becklin, J. Frogel, D. Kleinmann, G. Neugebauer, E. Ney, D. Strecker, *ibid.*, p. 15; W. Kunkel and H. Bradt, *ibid.*, p. 7.

34. G. Fritz, A. Davidsen, J. Meekins, H. Friedman, *Astrophys. J. Lett.* 164, 81 (1971); W. Forman, E. Kellogg, H. Gursky, H. Tananbaum, R. Giacconi, *Astrophys. J.* 178, 309 (1972).

35. H. J. Rood, T. L. Page, E. C. Kintner, I. R. King, *Astrophys, J. Lett.* 172, 124 (1972).

36. C. A. Muller, *Bull. Astron. Inst. Neth.* 14, 339 (1959).

37. J. F. Meekings, G. Fritz, T. A. Chubb, H. Friedman, R. C. Henry, *Nature* 231, 107 (1971).

38. H. Gursky, E. Kellogg, S. Murray, C. Leong, H. Tananbaum. R. Giacconi, *Astrophys. J. Lett.* 167, 181 (1971).

39. R. C. Henry, *ibid.* 172, 97 (1972); J. Holberg, S. Bowyer, M. Lampton, *Astrophys. J.* in press.

40. K. Brecher and E. R. Burbidge, *Nature* 237, 440 (1972).

41. J. E. Felten and P. Morrison, *Astrophys. J.* 146, 616 (1966).

42. M. Schmidt, *ibid.* 162, 371 (1970).

---

*The HEAO program was indeed reinstated by NASA, but with somewhat less ambitious capabilities, as noted by the author.—Ed.

43. A. Sandage, paper presented at the Mayall Symposium, Tucson, Arizona (1971).

44. R. Cowsik and E. J. Kobetich, *Astrophys. J.* 177, 585 (1972).

45. T. Gold and F. Hoyle, in *Paris Symposium on Radio Astronomy*, R. N. Bracewell, Ed. (Stanford Univ. Press, Stanford, Calif., 1959), p. 104.

46. The decay scheme was $n \rightarrow p + e + \bar{\nu}$, where $n$ is a neutron, $p$ a proton, $e$ an electron, and $\bar{\nu}$ an antineutrino.

47. J. I. Vette, D. Gruber, J. L. Matteson, L. E. Peterson, *Astrophys. J. Lett.* 160, 161 (1970).

48. J. R. Arnold, L. E. Peterson, A. E. Metzger, J. I. Trombka, *Proceedings of IAU Symposium No. 55*, H. Bradt, Ed. (Reidel, Dordrecht, in press); S. V. Golenetskii, E. P. Mazets, V. N. Ill'inskii, R. L. Aptekar, M. M. Bredov, Yu. A. Gur'yan, V. N. Panov, *Astrophys. Lett.* 9, 69 (1971); R. R. Daniel, G. Joseph, P. J. Lavaklare, *Astrophys. Space Sci.* 18, 462 (1972).

49. F. W. Stecker, *Astrophys. J.* 157, 507 (1969); *Nature* 229, 105 (1971).

50. M. J. Rees and J. Silk, *Astron. Astrophys.* 3, 452 (1969).

51. J. Silk, *Space Sci. Rev.* 11, 671 (1970).

52. R. L. Brown, *Lett. Nuovo Cimento* 4, 941 (1970).

53. F. W. Stecker, D. L. Morgan, Jr., J. Bredekamp, *Phys. Rev. Lett.* 27, 1469 (1971).

54. D. D. Clayton and J. Silk, *Astrophys. J. Lett.* 158, 43 (1969).

55. H. A. Mayer-Hasselwander, E. Pfeffermann, K. Pinkau, H. Rothermel, M. Sommer, *ibid.* 175, 23 (1972); G. H. Share, R. L. Kinzer, N. Seeman, preprint.

56. W. L. Kraushaar, G. W. Clark, G. P. Garmire, R. Borken, P. Higbe, C. Leong, T. Thorsos, *Astrophys. J.* 177, 341 (1972).

57. K. Brecher and P. Morrison, *Phys. Rev. Lett.* 23, 802 (1969).

58. C. S. Bowyer and G. B. Field, *Nature* 223, 573 (1969).

59. R. C. Henry, G. Fritz, J. F. Meekins, H. Friedman, E. T. Byram, *Astrophys. J. Lett.* 153, 11 (1968).

60. A. N. Bunner, P. L. Coleman, W. L. Kraushaar, D. McCammon, T. M. Palmieri, A. Shilepsky, M. Ulmer, *Nature* 223, 1222 (1969).

61. T. M. Palmieri, G. A. Burginyon, R. J. Grader, R. W. Hill, F. D. Seward, J. P. Stoering, *Astrophys. J.* 169, 33 (1971); D. J. Yentis, R. Novick, P. Van den Bout, *ibid.* 177, 365 (1972); R. C. Henry, G. Fritz, J. F. Meekins, T. Chubb, H. Friedmann, *Astrophys. J. Lett.* 163, 73 (1971); A. N. Bunner, P. L. Coleman, W. L. Kraushaar, D. McCammon, *ibid.* 167, 3 (1971); S. Hayakawa *et al.*, *Astrophys. Space Sci.* 12, 104 (1971).

62. A. Davidsen, S. Shulman, G. Fritz, J. F. Meekins, R. C. Henry, H. Friedman, *Astrophys. J.* 177, 629 (1972).

63. S. A. Ilovaisky and C. Ryter, *Astron. Astrophys.* 15, 224 (1971).

64. J. P. Ostriker, M. J. Rees, J. Silk, *Astrophys. Lett.* 6, 179 (1970).

65. D. McCammon, A. N. Bunner, P. L. Coleman, W. L. Kraushaar, *Astrophys. J. Lett.* 168, 33 (1971).

66. L. Spitzer, *Astrophys. J.* 124, 20 (1956).

67. In this expression $n_e$ is the electron density per cubic centimeter and $l$ is the path length.

68. J. Gott and J. Gunn, *Astrophys. J. Lett.* 169, 563 (1971).

69. D. S. DeYoung, *ibid.* 173, 7 (1972).

70. C. S. Bowyer, G. B. Field, J. E. Mack, *Nature* 217, 32 (1968).

71. E. A. Boldt, U. D. Desai, S. S. Holt, P. Serlemitsos, in *Non-Solar X- and Gamma-Ray Astronomy*, L. Gratton, Ed. (Reidel, Dordrecht, 1970), pp. 309–314.

72. G. Ducros, R. Ducros, R. Rocchia, A. Tarrius, preprint.

73. P. Gorenstein, E. M. Kellogg, H. Gursky, *Astrophys. J.* 156, 315 (1969).

74. D. W. Green, B. G. Wilson, A. J. Baxter, in *Space Research IX*, K. S. Champion *et al.*, Eds. (North-Holland, Amsterdam, 1969), pp. 222–225.

75. M. Matsuoka, M. Oda, Y. Ogawara, S. Hayakawa, T. Kato, *Can. J. Phys.* 46, S466 (1968).

76. F. Seward, G. Chodil, H. Mark, C. Swift, A. Toor, *Astrophys. J.* 150, 845 (1967).

77. J. A. M. Bleeker and A. J. M. Deerenberg, *Astrophys. J.* 159, 215 (1970).

78. A. E. Metzger, E. C. Anderson, M. A. Van Dilla, J. R. Arnold, *Nature* 204, 766 (1964).

79. D. A. Schwartz, H. S. Hudson, L. E. Peterson, *Astrophys. J.* 162, 431 (1970).

80. E. A. Boldt, U. D. Desai, S. S. Holt, *ibid.* 156, 427 (1967); E. A. Boldt, U. D. Desai, S. S. Holt, P. Serlemitsos, *Nature* 224, 677 (1969).

81. A. J. Baxter, B. G. Wilson, D. W. Green, *Can. J. Phys.* 47, 2651 (1969).

82. S. Hayakawa *et al.*, in *Non-Solar X- and Gamma-Ray Astronomy*, L. Gratton, Ed. (Reidel, Dordrecht, 1970), pp. 121–129.

83. A. Toor, F. D. Seward, L. R. Cathey, W. E. Kunkel, *Astrophys. J.* 160, 209 (1970).

84. C. Cunningham, D. Groves, R. Price, R. Rodrigues, C. Swift, H. Mark, *ibid.*, p. 1177.

85. R. Rothenflug, R. Rocchia, D. Boclet, P. Durouchoux, in *Space Research VIII*, A. P. Mitra *et al.*, Eds. (North Holland, Amsterdam, 1968), pp. 423–429.

# Observations of Gamma-Ray Bursts of Cosmic Origin

Ray W. Klebesadel, Ian B. Strong, and Roy A. Olson

Abstract. *Sixteen short bursts of photons in the energy range 0.2-1.5 MeV have been observed between 1969 July and 1972 July using widely separated spacecraft. Burst durations ranged from less than 0.1 s to ~30 s, and time-integrated flux densities from ~ $10^{-5}$ ergs $cm^{-2}$ to ~$2 \times 10^{-4}$ ergs $cm^{-2}$ in the energy range given. Significant time structure within bursts was observed. Directional information eliminates the Earth and Sun as sources.*

## I. Introduction

On several occasions in the past we have searched the records of data from early *Vela* spacecraft for indications of gamma-ray fluxes near the times of appearance of supernovae. These searches proved uniformly fruitless. Specific predictions of gamma-ray emission during the initial stages of the development of supernovae have since been made by Colgate (1968). Also, more recent Vela spacecraft are equipped with much improved instrumentation. This encouraged a more general search, not restricted to specific time periods. The search covered data acquired with almost continuous coverage between 1969 July and 1972 July, yielding records of 16 gamma-ray bursts distributed throughout that period. Search criteria and some characteristics of the bursts are given below.

## II. Instrumentation

The observations were made by detectors on the four *Vela* spacecraft, *Vela 5A, 5B, 6A,* and *6B,* which are arranged almost equally spaced in a circular orbit with a geocentric radius of ~$1.2 \times 10^5$ km.

On each spacecraft six 10 $cm^3$ CsI scintillation counters are so distributed as to achieve a nearly isotropic sensitivity. Individual detectors respond to energy depositions of 0.2-1.0 MeV for *Vela 5* spacecraft and 0.3-1.5 MeV for *Vela 6* spacecraft, with a detection efficiency ranging between 17 and 50 percent.

From *Astrophysical Journal*, vol. 182, pp. L85-L88, 1973, The University of Chicago Press. Reprinted with permission.

The scintillators are shielded against direct penetration by electrons below ~0.75 MeV and protons below ~20 MeV. A high-$Z$ shield attenuates photons with energy below that of the counting threshold. No active anticoincidence shielding is provided.

Normalized output pulses from the six detectors are summed into the counting and logics circuitry. Logical sensing of a rapid, statistically significant rise in count rate initiates the recording of discrete counts in a series of quasi-logarithmically increasing time intervals. This capability provides continuous coverage in time which, coupled with isotropic response, is unique in observational astronomy. A time measurement is also associated with each record.

The data accumulations include a background component due to cosmic particles and their secondary effects. The observed background rate, which is a function of the energy threshold, is ~ 150 counts per second for the *Vela 5* spacecraft and ~20 counts per second for the *Vela 6* spacecraft.

## III. Observations

Since these detectors are susceptible to stimulation by energetic particles, the following evidence is offered in support of the interpretation that the signals reported here are due to fluxes of photons within the quoted energy range. Other Vela detectors with high sensitivity to energetic charged particles and neutrons recorded no deviation from the steady counting rate induced by cosmic particle fluxes at the time of any of the observed bursts. It has been noted, furthermore, that the detailed time structure of each burst is reproduced at all spacecraft recording the event, even though the radiation must, in most cases, have traversed an appreciable portion of the geomagnetic field. Simple calculations show that electron energies of many GeV and proton energies of many MeV would be required to produce this degree of rigidity, and fluxes of such particles would create observable effects in the other instruments on the spacecraft. Additionally, no difference in the time of arrival of the stimulating signals at two different spacecraft has been found which exceeds 0.8 s, the maximum transit time for light, even though the search allowed a deviation from simultaneity as great as 4 s.

A count-rate record is generated only in response to a rapid rise in count rate to a level significantly above background. The frequency with which individual records are generated is relatively high for *Vela 5* spacecraft. Modifications to *Vela 6* detectors reduced this frequency, at some cost in sensitivity, to an insignificant level. Only 47 such records have been generated by both *Vela 6* spacecraft over a 2-year period, 22 of which are responses, in coincidence, to the bursts reported here. Present processing requires that at least two spacecraft record the burst with a deviation from simultaneity of 4 s or less. Sixteen events have been observed to meet these criteria, two of which were recorded by all four spacecraft. Absence of consistent response from all four spacecraft can be attributed in most cases to an inappropriate mode of operation or to marginal signal levels.

These bursts display a wide variety of characteristics. Time durations range from less than a second to about 30 s. Some count-rate records have a number of clearly resolved peaks while others do not appear to display any significant structure. The time-integrated flux density in the measured energy interval ranges from the minimum identifiable level of $\sim 10^{-5}$ ergs cm$^{-2}$ to more than $2 \times 10^{-4}$ ergs cm$^{-2}$. Instantaneous flux densities have exceeded $4 \times 10^{-4}$ ergs cm$^{-2}$ s$^{-1}$. An indication of the spectral distribution of the incident flux may be derived from the ratio of the response in the two energy intervals in those cases where both *Vela 5* and *Vela 6* spacecraft recorded the burst.

Allowing for differing energy thresholds and statistical fluctuations, the integrated flux for a particular event is independent of the recording spacecraft. Differences in the time of arrival of the signals at the various spacecraft imply that the spacecraft are not equidistant from any given source. Inverse-square law considerations thereby place the sources at a distance of at least 10 orbit diameters, or several million kilometers.

Arrival-time differences have been derived approximately in all cases, and fairly accurately ($\pm 0.05$ s) for a number of cases. For a two-spacecraft coincidence the transit delay defines a circle on the celestial sphere on which the source position must lie. For three spacecraft we can define intersecting circles, whose points of intersection represent the source position and its mirror image in the orbital plane of the spacecraft, a presently unresolved ambiguity. Nevertheless, it has been possible by this technique to rule out the Sun as a source. Also, in none of the 16 cases was there found any close correlation with any recorded indications of solar activity.

One event has been observed which almost certainly was associated with a solar outburst. It differs distinctly from the 16 bursts reported here, and will be described in detail at a later date.

A burst observed on 1970 August 22 is presented as an example. Figure 1 shows the count rate as a function of time. Each plot is presented in two parts. On the left, on a linear time scale, are plotted 10 measurements of count rate made at 4-minute intervals for the time immediately preceding the burst. These establish a background count rate. The record of the burst is plotted on the right on a logarithmic time scale. All the *Vela 5A* data have had a uniform 100 counts per second (a major fraction of the background) subtracted before plotting in order to facilitate comparison of time structure.

The initial part of the burst (extending to $\sim 4$ s) has an integrated flux density of $\sim 8 \times 10^{-5}$ ergs cm$^{-2}$ in the range 0.2–1.0 MeV, and $\sim 6 \times 10^{-5}$ ergs cm$^{-2}$ in the range 0.3–1.5 MeV. Within these 4 s there appears structure common to the records of all three spacecraft. Although the exact statistical significance of this structure has not yet been firmly established, it has been used to adjust these three records in time, relative to the initiation of the recordings. Exclusion of the Sun as the source, based on directional resolution, is unaffected by this correction.

In addition to the initial structure, all three records show a distinct peak centered around 6.5 s. For each record this peak is statistically significant to about 6 standard deviations. It represents integrated flux densities of $10^{-5}$ ergs cm$^{-2}$ and $4 \times 10^{-6}$ ergs cm$^{-2}$ in the lower and higher energy ranges, respectively. The spectrum is clearly softer than that of the initial part of the burst.

## IV. Discussion

A search was made for reports of a nova or supernova within a reasonable time ($\sim$ several weeks) of each gamma-ray burst. No reported novae were related in time or direction to any of the bursts. Only two reported supernovae reached maximum apparent magnitude within a few days of an observed burst. In both cases, however, reports of prediscovery observations were later made which preceded the gamma-ray burst by at least several days. In addition, the source positions derived from preliminary timing data are inconsistent with the locations of the supernovae.

The lack of correlation between gamma-ray bursts and reported supernovae does not conclusively argue against such an association, since it is possible that there are supernovae, not necessarily bright in the optical region ("theoreticians' supernovae"), whose rate of occurrence may exceed those which are optically visible (see, e.g., Thorne 1969). A source at a distance of 1 Mpc would need to emit $\sim 10^{46}$ ergs in the form of electromagnetic radiation between 0.2 and 1.5 MeV in order to produce the level of response observed here. Since this represents only a small fraction ($<10^{-3}$) of the energy usually associated with supernovae, the energy observed is not inconsistent with a supernova as a source.

The authors wish to acknowledge the interest shown in the past by Edward Teller, Stirling Colgate, and A. G. W. Cameron who have on a number of occasions encouraged us to look for bursts of energetic photons.

We also wish to thank J. H. Coon and all of our colleagues in the Space Science Group at Los Alamos who have helped with this work. The detector electronics were the responsibility of the Space Electronics section at Los Alamos, under the direction of J. P. Glore. Logics were developed by the Satellite Systems Division at Sandia Laboratories; in particular we wish to mention R. E. Spalding, G. J. Dodrill, and J. G. Mitchell.

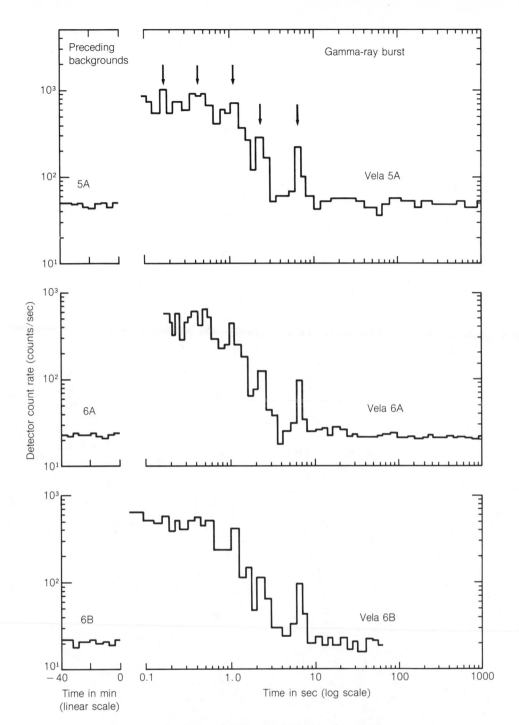

**Figure 1.** Count rate as a function of time for the gamma-ray burst of 1970 August 22 as recorded at three Vela spacecraft. Arrows indicate some of the common structure. Background count rates immediately preceding the burst are also shown. *Vela 5A* count rates have been reduced by 100 counts per second (a major fraction of the background) to emphasize structure.

This research was performed as part of the Vela Satellite Program, which is jointly sponsored by the U.S. Department of Defense and the U.S. Atomic Energy Commission. The program is managed by the U.S. Air Force, and satellite operation activities are under the jurisdiction of the Air Force Satellite Control Facility, Sunnyvale, California.

## References

Colgate, S. A. 1968, *Canadian J. Phys.*, 46, S476.
Thorne, K. S. 1969, in *Supernovae and Their Remnants*, ed. Peter J. Brancazio and A. G. W. Cameron (New York: Gordon & Breach).

Possible prehistoric record of the Crab
Nebula supernova. This pictograph is on
the ceiling of a shallow overhang in Chaco
Canyon National Monument, New Mexico.
It may represent the close conjunction of
the Crab Nebula supernova and the
crescent moon shortly before sunrise on
the morning of July 5, 1054 A.D. [National
Aeronautics and Space Administration.]

# PROBLEMS IN MODERN ASTRONOMY

"In Nine Centuries, Search Unravels Many Parts of the Crab Nebula Mystery" tells how astronomers discovered the source of the enormous energy that is released into space by the Crab Nebula. It also explains how old oriental records enable us to date the explosion that created this nebula. Since the article was written, the editors and their colleagues have found a number of ancient rock drawings in the western United States that reinforce the suggestion that American Indians may have observed and recorded this exploding star in the year 1054 A.D.

". . . the overwhelming majority of all the quasars that ever existed must have evolved by now into less luminous objects, perhaps ordinary galaxies. One can estimate that only about 35,000 quasars exist today." Astronomer Maarten Schmidt and *Scientific American* editor Francis Bello provide the best discussion of these most remarkable objects of distant space that we have yet seen in "The Evolution of Quasars." Several quasars with even larger red shifts than those discussed in this article have since been discovered, but the authors' conclusions remain valid.

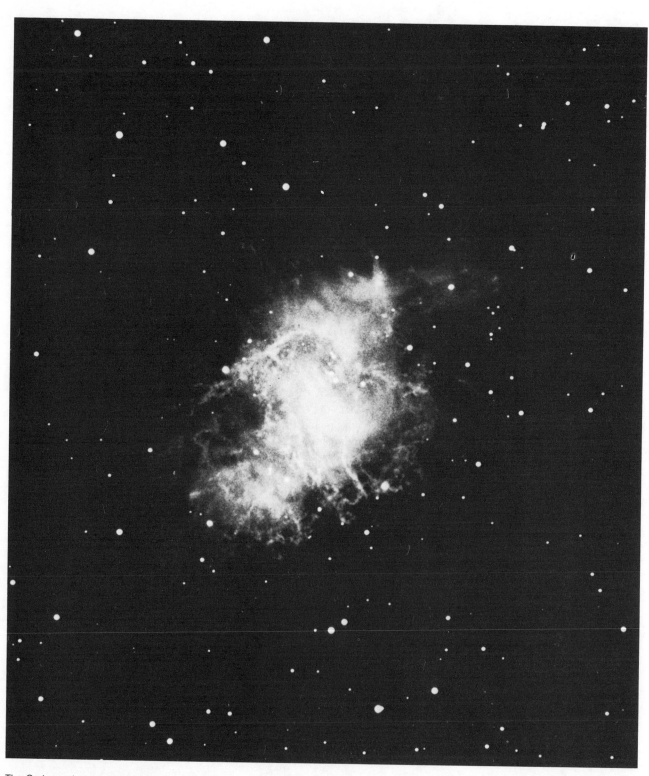

The Crab as photographed today from the earth. It is a vast and
rapidly expanding gas cloud, emitting radio waves, visible light
and x rays, powered by a spinning neutron star.

# In Nine Centuries, Search Unravels Many Parts of the Crab Nebula Mystery

Stephen P. Maran

Shortly before dawn on a July morning nine centuries ago, the Chinese astronomer Yang Wei–Tê glanced at a brilliant yellow star in the east, where day would soon break over Khaifêng, capital of the Sung dynasty. But this star did not belong there! The map of the constellations was very clear on this point, and the Chinese astronomers did not make mistakes in such matters. During nearly 1,300 years, they had unfailingly recorded all 17 appearances of the comet that, much later, would be named for Edmund Halley.

Yang's new star was as bright as the planet Venus, which outshines everything in the sky except the sun and moon. In fact, when the sun did rise, Yang found the star still visible. Although it soon began to fade, he reported that it could be seen in broad daylight for more than three weeks. After that it remained perceptible among the stars for some 600 nights. Then finally it passed beyond the limits of vision.

Astronomers in Peking and Japan recorded the star too, and so, we think, did North American Indians. At least they made drawings of a big round object near a crescent moon, and computations show that they probably were picturing the same event.

Yang may not have known of these other observations, but he did know his politics. His position as chief computer of the calendar gave him access to high authorities of the imperial court, before whom he was soon prostrated, begging leave to report the occurrence of the "guest star." The star, Yang explained, was an indication of the greatness of the present emperor. This explanation was favorably received, and the Bureau of Historiography was instructed to record Yang's report for posterity. Thus was documented the first clue in the tale of the Crab Nebula (opposite page), which has become an outstanding detective story of modern science.

In 1054 when Yang saw his star no one knew of nebulas—those spots of light in the sky that look like glowing clouds through a telescope. There was therefore no Crab Nebula mystery—just the first of many clues to a riddle that would develop as men studied the skies through the centuries. Like all good scientific problems, this riddle is open ended. Each new answer to some part of it suggests new questions. Yet in the last two years we have managed to explain a lot about the Crab Nebula. We know now that it was born when a great star larger than our sun exhausted its nuclear fuel and collapsed. As the stellar matter condensed towards the center, gravitational force upon it increased tremendously. The pressure rose so high that the individual atoms themselves collapsed; their constituent negative electrons and positive protons combined to form electrically neutral particles, neutrons. As the star collapsed, it spun ever faster.

A by-product of the great implosion was a splash. While most of the matter was drawing together in the star, some was flung out into surrounding space. It persists there as a rapidly expanding cloud of gas.

Electrons from the spinning star pass out into this gas cloud. Encountering magnetic fields, they release energy as radio waves, visible light and x rays. Here on earth we can detect this radiation, and it allows us now to understand this great, glowing, expanding gas cloud powered by a giant flywheel at the center. But the first step to follow Yang's sighting was relatively simple: Someone had to *find* the nebula.

After Yang's "guest star" faded beyond his view, no further clues presented themselves until 1731. By then telescopes were in fairly wide use and one belonged to an English doctor and amateur astronomer, John Bevis. Wholly unaware of the Chinese observations, Bevis happened to point his telescope at the same place where Yang's brilliant star had appeared and found the second clue. He saw not a dim, pointlike star but a fuzzy patch of light—in fact a "nebula."

Still another century later, in the mid-1840s, at Birr, Ireland, the third Earl of Rosse examined Bevis' nebula with the largest telescope the world had seen; its big mirror measured six feet across. He discerned clue three—filaments in the patch. Rosse's drawing of what he saw has been compared to a picture of a bug. He chose to call it a "Crab," and so the nebula was christened.

By the last decade of the 19th century, astronomical technology had so improved that the nebula could be photographed. In 1921 John Duncan of the Mount Wilson Ob-

From *Smithsonian*, vol. 1, no. 3, pp. 50–57, 1970. Reprinted with permission.

First report of a nova made by Chinese about 1300 B.C. Their followers recorded the 1054 supernova that gave birth to the Crab Nebula. [From *Science and Civilization in China*, volume III, by Joseph Needham (1959), Cambridge University Press.]

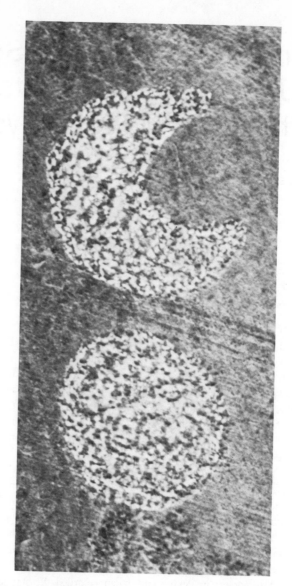

This drawing from Indian ruins in Navaho Canyon, Arizona, has been interpreted, along with another found in a cave, as the 1054 supernova. [William C. Miller, Hale Observatories.]

servatory in California compared two photographs of the Crab taken 11½ years apart, using the 60-inch telescope in the San Gabriel Mountains. On the more recent picture the filaments were a trifle farther from the center of the nebula than on the earlier exposure. This showed that the Crab was *expanding*. Duncan's finding, fully confirmed by other astronomers and measurement techniques, was the fourth great clue in the story: Where a star once flared briefly, a cloud of shining gas, six light years in diameter, was expanding into space at the rate of almost 700 miles per second. For comparison, the "escape velocity" that an Apollo spacecraft must attain to leave the earth on a lunar excursion is a paltry seven miles per second.

Distance dims the brightness of any light. We know now that the Crab Nebula is—and Yang's star was—some 6,000 light-years or 35,000,000 billion miles from the earth, whereas Venus, even at the most distant part of its orbit, is about 160 million miles away. Yet to Yang it seemed that the star was fully as bright as Venus. Considering that dust in space also dims the light of distant objects, the true brightness of Yang's star must have equaled the combined luminosity of at least 200 million suns.

A stellar outburst of this fantastic magnitude is a rare phenomenon indeed. In a spiral galaxy or "island universe" (so-called because it is a self-contained star system) such as our own Milky Way of some hundred billion stars, it will occur at most three times in a century. Astronomers call such an event a "supernova" and identify it with the death throes that occur when a massive star burns itself out.

The contemporary astronomer Hong-Yee Chiu has compared this process to a fire in a tile-roofed, wooden house: When the burning walls are nearly consumed, the roof falls in, and its impact causes one last splash of flame. In the same way, when the star exhausts its nuclear fuel, the exterior portion collapses, causing a great thermonuclear explosion. The tremendous brightness of Yang's star and the great speed of the expanding gas in the Crab show that the nebula is the result of a supernova.

What happens in a supernova outburst? During the 1930s, two famous physicists, Lev D. Landau in the Soviet Union and J. Robert Oppenheimer, in collaboration with G. M. Volkoff, in the United States, speculated about a new state of matter—and a new kind of star. You would get a "neutron star" if somehow you could take all the matter of a star like the sun (diameter 864,000 miles) and squeeze it into a ball just 10 miles across. The matter would then be as dense as the nucleus of an atom; a small matchbox would contain as much material as 17 trillion quarts of water. Heavy stuff indeed! If that same matchbox-sized chunk of neutron star could be "canned" and dropped on the earth, it would go right through as if a housewife dropped her steam iron through a wet Kleenex stretched on a hoop.

The most likely possibility for such a compression was that same collapse of a massive star that makes a supernova explosion. Theory suggested that the explosion might actually be the side effect of the *implosion*. A fraction of the star's mass would splash off as an expanding nebula, but most of it would remain in the tiny neutron star. Many physicists did not agree, however. So the concept of neutron stars, once raised, was soon neglected, until revived in the late 1950s by a few true believers.

Three more clues to the riddle of the Crab appeared as astronomers discovered and studied its strange radiations. The nebula was one of the principal objects of interest in the early history of *radio* astronomy. In 1948, John G. Bolton, using antennas atop a 400-foot cliff on the Australian coast near Sydney, detected radio waves from the constellation Taurus in which one finds the Crab. In the following year, Bolton and associates took more refined observations and pinned the "Taurus A" radio source down to the Crab Nebula. This work in Australia predated construction of the "big dish" radio telescopes that are a familar part of modern observatories and tracking stations. In fact the Crab was the first radio source identified with an object outside the solar system.

The Australian radio astronomers realized an incredible implication of their measurements. It had been known that stars and other natural objects emitted radiation by thermal processes; that is, the kind of radiation depended on the temperature of the object. Picture glowing molten iron fresh from the blast furnace. At first blue or white hot, it turns red as it cools. If you measure the color you get a direct indication of the temperature. In like manner, the colors of the stars indicate their surface temperatures, and these "color temperatures" have been verified by several techniques.

It was natural, therefore, to apply the theory of thermal processes to the radio measurements of the Crab. But the great strength of the radio signals implied a temperature of *two million degrees*. The Australians pointed out that this could not be reconciled with the much lower temperature indicated by the visible light of the nebula. This conclusion was clue number five. Perhaps some strange process, unlike those present in the stars, was at work in the Crab.

Then came the development of x-ray astronomy. Our atmosphere prevents x rays from reaching our planet. But after World War II, rockets could lift devices above the atmosphere to detect these radiations. The detectors found a pre-

viously unknown x-ray source somewhere near the Crab. The early instruments could not pinpoint the source, so its association with the nebula was uncertain. But in 1964, the celestial stage was set for a brilliant experiment, conceived by Herbert Friedman and his co-workers at the Naval Research Laboratory. On July 7, the moon was due to pass in front of the Crab. Friedman timed a launch so that the rocket would be aloft during this event, known as an "occultation."

Results of the occultation experiment, clue six, were conclusive: As the edge of the moon passed in front of the nebula, the x-ray signal faded out. This fading demonstrated that the Crab is a powerful source of x rays; it is presently known as the strongest such source in our galaxy. The same experiment also proved that the x rays come from an extended region in space, not just from a star. If the x rays were emitted by a star, they would have vanished quickly as the moon's edge occulted their source. Instead, they slowly faded as the moon passed across a region about two light-years in extent at the center of the nebula.

Clue number seven was the 1953 discovery that the visible light from the Crab was like light that comes out of electron synchrotrons—devices with doughnut-shaped chambers for accelerating electrons with which to bombard atomic nuclei. Scientists at the General Electric Company got one going in 1947, and when they looked through a window into the cavity containing the electron beam, they saw a strange light. When they studied it through a Polaroid filter, they found that this "synchrotron radiation" was polarized. Rotating the filter blanked it out.

A Soviet astrophysicist, I. S. Shklovsky suggested that the Crab was radiating like an accelerator. Its fast electrons were emitting synchrotron radiation.

The Russian theory, dubbed the "synchrotron model," proposed that relatively slow-moving electrons produced the nebula's radio emission while faster ones contributed to its visible light. There was one obvious consequence of this theory: The light from the Crab Nebula, like that from the G. E. synchrotron, should fade as the observer rotated his Polaroid filter. Later in 1953 this prediction was fully confirmed.

There was, however, one serious flaw to the synchrotron model theory. It very nicely accounted for the properties of the radio and light waves from the Crab. And later it explained x rays and infrared radiation from the nebula. But fast-moving particles that lose energy to synchrotron radiation slow down rapidly. The most energetic electrons, which produce x rays, should die out in less than a year. *They could not possibly have lasted since an explosion seen in 1054.*

This stumbling block in the theoretical explanation of the nebular radiation turned out to be a vital clue (number eight) in the story. Some unknown object or mechanism must supply these particles continuously.

### Strange Pulses from the Sky

By the late 1960s, radio astronomy had come of age and a surprising discovery in 1968 supplied the last two clues—so far. A general catalog of radio sources listed nearly 1,300 celestial emitters. An "antenna farm" of the National Radio Astronomy Observatory continuously monitored the sky from Green Bank, West Virginia. Jodrell Bank, the great radio observatory in England, was virtually a household word. Radio waves had been recorded from many sources. Perhaps the most common attribute of these radio sources was their continuous output at radio frequencies. But the announcement in 1968 of a cosmic radio source that "beeped" at intervals of about 1.337 seconds, repeating with a precision of better than a hundred thousandth of one percent, astonished the scientific community.

This "pulsar" was reported by Antony Hewish and his Cambridge University associates in the February 24 issue of *Nature*. Within two years, 50 pulsars were known and more than 400 articles about them had appeared in technical journals. The property that distinguishes pulsars from every other sort of natural object is their repetitive emission of regular short pulses. So extreme is the regularity of the pulses that they suggest man-made radar and the telemetry beeps of the early Soviet Sputniks. Half humorously, the Cambridge discoverers named the first pulsar "LGM-1" for "Little Green Men."

The Cambridge astronomers were very quiet about pulsars from the moment of discovery to the publication of it in *Nature*. Such an announcement had to be very carefully checked before being published. They also knew that once the existence of pulsars was revealed, American scientists operating a 1,000-foot bowl-shaped radio antenna cradled among the hills at Arecibo, Puerto Rico, would be much better able to explore them than could the British with the sophisticated, but less sensitive, equipment at Cambridge. At last, though, they had refined and confirmed their observations, and the announcement gave Arecibo its chance.

Extensive investigations by many astronomers, notably Frank Drake's Arecibo group, established the basic properties of the pulsed radio sources, showing among other things that these pulses were *not* produced by intelligent life. On the other hand (as Drake told the National Academy of Sciences) "stupid life" might not be excluded.

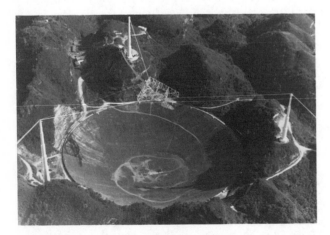

The big bowl at Arecibo, Puerto Rico, a radio telescope, explored pulsars after Cambridge astronomers had announced their existence. [Arecibo Observatory.]

Physicists stretched their imaginations further to explain pulsars. Clearly, their first task was to uncover the "clock mechanism." In other words, what celestial process spaced the pulses so perfectly that they rivaled the best man-made clocks? There were three basic explanations: vibration, rotation and orbital motion. A star might vibrate at a precise rate like a tuning fork or bell, producing a pulse at each vibration.

The rotation theory, on the other hand, would ascribe the pulsar radiation to a bright spot on a rapidly spinning star; as it turns past our line of sight, we perceive the bright spot as a pulse. Finally, the regular pulses could come from the orbital movements of a double star, each member whirling about the other; this idea can be compared to the annual motion of the earth around the sun.

The first group of pulsars had periods between pulses ranging from nearly two seconds down to a quarter second. Periods so short meant that pulsars are small and compact. A star like the sun cannot vibrate as rapidly as once per second and, if it spun that fast, centrifugal force would disrupt it and fling matter out into space at the stellar equator. This consideration eliminated the vibration theory. Orbital motion could be eliminated too. Calculations show that two ordinary stars cannot orbit with a one-second interval because the distance between them would have to be less than the size of the stars themselves.

Thomas Gold of Cornell University, however, suggested that pulsars were small, rapidly spinning neutron stars. Just as a figure skater with arms outstretched spins faster and faster as he brings his arms to his sides, so a supernova—a collapsing star—would rotate ever more rapidly as its particles came together to form a neutron star. It could easily reach a rate of one turn per second, or faster. Further, it would be so dense that centrifugal force could not disrupt it, and the implosion would amplify its magnetic field.

According to Gold, the end product of a supernova would be a rapidly spinning neutron star with a powerful magnetic field. Electrons, continuously streaming off a "sore spot" on the stellar surface, would travel through the magnetic field into surrounding space, generating a beam of synchrotron-like radiation. In short, a pulsar was a kind of lighthouse in the sky.

At first, Gold's idea did not attract very much attention. But at the Green Bank, West Virginia, observatory, David Staelin and E. C. Reifenstein III were hunting for new pulsars with a 300-foot antenna. They found two. One of them, promptly confirmed by the observers in Puerto Rico, was located in the Crab Nebula, beating at the tremendous rate of 30 times per second! Only a neutron star could vibrate, rotate or orbit so rapidly. The news of this discovery was soon relayed to Russia by transatlantic telephone. The American astronomers' names may not have been transliterated very accurately, however; for it is reported that an audience of physicists at Moscow University was told, "Now the Americans have a Stalin, and he has found a neutron star!"

The discovery of the neutron star in the Crab Nebula (clue nine) was unique. For the first time in astronomy a theoretical prediction for a wholly new class of objects really came true. It was too late for congratulations, however. Of the chief authors of the neutron star concept, Oppenheimer had died in February 1967, a year before the announcement of the first pulsar, and Landau, who died in the following year, had never fully recovered from severe brain damage that resulted from a 1962 accident.

## A Gradually Slowing Neutron Star

Before long, the observers at Arecibo noticed a striking property of the Crab Nebula pulsar. Unlike the other pulsars, as thus far observed, the Crab did not have a precisely constant interval between pulses. In fact, accurate timings showed that the period of the Crab pulsar was lengthening by 37 billionths of a second per day. Imagine a wristwatch that erred by only that amount! Gold realized the large implications of the minuscule change. It was already accepted that the pulsar period of 1/30 second could only describe the rotation of a neutron star. Then the tiny lengthening of the pulse spacing must mean that the spinning neutron star was slowly running down. Think of this star as a giant flywheel, ten miles in diameter, set in motion by the supernova outburst, turning 30 times per second and possessing the mass of the sun. For a wheel like this, a slowdown of even 37 billionths of a second per day represents an enormous loss of energy—more than enough to power the whole Crab Nebula—x rays, light, radio emission and all. The observation of a gradual slowing of the rapidly spinning neutron star was our tenth and final clue: The explosion observed in 1054 A.D. had set a cosmic flywheel spinning, and its spin is energizing the Crab Nebula today.

So this most recent astronomical discovery of the pulsars tells us the "who" of a whodunit that started with a bang and a flash seen in 1054. The flash was from a great collapse, the implosion of a star that crushed itself and its atoms under the force of its own gravity. Electrons streaming from this rotating neutron star pass out into the nebula formed by the supernova explosion, where they encounter magnetic fields and produce synchrotron radiation.

But the term "final clue" is a deliberate misnomer because future astronomers must find out much more about the Crab. Precisely *how* does the spinning, magnetized neutron star accelerate electrons to the enormous velocities required by the synchrotron theory? Just where do the radio pulses come from? Do they arise in the distant surroundings of the star, as has been suggested by Franco Pacini of the University of Rome? Or do the pulses arise at the magnetic poles on the very surface of the star?

The Crab pulsar is unique among the objects of its class. Climaxing the long search for visible light from the pulsars three University of Arizona astronomers detected light pulses from the nebula, and another Friedman rocket shot found pulsed x rays as well. At the Kitt Peak National Observatory near Tucson, Roger Lynds, D. E. Trumbo and I determined that the light pulses arise from the same star in the nebular center that the Mount Wilson astronomers, Walter Baade and Rudolph Minkowski had proposed in 1942 as the most likely candidate for the remnant star of the explosion. Because the Mount Wilson photographs were long exposures, only in 1969, in the course of the pulsar studies, did we realize that

(unlike all other stars) this star does not shine continuously. Special photographs (below) show that it flashes at the same rate as the radio pulses.

Why do we find no evidence for light or x rays* from any pulsar but the one in the Crab? What causes the strange ripples of light that occasionally travel across the nebular gas? Do the pulsars produce the cosmic rays that pervade space, as Gold believes? What caused a recent abrupt change in the spin rate of the Crab neutron star? Was it a "starquake" or the eruption of a stellar "volcano"? It seems that we have a great deal more to learn from the Crab Nebula, and if experience since 1054 is any guide, the new facts will be developed by scientists of many nations, most of whom will be surprised by what they find.

Flashing, neutron star. A series of photographs synchronized with the radio pulse shows that the star "turns on" in visible light when radio emission is at a maximum and goes off at other times. These equally spaced pictures show the star on in first and third, off in second and fourth. [Kitt Peak National Observatory photographs by Hong-Yee Chiu, Roger Lynds and Stephen P. Maran.]

---

*There is now evidence for x-ray and gamma-ray emission from a few other pulsars, but among the more than 150 known radio pulsars, only the Crab pulsar has been observed to emit visible light pulses.—Ed.

# The Evolution of Quasars

Maarten Schmidt and Francis Bello

Since light has a finite velocity the astronomer can never hope to see the universe as it actually exists today. Far from being a handicap, however, the finite velocity of light enables him to peer back in time as far as his instruments and ingenuity can carry him. If he can correctly interpret the complex messages coded in electromagnetic radiation of various wavelengths, he may be able to piece together the evolution of the universe back virtually to the moment of creation. According to prevailing theory, that moment was some 10 billion years ago, when the total mass of the universe exploded out of a small volume, giving rise to the myriad of galaxies, radio galaxies and quasars (starlike objects more luminous than galaxies) whose existence has been slowly revealed during the past half-century.

Optical observations have shed little light on the evolution of ordinary galaxies because even with the most powerful optical telescopes such galaxies cannot be studied in detail if they are much farther away than one or two billion light-years. The astronomer sees them as they looked one or two billion years ago, when they were already perhaps eight or nine billion years old.

Quasars, on the other hand, provide a direct glimpse of the universe as it existed eight or nine billion years ago, only one or two billion years after the "big bang" that presumably started it all.

Some 50 years ago the first large telescopes had shown that the light from distant galaxies is shifted toward the red end of the spectrum; the more distant the galaxy, the greater its red shift and the higher its velocity of recession. Like raisins in an expanding cosmic pudding, all the galaxies are receding from one another. From the observed velocities of recession one can compute that some 10 billion years ago all the matter in the universe was jammed into a tiny volume of space.

The term quasar, a contraction of "quasi-stellar radio source," was originally applied only to the starlike counterparts of certain strong radio sources whose optical spectra exhibit red shifts much larger than those of galaxies. Before long, however, a class of quasi-stellar objects was discovered with large red shifts that have little or no emission at radio wavelengths. "Quasar" is now commonly applied to starlike objects with large red shifts regardless of their radio emissivity.

This article is based on the hypothesis that the quasar red shifts are cosmological, that is, they are a consequence of the expansion of the universe and thus directly related to the distance of the object. On that hypothesis quasars are very remote objects. According to a contrary hypothesis, which will be discussed toward the end of the article, quasars are relatively close objects.

A recent study of quasars carried out with the aid of the 200-inch Hale telescope on Palomar Mountain has provided evidence that these extremely luminous objects evolved quite rapidly when the universe was young. The study indicates that quasars were about 100 times more plentiful when the sun and the earth were formed some five billion years ago than they are today. They were perhaps more than 1,000 times more plentiful at a still earlier epoch, say eight or nine billion years ago. Earlier than that, however, there may have been fewer quasars, perhaps because conditions in the universe had not yet favored their development.

The study embraced all the quasars in two areas of the sky representing a thousandth of the total celestial sphere. By extrapolation one can say with reasonable confidence that a complete sky survey with the largest telescopes should reveal on the order of 15 million quasars. The overwhelming majority are so far away that they almost certainly burned themselves out in the billions of years required for their light to reach us. All of them, of course, can still be studied telescopically, given the time and the inclination. If, however, it were possible to conduct an instantaneous survey of the universe, one might find that only about 35,000 quasars are in existence and radiating with their characteristic intensity at the present time. These find-

BRIGHTEST QUASAR and first member of its class to be recognized is 3C 273, indicated by the reticle in the photograph on the opposite page. The negative print was made from a 1-by-1⅜-inch portion near the edge of a 14-inch square plate taken with the 48-inch Schmidt telescope on Palomar Mountain as part of the National Geographic Society–Palomar Sky Survey. In 1962 the starlike object was found to coincide with the position of a strong radio source designated No. 273 in the third catalogue compiled by radio astronomers at the University of Cambridge. The optical magnitude of 3C 273 is 13. In the entire sky there are at least a million stars of that magnitude. A study of 3C 273's strange spectrum revealed, however, that its light was shifted toward the red end of the spectrum by an amount that indicated the object was receding at about one-sixth the velocity of light. This implied that it was not a nearby star but an object between one billion and two billion light-years away. A galaxy at the same distance would appear at least four magnitudes fainter, which means that 3C 273 is intrinsically at least 40 times brighter. The term "quasi-stellar radio source," or quasar, was coined to describe 3C 273 and other starlike objects exhibiting a large red shift.

From *Scientific American*, vol. 224, pp. 55–69, May 1971. Copyright © 1971 by Scientific American, Inc. All rights reserved. Reprinted with permission.

ings are in conflict with the "steady state" hypothesis, which holds that the universe has always looked exactly the way it does today. That hypothesis postulates that new matter is continuously being created to maintain the expanding universe at a constant density.

After 10 years of intensive study by optical and radio astronomers quasars remain among the most puzzling of all celestial objects. Assuming that they are at cosmological distances, one can easily show that many quasars are from 50 to 100 times brighter than entire galaxies

containing hundreds of billions of stars. Unlike the light output of normal galaxies, the light output of some quasars has been observed to change significantly in a matter of days. The only explanation is that some variable component of a quasar, if not the entire quasar, may be not much larger than the solar system.

## The Discovery of Quasars

Before 1960 radio astronomers had identified and catalogued hundreds of radio sources: invisible objects in the universe that emit radiation at radio frequencies. From time to time optical astronomers would succeed in identifying an object—usually a galaxy—whose position coincided with that of the radio source. Thereafter the object was called a radio galaxy. The large majority of radio sources remained unidentified, however, and the general belief was that the source of the emission was a galaxy too far away, or at least too faint, to be recorded on a photographic plate.

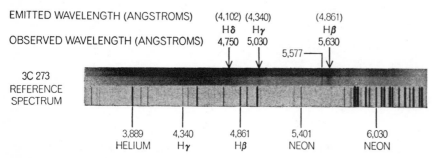

PORTION OF SPECTRUM OF QUASAR 3C 273 shows three prominent emission lines centered at 4,750, 5,030 and 5,630 angstroms, corresponding to the hydrogen emission lines delta, gamma and beta in the Balmer series. The upper and lower halves of the spectrogram were given different exposures to facilitate study. The three emission lines are produced by hydrogen atoms in various states of excitation. Two of the three lines, H gamma and H beta, also appear in the reference spectrum at their normal emission wavelength: 4,340 and 4,861 angstroms. The normal wavelength for H delta is 4,102 angstroms. The red shift, z, is obtained by subtracting the normal wavelength from the observed wavelength and dividing the difference by the normal wavelength. For 3C 273 z is .158, indicating the quasar is receding at nearly a sixth the speed of light. The sharp line at 5,577 often appears in spectra of astronomical objects and serves as a convenient reference point; it is produced by excited oxygen atoms in the upper atmosphere. This spectrogram and others in this article were made by one of the authors (Schmidt), who provided the original interpretation of 3C 273's spectrum.

In 1960 Thomas Matthews and Allan Sandage first discovered a starlike object at the position given for a radio source in the Third Cambridge ("3C") Catalogue, compiled by Martin Ryle and his colleagues at the University of Cambridge. The radio object 3C 48 coincided in position with a 16th-magnitude star whose spectrum exhibited broad emission lines that could not be identified. Not only did the object emit much more ultraviolet radiation than an ordinary star of the same magnitude but also its brightness varied by more than 40 percent in a year.

Object 3C 48 was thought to be a unique kind of radio-emitting star in our own galaxy until 1963, when the strong radio source 3C 273 was identified with a starlike object of 13th magnitude and one of the authors (Schmidt) recognized that most of the puzzling lines in its spectrum could be explained as the Balmer series of hydrogen lines, shifted in wavelength toward the red by 15.8 percent, or .158 [see illustration on page 346 and upper illustration at left]. Red shifts are commonly expressed as a fraction or percentage obtained by dividing the measured displacement of a line by the wavelength of the undisplaced line. With this clue it was immediately evident that the lines in the spectrum of 3C 48 had a red shift of .367 [see "Quasistellar Radio Sources," by Jesse L. Greenstein; SCIENTIFIC AMERICAN, December, 1963].

Such large red shifts, equivalent to a significant fraction of the velocity of

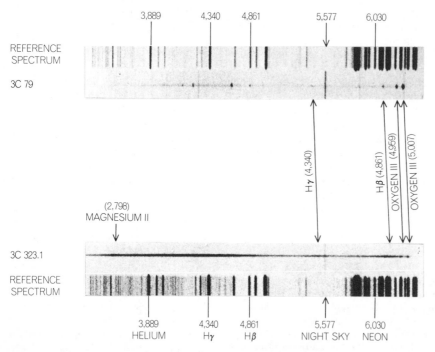

CONTRAST BETWEEN QUASAR AND RADIO GALAXY is shown by these two spectra. The spectrum at the top is that of the strong radio galaxy 3C 79, which has a red shift of .256. The bottom spectrum is that of quasar 3C 323.1, whose red shift is just slightly greater: .264. The radio galaxy produces a substantial number of sharp emission lines. Four of the lines in the right half of its spectrum are identified and compared with their much broadened counterparts as they appear in the spectrum of the quasar. Quasars characteristically emit strongly in the ultraviolet part of the spectrum. A common emitting ion is singly ionized magnesium, designated magnesium II, which has an emission wavelength of 2,798 angstroms.

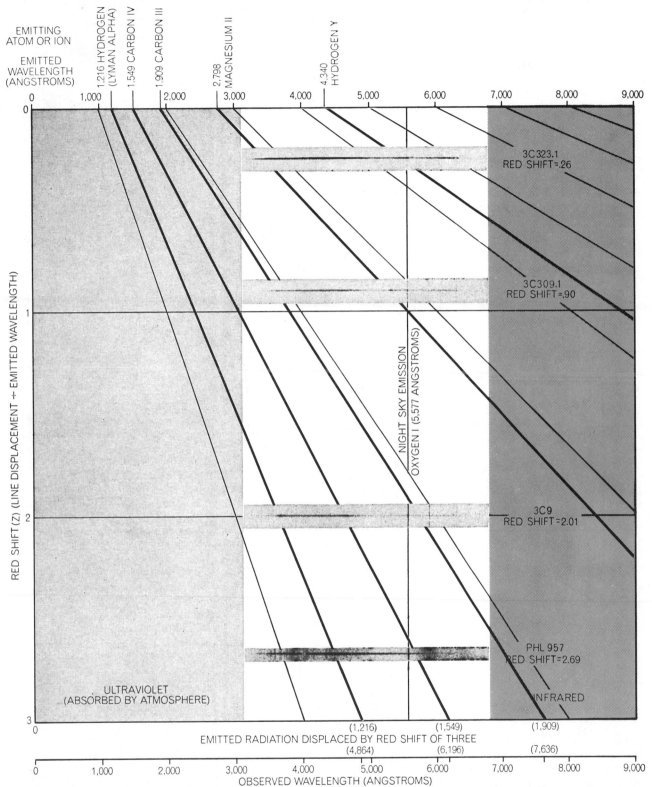

FOUR QUASAR SPECTRA are positioned on a diagram that shows how radiation emitted at one wavelength billions of years ago is "stretched" on its long journey through space by the presumed expansion of the universe. At least two emission lines are needed to establish the red shift of an astronomical object. A single line could represent any line shifted by any arbitrary amount. Here the heavy slanting lines correspond to the radiation emitted by hydrogen (Lyman alpha), carbon IV, carbon III, magnesium II and hydrogen (gamma). The roman numerals are one greater than the number of electrons missing from the atom. At a red shift, $z$, of 1 the Lyman-alpha line is observed at 2.432 angstroms; at a red shift of 2 the line is observed at 3,648 angstroms; at a red shift of 3 it would appear at 4,864 angstroms. Thus when $z$ equals 2 the initial wavelength is stretched exactly three times; when $z$ equals 3, four times and so on. The quantity $1 + z$ expresses how much the universe has expanded between the emission of a photon and its observation. Only two quasars are known with a red shift greater than 2.5; one of them is PHL 957, whose spectrum appears here. Its spectrum was made with an image-tube spectrograph; the other three spectra were recorded directly on photographic film. The photons that produced the spectrum of PHL 957 left the quasar when the universe was only about 13 percent of its present age.

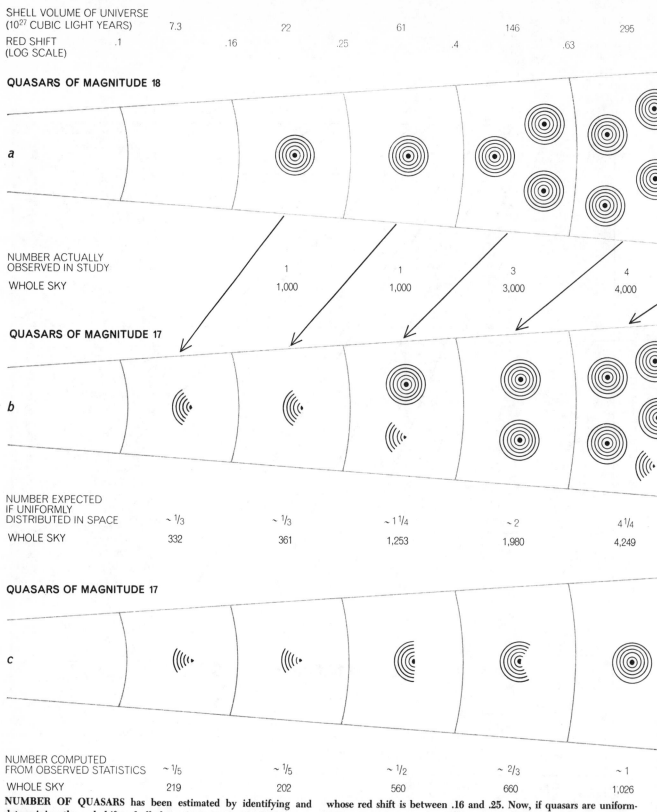

| SHELL VOLUME OF UNIVERSE (10²⁷ CUBIC LIGHT YEARS) | | 7.3 | | 22 | | 61 | | 146 | | 295 |
|---|---|---|---|---|---|---|---|---|---|---|

| RED SHIFT (LOG SCALE) | .1 | | .16 | | .25 | | .4 | | .63 | |
|---|---|---|---|---|---|---|---|---|---|---|

**QUASARS OF MAGNITUDE 18**

a

| NUMBER ACTUALLY OBSERVED IN STUDY | | 1 | 1 | 3 | 4 |
|---|---|---|---|---|---|
| WHOLE SKY | | 1,000 | 1,000 | 3,000 | 4,000 |

**QUASARS OF MAGNITUDE 17**

b

| NUMBER EXPECTED IF UNIFORMLY DISTRIBUTED IN SPACE | ~ 1/3 | ~ 1/3 | ~ 1 1/4 | ~ 2 | 4 1/4 |
|---|---|---|---|---|---|
| WHOLE SKY | 332 | 361 | 1,253 | 1,980 | 4,249 |

**QUASARS OF MAGNITUDE 17**

c

| NUMBER COMPUTED FROM OBSERVED STATISTICS | ~ 1/5 | ~ 1/5 | ~ 1/2 | ~ 2/3 | ~ 1 |
|---|---|---|---|---|---|
| WHOLE SKY | 219 | 202 | 560 | 660 | 1,026 |

**NUMBER OF QUASARS** has been estimated by identifying and determining the red shifts of all the quasars in sample fields representing one-thousandth of the whole sky. The sample consisted of 20 quasars with an optical, or apparent, magnitude of about 18. It was clear from their red shifts, however, that some are much farther away than others and therefore are intrinsically brighter, as depicted in *a*. The red-shift intervals have been chosen so that the quasars in any given "shell" of the universe are on the average one magnitude (2.5 times) brighter in absolute luminosity than those in the next shell inward. Thus the four quasars in the sample box representing the most remote shell (red shift: 1.58 to 2.51) are each 100 times more luminous than the single quasar in the box whose red shift is between .16 and .25. Now, if quasars are uniformly distributed in space, and if there are 4,000 of maximum luminosity in the most remote shell, one would expect to find a proportional number in the next shell inward, whose volume is only two-thirds that of the outer shell. Two-thirds times 4,000 is 2,667. Thus diagram *b* shows that in the red-shift interval between 1 and 1.58 one would expect to find 2,667 quasars of maximum luminosity in the whole sky (or, proportionately, 2⅔ quasars in the small area actually sampled). In photographs these 2,667 should appear one magnitude brighter (magnitude 17) than the 4,000 of the same intrinsic brightness that are farther away. Using the same assumptions, one can estimate the number of quasars in still nearer shells.

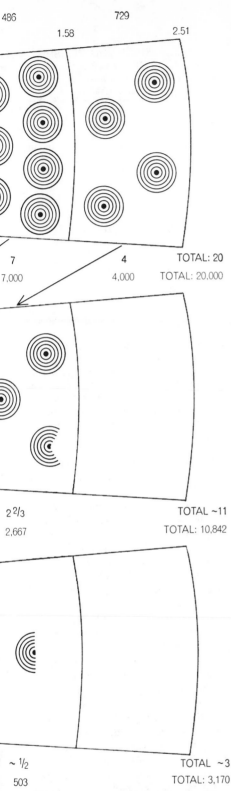

The total number is 10,842, distributed as shown in *b*. One concludes, therefore, that if quasars are uniformly distributed in space, one should observe about twice as many 18th-magnitude quasars as 17th-magnitude quasars. In actuality, however, surveys show that the number of quasars goes up by a factor of about six per magnitude. To satisfy this observation there can be only about 3,000 quasars of the 17th magnitude in the whole sky. An appropriate red-shift distribution for that approximate number is shown in *c*.

light, ruled out the possibility that 3C 273 and 3C 48 were stars in our galaxy. It was proposed that the red shifts are cosmological, which implies that the two objects have to be billions of light-years away and therefore extremely luminous to look as bright as they do in our night sky. They soon became known as quasars. Within the next few years quasars with even larger red shifts were discovered, including some with red shifts of more than 2, or more than six times the largest red shift ever observed for an ordinary galaxy. On the cosmological hypothesis, a red shift of 2 suggests that the light from the object has been traveling for about 80 percent of the age of the universe.

## The Quiet Quasars

Several hundred radio sources have now been identified with starlike objects. Most of the identifications are made on the basis of positions provided by two or more radio telescopes spaced from several hundred meters to several thousand kilometers apart, used as an interferometer. The technique yields a precise measure of the difference in the time required for radio waves from the source to reach each telescope of the interferometer. One can then locate the source with an accuracy of between one second and 15 seconds of arc. Once the search has been narrowed to an optical candidate the final test is to see if its spectrum shows a red shift. More than half of the objects identified on the basis of their radio position usually turn out to be quasars. The spectroscopic test is unambiguous because the maximum red shift ever recorded for a star is .002; the smallest red shift for a quasar identified on the basis of its radio emission is .158 (for 3C 273).

It was noticed early that quasars usually emit rather strongly in the ultraviolet part of the spectrum. In 1964, when radio positions were known with considerably less accuracy than they are today, Ryle and Sandage conceived the idea of using ultraviolet strength as a clue in searching for optical counterparts of radio sources. They used a technique in which a single photographic plate of a star field was exposed to blue light and then was shifted slightly and exposed to ultraviolet radiation. By visual examination it was possible to readily distinguish strong ultraviolet emitters from normal stars.

In 1965 Sandage noted that objects with excess ultraviolet emission were much more plentiful than known radio sources in typical star fields. He soon discovered that some of these "blue stellar objects" exhibit red shifts that qualify them as quasars even though no radio emission has been detected from them. Most of the other strong ultraviolet emitters turn out to be white-dwarf stars in our own galaxy. Thus only a small fraction of quasars are strong radio emitters. The rest are radio quiet, or virtually so. It may be that a typical quasar is a strong radio emitter for only a small part of its life-span. Alternatively, it may be that relatively few quasars are born to be strong radio sources.

Two years ago Sandage and Willem J. Luyten published photometric analyses of 301 blue objects in seven survey fields. They counted quasar candidates tentatively selected from these blue objects and estimated that in one square degree of the sky (roughly equal to five times the area of the moon) there is, on the average, .4 quasar brighter than magnitude 18.1. They also estimated that there are five quasars per square degree brighter than magnitude 19.4, and that tentatively there are as many as 100 brighter than magnitude 21.4. Over the entire sky they estimated there may be 10 million quasars of the 22nd magnitude or brighter.

The greater the magnitude, of course, the dimmer the object; every increase of five magnitudes (say from the 18th magnitude to the 23rd) corresponds to a decrease by a factor of 100 in brightness. The number of quasars increases steeply with magnitude, by a factor of about six per magnitude. This steep increase is incompatible with a uniform distribution of quasars in space, as we shall see.

## Counting Verified Quasars

The objects isolated by Sandage and Luyten are defined as "faint blue objects [with] an ultraviolet excess." For a detailed statistical study one has to obtain the spectrum of each "faint blue" candidate individually to establish whether or not it is really a quasar. One of the authors (Schmidt) began this task about four years ago, working with several of the star fields examined by Sandage and Luyten. The ultimate goal of the study is to establish how quasars are distributed by red shift (distance) and luminosity.

Of 55 faint blue objects investigated in two of the Sandage-Luyten fields, 32 turned out to have negligible red shifts and therefore could be rejected as being dwarf stars within our own galaxy. The 23 remaining objects exhibited spectra characteristic of quasars, and all but one of the spectra contained the minimum of two lines needed for establishing a red shift. A single line could represent almost any emitting atom red-shifted by any arbitrary amount. When a spectrum contains two lines, however, it is almost always possible to assign a unique red-shift value that identifies a reasonable emission wavelength for each line [see *illustration on page 349*]. Unfortunately the spectra of some quasars show only a single clear line, thereby frustrating efforts to establish their red shift. Although the red shifts assigned to several of the objects are still tentative, the overall distribution must be essentially correct. The red shifts range from .18 to 2.21. None of the 23 quasars appears in any of the catalogues of strong radio sources.

At this stage it will be most useful in our discussion to concentrate on the quasars of 18th magnitude in the sample. There are 20 such quasars. Since the 20 objects exhibit a variety of red shifts, however, we know they must lie at vastly different distances and therefore must differ greatly in *absolute* luminosity even though they look equally bright to an observer.

To express these differences in absolute luminosity one can classify the objects by red shift in such a way that each red-shift category represents a step of one magnitude in absolute luminosity. The relation between the red shift and the magnitude of a standard source de-pends on the properties of the universe. In the cosmological model followed in this study a quasar of 18th magnitude whose red shift falls in the range between .25 and .4 is intrinsically brighter by one magnitude than a nearer object whose red shift lies between .16 and .25. Six red-shift categories, each corresponding to a step of one magnitude in absolute luminosity, are enough to cover the range of red shifts actually exhibited by the 20 objects. The brightest members of the group are five magnitudes, or 100 times, brighter than the least luminous.

When the 20 quasars were grouped by red shift in this way, their distribution was found to be similar to the red-shift distribution of radio quasars of the same optical magnitude. Taking this into account and rounding things off somewhat, the following distribution for the red shifts of 18th-magnitude quasars was adopted for the subsequent analysis:

| Red shift | 1.58–2.51 | 20 percent |
| Red shift | 1.00–1.58 | 35 percent |
| Red shift | .63–1.00 | 20 percent |
| Red shift | .40– .63 | 15 percent |
| Red shift | .25– .40 | 5 percent |
| Red shift | .16– .25 | 5 percent |

The Sandage-Luyten survey had shown that in the entire sky there are, in round numbers, 20,000 quasars of apparent magnitude 18—just 1,000 times as many as in the new detailed sample.

| RED SHIFT (Z) | SHELL VOLUME OF UNIVERSE ($10^{27}$ CUBIC LIGHT YEARS) | SHELL VOL OF UNIVER X$(1 + Z)^6$ |
|---|---|---|
| 1.58 —2.51 | 729 | 593,000 |
| 1.00 —1.58 | 486 | 74,600 |
| .63 —1.00 | 295 | 10,900 |
| .40 — .63 | 146 | 1,800 |
| .25 — .40 | 61 | 336 |
| .16 — .25 | 22 | 68 |
| .10 — .16 | 7.3 | 15 |
| .06 — .10 | 2.1 | 3.4 |
| .04 — .06 | .59 | .8 |
| .025— .04 | .16 | .19 |
| .016— .025 | .04 | .05 |

**TOTAL QUASAR POPULATION** of universe is estimated to be on the order of 14 million, of which more than 99.7 percent are evidently fainter than the 18th magnitude and have red shifts greater than .4. From the 13th to 18th visual magnitude the number of quasars increases by a factor of five or six

If the 20,000 are distributed according to the percentages listed above, one finds that the number in each red-shift category, starting with the highest, is as follows: 4,000, 7,000, 4,000, 3,000, 1,000 and 1,000. It is clear that in a random sample of 18th-magnitude quasars more than half are extremely distant (red shift greater than 1) and therefore belong to the most luminous members of their class. A red shift of 1 corresponds to looking back two-thirds of the time that has elapsed since the universe began its expansion.

Proceeding to the next stage of the analysis, one would like to estimate the number of quasars whose apparent magnitude is either brighter or fainter than 18 and how they are distributed according to red shift. To do this one must know the volumes of the successive shells of the universe in which we have placed our 18th-magnitude quasars. These volumes depend on the cosmological model followed. Our unit of volume, $10^{27}$ cubic light-years, or a cube of which each side is a billion light-years, is co-moving, which means that no matter what the red shift, the unit of volume expands with the universe into our "local" unit of $10^{27}$ cubic light-years.

The 4,000 brightest and most distant quasars (red shift 1.58 to 2.51) occupy a shell with a volume of 729 × $10^{27}$ light-years. The problem now is to use this in-

| RED SHIFT (Z) | SHELL VOLUME OF UNIVERSE ($10^{27}$ CUBIC LIGHT YEARS) | NUMBER OF QUASARS IN WHOLE SKY | | |
|---|---|---|---|---|
| | | MAGNITUDE 17 CORRECTED DISTRIBUTION ~$(1 + Z)^6$ | MAGNITUDE 17 IF UNIFORMLY DISTRIBUTED | MAGNITUDE 18 DERIVED FROM OBSERVATION |
| 1.58–2.51 | 729 | | | 4,000 |
| 1.00–1.58 | 486 | 503 | 2,667 | 7,000 |
| .63–1.00 | 295 | 1,026 | 4,249 | 4,000 |
| .40– .63 | 146 | 660 | 1,980 | 3,000 |
| .25– .40 | 61 | 560 | 1,253 | 1,000 |
| .16– .25 | 22 | 202 | 361 | 1,000 |
| .10– .16 | 7.3 | 219 | 332 | |
| | | 3,170 | 10,842 | 20,000 |

**DISTRIBUTION OF QUASARS** according to red shift is shown for 20,000 quasars of 18th optical magnitude (*column at far right*), based on a representative sample of 20 quasars. The adjacent columns present two different estimates of the total number of quasars of the 17th magnitude. The method of making the estimates is explained in the illustration on the preceding two pages, where the same numbers appear in the diagrams labeled *b* and *c*. Observation shows that the number of quasars goes up by a factor of about six per magnitude rather than the factor of two expected if quasars were uniformly distributed throughout space. One can obtain the observed distribution by multiplying the shell volume of the universe by $(1 + z)^6$, where z is the red shift and the exponent 6 is an experimentally determined value that yields the desired increment per magnitude. The table at top of these two pages shows the computed number and red shift of all quasars from magnitude 13 through 23.

APPROXIMATE VISUAL MAGNITUDE

| 14 | 15 | 16 | 17 | 18 | 19 | 20 | 21 | 22 | 23 |
|---|---|---|---|---|---|---|---|---|---|
| — | — | — | —— | 4,000 | 56,000 | 217,000 | 1,000,000 | 2,000,000 | 9,000,000 |
| — | — | — | 503 | 7,000 | 27,000 | 124,000 | 200,000 | 1,000,000 | ——— |
| — | — | 74 | 1,026 | 4,000 | 18,000 | 32,000 | 200,000 | ——— | ——— |
| — | 12 | 169 | 660 | 3,000 | 5,000 | 27,000 | —— | —— | —— |
| 2 | 32 | 123 | 560 | 1,000 | 5,000 | —— | —— | —— | —— |
| 6 | 25 | 113 | 202 | 1,000 | ——— | — | — | — | — |
| 6 | 25 | 44 | 219 | —— | ——— | — | — | — | — |
| 6 | 10 | 50 | —— | —— | — | — | — | — | — |
| 2 | 12 | — | —— | —— | — | — | — | — | — |
| 3 | — | — | —— | — | — | — | — | — | — |
| — | — | — | —— | —— | — | — | — | — | — |
| 25 | 116 | 573 | 3,170 | 20,000 | 111,000 | 400,000 | 1,400,000 | 3,000,000 | 9,000,000 |

for each decline of one magnitude in brightness. Beyond the 18th magnitude, however, the increase is slower because the table contains no entries for quasars with red shifts greater than 2.51. In fact, only two quasars with larger red shifts are known, which suggests that there is a genuine paucity of such objects.° Any quasar with a red shift of 2.5 is so distant that its light has been traveling through space for more than 85 percent of the age of the universe. The light from the more than 13.5 million quasars with a red shift greater than 1 has been en route for at least 6.8 billion years, assuming that the universe is on the order of 10 billion years old. Because the lifetime of a quasar is probably well under a billion years, the overwhelming majority of all the quasars that ever existed must have evolved by now into less luminous objects, perhaps ordinary galaxies. One can estimate that only about 35,000 quasars exist today.

formation to compute how many quasars of the same absolute luminosity would appear in the shell immediately within the outermost one, whose red shift corresponds to between 1 and 1.58. That shell, according to the cosmological model selected, has a volume of 486 × 10²⁷ cubic light-years, or two-thirds of the volume of the outer shell. Now we introduce a supposition. If quasars were uniformly distributed in space, the inner shell would contain two-thirds times 4,000, or 2,667, quasars exactly like those in the outer shell. If quasars of that intrinsic luminosity were moved one shell closer to us, their apparent luminosity, as we observe them, would therefore be one magnitude brighter, that is, magnitude 17 instead of magnitude 18 [see illustration on pages 350 and 351]. Remember that the red-shift intervals were chosen specifically so that each step would correspond to a one-magnitude change in brightness.

A similar computation is now performed for the 18th-magnitude quasars in each of the other red-shift categories. In each case one computes the number expected in the shell within the preceding one, assuming as before that quasars are uniformly distributed in space. This calculation yields the following additional numbers: 4,249, 1,980, 1,253, 361 and 332. When these are added to the number 2,667 previously computed,

one obtains a total of 10,842 quasars of apparent magnitude 17, or roughly half as many quasars as one expects to find of magnitude 18 (assuming uniform distribution).

We recall that the Sandage-Luyten survey shows that the number of quasarlike objects increases not by a factor of two per magnitude (from 10,842 to 20,000 in the exercise just completed) but by a factor of about six. In other words, their statistics would predict only some 3,000 or 4,000 quasars of apparent magnitude 17 rather than 10,842.

What the factor of six tells us, of course, is that there are more faint quasars than one would expect to find if space were uniformly filled with quasars. The only plausible explanation is that the density of quasars must increase with increasing distance, that is, as we look back farther in time. To arrive at a distribution law that satisfies the observational evidence, let us assume that the density is proportional to some power, $n$, of the scale of the universe. The scale, or size, of the universe is inversely proportional to the amount by which light has been "stretched" by the expansion of the universe. Thus if the Lymanalpha line emitted at 1,216 angstroms is observed at 3,648 angstroms, one can say that the universe has expanded by a factor of three since the radiation left the emitter. Since the red shift, $z$, in this

case is 2 (3,648 minus 1,216 divided by 1,216) it is evident that the scale of the universe is given not by $z$ but by $1 + z$. The density law we are seeking is therefore $(1 + z)^n$.

The value of $n$ is simply obtained by trial and error to yield about 3,000 quasars of magnitude 17 [see illustration on opposite page]. Quite by accident the value of $n$ turns out to be 6. It is only a coincidence that $n$ is 6 and that the increase in the number of quasars per magnitude is also six. With this density law it is a simple matter to extend the distribution table downward from magnitude 17 and upward from magnitude 18 [see illustration above].

Along the bottom of the table one can read off the number of quasars expected in the entire sky for each magnitude. For the five magnitudes brighter than magnitude 18 the expected quasar population decreases steadily at each step from 3,170 (17th magnitude) to 573 (16th) to 116 (15th) to 25 (14th) and finally to five (13th). For the five magnitudes fainter than 18 the expected population rises steeply at each step from 111,000 (19th magnitude) to 400,000 (20th) to 1.4 million (21st) to three million (22nd) and finally to nine million (23rd). The total estimated quasar population from magnitude 13 to magnitude 23 inclusive is thus about 14 million.

The table does not list entries for

---

°Additional high red-shift quasars have since been found. —Ed.

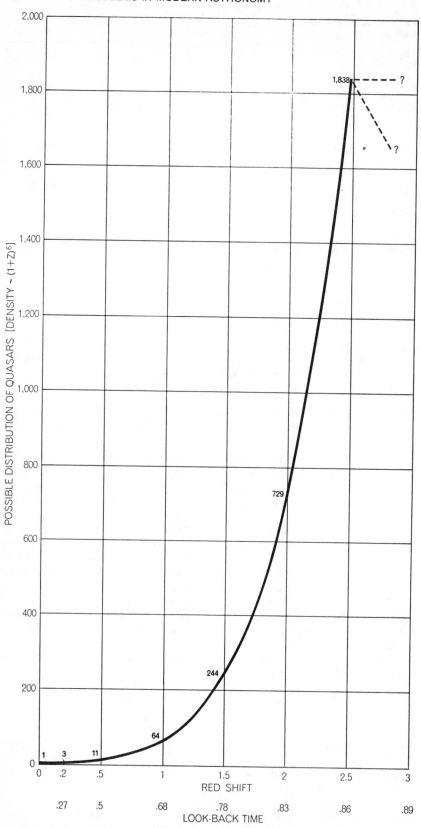

CHANGE IN QUASAR DENSITY WITH TIME can be derived from the table on the preceding two pages. The curve shows that the number of quasars rises steeply with increasing red shift, which is equivalent to looking back in time. Thus if one looks back 68 percent of the age of the universe, one would find more than 60 quasars in the volume of space that now contains one quasar. Looking back 83 percent of the age of the universe, one would find more than 700 quasars in the same volume. The maximum density may have existed when the universe had reached only about 14 percent of its present age. The scarcity of quasars with a red shift greater than 2.5 suggests that their density was no greater at earlier epochs.

quasars with red shifts greater than 2.5. Actually two quasars with larger red shifts are known: one, PHL 957, has a red shift of 2.69; the other, 4C 05.34, has a red shift of 2.88. Their magnitudes are respectively 17 and 18. If the density law $(1 + z)^6$ continued to hold, one would expect a great many 19th-magnitude quasars with red shifts larger than 2.5. Their scarcity suggests that the density does not increase beyond 2.5 and that it may actually decrease.[°]

The probable scarcity of quasars with red shifts greater than 2.5 implies that the largest telescopes are able to look back in time to the epoch when quasars made their first appearance in the universe. Depending somewhat on the cosmological model selected, one can say that the light from a quasar with a red shift of 2.5 began its journey through space some 8.6 billion years ago, or some 1.5 billion years after the big bang that hypothetically created the universe as we know it. Within the next few billion years the great majority of quasars were born and began their brief but brilliant career [see illustration at left].

One can estimate that the universe at present contains only some 35,000 quasars. All the rest have presumably evolved into less remarkable objects, perhaps ordinary galaxies; we know of their existence because the signals they emitted billions of years ago are only now reaching our telescopes. The quasars of the lowest intrinsic luminosity (those at the bottom of the table on the preceding two pages) are no brighter than large galaxies. It is therefore uncertain whether all of them are quasars or whether some are compact galaxies of one kind or another. To avoid such confusion one could consider leaving out the quasars listed in the two lowest (least luminous) categories all across the table. The remaining "high luminosity" quasars would then number about 1.5 million for the entire sky, and the number existing at the present time would drop to only 3,500.

Another way to look at the quasar population developed in this analysis is to compare the number of quasars with the number of galaxies in a given volume of space. A volume of $10^{27}$ cubic light-years in our neighborhood contains about 20 quasars, of which two are objects of high luminosity. In very round numbers the same volume of space contains probably between one million and 10 million galaxies.

The study described above involved quasars selected solely on the basis of their optical properties; their radio

[°]A few quasars with red shifts larger than 3.0 have since been found.—Ed.

emission, if any, is negligible. It is therefore important to ask if quasars selected on the basis of their radio luminosity also show an increase in density with increasing distance. The 3C catalogue mentioned above is a comprehensive listing of all radio sources in the northern half of the sky with a certain minimum radio intensity. (The minimum value is nine "flux units" at 178 megahertz, or $9 \times 10^{-26}$ watt per square meter per hertz.) By the late 1960's 44 of the 300-odd extragalactic radio sources in the 3C catalogue had been optically identified as quasars. Of these 44 objects 33 had optical magnitudes of 18.5 or greater, and there was reason to believe that the 33 represented essentially all the 3C quasars down to that limiting magnitude.

## Radio-bright Quasars

The analysis of the distribution of the 33 objects is complicated because both a radio limitation and an optical limitation were involved in their selection. That is to say, to appear in the group of 33 quasars an object had to radiate strongly in two widely separated parts of the spectrum: the radio region and the optical region. The analysis made by one of the authors (Schmidt) went as follows:

From the red shift the distance to each object was computed on the basis of some particular model of the expanding universe. This distance equaled the radius of the volume of space within which the object was actually observed. One can then ask how far the object could be moved outward before one of two things happen: either its apparent magnitude drops below 18.5 or its radio flux falls below nine units. This distance defines the radius of the maximum volume beyond which the object could not lie and still remain a member of its original class.

For each object one can express the ratio of the two volumes, actual volume over maximum volume, as a decimal fraction. A priori, if the 33 objects were uniformly distributed, one would expect the average value of this fraction to be .5. Thus one would expect half of the values to be less than .5 and the other half to lie between .5 and 1. Actually only six of the objects yield values below .5 whereas 27 give higher values. In other words, radio quasars tend to occupy the outer reaches of the volume within which they can be observed. This tells us that their density increases with distance. When the density law is worked out in detail, it is found to lie

between $(1 + z)^5$ and $(1 + z)^6$. That is remarkably similar to the density law obtained for optically selected quasars, which on the average show negligible radio emission. The conclusion is that quasars have a density distribution that is only slightly or not at all dependent on their radio properties. This still leaves unsettled, however, the two possibilities already mentioned: either most quasars pass through a brief evolutionary stage during which they emit strongly at radio wavelengths or else only a small fraction of all quasars are destined to evolve into strong radio emitters.

## Other Quasar Hypotheses

A number of astronomers and theorists originally found it difficult to accept the idea that the red shifts of quasars are cosmological. They did not see how it was possible for an object to emit as much light as 100 galaxies and yet vary in intensity by 10 percent or more in a few days. They proposed, as one alternative, that quasars might be much nearer and smaller objects ejected at high velocity from the center of our own galaxy. This is sometimes called the local-Doppler hypothesis because the red shift is a Doppler shift and the objects are of local origin. Being only a few million light-years away, rather than billions of light-years, their actual energy output would be much less.

This hypothesis has encountered the difficulty that quasars are much more numerous than anyone suspected in the early 1960's. As we have just seen, recent estimates run into the millions, and on the most conservative basis one can hardly assume fewer than a million quasars. It may be estimated that the mass of the typical quasar, on the basis of the local-Doppler hypothesis, would have to be at least 10,000 suns. The ejection of a million objects, each of 10,000 solar masses, from the center of our galaxy would require that the mass of all the stars in the galactic nucleus be completely converted into energy. One must also explain why the only quasars ever observed are those ejected by our own galaxy. If any quasar-like objects had been ejected by any of the scores of galaxies in our immediate neighborhood, some of them should be observed to be heading *toward* us and thus should exhibit a blue shift rather than a red shift. Yet no quasar-like object with a blue shift has ever been detected. The local-Doppler explanation, on the whole, must be regarded as being quite unlikely.

A totally different explanation for the

red shift of quasars seemed attractive at first. According to this hypothesis quasars are objects in which a substantial mass is compressed into an extremely small volume. Light emitted from such an object would have to overcome an immense gravitational potential and would be red-shifted just as it is in quasars. The physical conditions that the hypothesis must account for can be rather precisely calculated. It is possible to compute, therefore, how large an emitting envelope of gas is needed, and what its density and temperature must be, to produce the spectral lines actually observed in quasars.

But if one assumes, to take an extreme case, that the highly condensed mass is comparable to the mass of the sun, its emitting envelope would not exhibit the required luminosity unless it were within 10 kilometers of the observer! The object has to be more distant, of course, and that will require a larger mass. The masses computed are large, and thus tend to create inadmissible side effects. For example, at a distance of 30,000 light-years the mass would have to be $10^{11}$ suns; it would rival the mass of our own galaxy, whose center is at the same distance. If the mass is raised still further to $2 \times 10^{13}$ suns, the minimum distance can be raised to 10 million light-years. In that case, in order not to raise the observed average density of the universe, a million such quasars would have to be distributed out to a distance of at least a billion light-years, at which point they would hardly qualify any longer as local objects.

One other "anticosmological" hypothesis should be mentioned for the sake of completeness: the hypothesis that the cause of the quasar red shift is simply unknown and thus lies outside present-day physics. Since no arguments can be made against such a metaphysical hypothesis it cannot be excluded.

## The Cosmological Hypothesis

An attractive feature of the cosmological hypothesis is that the quasar red shift comes "free," without requiring the introduction of bizarre physical conditions to explain the shift. The quasars exhibit a red shift simply because they are being carried along by the expansion of the universe. The extraordinary luminosity of quasars, together with their short-term variability, originally constituted the strongest objection to the cosmological hypothesis. In the past five years, however, short-term luminosity fluctuations of considerable magnitude have been observed in the nuclei of two rather special kinds of galaxy: N-type galaxies and Seyfert galaxies. These nuclei are starlike and resemble quasars in producing an excess of ultraviolet radiation. Moreover, there is general agreement that their red shifts, even though they are modest in the case of Seyfert galaxies, are cosmological in origin.

Recently it has been found that both quasars and Seyfert galaxies radiate strongly in the infrared region of the spectrum. Indeed, the infrared luminosity of the nearby Seyfert radio galaxy 3C 120 is $10^{46}$ ergs per second, which is equal to the infrared luminosity of many quasars when their luminosity is calculated on the assumption of their being at cosmological distances. In other words, we now have examples of objects whose extraordinary energy output is as difficult to explain as the output of qua-

sars (regarded as cosmological objects) and whose output varies over time scales that are just as brief as the time scales for the variation of quasars. Therefore the cosmological hypothesis cannot be ruled out on the basis of the difficulties encountered in explaining the quasars' rapidly varying high luminosity, because the same difficulties hold for galaxies whose properties and distances are not in question.

Support for the cosmological hypothesis has recently been obtained by James E. Gunn of the Hale Observatories. He found that the image of the quasar PKS 2251 + 11 (red shift .323) is superposed on the image of a small, compact cluster of galaxies. Gunn was able to determine the red shift of the brightest galaxy in the cluster and found a value of .33 ± .01. The coincidence in direction and red shift makes it very likely

that the quasar is associated with the cluster of galaxies, thus confirming the cosmological nature of its red shift.

As for the ultimate source of the tremendous energy observed in quasars, there has been no lack of hypotheses, among them stellar collisions, the gravitational collapse of massive stars, supernova explosions, conversion of gravitational energy into particle energy by magnetic fields, matter-antimatter annihilation and the rotational energy of a very compact mass (as proposed for pulsars). There is also no agreement about the radiation mechanism, particularly in the infrared, where much of the output is radiated. Similar problems exist for nuclei of galaxies, notably for those of Seyfert galaxies. The solution of these problems constitutes one of the main challenges to present-day astronomy.

What does life on other worlds look like?
Most views on this subject are based on the
assumption that life elsewhere must in a
major sense resemble life on earth. This
assumption might well be false because the
conditions that would strongly affect the
development of life may differ greatly from
one world to another. Hence, we must be
prepared for the probability that extraterrestrial beings will not resemble anything
familiar to us, a point made by Louise
Zingarelli, whose drawing is shown here.

PART XII

# EPILOGUE

What lies in the future as we continue the exploration of space? Surely the most profound question that occurs to us is the possibility that life exists on other worlds, and that it may include intelligent beings with whom we can communicate. Several avenues of communication are possible. In 1972, a plaque with pictorial and mathematically coded information was launched aboard *Pioneer 10,* the spacecraft that flew past Jupiter. This probe's journey will eventually carry it out of the solar system. However, it seems doubtful that anyone will ever find it and decipher the plaque, for as artist Linda Salzman Sagan and astronomers Carl Sagan and Frank Drake remark in "A Message from Earth," "We do not know the likelihood of the Galaxy being filled with advanced technological societies capable of and interested in intercepting such a spacecraft." (These authors use the abbreviation *pc* for *parsec,* the astronomical unit of distance that equals 3.26 light years.) Another avenue of communication is by means of radio. On November 16, 1974, at Drake's direction, a powerful, coded signal was beamed toward the globular cluster Messier 13 by the 1,000-foot radio telescope at Arecibo, Puerto Rico. This radio beam will strike the hundreds of thousands of stars in the cluster about 24,000 years from now. Should one of these stars have a planet inhabited by an advanced civilization, perhaps the message will be received and a reply sent back to Earth, which will take an equal amount of time to get here. If humans are intelligent enough to survive for the next 24,000 or 48,000 years, could they not also devise a more efficient interstellar communication system? Or discover one that has already been put into operation by other beings? Several books have been published recently that deal with how we may communicate with life on other worlds. Two of the best (both available in paperback) are *The Galactic Club: Intelligent Life in Outer Space* by Ronald N. Bracewell (W. H. Freeman and Company, 1975) and *Interstellar Communication: Scientific Perspectives* by Cyril Ponnamperuma and A. G. W. Cameron (Houghton Mifflin Company, 1974).

# 44

# A Message from Earth

Carl Sagan, Linda Salzman Sagan, and Frank Drake

Pioneer 10 is the first spacecraft that will leave the solar system. Scheduled for a launch no earlier than 27 February 1972, its 630- to 790-day-long flight will take it within two planetary radii of Jupiter, where, in a momentum exchange with the largest planet in the solar system, the spacecraft will be accelerated out of the solar system with a residual velocity at infinity of 11.5 km/sec.* The spacecraft is designed to examine interplanetary space between the earth and Jupiter, perform preliminary reconnaissance in the asteroid belt, and make the first close-up observations of Jupiter and its particles and fields environment.

It seemed to us appropriate that this spacecraft, the first man-made object to leave the solar system, should carry some indication of the locale, epoch, and nature of its builders. We do not know the likelihood of the Galaxy being filled with advanced technological societies capable of and interested in intercepting such a spacecraft. It is clear, however, that such interception is a very long term proposition. With a residual interstellar velocity of 11.5 km/sec, the characteristic time for Pioneer 10 to travel 1 parsec (pc)—slightly less than the distance to the nearest star—is some 80,000 years. From the simplest collision physics, it follows that the mean time for such a spacecraft to come within 30 astronomical units (1 A.U. = $1.5 \times 10^{13}$ cm) of a star is much longer than the age of the Galaxy. Consequently there is a negligible chance that Pioneer 10 will penetrate the planetary system of a technologically advanced society. But it appears possible that some civilizations technologically much more advanced than ours have the means of detecting an object such as Pioneer 10 in interstellar space, distinguishing it from other objects of comparable size but not of artificial origin, and then intercepting and acquiring the spacecraft.

But if the intercepting civilization is not within the immediate solar neighborhood, the epoch of such an interception can only be in the very distant future. Accordingly, we cannot see any conceivable danger in indicating our position in the Galaxy, even in the eventuality, which we consider highly unlikely, that such advanced societies would be hostile. In addition we have already sent much more rapidly moving indications of our presence and locale: the artificial radio-frequency emission which we use for our own purposes on Earth.

Erosional processes in the interstellar environment are largely unknown, but are very likely less efficient than erosion within the solar system, where a characteristic erosion rate, due mainly to micrometeoritic pitting, is of the order of 1 Å/year. Thus a plate etched to a depth $\sim 10^{-2}$ cm should survive recognizably at least to a distance $\sim$ 10 pc, and most probably to $>>$ 100 pc. Accordingly, Pioneer 10 and any etched metal message aboard it are likely to survive for much longer periods than any of the works of man on Earth.

With the support of the Pioneer Project Office at NASA's Ames Research Center in Mountain View, California, and of NASA Headquarters in Washington, D.C., it was agreed to prepare a message on a 6- by 9-inch surface of 6061 T6 gold-anodized aluminum plate, 50/1000 inch thick. The mean depth of engraving is 15/1000 inch. The plate is mounted in an exterior but largely protected position on the antenna support struts, behind the ARC plasma experimental package, on the Pioneer 10 spacecraft.

The question of the contents of such a message is not an easy one. The message finally agreed upon (Fig. 1) is in our view an adequate but hardly ideal solution to the problem. A time interval of only 3 weeks existed between the formulation of the idea of including a message on Pioneer 10, achieving NASA concurrence, devising the message, and delivering the draft message for engraving. We believe that any such message will be constrained, to a greater or lesser degree, by the limitations of human perceptual and logical processes. The message inadvertently contains anthropocentric content. Nevertheless we feel that an advanced technical civilization would be able to decipher it.

At top left is a schematic representation of the hyperfine transition of neutral atomic hydrogen. A transition from anti-

---

From *Science*, vol. 175, pp. 881–884, 25 February 1972. Copyright 1972 by the American Association for the Advancement of Science. Reprinted with permission.

*Launched March 3, 1972, a 21-month-long flight took it to within 130,000 km of the Jovian cloud tops. It is now en route to interstellar space.—Ed.

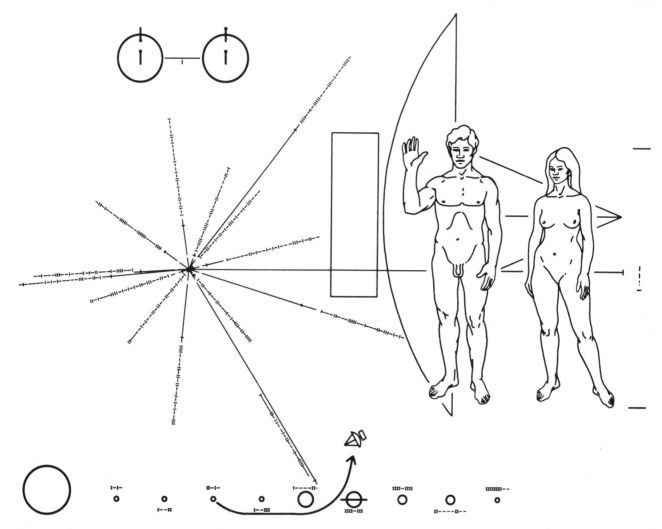

**Figure 1.** The engraved aluminum plate carried aboard Pioneer 10. It contains information on the position, epoch, and nature of the spacecraft. [NASA.]

parallel nuclear and electronic spins to parallel nuclear and electronic spins is shown above the binary digit 1. So far the message does not specify whether this is a unit of length (21 cm) or a unit of time $[(1420\ \text{Mhz})^{-1}]$. This fundamental transition of the most abundant atom in the Galaxy should be readily recognizable to the physicists of other civilizations. As a cross-check, we have indicated the binary equivalent of the decimal number 8 along the right-hand margin, between two tote marks corresponding to the height of the human beings shown. The Pioneer 10 spacecraft is displayed behind the human beings and to the same scale. A society that intercepts the spacecraft will of course be able to measure its dimensions and determine that 8 by 21 cm corresponds to the characteristic dimensions of the spacecraft.

With this first unit of space or time specified we now consider the radial pattern at left center. This is in fact a polar coordinate representation of the positions of some objects about some origin, with this interpretation being a probable, but not certain, initial hypothesis to scientists elsewhere. The two most likely origins in an astronomical interpretation would be the home star of the launch civilization and the center of

the Galaxy. There are 15 lines emanating from the origin, corresponding to 15 objects. Fourteen of these objects have a long binary number attached, corresponding to a 10-digit number in decimal notation. The large number of digits is the key that these numbers indicate time intervals, not distances or some other quantity. A civilization at our level of technology (as evinced from the Pioneer 10 spacecraft itself) will not know the distances to galactic objects useful for direction-finding to ten significant figures; and, even if we did, the proper motion of such objects within the Galaxy would render this degree of precision pointless. There are no other conceivable quantities that we might know to ten significant figures for relatively distant cosmic objects. The numbers attached to the 14 objects are therefore most plausibly time intervals. From the unit of time, the indicated time intervals are all $\sim 0.1$ second. For what objects might a civilization at our level of advance know time periods $\sim 0.1$ second to ten significant figures? Pulsars are the obvious answer. Since pulsars are running down at largely known rates they can be used as galactic clocks for time intervals of hundreds of millions of years. The radial pattern therefore must indicate the positions (obtained by us from the observed dispersion measures) and periods at the launch epoch of 14 pulsars, plus one additional object which is the most distant.

The problem thus reduces to searching the astronomical records to find a locale and epoch within the galaxy at which 14 pulsars were in evidence with the denoted periods and relative coordinates. Because the message is so overspecified, and because the pulsar periods are given to such precision, we believe that this is not an extremely difficult computer task, even with time intervals $>> 10^6$ years between launch and recovery. The pulsars utilized, with their periods in seconds and in units of the hydrogen hyperfine transition, are indicated in Table 1. The hyperfine period of $(1.420405752 \times 10^9$ $\text{sec}^{-1})^{-1}$, a fraction of a nanosecond, is just small enough so that all the known digits of the pulsar periods can just be written to the left of the decimal point. Accordingly decimals and fractions are entirely avoided with no loss of accuracy and without many noninformative digits. The presence of several consecutive terminal zeros (Table 1), particularly in pulsars 1240 and 1727, imply that, for these two pulsars, we have given a precision greater than we now have. The problem of which end of a number is the most significant digit is expressed automatically in this formulation, since all binary numbers start with a 1 but end in a 1 or a 0. The binary notation, in addition to being the simplest, is selected in order to produce a message that can suffer considerable erosion and still be readable. In principle, the reader only need determine that there were two varieties of symbols present, and the spacings alone will lead to a correct reconstruction of the number.

Those radial lines for which the earth-pulsar distance is not accurately known are shown with breaks. All three spatial coordinates of the pulsars are indicated. The $(r, \theta)$ coordinates are given in the usual polar projection. The tick marks near the ends of the radial lines give the $z$ coordinate normal to the galactic plane, with the distances measured from the far end of the line. The reconstruction of pulsar periods will indicate that the origin of $(r, \theta)$ coordinates is not the center of the Galaxy. Accordingly the long line extending to the right, behind the human beings, and which is not accompanied by a

pulsar period, should be identifiable as the distance to the galactic center. Since the tick mark of this line is precisely at its end, this should simultaneously confirm that the ticks denote the galactic $z$ coordinate, and that the longest line represents the distance from the launch planet to the galactic center. The tick marks were intended to be asymmetric about the radial distance lines, in order to give the sign of the galactic latitude or $z$ coordinate. In the execution of the message this convention was inadvertently breached. But the sign of the $z$ coordinate should be easily deducible without this aid. There is an initial ambiguity about whether the $(r, \theta)$ presentation is from the North or South Galactic Pole, but this ambiguity would be resolved as soon as even one pulsar was identified.

The 14 pulsars denoted have been chosen to include the shortest period pulsars which give the greatest longevity and the greatest luminosity; they are, therefore, the pulsars of greatest use in this problem where interception of the message occurs only in the far future. They are also selected to be distributed as evenly as possible in galactic longitude. Included are both pulsars in the vicinity of the Crab Nebula; the second (PSR 0525) has the longest known period. Fourteen pulsars were included to provide redundancy for any position and time solutions, but also to allow for the good possibility that pulsar emission is highly beamed and that not all pulsars are visible at all view angles. We expect that some of the 14 would be observable from all locales. In addition a very advanced civilization might have information on astronomy from other locales in the Galaxy. If the spacecraft is intercepted after only a few tens of millions of years (having traveled several hundred parsecs), all 14 pulsars may still be detectable.

The reconstruction of the epoch in which the message was devised should be performable to high precision: With 14 periods, almost all of which are accurate to nine significant figures in decimal equivalent, a society which has detailed records of past pulsar behavior should be able to reconstruct the epoch of launch to the equivalent of the year 1971. If past records of pulsar "glitches" (discontinuities in the period) are not kept or reconstructable from the physics it should still be possible to reconstruct the epoch to the nearest century or millennium.

Fortuitously, two of the pulsars are very near Earth. If either is correctly identified, it can be used to place the position of our solar system in the galaxy to approximately 20 pc, thereby specifying our location to approximately 1 in $10^3$ stars.

To specify our position to greater accuracy, we have included a schematic solar system at the bottom of the diagram. Because of the limited plate dimensions, the solar system was engraved with the planets not in the solar equatorial plane. (If this were an accurate representation of our solar system it would identify it very well indeed!) Relative distances of the planets are indicated in binary notation above or below each planet. The serifs on the binary "ones" are presented to stress that the units are different from those of pulsar length and period. The numbers represent the semimajor axes of the planetary orbits in units of one-tenth the semimajor axis of the orbit of Mercury, or 0.0387 A.U., approximately. There is no way for this unit of length to be deciphered in the message, but the schematic sizes and relative distances—given to three

TABLE 1

*The 14 selected pulsars.*

| Pulsar | Period (1970/1971 epoch) (second) | Period (units of H hyperfine transition) |
|--------|-----------------------------------|------------------------------------------|
| 0328 | $7.145186424 \times 10^{-1}$ | $1.014906390 \times 10^9$ |
| 0525 | $3.745490800$ | $5.320116676 \times 10^9$ |
| 0531 | $3.312964500 \times 10^{-2}$ | $4.705753832 \times 10^7$ |
| 0823 | $5.306595990 \times 10^{-1}$ | $7.537519468 \times 10^8$ |
| 0833 | $8.921874790 \times 10^{-2}$ | $1.267268227 \times 10^8$ |
| 0950 | $2.530650432 \times 10^{-1}$ | $3.594550429 \times 10^8$ |
| 1240 | $3.880000000 \times 10^{-1}$ | $5.511174318 \times 10^8$ |
| 1451 | $2.633767640 \times 10^{-1}$ | $3.741018705 \times 10^8$ |
| 1642 | $3.876887790 \times 10^{-1}$ | $5.506753717 \times 10^8$ |
| 1727 | $8.296830000 \times 10^{-1}$ | $1.178486506 \times 10^9$ |
| 1929 | $2.265170380 \times 10^{-1}$ | $3.217461037 \times 10^8$ |
| 1933 | $3.587354200 \times 10^{-1}$ | $5.095498540 \times 10^8$ |
| 2016 | $5.579533900 \times 10^{-1}$ | $7.925202045 \times 10^8$ |
| 2217 | $5.384673780 \times 10^{-1}$ | $7.648421610 \times 10^8$ |

significant figures in decimal equivalent—of the planets in our solar system, as well as the schematic representation of the rings of Saturn seen edge-on, should easily distinguish our solar system from the few thousand nearest stars if they have been surveyed once. Also indicated is a schematic trajectory of the Pioneer 10 spacecraft, passing by Jupiter and leaving the solar system. Its antenna is shown pointing approximately back at Earth. The cross-correlation between this stage of solar system exploration and the instrumentation and electronics of the Pioneer 10 spacecraft itself should specify the level of contemporary human technology with some precision.

The message is completed by a representation at right of a man and woman drawn to scale before a schematic Pioneer 10 spacecraft. The absolute dimensions of the human beings are specified in two ways: by comparison with the Pioneer 10 spacecraft and in units of the wavelength of the hyperfine transition of hydrogen, as described above. It is not clear how much evolutionary or anthropological information can be deduced from such a sketch drawing. Ten fingers and ten toes may provide a clue to man's arboreal ancestry, and the fact that the distance of Mercury from the sun is given as 10 units may be a clue to the development of counting. It seems likely, if the interceptor society has not had previous contact with organisms similar to human beings, that many of the body characteristics shown will prove deeply mysterious. We rejected many alternative representations of human beings for a variety of reasons; for example, we do not show them holding hands lest one rather than two organisms be deduced. With a set of human representations to this degree of detail, it was not possible to avoid some racial stereotypes, but we hope that this man and woman will be considered representative of all of mankind. A raised outstretched right hand has been indicated as a "universal" symbol of good will in many human writings; we doubt any literal universality, but included it for want of a better symbol. It has at least the advantage of displaying an opposable thumb.

Among the large number of alternative message contents considered and rejected were radioactive time markers (rejected because of interference with the Pioneer radiation detectors), star map position indicators (rejected because of stellar proper motions and serious data-handling problems in decoding), and schematic representations of the vascular, neurological, or muscular apparatus of human beings or some indication of the number of cortical neural connections (rejected because of the ambiguity of the envisioned representations). It is nevertheless clear that the message can be improved upon; and we hope that future spacecraft launched beyond the solar system will carry such improved messages.

This message then is a first attempt to specify our position in the Galaxy, our epoch and something of our nature. We do not know if the message will ever be found or decoded; but its inclusion on the Pioneer 10 spacecraft seems to us a hopeful symbol of a vigorous civilization on Earth.

## Note

1.  We thank the Pioneer Project Office at Ames Research Center, especially Charles Hall, the Program Manager, and Theodore Webber; and officials at NASA Headquarters, particularly John Naugle, Ishtiaque Rasool, and Henry J. Smith, for supporting a small project involving rather longer time scales than government agencies usually plan for. The initial suggestion to include some message aboard Pioneer 10 was made by Eric Burgess and Richard Hoagland. A redrawing of the initial message for engraving was made by Owen Finstad; the message was engraved by Carl Ray. We are grateful to A. G. W. Cameron for reviewing this message and for suggesting the serifs on the solar system distance indicators; and to J. Berger and J. R. Houck for assistance in computer programming.

# BIBLIOGRAPHIES AND SOURCES OF ILLUSTRATIONS
## for articles originally published in SCIENTIFIC AMERICAN

## ARTICLE 2

### Bibliography

ON THE REVOLUTIONS OF THE HEAVENLY SPHERES. Nicolaus Copernicus, translated by Charles Glenn Wallis in *Great Books of the Western World: Vol. 16,* edited by Robert M. Hutchins. Encyclopaedia Britannica, Inc., 1952.

TYCHO BRAHE: A PICTURE OF SCIENTIFIC LIFE AND WORK IN THE SIXTEENTH CENTURY. J. L. E. Dreyer. Dover Publications, Inc., 1963.

THE COPERNICAN REVOLUTION: PLANETARY ASTRONOMY IN THE DEVELOPMENT OF WESTERN THOUGHT. Thomas S. Kuhn. Harvard University Press, 1966.

THREE COPERNICAN TREATISES. Edward Rosen. Dover Publications, Inc., 1971.

A FRESH LOOK AT COPERNICUS. Owen Gingerich in *The Great Ideas Today,* edited by Robert M. Hutchins and Mortimer J. Adler. Encyclopaedia Britannica, Inc., 1973.

### Sources of Illustrations

PAGE 8    Houghton Library, Harvard University
PAGE 10    Charles Eames, University of Uppsala Library
PAGES 12, 14, 16, 18    Vatican Library

## ARTICLE 4

### Bibliography

RADIO INTERFEROMETRY AT ONE-THOUSANDTH SECOND OF ARC. M. H. Cohen, D. L. Jauncey, K. I. Kellermann and B. G. Clark in *Science,* Vol. 162, No. 3849, pages 88–94; October 4, 1968.

LONG-BASELINE INTERFEROMETRY. Bernard F. Burke in *Physics Today,* Vol. 22, No. 7, pages 54–63; July, 1969.

JOINT SOVIET-AMERICAN RADIO INTERFEROMETRY. K. I. Kellermann in *Sky and Telescope,* Vol. 42, No. 3, pages 132–133; September, 1971.

HIGH-RESOLUTION OBSERVATIONS OF COMPACT RADIO SOURCES AT 6 AND 18 CENTIMETERS. K. I. Kellermann, D. L. Jauncey, M. H. Cohen, B. B. Shaffer, B. G. Clark, J. Broderick, B. Rönnäng, O. E. H. Rydbeck, L. Matveyenko, I. Moiseyev, V. V. Vitkevitch, B. F. C. Cooper and R. Batchelor in *The Astrophysical Journal,* Vol. 169, No. 1, Part 1, pages 1–24; October 1, 1971.

### Sources of Illustrations

PAGE 29    Jet Propulsion Laboratory, California Institute of Technology
PAGES 36, 38    Hale Observatories

## ARTICLE 13

### Bibliography

MAGNETISM AND THE COSMOS. Edited by W. R. Hindmarsh, F. J. Lowes, P. H. Roberts and S. K. Runcorn. Oliver & Boyd, 1967.

THE MOON: OUR NEAREST CELESTIAL NEIGHBOUR. Zdeněk Kopal. D. Reidel Publishing Company, 1969.

LUNAR PORTABLE MAGNETOMETER EXPERIMENT. P. Dyal, C. W. Parkin, C. P. Sonett, R. L. DuBois and G. Simmons in *Apollo 14 Preliminary Science Report.* NASA SP-272. U.S. Government Printing Office, July, 1971.

THE APOLLO 12 MAGNETOMETER EXPERIMENT: INTERNAL LUNAR PROPERTIES FROM TRANSIENT AND STEADY MAGNETIC FIELD MEASUREMENTS. P. Dyal and C. W. Parkin in *Proceedings of the Second Lunar Science Conference: Houston, Texas, January 11-14, 1971.* The M.I.T. Press, 1971.

### Source of Illustrations

PAGES 90, 92   NASA

## ARTICLE 14

### Bibliography

SCIENCE, Vol. 175, No. 4019; January, 21, 1972.

MARS: THE VIEW FROM MARINER 9. Carl Sagan in *Astronautics & Aeronautics,* Vol. 10, No. 9, pages 26-41; September, 1972.

ICARUS, Vol. 17, No. 2; October, 1972.

### Source of Illustrations

PAGES 104, 105, 106, 107 (*bottom*), 110-111, 112, 113, 114, 115, 116, 117, 118, 119   Jet Propulsion Laboratory, California Institute of Technology

## ARTICLE 21

### Bibliography

ASTROPHYSICS. NUCLEAR TRANSFORMATIONS, STELLAR INTERIORS AND NEBULAE. Lawrence H. Aller. The Ronald Press Company, 1954.

THE PERIOD-LUMINOSITY RELATION OF THE CEPHEIDS. W. Baade in *Publications of the Astronomical Society of the Pacific,* Vol. 68, No. 400, pages 5-16; February, 1956.

### Source of Illustrations

PAGES 153, 154-155   Hale Observatories

## ARTICLE 24

### Bibliography

DARK NEBULAE, GLOBULES AND PROTOSTARS. Edited by Beverly T. Lynds. University of Arizona Press, 1971.

THEORIES OF STAR FORMATION. D. McNally in *Reports on Progress in Physics,* Vol. 34, page 71; 1971.

LIGHT MOLECULES AND DARK CLOUDS. David Buhl in *Mercury,* Vol. 1, No. 5, page 4, Vol. 1, No. 6, pages 4-8, 1972.

MOLECULES IN THE GALACTIC ENVIRONMENT. Edited by M. A. Gordon and L. E. Snyder. Wiley-Interscience, 1973.

### Sources of Illustrations

PAGES 176, 182   Hale Observatories
PAGES 177, 181, 183, 184, 186-187   Bart J. Bok

## ARTICLE 25

### Bibliography

EVOLUTION OF STARS AND GALAXIES. Walter Baade. Harvard University Press, 1963.

STELLAR EVOLUTION: COMPARISON OF THEORY WITH OBSERVATION. Icko Iben, Jr., in *Science,* Vol. 155, No. 3764, pages 785-796; February 17, 1967.

ENERGY PRODUCTION IN STARS. Hans A. Bethe in *Science,* Vol. 161, No. 3841, pages 541-547; August 9, 1968.

STELLAR ABUNDANCES AND THE ORIGIN OF THE ELEMENTS. Albrecht O. J. Unsöld in *Science,* Vol. 162, No. 3871, pages 1015-1025; March 7, 1969.

### Source of Illustrations

PAGES 188, 192   Hale Observatories

## ARTICLE 30

### Bibliography

INTERSTELLAR MOLECULES AND DENSE CLOUDS. D. M. Rank, C. H. Townes and W. J. Welch in *Science,* Vol. 174, No. 4014, pages 1083-1101; December 10, 1971.

INTERSTELLAR MOLECULES. B. E. Turner in *Galactic and Extra-Galactic Radio Astronomy,* edited by G. L. Verschuur and K. I. Kellermann. Springer-Verlag, Inc., pages 199-255, 1974.

MOLECULES IN THE GALACTIC ENVIRONMENT. Edited by M. A. Gordon and L. E. Snyder. Wiley-Interscience, 1973.

### Source of Illustrations

PAGE 223   Lick Observatory

## ARTICLE 33

### Bibliography

THE ROTATION AND RADIAL VELOCITY OF THE CENTRAL PART OF THE ANDROMEDA NEBULA. F. G. Pease in *Proceedings of the National Academy of Sciences of the United States of America*, Vol. 4, No. 1, pages 21–24; January 15, 1918.

ROTATION OF THE ANDROMEDA NEBULA FROM A SPECTROSCOPIC SURVEY OF EMISSION REGIONS. Vera C. Rubin and W. Kent Ford, Jr., in *The Astrophysical Journal*, Vol. 159, No. 2, Part 1, pages 379–403; February, 1970.

STELLAR MOTION NEAR THE NUCLEUS OF M31. Vera C. Rubin, W. Kent Ford, Jr., and C. Krishna Kumar in *The Astrophysical Journal*, Vol. 181, No. 1, Part 1, pages 61–77; April, 1973.

### Sources of Illustrations

PAGES 263, 264, 265 (*top*), 266    Hale Observatories
PAGE 269    Vera C. Rubin

## ARTICLE 34

### Bibliography

MULTIPLE GALAXIES. F. Zwicky in *Ergebnisse der exakten Naturwissenschaften*, Vol. 29, pages 344–385; 1956.

GRAVITATIONSEFFEKTE BEI DER BEGEGNUNG ZWEIER GALAXIEN. J. Pfleiderer in *Zeitschrift für Astrophysik*, Vol. 58, No. 1, pages 12–22; August, 1963.

ATLAS OF PECULIAR GALAXIES. Halton Arp. California Institute of Technology, 1966.

GALACTIC BRIDGES AND TAILS. Alar Toomre and Juri Toomre in *The Astrophysical Journal*, Vol. 178, pages 623–666; December 15, 1972.

TIDAL INTERACTION OF GALAXIES. T. M. Eneev, N. N. Kozlov and R. A. Sunyaev in *Astronomy and Astrophysics*, Vol. 22, No. 1, pages 41–60; January, 1973.

### Sources of Illustrations

PAGE 270    F. Zwicky, Hale Observatories
PAGES 274 (*top*), 280    H. C. Arp, Hale Observatories
PAGE 278 (*top*)    Hale Observatories
PAGE 278 (*bottom*)    S. van den Bergh, David Dunlap Observatory

## ARTICLE 43

### Bibliography

QUASI-STELLAR OBJECTS. Geoffrey Burbidge and Margaret Burbidge. W. H. Freeman and Company, 1967.

SPACE DISTRIBUTION AND LUMINOSITY FUNCTIONS OF QUASI-STELLAR RADIO SOURCES. Maarten Schmidt in *The Astrophysical Journal*, Vol. 151, No. 2, pages 393–409; February, 1968.

ON THE NATURE OF FAINT BLUE OBJECTS IN HIGH GALACTIC LATITUDES, II: SUMMARY OF PHOTOMETRIC RESULTS FOR 301 OBJECTS IN SEVEN SURVEY FIELDS. Allan Sandage and Willem J. Luyten in *The Astrophysical Journal*, Vol. 155, No. 3, pages 913–918; March, 1969.

SPACE DISTRIBUTION AND LUMINOSITY FUNCTIONS OF QUASARS. Maarten Schmidt in *The Astrophysical Journal*, Vol. 162, No. 2, Part 1, pages 371–379; November, 1970.

### Source of Illustrations

PAGES 346, 348    M. Schmidt

# INDEX

8172